Developments in Precambrian Geology 3

ARCHEAN GREENSTONE BELTS

DEVELOPMENTS IN PRECAMBRIAN GEOLOGY
Advisory Editor B.F. Windley

Further titles in this series

1. B.F. WINDLEY and S.M. NAQVI (Editors)
   Archaean Geochemistry
2. D.R. Hunter (Editor)
   Precambrian of the Southern Hemisphere

DEVELOPMENTS IN PRECAMBRIAN GEOLOGY 3

ARCHEAN GREENSTONE BELTS

KENT C. CONDIE

Department of Geoscience, New Mexico Institute of Mining and Technology, Socorro, New Mexico, U.S.A.

ELSEVIER SCIENTIFIC PUBLISHING COMPANY
Amsterdam – Oxford – New York 1981

ELSEVIER SCIENTIFIC PUBLISHING COMPANY
335 Jan van Galenstraat
P.O. Box 211, Amsterdam, The Netherlands

Distributors for the United States and Canada:

ELSEVIER NORTH-HOLLAND INC.
52, Vanderbilt Avenue
New York, N.Y. 10017

Library of Congress Cataloging in Publication Data

Condie, Kent C
Archean greenstone belts.

(Developments in Precambrian geology ; v. 3)
Bibliography: p.
Includes index.
1. Geology, Stratigraphic--Archaean. 2. Rocks, Metamorphic. I. Title. II. Series.
QE653.C65 551.7'12 80-10317
ISBN 0-444-41854-7

ISBN 0-444-41854-7 (Vol. 3)
ISBN 0-444-41719-2 (Series)

Printed in The Netherlands

PREFACE

This book presents a summary of data and interpretations related to the origin of Archean greenstone belts and associated granitic terranes. Although most of the published literature relates chiefly to greenstone belts, I treat both greenstone and granitic components of Archean low-grade terranes because the origin and tectonic setting of one component cannot be regarded as independent of the other. Results are presented from numerous fields including volcanology, sedimentology, stratigraphy, metamorphic petrology, structure, geophysics, and geochemistry. The approach is not chiefly a descriptive one, but represents a combination of interpretation and description. No attempt is made to review the details of stratigraphic nomenclature in specific areas nor to summarize other aspects of the geology which are chiefly of local interest. Examples of typical greenstone successions and representative geologic histories, however, are presented to illustrate similarities and differences and overall characteristics. Extensive references are given for those wishing more detailed information in given areas.

The first chapter deals with the general features, geographic distribution, and geochronology of Archean granite-greenstone terranes and Chapter 2 discusses greenstone belt stratigraphy, upper and lower greenstone successions, and provinciality. Chapter 3 deals with volcanic and hypabyssal rocks and Chapter 4 with sedimentary rocks in greenstone belts. Granitic rocks are discussed in Chapter 5 and structure and metamorphism in Chapter 6. Mineral deposits in granite-greenstone terranes are briefly reviewed in Chapter 7 and evidences for Archean life in greenstone successions are discussed in Chapter 8. Chapter 9 is a summary of recent geochemical and isotopic studies related to the origin and source of Archean magmas. Lastly, in Chapter 10, a discussion is presented of the origin and development of the early crust and lithosphere, the Archean thermal regime, the possible role of plate tectonics in the Archean, the relation of high-grade to low-grade Archean terranes, and a review of models for the origin of greenstone belts.

The book is intended as a reference book for both academic and industrial geoscientists. Although not primarily designed as a text, it could be used as such in an upper division or graduate course on the Archean. James M. Robertson and Stephen White are acknowledged for reading and criticizing some chapters. Carolyn Condie helped with editing and Rose Mary Richards carefully typed various versions of the manuscript. I would like to thank the many authors who supplied original figures to use in the book and also acknowledge publishers and authors for their permission to publish the figures.

Kent C. Condie
Socorro, New Mexico
August, 1979

CONTENTS

Chapter 1

ARCHEAN GRANITE-GREENSTONE TERRANES

INTRODUCTION

Archean rocks, which are rocks $\gtrsim$ 2.5 billion years (b.y.) in age, are exposed in small areas on all of the continents (Fig. 1-1). They comprise crustal provinces which are roughly equidimensional in plan view and range in size from < 0.1 to 2.6×10^6 km^2 in area, with most falling between 0.25 and 0.5×10^6 km^2. A crustal province is herein defined as a segment of the crust which records dominantly a singular range of radiometric ages and commonly exhibits a similar structural style (Condie, 1976a). Archean provinces contain rocks which range in age from 2.5 to 3.8 b.y.[1] Three rock associations occur in Archean provinces, which are in order of relative abundance: the granite-greenstone association, the high-grade association, and the cratonic basin association. The granite-greenstone association is characterized by supracrustal successions comprised dominantly of mafic volcanic rocks, known as greenstone belts, engulfed in a sea of granitic rocks. This association dominates in Archean provinces in North America, southern Africa, and Australia. The high-grade association, which dominates in Archean provinces in central and northern Africa, Greenland, and in the Soviet Union, is characterized by gneiss-migmatite-granulite complexes, layered igneous intrusions, and high-grade supracrustal remnants. The cratonic basin association has thus far been described only in the Kaapvaal province in South Africa (Anhaeusser, 1973a) and is characterized by a succession composed dominantly of quartzites, shales and carbonates with smaller amounts of volcanic rock. This association may have been more widespread during the Archean, however, as evidenced by the widespread distribution of inclusions of these rock types at higher metamorphic grades in the Archean high-grade terranes. Although, in detail, structural trends in Archean provinces are complex and reflect polyphase deformation, an overall structural grain is exhibited by large portions of some provinces (Fig. 1-1).

Boundaries of Archean provinces fall into one or a combination of three categories: rapid increases in metamorphic grade, faults or shear zones, and

[1] All radiometric ages in this book are calculated with reference to the following decay constants (after Steiger and Jäger, 1977): $^{238}U = 0.155 \times 10^{-9}$ yr^{-1}; $^{235}U = 0.985 \times 10^{-9}$ yr^{-1}; $^{232}Th = 0.0495 \times 10^{-9}$ yr^{-1}; $^{40}K = 5.81 \times 10^{-10}$ yr^{-1}; $^{87}Rb = 1.42 \times 10^{-11}$ yr^{-1}; $^{147}Sm = 6.54 \times 10^{-12}$ yr^{-1}.

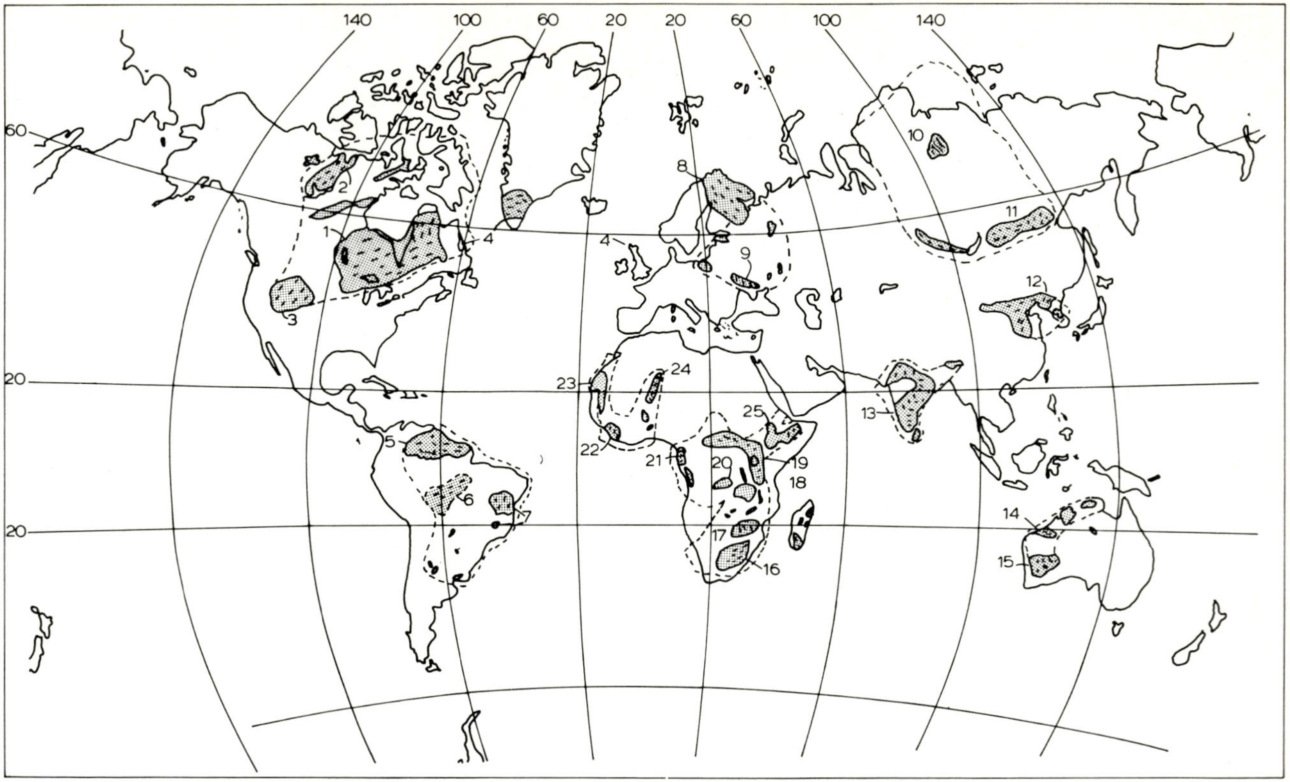

Fig. 1-1. Distribution of Archean provinces (shown in gray). Bold dashed lines outline areas probably underlain by Archean terranes. Structural trends are indicated where available. Key to major provinces: *1* = Superior (B), *2* = Slave (G), *3* = Wyoming (G), *4* = North Atlantic (H), (Nain, Godthaab, Lewisian), *5* = Guiana (H), *6* = Guaporé (H), *7* = São Francisco (B), *8* = Kola (B), *9* = Ukrainian (B), *10* = Anabar (H), *11* = Aldan (H), *12* = Chinese (H), *13* = Indian (B), *14* = Pilbara (G), *15* = Yilgarn (B), *16* = Kaapvaal (G), *17* = Rhodesian (G), *18* = Zambia (H), *19* = Central African (B), *20* = Kasai (H), *21* = Cameroons (H), *22* = Liberian (B), *23* = Maritanian (H), *24* = Ouzzalian (H), *25* = Ethiopian (H). Symbols: G = granite-greenstone terrane; H = high-grade terane; B = both granite-greenstone and high-grade terranes.

unconformities with younger sedimentary terranes. Increases in metamorphic grade and tectonic contacts characterize boundaries with Proterozoic mobile belts. Such mobile belts are orogenic-metamorphic belts which partially surround and cross-cut most Archean provinces (Anhaeusser et al., 1969; Condie, 1976a; Kröner, 1977a). Metamorphic and/or tectonic contacts have been described in Canada along the eastern borders of the Superior Province (the Grenville Front) (Wynne-Edwards, 1972) and the Slave Province (the Thelon Front) (Gibb and Thomas, 1977) (Figs. 1-6 and 1-9). They have also been described in Africa along the southern margin of the Rhodesian Province (Mason, 1973) and along the eastern margin of the Central African Province (the Mozambiquian Front) (Sanders, 1965; Hepworth, 1972). Fault boundaries may exhibit normal, reverse, or transcurrent motions and often have associated mylonitic zones. A diagrammatic cross-section of a sheared contact between the Archean Wyoming Province and the Proterozoic Churchill Province in North America is shown in Fig 1-2. The boundary between the two provinces is a near-vertical shear zone (the Mullen Creek–Nash Fork shear zone) separating Archean gneisses from Proterozoic granites and metamorphic rocks. The shear zone also displaces deformed miogeoclinal metasedimentary rocks that rest unconformably on the Archean gneisses. Some boundaries, such as much of the Grenville and Mozambiquian Fronts, are defined by rapid increases in metamorphic grade from the greenschist to the granulite facies. Such changes may occur over a few kilometers distance. Tectonic and metamorphic boundaries are also characterized by geophysical anomalies (Goodwin et al., 1972). Paired negative and positive Bouguer gravity anomalies, for instance, characterize the Grenville and Thelon Fronts (Thomas and Tanner, 1975; Gibb and Thomas, 1977) and the Nelson River shear zones (Innes, 1960) in Canada. Seismic data indicate that the crust thickens by 5—10 km over a distance of 50—70 km beneath the Grenville Front (Mereu and Jobidon, 1971). Crustal thickening is also recognized along the western boundary of the Superior Province (Mereu and Hunter, 1969; Hajnal and Mclure, 1977).

Although the distribution of Archean provinces shown in Fig. 1-1 suggests that they comprise a small portion of the total continental crust, the widespread occurrence of Archean radiometric dates in Proterozoic mobile belts suggests that the Archean crust was originally much more extensive (Condie, 1976a; Kröner, 1977a, b). Inliers of Archean crust have been described in some Proterozoic mobile belts, some of the largest of which occur in the Churchill Province in Canada (Fig. 1-1) (Goodwin, 1974). Archean inliers have been known for some time in Proterozoic mobile belts in southern Africa (Cahen and Snelling, 1966; Kröner, 1977a, b) and have recently been described in the Proterozoic mobile belt between the Yilgarn and Pilbara Provinces in western Australia (Horwitz and Smith, 1978). Mafic dike distributions in the Yilgarn and Pilbara Provinces also indicate that these provinces were probably contiguous by late Archean time. Available evidence

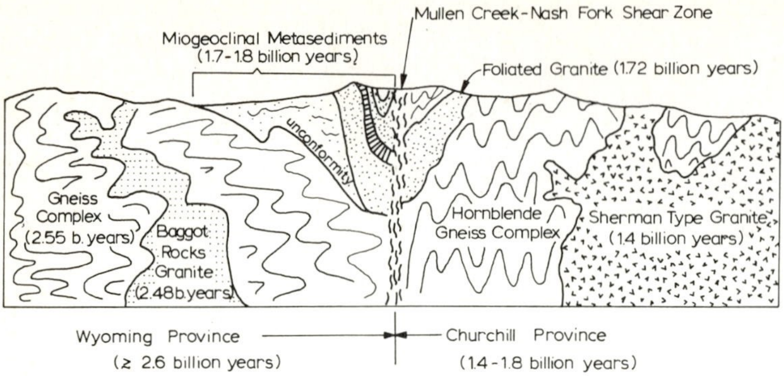

Fig. 1-2. Diagrammatic cross-section of the boundary between the Wyoming and Churchill Provinces in southeastern Wyoming (after Hills et al., 1968).

suggests that the Archean crust may have underlain much of the area now occupied by Proterozoic mobile belts as indicated in Fig. 1-1.

The study of Archean rocks has been extremely fruitful in enhancing our understanding of the early stages of the earth's history. Many basic problems, however, remain to be solved. Although it is unlikely that fragments of the earth's earliest crust are preserved, any origin for the observed Archean crust must be a natural consequence of earlier crustal processes (Condie, 1979a). Hence, by gaining an understanding of the development of the crust between 2.5 and 3.8 b.y., we can provide an important constraint for the origin and early development of the crust. What was the composition of the earth's early crust? Mafic, granitic, andesitic, and anorthositic compositions have been proposed (Condie, 1979a). The oldest known rocks on earth (~ 3.8 b.y.; Moorbath et al., 1973) comprise the Isua greenstone belt in Greenland and are mafic and felsic volcanics, iron formation and clastic sediments reflecting volcanic provenance (Allaart, 1976). Even these rocks, however, are probably younger than the earliest crust which may have formed prior to 4.0 b.y. ago.

One of the most important problems in Archean terranes, which will be discussed further in Chapter 10, is that of understanding the relationship of the granite-greenstone to the high-grade terranes. Do they represent different tectonic settings or do they reflect different levels of erosion in the Archean crust (Windley, 1973, 1976)? Did both oceanic and continental crust exist in the Archean and if so, did they reflect plate tectonic processes similar to the present? Just when did plate tectonics begin? Some investigators favor the onset of plate tectonics with the formation of the first crust (Condie, 1979a). Others suggest that plate tectonics did not begin until about 1 b.y. ago when the lithosphere had cooled sufficiently to act as a brittle solid (Wynne-Edwards. 1976; Baer, 1977). Still another possibility is that plate tectonic processes were episodic, dominating prior to 2.5 b.y. and again after about 1 b.y. (Engel and Kelm, 1972). An understanding of the tectonic settings reflected by the three major Archean rock associations together with paleomagnetic studies of these rocks is necessary to resolve these problems.

Another question of interest is whether or not greenstone belts are limited to the Archean. This question depends, in part, on how one defines a greenstone belt. If a rather generalized definition is accepted in which greenstone belts are considered as supracrustal successions in which mafic volcanics dominate, greenstone belts are not limited to the Archean. Recent radiometric dating has shown that the Snow Lake—Flin Flon belt in southeastern Manitoba, which was long considered to represent a typical Archean greenstone belt, is 1.7—1.8 b.y. in age (Bell et al., 1975; Moore, 1977). Similar successions of Proterozoic age are known from the southwestern United States (1.7—1.8 b.y.) (Anderson and Silver, 1976), the Birrimian in West Africa (~ 2.0 b.y.) (Burke and Dewey, 1972), and the Grenville Province in eastern Canada (1.3 b.y.) (Sethuraman and Moore, 1973). Other occurrences of probable Proterozoic age are known on most continents. If a more restrictive definition of greenstone belt is used, which includes the presence of relatively large amounts of ultramafic and komatiitic volcanics, post-Archean examples are rare or absent. Such a definition, however, eliminates many Archean greenstones which do not contain significant amounts of ultramafic volcanics. In the author's opinion, it would appear that greenstone belts are not limited to the Archean but formed also in the Proterozoic and perhaps in the Phanerozoic.

GENERAL FEATURES OF ARCHEAN GRANITE-GREENSTONE TERRANES

Archean granite-greenstone terranes are composed in large part of granitic and gneissic rocks (80—90%) which surround and, in part, intrude greenstone belts which comprise the remainder. The most striking feature of these terranes is their world-wide similarities (Anhaeusser et al., 1969; Anhaeusser, 1973a, 1975; Condie, 1976a; Windley, 1977). Greenstone belts are linear- to irregular-shaped, synformal supracrustal successions which range in width from 5 to 250 km and in length up to several hundred kilometers. Most belts range from 10 to 50 km wide and 100—300 km long. They contain exposed stratigraphic thicknesses ranging from 10 to 20 km. Although the oldest known greenstone belts are 3.5—3.8 b.y. in age (Isua in Greenland, Barberton in South Africa, and the older greenstone belts in Rhodesia), most greenstone belts appear to have formed between 2.6 and 2.7 b.y. An idealized map of a typical greenstone belt is given in Fig. 1-3 and shows some of the main features. Most greenstone belts are faulted synforms with fold axes and major faults paralleling the synformal axis. The keel-shaped outline is produced by diapiric, intrusive plutons. Greenstone belts are typically metamorphosed to the greenschist or amphibolite facies and metamorphic grade may increase near contacts with plutons. Primary textures and structures are often well-preserved in greenstone

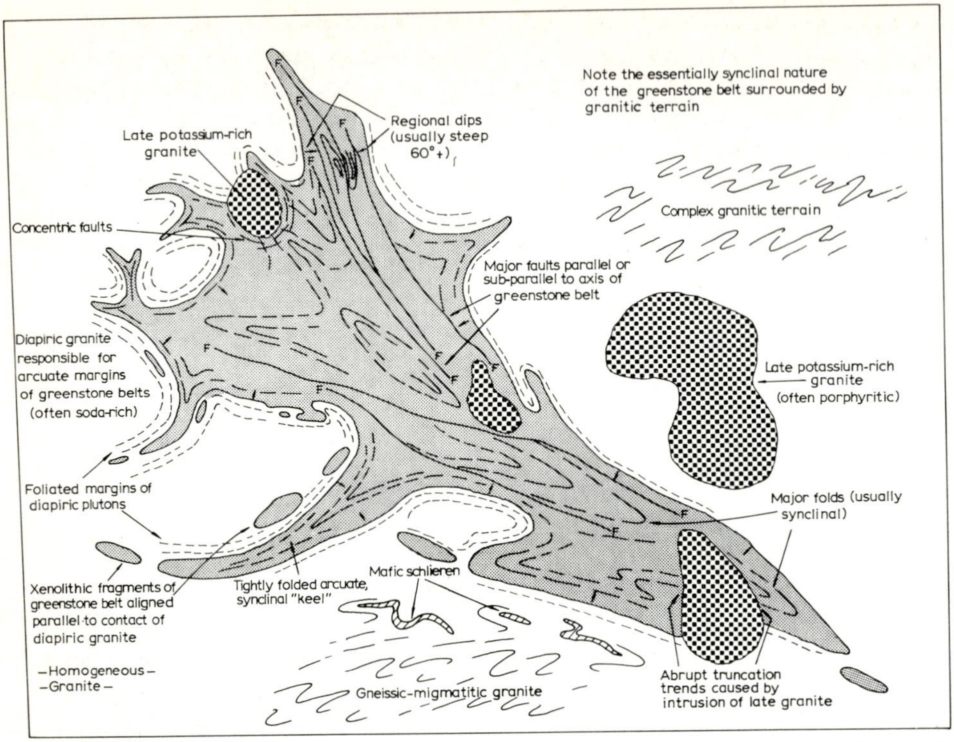

Fig. 1-3. Idealized map of a typical Archean greenstone belt (after Anhaeusser et al., 1969).

successions. Surrounding granitic terranes are comprised of gneissic complexes, diapiric intrusives, batholiths, and late, discordant plutons. The gneissic complexes are highly deformed and contain inclusions of greenstone belts, some of which may represent fragments of earlier greenstone belts. Diapiric plutons are foliated with the degree and dip of foliation increasing near their margins. Such foliation is broadly concordant to that in adjacent greenstone belts. Late, discordant plutons are intrusive into greenstone belts and into gneissic terranes. Structural studies in granite-greenstone terranes indicate the dominance of vertical forces although in some, horizontal forces also may have been important.

Greenstone successions are composed chiefly of pillowed, mafic volcanic rocks. Calc-alkaline volcanic rocks increase in abundance with stratigraphic height in some successions. Some greenstone belts contain an abundance of ultramafic and komatiitic lavas in their lower parts. Sediments comprise a minor but important part of greenstone belts generally being most abundant in upper stratigraphic levels. They are dominantly graywacke-argillite with smaller amounts of chert and other clastic sediments. The earliest evidences of life occur in chert horizons in Archean greenstone belts. Many mineral deposits occur in Archean greenstone belts among which the most important are Cu, Ni, Fe, Au, and Cr.

Greenstone-granite terranes have been likened by some to Phanerozoic orogenic belts. However, several notable differences have been pointed out by Engel and Kelm (1972). Length-to-width ratios of Phanerozoic belts, although variable, generally exceed 100:1 whereas Archean granite-greenstone terranes rarely exceed 5:1. The original length-to-width ratios of the Archean belts, however, may have been much greater if the continents were part of one or two supercontinents in late Archean time as the interpretation of paleomagnetic data by some investigators suggests (Piper, 1976b). Another difference between Archean and Phanerozoic orogenic belts is the size of folds. Wavelengths and amplitudes of most folds in granite-greenstone terranes are small compared to Phanerozoic counterparts. Other differences include the near-absence of blueschist-facies metamorphism and the abundance of ultramafic lavas in Archean greenstone belts. Differences and similarities between Archean and Phanerozoic orogenic provinces are important in reconstructing Archean tectonic settings as discussed in Chapter 10.

GENERAL FEATURES OF ARCHEAN HIGH-GRADE TERRANES

To discuss the significance of granite-greenstone terranes in terms of crustal evolution and tectonic setting, it is necessary also to consider Archean high-grade terranes. Recent detailed studies in southwest Greenland (McGregor, 1973; Windley et al., 1973; Bridgwater et al., 1976, 1978) and Scotland (Sheraton et al., 1973) have been informative in enhancing our understanding of these terranes. High-grade areas are composed chiefly of quartzofeldspathic gneiss-migmatite terranes ($\gtrsim$ 80%) with varying amounts of granulite-facies rocks. In addition, varying but generally minor amounts of supracrustal rocks, layered igneous complexes, and mafic dikes are found. The metamorphic grade ranges from middle amphibolite to upper granulite facies, and only rarely are primary textures preserved in volcanic or sedimentary supracrustal rocks. Metamorphic mineral assemblages indicate that some granulite-facies terranes were buried to 30—40 km depth at the time of metamorphism (Chapter 6). High-grade terranes are characterized by complex polyphase deformation which penetrates all rocks. Folds are characterized by interference patterns up to several kilometers across. Unlike granite-greenstone terranes, the dominant stress regimes appear to have been sub-horizontal and tangential producing major thrusts and recumbent folds (Bridgwater et al., 1974). Although some high-grade terranes (i.e., southwest Greenland and Labrador, ~ 3.8 b.y.) are distinctly older than most greenstone-granite terranes, they appear as a whole to range in age throughout the Archean with some of the youngest occurrences (~ 2.5 b.y.) found in the Archean Chinese Province (Fig. 1-1).

At least some high-grade gneiss-migmatite complexes differ from the gneiss-migmatite complexes found in granite-greenstone terranes in that they are more K_2O-rich (Bridgwater et al., 1976). Supracrustal rocks occur infolded in gneiss terranes and reflect a variety of progenitors. The Isua succession in southwest Greenland is composed of mafic and ultramafic volcanics, quartzite, carbonate-rich schists, felsic tuffs, conglomerate, and banded iron formation (Allaart, 1976). This succession is not unlike some greenstone successions. More typical supracrustal remnants, however, are quartzite-carbonate-mica schist successions. Associated with these are amphibolites, anorthosites, and ultramafic rocks which represent metaigneous rocks. Large successions of layered mafic to felsic granulites are found in Scotland (Sheraton et al., 1973) and in the Peninsular gneisses in India (Ramiengar et al., 1978). Layered igneous complexes are distinctive components of most high-grade terranes (Windley and Bridgwater, 1971). They range from very small to large stratiform sheets such as the Fiskenaesset Complex in southwest Greenland (Windley et al., 1973) with a thickness up to 1.5 km and a strike length of at least 60 km. Layered intrusions found in high-grade terranes differ from those found in granite-greenstone terranes by the presence of primary hornblende, the presence of magnetite throughout the intrusion, the absence of enrichment of alkalies in late liquids, and the presence of calcic plagioclase (An_{80}—An_{100}) throughout. These features appear to reflect crystallization in the presence of abundant water as opposed to the relatively dry crystallization characterizing layered igneous complexes in granite-greenstone terranes (Windley and Smith, 1976).

The geologic studies of high-grade terranes in southwest Greenland indicate a sequence of events as follows (see Fig. 1-20) (after Bridgwater et al., 1974; Myers, 1976): (1) formation of basement gneisses (the Amitsoq gneisses) at ~ 3.8 b.y.; (2) deposition of supracrustal rocks; (3) interfolding, thrusting, and metamorphism of gneisses and supracrustals (3.55—3.65 b.y.); (4) intrusion of Ameralik dikes; (5) deposition of sediments and minor volcanism producing the Malene supracrustal succession; (6) intrusion of layered igneous complexes; (7) intrusion of tonalite and granodiorite (the Nûk gneisses) at 2.8—2.9 b.y.; (8) intense polyphase deformation and middle to high-grade metamorphism at ~ 2.8 b.y.; and (9) intrusion of the Qôrqut granite at ~ 2.5 b.y.

ARCHEAN CRATONIC BASIN ASSOCIATIONS

As mentioned above, the only well-documented example of an Archean cratonic basin is the Kaapvaal basin in southern Africa (Anhaeusser, 1973a; Vajner, 1976). The rocks in this basin are only slightly deformed and exhibit very low grades of metamorphism. Primary textures and structures are well-preserved in sediments and indicate stable-shelf, near-shore deposition. This cratonic succession differs from most typical post-Archean successions in that a large proportion of volcanic rocks are interlayered with the sediments.

The nature of these volcanics is poorly known, but available data suggest that they represent a bimodal felsic and mafic association. The axis of deposition of the Kaapvaal basin, which lies unconformably on the Kaapvaal granite-greenstone province, migrated northwestward over a distance of about 600 km between 3.0 and 1.8 b.y. ago (Anhaeusser, 1973a; Pretorius, 1974). The earliest rocks deposited along the southeastern margin of the province were dominantly quartzites, shales, and carbonates with associated mafic and felsic volcanics that formed the Pongola Supergroup (Von Brunn and Hobday, 1976). As the basin migrated northwestward between 2.75 and 2.9 b.y., it filled with quartzites, shales, conglomerates and associated volcanics of the Witwatersrand Supergroup attaining a maximum thickness of 14 km. Between 2.8 and 1.8 b.y., it continued to move to the northwest collecting similar sediments with increasing amounts of carbonate and decreasing amounts of volcanic rock.

Other cratonic basins may have existed during the Archean as evidenced by the large proportion of quartzite and mica schist in high-grade terranes of the Indian Province (the Sargur supracrustal successions; Viswanatha and Ramakrishnan, 1975) and the Aldan and Anabar Provinces in Siberia (Salop, 1968; Salop and Travin, 1972).

THE BASEMENT PROBLEM

One of the major problems in Archean granite-greenstone terranes is that of the nature of the basement upon which greenstone belts were erupted. Because contacts between greenstone successions and surrounding granitic terranes are often poorly exposed or faulted, it is not possible to determine the relative age relationships. Some contacts are clearly intrusive. It is not clear, however, if intrusive plutons, and in particular tonalitic diapirs, represent new additions of granitic magma to the crust of remobilization and diapiric intrusion of basement gneisses upon which greenstones were erupted. There are now many examples of unconformable relationships between gneisses and overlying greenstone successions (Windley, 1973; Shackleton, 1973a; Baragar and McGlynn, 1976). Some of the best documented cases are as follows: Steeprock Lake (Jolliffe, 1966), Cross Lake (Rousell, 1965), and Oxford Lake—Gods Lake (Elbers, 1973) in the Superior Province in Canada; Ross Lake (Davidson, 1972; Baragar and McGlynn, 1976), Benjamin Lake (Heywood and Davidson, 1969) and Point Lake (Stockwell, 1933) in the Slave Province in Canada (Fig. 1-9); beneath greenstone belts along the southern margin of the Karnataka subprovince in India (Nautiyal, 1966); in the Shabani belt in Rhodesia (Bickle et al., 1975; Wilson et al., 1978); and perhaps two unconformities in the Lawlers-Mt. Keith greenstone succession in the Yilgarn Province of Western Australia (Naldrett and Turner, 1977). Unconformities separate greenstone belts of more than one age in some areas (Chapter 2).

Existing evidence clearly indicates that some greenstone successions, in part or entirely, were erupted on older gneissic complexes. However, the fact that such complexes contain remnants of greenstones indicates that they, in turn, were preceded by an earlier episode (or episodes) of greenstone volcanism.

GEOPHYSICAL CHARACTERISTICS

Most geophysical data related to Archean granite-greenstone terranes come from the Superior Province of Canada (Goodwin et al., 1972). In general, there is a good correlation between surface geology and Bouguer gravity and magnetic anomalies. Positive Bouguer anomalies occur over greenstone belts, granulites, and mafic-ultramafic intrusions and negative anomalies characterize granitic terranes (Innes, 1960; Tanner, 1969). Such a distribution is predictable based on density contrasts of surface rocks. Detailed gravity studies in the northwestern Superior Province and in the Kaapvaal Province in Southern Africa indicate that greenstone belts are of shallow depth bottoming out between 3 and 6 km (Burley et al., 1970; Brisbin, 1971; Dusanowskyj, 1977). The fact that stratigraphic thicknesses in greenstone belts are considerably greater (10—20 km) suggests that such thicknesses may not represent true stratigraphic thicknesses. Gravity studies of granitic plutons in granite-greenstone terranes also indicate rather shallow depths for most plutons of the order of 3—10 km (Szewczyk and West, 1976; West, 1976).

Magnetic anomalies in granite-greenstone terranes exhibit broad swirling patterns (Fig. 1-4) and have amplitudes of a few hundred gammas (MacLaren and Charbonneau, 1968). Unexpectedly, magnetic highs correlate with granitic terranes and lows with greenstone belts. This peculiar correlation also exists after filtering of high-frequency anomalies and removal of the core-generated magnetic field (Bhattacharya and Morely, 1965; McGrath and Hall, 1969). Data indicate that the principal magnetization lies within the upper part of the crust. Since the surface granitic rocks do not exhibit large magnetizations, they must be underlain by rocks (probably granitic, to satisfy gravity data) that are more highly magnetized (Hall, 1968).

Heat flow measurements in Archean granite-greenstone terranes are not numerous. Most measurements come from the Superior and Yilgarn Provinces (Hyndman et al., 1968; Cermak and Jessop, 1971; Slack, 1974; Rao and Jessop, 1975; Allis, 1979; Jessop and Lewis, 1978). Most values fall in the range of 25—50 mW/m^2, averaging about 40 mW/m^2. This is similar to Precambrian shields as a whole (Slack, 1974; Condie, 1976a). Employing the linear relationship between heat flow and heat productivity of near-surface rocks (Roy et al., 1968), it would appear that 25—30 mW/m^2 come from the lower crust and/or mantle (Cermak and Jessop, 1971; Rao and Jessop, 1975). Jessop and Lewis (1978) reported an average reduced heat

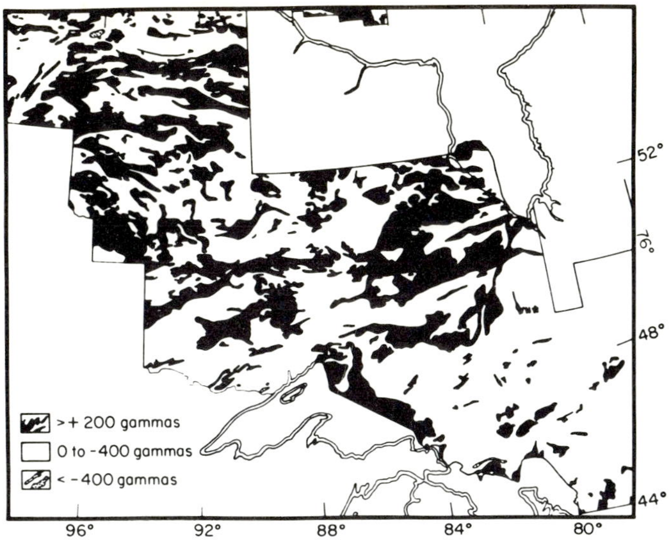

Fig. 1-4. Magnetic anomaly patterns in the northwestern part of the Superior Province (after Morley et al., 1967). Magnetic heighs (in black) overlie areas of granite and gneiss and lows (in white) overlie greenstone belts.

flow value for the Superior Province of 21 mW/m^2 and most heat generation values for the Archean crust in this province are $< 3\,\mu$W/m^3. Allis (1979) reported a correlation between Bouguer gravity anomaly and heat flow in the Superior Province. The slope of the regression line (1 mW/m$^2 \simeq -4$ mgal) is that which is expected from the relation between heat productivity and density of surface rocks in the Superior Province. Results are interpreted to reflect an exponential decrease in heat productivity to about 20 km depth followed by a rapid decrease in heat productivity (Fig. 1-5). The results also indicate a slight increase in heat productivity beneath greenstone belts.

Seismic surveys of the crust in Archean granite-greenstone terranes reveal that the crust ranges from 30 to 40 km thick (Goodwin et al., 1972). Again, almost all of the data come from the Superior Province (Hall and Hajnal, 1973; Berry, 1973; Wright and West, 1976; West et al., 1977; Young and West, 1977; Wright, 1977). Crustal models range from one to three layers. The Conrad discontinuity appears to be missing or of only local extent in some areas, while it is present in others (i.e., the western part of the Superior Province). Young (1979) has recently documented the presence of a crustal low-velocity zone 10—20 km deep and ~ 8 km wide in the northwestern Superior Province. A high-conductivity zone has also recently been described at similar depths in this area (Koziar and Strangway, 1978). The latter authors interpret the high-conductivity zone to reflect the presence of small amounts of pore-space water at mid-crustal depths. P_n velocities are about 8 km/s in the central and western parts of the Superior Province

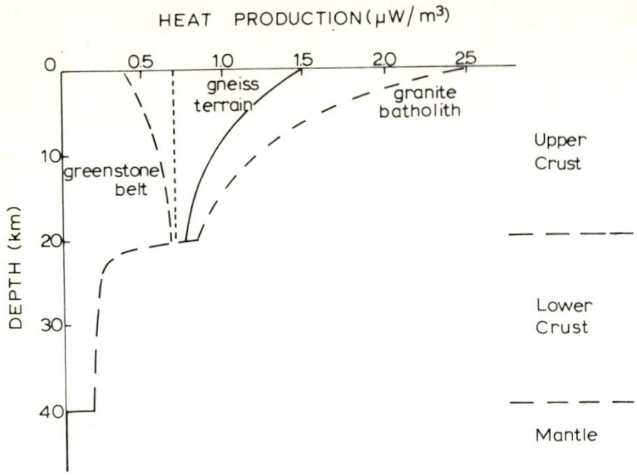

Fig. 1-5. Ideal heat productivity as a function of depth for three types of surface rock in Archean granite-greenstone terranes (from Allis, 1979).

rising to 8.1 km/s near Lake Superior and 8.3 km/s beneath Hudson Bay (Berry, 1973). Little evidence exists for an upper mantle low-velocity zone, and if one exists it cannot exceed 50 km in thickness (Barr, 1967). A high-velocity layer in the upper mantle has been recognized in the western part of the Superior Province (Gurbuz, 1970).

MAJOR GRANITE-GREENSTONE PROVINCES

North America

Superior Province

The Superior Province is the largest Archean province covering an area of about 2.6×10^6 km^2 in southeastern Canada (Fig. 1-1). Most whole-rock Rb-Sr isochron and U-Pb zircon ages from this province fall between 2.6 and 2.7 b.y. with a smaller number (principally from granitic rocks) falling between 2.8 and 3.0 b.y. Rocks 3.5—3.6 b.y. in age are recorded from the Minnesota River Valley along the extreme southwestern edge of the province (Goldich et al., 1970). The Superior Province may have been considerably larger prior to 1.8 b.y. as evidence by the numerous relict Archean ages in the surrounding Churchill Province. Recent Sm-Nd dates of sample composites from the Churchill Province also indicate that much of this province was formed during the Archean (McCulloch and Wasserburg, 1978). Based on tectonic features, lithologic associations, and metamorphic grade, the province has been divided into ten subprovinces (Stockwell et al., 1970; Goodwin, 1978) (Fig. 1-6). The subprovinces in the western part of the province occur as large belts of alternating volcanic-plutonic and sedimentary-plutonic rocks. To avoid confusion with the more conventional use of the

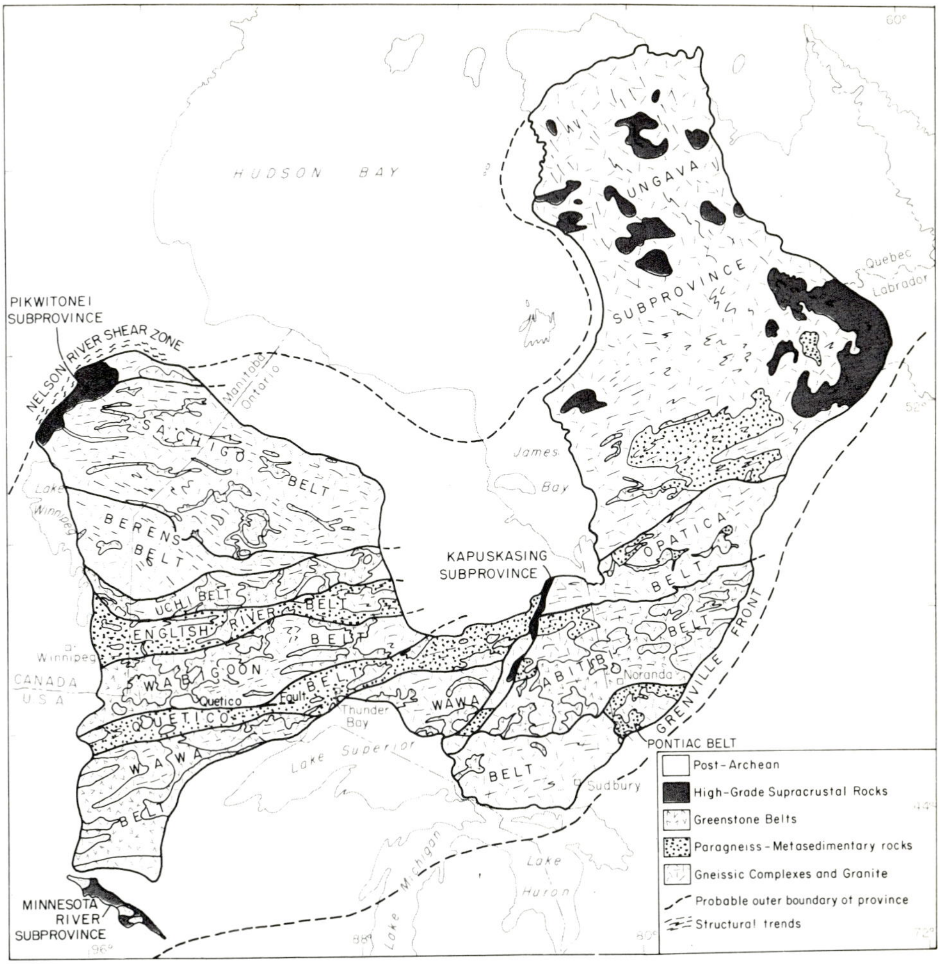

Fig. 1-6. Geologic map of the Superior Province, Canada and U.S.A. (after Goodwin et al., 1972; Goodwin, 1974, 1978).

term greenstone belt, these belts, which average from 50 to 200 km wide and up to 1300 km in length, are referred to as *superbelts* (Goodwin, 1978). From north to south, the volcanic-plutonic superbelts are the Sachigo, Uchi, Wabigoon, and Wawa (Shebandowan) belts; the sedimentary-plutonic superbelts are the Berens, English River, and Quetico belts. The volcanic-plutonic superbelts are composed of typical granite-greenstone terranes. The average abundances of rock types are summarized in Table 1-1. Metamorphic grade is typically greenschist facies. Unlike these superbelts, the sedimentary-plutonic belts are composed chiefly of paragneiss, granitic plutons, and sediments (chiefly graywacke-argillite) (Breaks et al., 1978). Metamorphis grade in the sedimentary-plutonic belts is typically amphibolite facies and may

TABLE 1-1

Average lithologic abundances (in percent) in superbelts of the Archean Superior Province (after Goodwin, 1978).

Superbelt type	Granitic plutons	Orthogneiss	Paragneiss	Volcanic rocks	Sedimentary rocks
Volcanic-plutonic	29	39	2	24	6
Sedimentary-plutonic	35	17	37	1	10

locally range upwards to the granulite facies. The English River and Quetico Superbelts can be further subdivided into domains in which paragneiss-metasediment and granitic plutons dominate, respectively.

Superbelt boundaries fall into one or a combination of three categories (Goodwin, 1978): lithologic transitions, transcurrent faults, or rapid transitions in metamorphic grade. An example of a lithologic transition is volcanic rocks of the Uchi Superbelt that interfinger with and pass into graywacke-argillite of the English River Superbelt to the south. Stratigraphic and structural studies of Ayres (1969) suggest an intertonguing relationship between clastic sediments of the Quetico Superbelt and volcanics and clastic sediments of the adjacent Wabigoon and Wawa Superbelts (Fig. 1-7). The greenstone belts evolve in much the same manner described in Chapter 2 with sediments and calc-alkaline volcanics becoming important towards the tops of greenstone successions. In the Wawa Superbelt, a second volcanic cycle is shown at the top marked by the reappearance of mafic volcanics. The boundary of the Quetico and Wabigoon Superbelts is accentuated throughout much of its length by the dextral, strike-slip Quetico fault. In the vicinity of DeCourcey and Smiley Lakes, however, this is marked by a change in metamorphic grade over a distance of 4—6 km (Fig. 1-8) (Kehlenbeck, 1976). Pelitic sediments in the contact zone increase in metamorphic grade in going south towards the gneisses of the Quetico belt. This increase in grade and eventual appearance of migmatites in the Quetico belt suggests a steeper geothermal gradient and perhaps a greater burial depth represented in the Quetico than in the Wabigoon Superbelt.

The superbelts in the Western part of the Superior Province are disrupted by the Kapuskasing fault zone south of James Bay (Fig. 1-6). This northeasternly trending fault zone defines the eastern boundary of the Kapuskasing subprovince. The western boundary of this subprovince is also, in part, fault bounded (Stockwell et al., 1970). The Kapuskasing subprovince is characterized by gravity and magnetic highs which disrupt the ENE trend of the western superbelts. The southern part of the subprovince is comprised of typical granite-greenstone or metasedimentary terranes while the northern

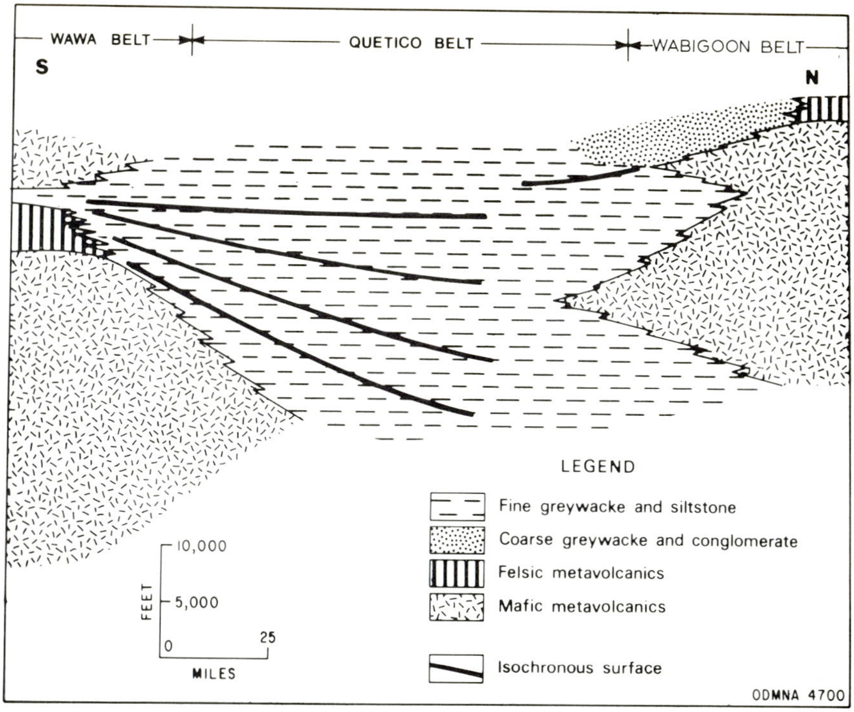

Fig. 1-7. Diagrammatic cross-section through the Wawa-Quetico-Wabigoon Superbelts prior to deformation and plutonism (after Ayres, 1969).

part, which lies in a fault-bounded horst approximately 200 km long and 10–20 km wide, contains granulite-facies rocks. This portion of the subprovince is thought to represent a deeply eroded continental rift zone similar to the East African rift zone (Innes, 1960; Innes et al., 1967).

East of the Kapuskasing subprovince, the alternating superbelt pattern seems to continue, although it is not as well developed as in the western Superior Province (Fig. 1-6). The Abitibi subprovince is generally considered as one vast greenstone belt with associated granites (Chapter 3). Individual greenstone successions completely interfinger in this belt. The Ungava subprovince is the largest and most poorly known portion of the Superior Province (Stockwell et al., 1970; Herd, 1978). It is composed dominantly of granitic rocks with a smaller amount of supracrustal rocks metamorphosed to the hornblende-granulite facies. The structural trends change in this subprovince from the typical ENE trends characteristic of the southern and western parts of the Superior Province to a northernly trend in the northern portion of the subprovince. Although the subprovince contains some recognizable greenstone belts in the southern part, the relationship of this large, mostly high-grade region, to the alternating superbelts to the south is largely unknown. It is possible that the Ungava subprovince represents a

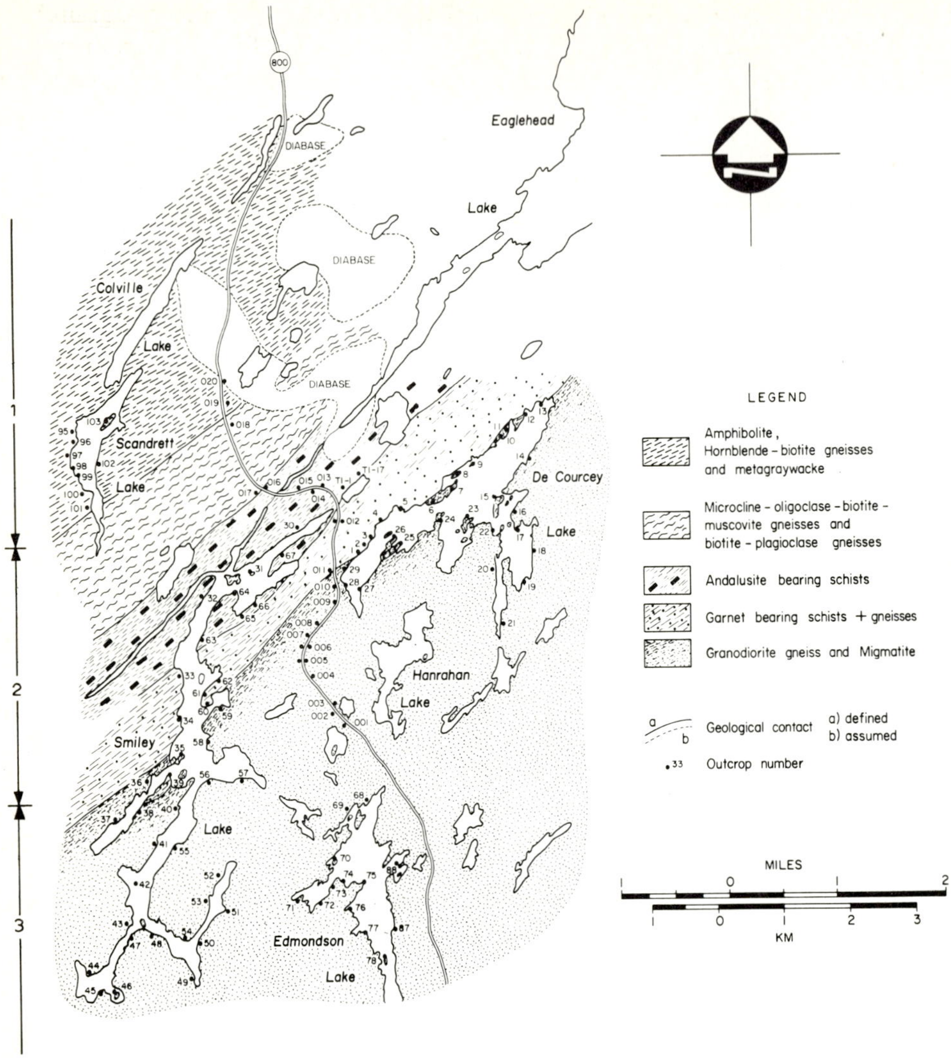

Fig. 1-8. Geologic map of the boundary zone of the Quetico and Wabigoon superbelts near De Courcey and Smiley Lakes, Ontario (after Kehlenbeck, 1976; reproduced by permission of the National Research Council of Canada). Key: *1* = Wabigoon belt, *2* = Transition zone, *3* = Quetico belt.

deeper erosion level of the granite-greenstone and metasedimentary superbelts. Another granulite-facies subprovince, the Pikwitonei, occurs along the northwestern edge of the Superior Province. This subprovince, which abuts the Churchill Province along the Nelson River shear zone, may represent an uplifted portion of an old craton ($\geqslant$ 3.0 b.y. in age) (Ermanovics and Davison, 1976). High-grade rocks also occur along the extreme southwestern margin of the Superior Province in the Minnesota River subprovince. A granulite-facies metamorphic terrane has been reported in this area (Himmelberg and Phinney, 1967). Archean granitic rocks have also recently been described from northern Wisconsin although their distribution and relation to other parts of the Superior Province are not as yet known (Van Schmus and Woolsey, 1975; Van Schmus, 1976; Van Schmus and Anderson, 1977).

Although radiometric ages are available from many localities in the Superior Province, only a few areas have been studied in detail. The oldest reported rocks from the province are the tonalite-trondhjemite gneisses in the vicinity of Morton and Montevideo in the Minnesota River subprovince. Existing Rb-Sr whole-rock isochron dates and U-Pb zircon dates suggest an age for this gneissic complex in the range of 3.5—3.7 b.y. (Goldich et al., 1970; Goldich, 1972; Goldich and Hedge, 1974). These gneisses are intrusive bodies and contain inclusions of amphibolite and pyroxene granulite that may represent remnants of still older greenstone belts (Grant, 1972). Preliminary results indicate that some of the granitic components in the gneisses formed at approximately 3.0 b.y. (Goldich and Hedge, 1974). A major regional metamorphism at 2.6 b.y. is recorded by mixed whole-rock and mineral Rb-Sr isochrons and is contemporary with the first deformation (F_1). The Sacred Heart and Ortonville quartz monzonite plutons yield Rb-Sr and U-Pb zircon dates of 2.6—2.7 b.y.; field relationships suggest they are late-syntectonic with F_1 (Grant, 1972). Minor structures (F_2) post-date F_1 and may be as young as 2.1 b.y. (Hanson and Himmelberg, 1967). Granitic plutons and a few mafic dikes were emplaced at 1.7—1.8 b.y. and partial resetting of Rb-Sr and K-Ar mineral systems reflect minor metamorphism at this time.

The geochronology of the Rainy Lake area along the Minnesota-Ontario boundary has been reported by Peterman et al. (1972). Both Rb-Sr and zircon dates confirm that graywacke sedimentation, magmatic activity (both greenstone and granitic), folding, and metamorphism occurred over an approximately 50 m.y. time interval of 2.65—2.7 b.y. Five events are reported from the western part of the English River Superbelt (Krogh et al., 1976; Goodwin, 1977; Wooden, 1978; Gower et al., 1978): (1) intrusion at ~ 3.0 b.y. of tonalitic gneisses that contain amphibolite inclusions which may represent remnants of older greenstone belts; (2) a major volcanic-plutonic episode at about 2.7 b.y.; (3) deformation and regional metamorphism at 2.62—2.63 b.y.; (4) emplacement of diapiric granitic plutons between 2.55 and 2.6 b.y.; and (5) emplacement of minor post-tectonic plutons at about 2.5 b.y. Late granitic plutons (2.45—2.5 b.y.) are also reported in other areas in the province (Jones et al., 1974).

Major magmatic events in the Superior Province fall into four groups (in b.y.): 3.5—3.7, 2.8—3.0, 2.6—2.7, and 2.45—2.50 (Fig. 1-20). As mentioned above, the 3.5—3.7 b.y. rocks have thus far been reported only from the Minnesota River Valley. The 2.8—3.0 b.y. event is chiefly, if not exclusively, a plutonic event (Wooden, 1978). How widespread this event may be is unknown currently. Data indicate that it may be more widespread in the sedimentary-plutonic than in the volcanic-plutonic superbelts. Although greenstone belts older than 2.7 b.y. have not been recognized in the Superior Province, inclusions of amphibolite, chert, and ultramafic rock in the 3.5—3.7 b.y. and 2.8—3.0 b.y. granitic terranes suggest the presence of one or more pre-2.7-b.y. periods of greenstone-belt volcanism. The major period of greenstone belt and granite formation occurred throughout most of the Superior Province at 2.6—2.7 b.y. Many greenstone belts and associated granitic rocks appear to have formed between 2.65 and 2.7 b.y. The most widespread plutonism, regional metamorphism, and deformation is between 2.6 and 2.65 b.y. In some areas, such as Rainy Lake and northeastern Minnesota (Jahn and Murthy, 1975; Arth and Hanson, 1975), the entire cycle of volcanism, sedimentation, burial, deformation, and plutonism appears to have occurred in 50—70 m.y. Minor, but widespread, post-tectonic granitic plutonism is recorded between 2.45 and 2.5 b.y. and widespread heating probably related to regional uplift is recorded by mineral ages at 2.1—2.3 b.y.

Slave Province

The Slave Province underlies an area of about 250,000 km^2 in northwestern Canada (Fig. 1-1). This province, which is roughly ovoid in shape, is comprised of about 40% greenstone belts and 60% granitic terrane (McGlynn and Henderson, 1970, 1972). Greenstone belts trend north to northeast in the western half of the province and west to northwest in the eastern half (Fig. 1-9). All of the greenstone successions in the province are grouped into the Yellowknife Supergroup (Henderson, 1970). These successions occur in three major areas (Fig. 1-9): along the western margin of the province, in a belt extending north from Great Slave Lake in the central part of the province, and in a large rather equidimensional area in the northeast. The areas are separated by complex granitic-gneiss terranes. Approximately 15—20% of the greenstone successions are composed principally of mafic volcanic rocks and the remainder of monotonous graywacke-argillite sequences. These proportions of volcanic and sediment are opposite to those observed in most Archean greenstone belts, thus rendering the Slave Province rather unique in this respect. The volcanic rocks occur in 18 or 19 greenstone belts mostly along the margins of the sedimentary sections and at the base of the succession (Baragar and McGlynn, 1976). Thicknesses of individual sections range from 300 m to 12 km averaging about 3 km. Unconformities between the Yellowknife Supergroup and gneissic basement are exposed at three locations in the province (Fig. 1-9).

Major folds in the greenstone belts are isoclinal and trend north to northwest. These are modified in many areas by later cross-folds. Sedimentary assemblages are generally more tightly folded than volcanic assemblages. Metamorphic grade in the province ranges from greenschist to amphibolite facies. The western border of the province is chiefly a fault contact with

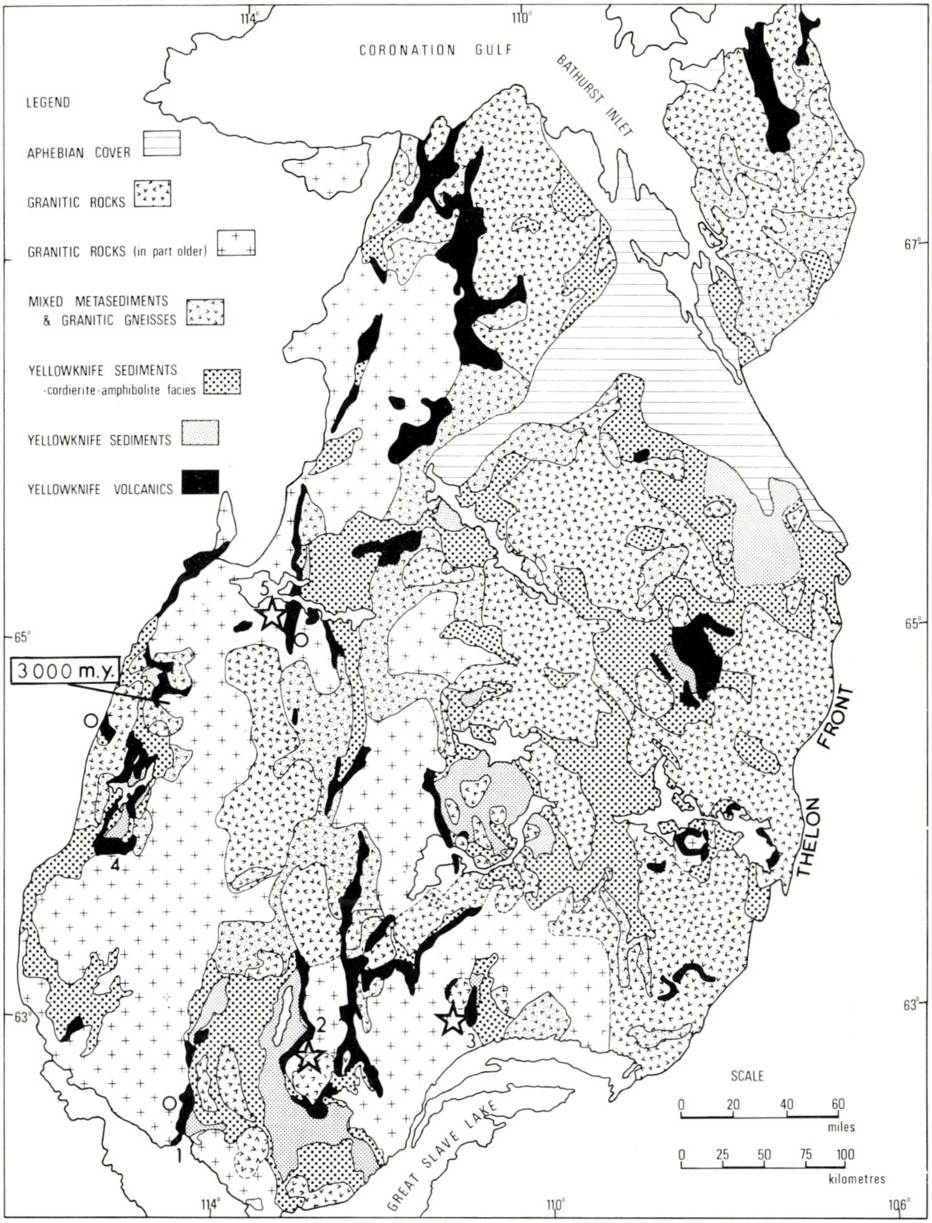

Fig. 1-9. Geologic map of the Slave Province, Canada (from Baragar and McGlynn, 1976). Stars represent locations of unconformities beneath the Yellowknife Supergroup. Location of 3000-m.y.-old gneisses are also shown. Greenstone belts: *1* = Yellowknife, *2* = Cameron River—Ross Lake, *3* = Benjamin Lake, *4* = Indin Lake, *5* = Point Lake.

gneissic terranes of the Proterozoic Bear Province. This latter province, however, appears to contain appreciable amounts of Archean rock along its eastern margin (Frith et al., 1977). Proterozoic sediments overlie the southeastern border and part of the northern part of the province in the Bathurst Inlet area. The Thelon Front forms the eastern margin of the province.

Segments of Archean crust have been identified in the Churchill Province between the Slave and Superior Provinces (Goodwin, 1974) (Fig. 1-1). Two major problems in Canadian geology at the present time are: (1) just how much of the Churchill Province represents reworked Archean crust, and (2) were the Slave and Superior Provinces originally part of the same province which are separated today by deformed Proterozoic supracrustal and granitic rocks which were deposited on and intruded into the Archean crust?

The oldest rocks in the Slave Province are tonalite-trondhjemite gneisses 2.95—3.0 b.y. in age (Frith et al., 1974, 1977). Remnants of amphibolite in these gneisses may represent, in part, fragments of older greenstone belts. So far, rocks of this age have been found only along the western margin of the province in the vicinity of Grenville Lake (Fig. 1-9). They appear to form at least in part, the basement upon which the Yellowknife Supergroup was deposited. The major period of greenstone belt development, deformation, pluton emplacement, and regional metamorphism falls within a small time interval of 2.55—2.60 b.y. (Green and Baadsgaard, 1971; Frith et al., 1977). Some gneisses record ages of 2.65 b.y. Post-tectonic quartz monzonites emplaced between 2.45 and 2.5 b.y. are of minor importance in the province. K-Ar mineral ages reflect one or more thermal (rifting?) events between 2.1 and 2.25 b.y. and/or regional uplift.

Wyoming Province

The Wyoming Province underlies an area of about 420,000 km^2 in the western United States and is exposed in the cores of the major mountain ranges in Wyoming and Montana and in a few minor areas in other states surrounding Wyoming (Fig. 1-10) (Condie, 1969a, 1976b). Because of inadequate exposure, the boundaries of the province are poorly known. The southern boundary is exposed in the Sierra Madre and Medicine Bow Mountains in southeastern Wyoming (Fig. 1-2). At these locations a shear zone separates Archean gneisses from Proterozoic metamorphic and granitic rocks (Hills et al., 1968; Divis, 1977). The western extent of the province is not known although Archean Rb-Sr isochrons from young basalts suggest that it extended at least as far west as central Idaho (Leeman, 1975).

Fig. 1-10. Geologic map of the Wyoming Province, U.S.A. (after Condie, 1976b). Greenstone belts: *1* = South Pass, *2* = Owl Creek, *3* = Western Granite Mountains, *4* = Rattlesnake Hills, *5* = Seminoe Mountains, *6* = Casper Mountain.

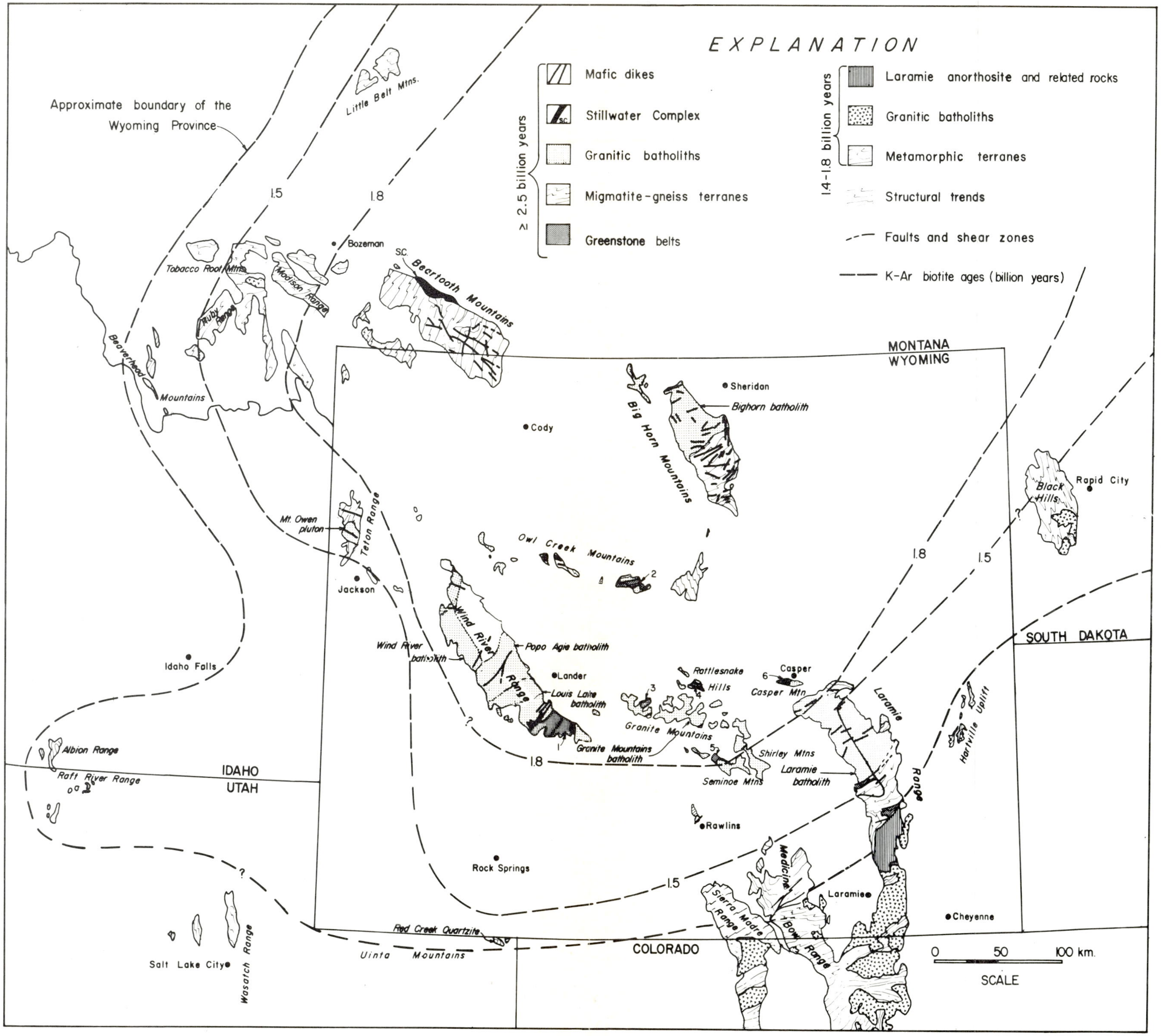
EXPLANATION
≥ 2.5 billion years
Mafic dikes
Stillwater Complex
Granitic batholiths
Migmatite-gneiss terranes
Greenstone belts
1.4-1.8 billion years
Laramie anorthosite and related rocks
Granitic batholiths
Metamorphic terranes
Structural trends
Faults and shear zones
K-Ar biotite ages (billion years)
Approximate boundary of the Wyoming Province
Little Belt Mtns.
Bozeman
Tobacco Root Mtns
Madison Range
Ruby Range
SC.
Beartooth Mountains
Beaverhead Mountains
MONTANA
WYOMING
Sheridan
Bighorn batholith
Big Horn Mountains
Cody
Rapid City
Black Hills
Mt. Owen pluton
Teton Range
Jackson
Owl Creek Mountains
Wind River batholith
Wind River Range
Popo Agie batholith
Lander
Louis Lake batholith
Rattlesnake Hills
Casper
Casper Mtn
Granite Mountains
Granite Mountains batholith
Shirley Mtns
Seminoe Mtns
Laramie batholith
Laramie Range
Hartville Uplift
SOUTH DAKOTA
Idaho Falls
Albion Range
Raft River Range
IDAHO
UTAH
Rawlins
Rock Springs
Medicine Bow Range
Sierra Madre Range
Laramie
Cheyenne
Red Creek Quartzite
Uinta Mountains
COLORADO
Salt Lake City
Wasatch Range
1.5
1.8
0 50 100 km.
SCALE

Of the exposed Wyoming Province, gneisses and migmatites comprise about 60% and granitic plutons 30%. Greenstone belt remnants, of which there is only one major example, the South Pass belt (Bayley et al., 1973), comprise most of the remainder of the province (Fig. 1-10). The Stillwater Complex occurs in the northern Beartooth Mountains. In southwestern Montana and northeastern Utah, remnants of some combination of quartzite, iron formation, mica schist, and marble occur in the gneiss terrane. The relationship of these supracrustal rocks to the greenstone belt remnants is unknown, but they appear to represent stable-shelf or miogeoclinal prisms along the margins of the Wyoming Province. Diabase dikes form a minor but widespread component throughout the province. Structurally, the Wyoming Province is very complex and no overall structural trend characterizes the province as a whole. Metamorphic grade ranges from the upper greenschist to the lower granulite facies with amphibolite-facies terranes being most widespread.

A large number of radiometric ages are available from the Wyoming Province as summarized by Condie (1969a, 1976b) and Reed and Zartman (1973). Discordant zircon dates from gneisses in the Beartooth Mountains indicate an age greater than 3.2 b.y. (Catanzaro and Kulp, 1964) and perhaps as old as 3.8 b.y. Rb-Sr isochron dates of 2.85—3.0 b.y. are recorded by granitic plutons of variable composition from the Teton, Granite, Bighorn, and Sierra Madre Mountains. Deformation and regional metamorphism appear to be of the same age in these areas. Amphibolite remnants in tonalitic-trondhjemitic gneisses of this age may represent fragments of early greenstone belts. Post-tectonic mafic dikes were emplaced in the Bighorn Mountains at about 2.85 b.y. (Stueber et al., 1976). As in the Slave Province, the most widespread period of greenstone belt development, deformation, plutonism, and regional metamorphism is between 2.6 and 2.7 b.y. The Stillwater Complex and pre- to post-tectonic mafic dikes were emplaced during this time. Post-tectonic granitic plutons were emplaced at 2.45—2.5 b.y. in the Teton and Medicine Bow Mountains. One or more diabase dike swarms were intruded in the central part of the province between 2.0 and 2.15 b.y. Mineral ages, especially around the outer parts of the province, recorded heating accompanying the Hudsonian orogeny between 1.6 and 1.8 b.y. The effect of heating diminishes inwards from the margins of the province as indicated by the contours of equal K-Ar biotite ages in Fig. 1-10. The relatively large number of high initial $^{87}Sr/^{86}Sr$ ratios (0.703—0.706) from the Wyoming Province (Fig. 10-22) indicate that the igneous rocks cannot have been derived directly from the mantle which would have had ratios of 0.700—0.701 at this time. Reworking of older crustal rocks ($\gtrsim$ 3.2 b.y.) is consistent with the high initial ratios in the 2.9- and 2.6-b.y. terranes. Relict Archean ages from Precambrian basement rocks between the Wyoming and Superior Provinces suggest that these two provinces were originally part of the same province.

Kaapvaal Province

Although the Kaapvaal Province underlies about 600,000 km^2 of southern Africa (Fig. 1-1), the chief exposures are concentrated in a narrow belt along the eastern margin of the province (Fig. 1-11). Most of the province is overlain by cratonic sediments of the Kaapvaal basin or by Phanerozoic sediments. Of the exposed rocks in the province, 91% are gneisses and granites and the remainder are chiefly greenstone belts (Anhaeusser, 1976a). The major greenstone belts are the Barberton, Murchison, and Pietersburg belts all of which exhibit a northeasterly trend similar to the dominant structural trend in the Limpopo belt to the north (Mason, 1973; Hunter, 1974a). In the western part of the province, the few greenstone remnants exposed exhibit a more northerly trend. The Barberton greenstone belt (Fig. 1-11) contains the most complete and one of the best preserved greenstone successions known (Anhaeusser, 1973a). This succession is the type locality for komatiites (Viljoen and Viljoen, 1969b). The nature of the contact of this succession with the adjacent Ancient Gneiss Complex in Swaziland is matter of much discussion. One school favors an unconformable relationship (Hunter, 1974a) and the other favors a tectonic contact with the Barberton volcanic succession being older (Anhaeusser, 1973a). Metamorphic grade in the Kaapvaal Province ranges from greenschist to amphibolite facies with some possible remnants of granulite-facies rocks in the Ancient Gneiss Complex (Hunter, 1973). The boundaries of the Kaapvaal Province are poorly exposed. On the north, the province grades into or is in fault contact with the Limpopo mobile belt (Kröner, 1977b). The only clear exposure of the contact is west of Prieska, where the province is in fault contact with the Namaqua mobile belt (Pretorius, 1974).

The oldest radiometric ages from the Kaapvaal Province come from the Barberton region. An Rb-Sr mineral isochron from a mafic volcanic in the lower Onverwacht Group suggests an age for this group of 3.4 b.y. (Jahn and Shih, 1974). A minimum age for the Middle Marker in the Swaziland Supergroup is given by an Rb-Sr isochron date of 3.28 b.y. (Hurley et al., 1972). A recent Sm-Nd isochron date of 3.45 ± 0.3 b.y. from Onverwacht volcanic rocks probably provides the most accurate age for the Barberton succession (Hamilton et al., 1979). Because of a complex deformational history, it has been difficult to date the Ancient Gneiss Complex. Reported isochron ages range from 3.1 to 3.3 b.y. with initial ratios of 0.7022—0.7060 (Hunter, 1974a; Davies and Allsopp, 1976). The results of Davies and Allsopp (1976) fall on a growth curve (with Rb/Sr = 0.53) which intersects the mantle growth curve at 3.3—3.4 b.y. They interpret these results to reflect production of gneisses at about 3.4 b.y. followed by repeated metamorphism and Sr homogenization. Hunter et al. (1978), however, believe the growth curve is fortuitous because field relations suggest the gneissic samples in each isochron

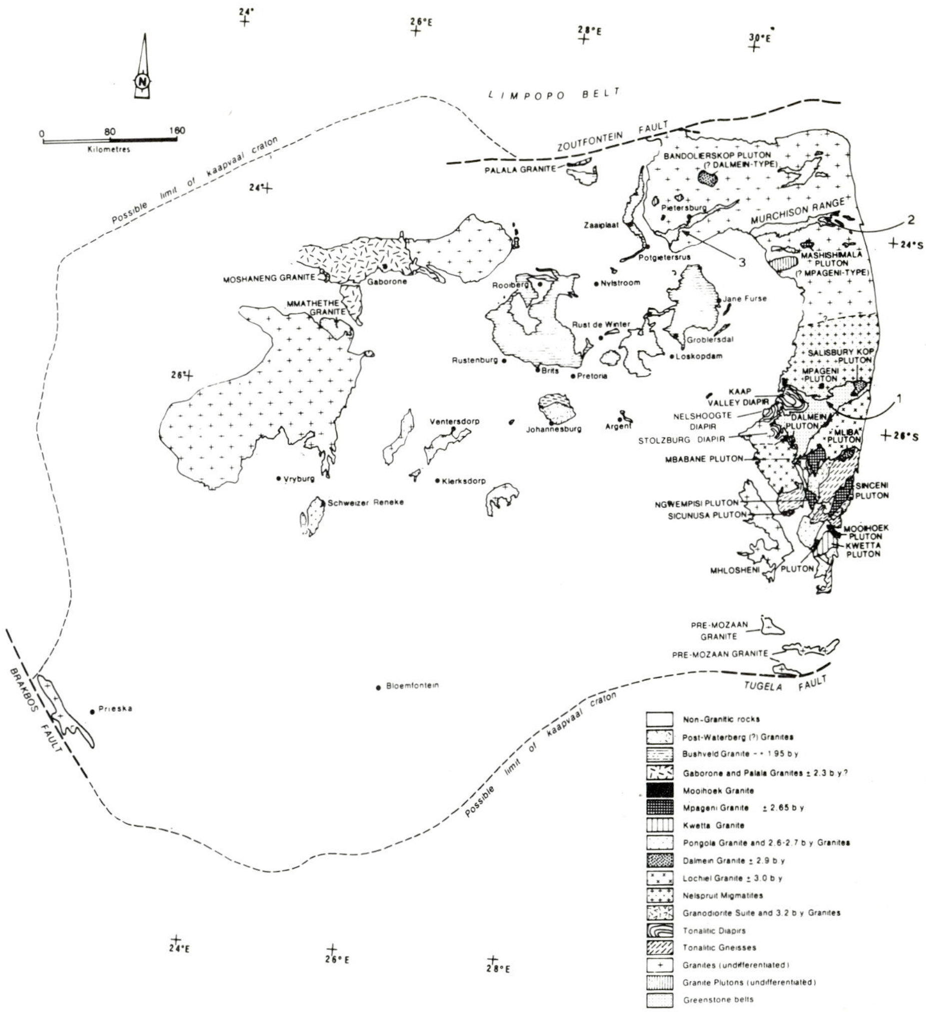

Fig. 1-11. Geologic map of the Kaapvaal Province, southern Africa (after Hunter, 1974a). Major greenstone belts: *1* = Barberton, *2* = Murchison, *3* = Pietersburg.

are not related to each other. They suggest that the Ancient Gneiss Complex formed prior to 3.4 b.y. (3.5—3.7 b.y.) and that after emplacement, its $^{87}Sr/^{86}Sr$ ratio grew along growth curves with low Rb/Sr ratios but that later metamorphism redistributed Rb and Sr increasing the Rb/Sr ratios. This interpretation would make the Ancient Gneiss Complex several hundred million years older than the Barberton greenstone belt, whereas the interpretation of Davies and Allsopp (1976) suggests they are about the same age. The tonalitic diapirs in the Barberton region were emplaced between 3.1 and 3.2 b.y. and this was also the probable time of major deformation and regional metamorphism. At about 2.8—2.9 b.y., the Lochiel batholith and associated granitic plutons of the Dalmein-type were emplaced and finally widespread granitic plutonism is recorded between 2.6 and 2.7 b.y. (Hunter, 1974a, b). The Usushwana stratiform complex was intruded at about 2.8 b.y. It is noteworthy also that the Kaapvaal basin began to develop at about 2.8—2.9 b.y. with the deposition of the Pongola Supergroup south of the Barberton region. Continued cratonic sedimentation and volcanism is recorded in the Dominion Reef and Witwatersrand Supergroups between 2.5 and 2.8 b.y. Cratonic sedimentation (± volcanism) continued until about 1.8 b.y. as the Kaapvaal basin migrated northward and another less widespread period of granitic plutonism is recorded at about 2.25 b.y. High initial strontium ratios in many of the Archean granites from the Kaapvaal Province (0.703—0.707) necessitate some crustal contribution during their formation (Hunter, 1974b).

Rhodesian Province

The Rhodesian Province is an oval-shaped granite-greenstone province underlying about 300,000 km^2 in eastern Rhodesia (Bliss and Stidolph, 1969) (Fig. 1-1). It is bounded on the north, east, and south by Precambrian mobile belts (Fig. 1-12) which, in part, contain reworked rocks of the Rhodesian Province. Contacts with these mobile belts, although poorly exposed, appear to represent increases in metamorphic grade over distances of many kilometers (Wilson, 1973a; Mason, 1973; Key et al., 1976). Approximately 83% of the rocks in the province are gneisses and various granitic rocks with the remainder being chiefly greenstone belts (Anhaeusser, 1976a). The structural grain in the province is complex and reflects polyphase deformation and diapiric plutonism (Macgregor, 1951). Three trends are recognized (Fig. 1-12) by Wilson (1973a): NNE, NW, and ENE. The Great Dyke cross-cuts the province with a NNE trend. This dike is a series of at least four stratiform igneous bodies that were emplaced at 2.46 b.y. (Davies et al., 1970; Hamilton, 1977). Metamorphic grade ranges from prehnite-pumpellyite to amphibolite facies; a few remnants of granulites have also been reported from the older gneisses in the Selukwe area (Stowe, 1973). Two ages of greenstone belts are now recognized in Rhodesia (Wilson et al., 1978); these are separated by an angular unconformity in the Shabani area. The details of this division are discussed in Chapter 2.

A large number of radiometric dates have been reported from the Rhodesian Province in recent years (Hawkesworth et al., 1975; Jahn and Condie, 1976; Moorbath et al., 1976, 1977a; Hamilton, 1977; Hickman, 1978). The oldest dated rocks in the Rhodesian Province are tonalitic-trondhjemitic gneisses occupying a roughly triangular area with Selukwe, Fort Victoria, and Shabani as the corners (Fig. 1-12) (Wilson et al., 1978). These gneisses are about 3.5 b.y. in age and contain infolded remnants of greenstone belts, referred to as the Sebakwian greenstone belts. The gneisses are cut by the 3.45-b.y. Mushandike granite in the Mashaba area (Hickman, 1974) and the

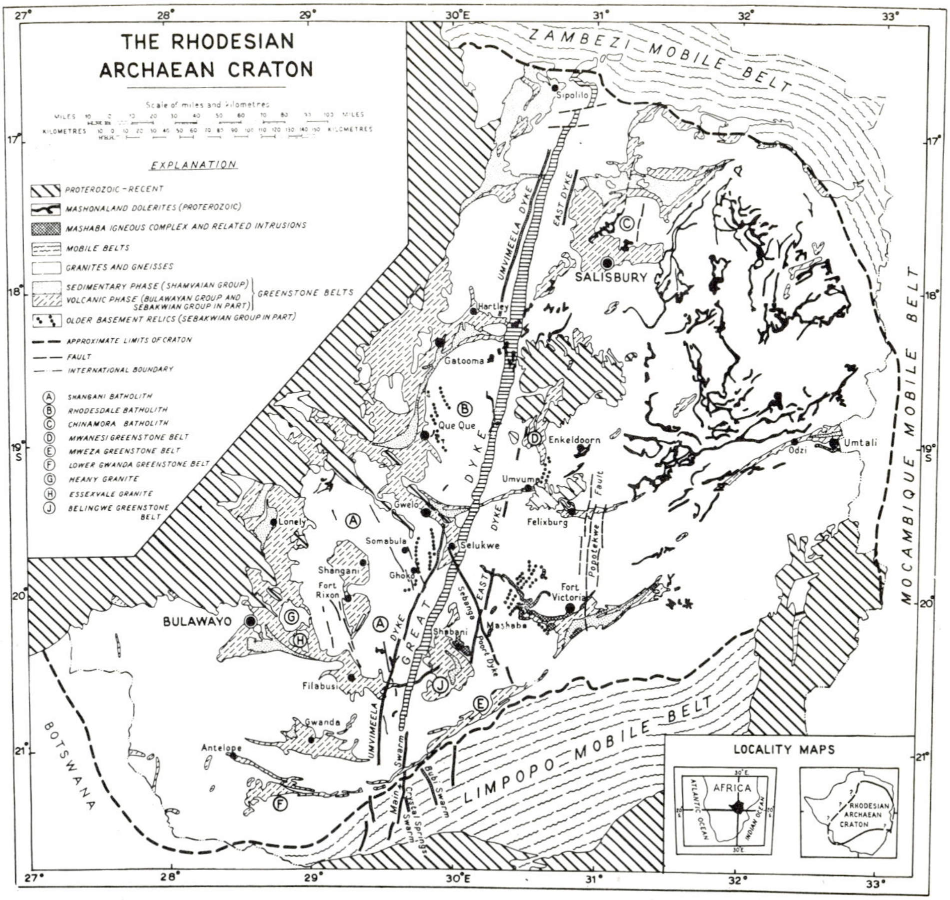

Fig. 1-12. Geologic map of the Rhodesian Province, southern Africa (after Wilson, 1973a).

Mont d'Or granite (3.35 b.y.) intrudes the Selukwe greenstone belt south of Selukwe (Moorbath et al., 1976). A conglomerate within this greenstone belt (in the Wanderer Formation) contains granitic boulders which indicate the existence of pre-3.5-b.y. sialic crust in this region. Rb-Sr isochron dates from the Mashaba tonalite (2.9 b.y.) and the Chingezi gneisses (2.82 b.y.) and K-Ar dates from other granites in the province suggest a period of granitic plutonism of unknown extent at 2.8—2.9 b.y. (Wilson et al., 1978; Hawkesworth et al., 1975). The major period of greenstone belt formation, granitic plutonism, metamorphism and deformation occurred between 2.6 and 2.7 b.y. This includes the formation of the Bulawayan and Shamvaian Groups, intrusion of mafic dikes and the Mashaba Complex, and emplacement of various tonalitic plutons. Numerous granites and quartz monzonites of the Chilimanzi suite were intruded between 2.55 and 2.6 b.y., some

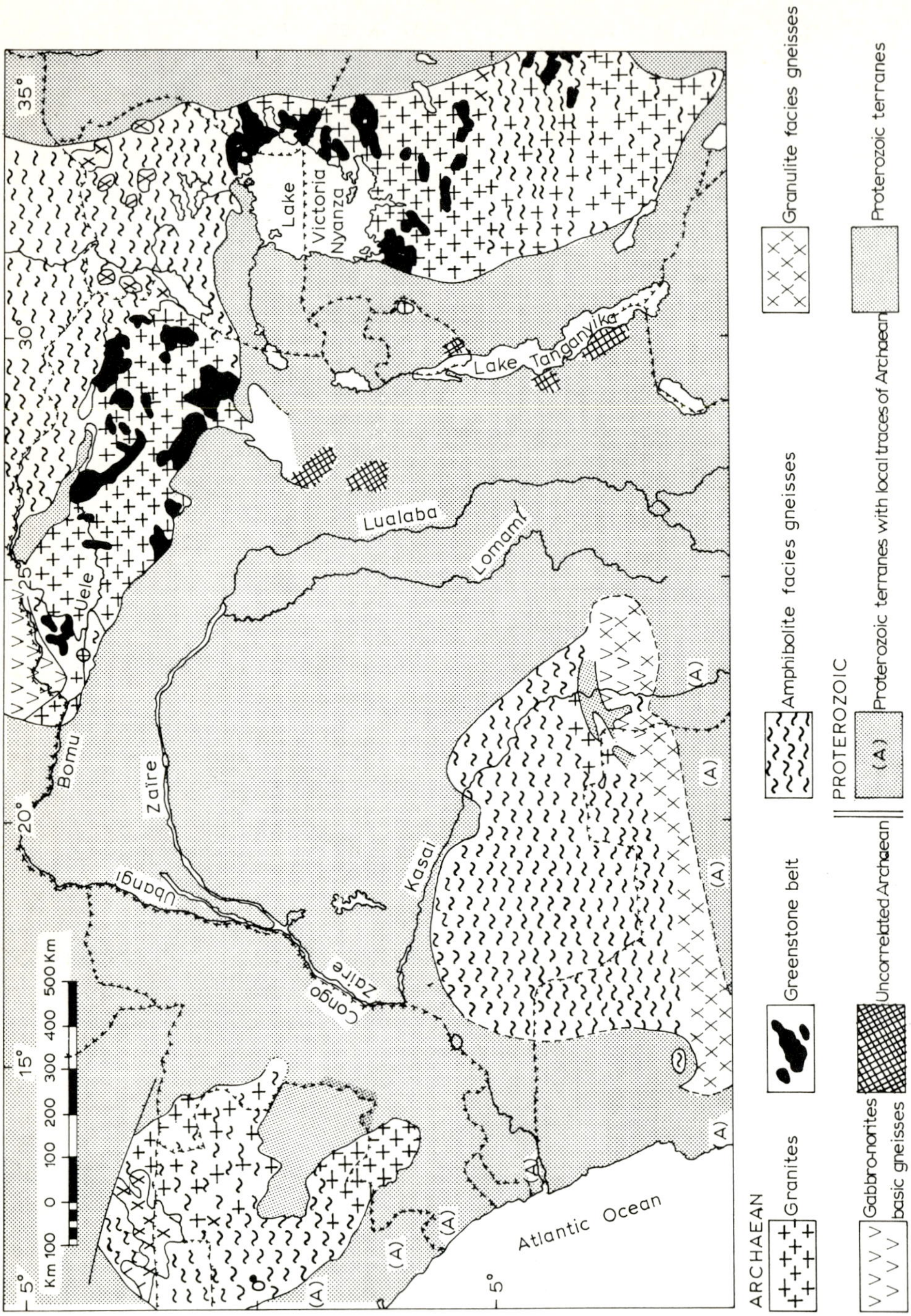

Km 100 0 100 200 300 400 500 Km
5°
15°
20°
25°
30°
35°
5°
Lake Victoria Nyanza
Lake Tanganyika
Lualaba
Lomami
Uele
Bomu
Zaïre
Ubangi
Congo
Zaïre
Kasai
Atlantic Ocean
(A)
ARCHAEAN
Granites
Greenstone belt
Amphibolite facies gneisses
Granulite facies gneisses
Gabbro-norites basic gneisses
Uncorrelated Archaean
PROTEROZOIC
Proterozoic terranes with local traces of Archaean
Proterozoic terranes

of which have somewhat high initial $^{87}Sr/^{86}Sr$ ratios (0.703—0.704) reflecting some reworking of sialic crust not older than about 3.0 b.y. (Hickman, 1978). The last major Archean event recorded in the Rhodesian Province is intrusion of the Great Dyke at 2.46 b.y. (Hamilton, 1977). The Mashonaland dolerites were intruded at about 1.85—1.9 b.y. (Wilson, 1973a).

Central African Province

The Central African Province underlies a poorly defined area of about 1.3×10^6 km^2 chiefly in Zaire, Uganda, Kenya, and Tanzania (Fig. 1-1). The province is now completely surrounded and, in part, engulfed and reworked by Proterozoic mobile belts (Kröner, 1977a). Remnants of Archean rock in these mobile belts suggest the province may have been much larger. The province is composed of mixed high-grade and granite-greenstone terranes. The principal granite-greenstone terrane lies in a northwesterly trending belt extending from northeastern Zaire across Lake Victoria into Kenya and Tanzania (Fig. 1-13) (Cahen et al., 1976). Greenstone belts at low metamorphic grades with well-preserved primary textures and structures occur in the Nyanzian-Kavirondian successions in Kenya (Haughton, 1963; Pallister, 1971) and in the Kibalian sequence in northeastern Zaire (Lavreau and Ledent, 1975; Cahen et al., 1976). The Nyanzian successions differ from many greenstone belts by the relatively large amounts of andesite and rhyolite present. The overlying Kavirondian succession is composed chiefly of graywacke-argillite, tuff, arkose, and conglomerate. High-grade terranes, which include gneissic complexes containing supracrustal rocks ranging in metamorphic grade from the amphibolite to the granulite facies, are represented by the West Nile Complex in Uganda and Zaire, the Dodoman System in Tanzania, and the Bomu Complex in Zaire (Hepworth, 1964, 1972; Hepworth and MacDonald, 1966; Cahen et al., 1976; Kröner, 1977b). Structural trends are complex and variable. In the Nyanzian and Kibalian greenstone-granite terranes they are dominantly N to WNW, while in the West Nile gneisses in Uganda they are more northerly (Tanner, 1973).

Accurate radiometric dates from the Central African Province that are tied closely to well-documented field relationships are rare. The oldest dates from the province of about 3 b.y. are from granitic rocks (Leggo, 1974), although as pointed out by Kröner (1977b), it is likely that sialic rocks of greater age are present. A lead model age (Cahen and Snelling, 1966) suggests the existence of 3.5-b.y.-old rocks in northeastern Zaire in the Bomu Complex. Granite emplacement, deformation, and high-grade (granulite-facies) metamorphism are recorded in Uganda and northeastern Zaire at 2.9—3.0 b.y. (Leggo, 1974; Cahen et al., 1976). Available data suggest the

Fig. 1-13. Geologic map of equatorial Africa showing the Kasai, Cameroons, and Central African Provinces (excluding Phanerozoic cover) (after Cahen et al., 1976).

major period of greenstone volcanism forming the Nyanzian, Kibalian and Dodoman assemblages is > 2.74 b.y. and probably as old or older than 3.1 b.y. (Old and Rex, 1971; Dodson et al., 1975; Lavréau and Ledent, 1975). The younger greenstone belts, as represented by the Kavirondian System and associated granites, appear to have formed between 2.6 and 2.7 b.y. Post-tectonic granitic plutonism is recorded in parts of the province at 2.45—2.5 b.y. and about 2.0 b.y.

Liberian Province

The Liberial Province underlies an area of about 150,000 km^2 in Sierra Leone and Liberia in West Africa (Fig. 1-1). The province is bounded by a Pan-African mobile belt on the south and by mobile belts of Eburnean age on the other sides (Hurley et al., 1971, 1976). Two divisions are recognized in the province: a granite-greenstone terrane comprizing most of the province and a high-grade terrane, the Kasila belt, on the south (Fig. 1-14) (Williams, 1978). Structural trends in both terranes generally lie between northwest and northeast. Most of the supracrustal rocks in the granite-greenstone terrane are metamorphosed to the upper-greenschist or lower-amphibolite facies while those in the Kasila belt record amphibolite to granulite grades. The boundary between the two provinces is a zone of high strain and thrusting. Field and geochronologic evidence suggest that the low-grade terrane on the north continues into the high-grade to the south (Williams and Williams, 1976). The type of sediments and amount of volcanic rocks in the two terranes, however, imply different sedimentary environments and tectonic settings (Rollinson, 1978) (Chapter 10). When the Atlantic basin is closed, the Liberian Province lies adjacent to the Archean Guiana Province in South America (Hurley and Rand, 1974; Hurley et al., 1976) and high-grade rocks of the Kasila belt match up with high-grade rocks of the Imataca Complex in Venezuela. Both terranes appear to have been part of the same province prior to the opening of the Atlantic basin in the Triassic.

The oldest radiometric dates from the Liberian Province are from the Kenema assemblage which is a tonalite-trondhjemite gneiss-migmatite complex underlying much of the province (Grant, 1973). Rb-Sr isochron results suggest that at least some of this assemblage is between 3.1 and 3.4 b.y. in age (Hurley et al., 1976). The major period of greenstone belt development, plutonism, metamorphism, and deformation, however, appears to be between 2.6 and 2.75 b.y. (Hurley et al., 1971, 1976; Hedge et al., 1975).

Australia

Yilgarn Province

The Yilgarn Province is a broadly triangular-shaped province underlying an area of about 650,000 km^2 in southwestern Australia (Fig. 1-1) (Gee, 1979). It is bounded on the northwest and on the south and southeast by

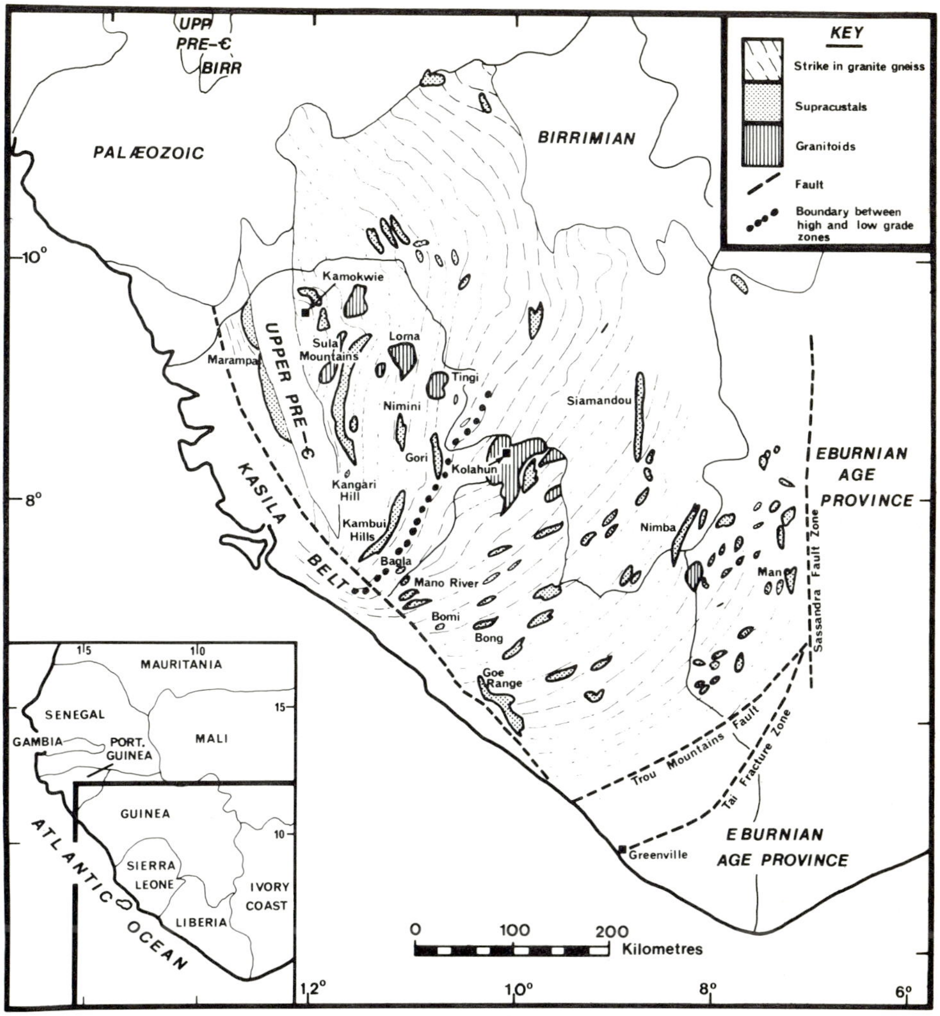

Fig. 1-14. Geologic map of the Liberian Province, West Africa (from Rollinson, 1978).

Proterozoic mobile belts (Rutland, 1973, 1976). The western margin of the province is bounded by the Darling fault of Phanerozoic age and the north-eastern margin is not exposed being unconformably overlain by Phanerozoic sediments. The Yilgarn Province has been divided into three subprovinces based on tectonic style, metamorphic grade, and lithologic abundances (Fig. 1-15) (Trendall, 1975; Rutland, 1976).[1] The largest, the Eastern Goldfields

[1] Yilgarn subprovinces are referred to as provinces by Trendall (1975). To avoid confusion with the more conventional use of the term province, they will herein be referred to as subprovinces. In turn, the subprovinces of I.R. Williams (1973, 1975a) will be referred to as regions.

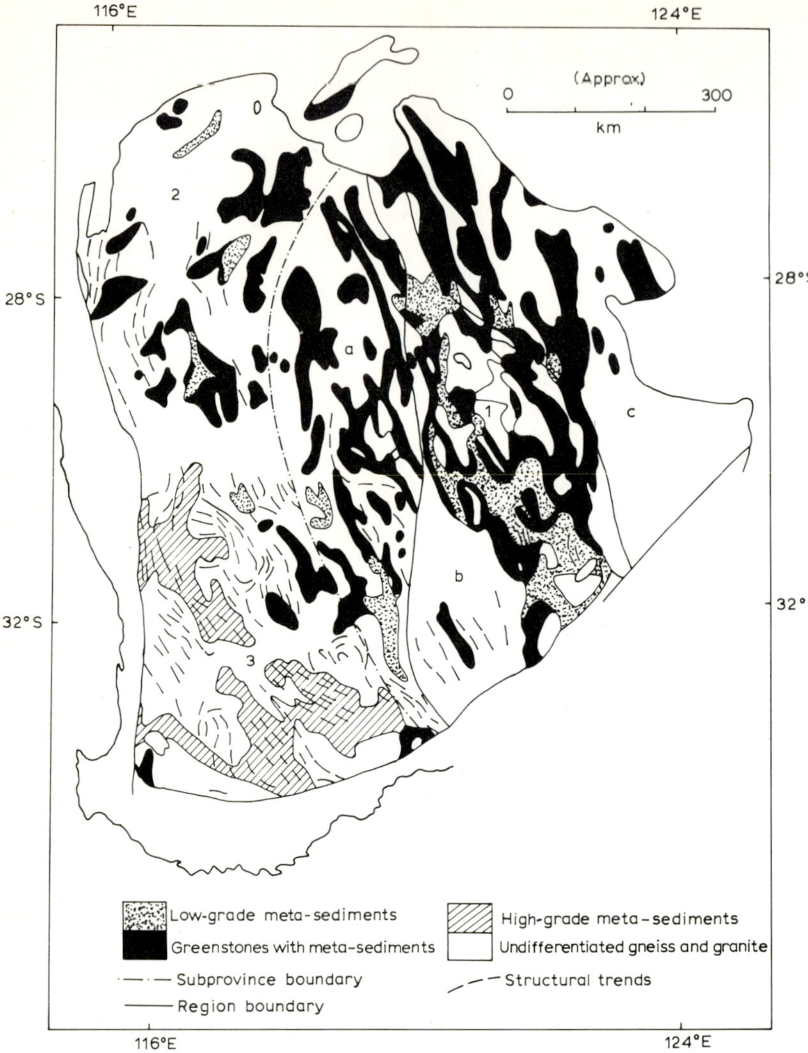

Fig. 1-15. Geologic map of the Yilgarn Province, southwest Australia (after Rutland, 1973). Subprovinces: *1* = Eastern Goldfields, *2* = Murchison, *3* = Southwestern (Wheat Belt). Regions of the Eastern Goldfields subprovince: *a* = Southern Cross, *b* = Kalgoorlie, *c* = Laverton.

subprovince, is composed of a typical granite-greenstone terrane exhibiting a NNW structural trend. It contains an interconnected network of greenstone belts occurring in large synformal structures. An average of 75% of the greenstone successions are comprised of volcanic rocks. This subprovince is particularly rich in gold and some base metal deposits. I.R. Williams (1973, 1975a) has further divided the Eastern Goldfields subprovince into three regions (Fig. 1-15). The Southern Cross region on the west is characterized by arcuate, synformal greenstone belts with a northerly trend. The Laverton region on the east contains greenstone belts that are somewhat more linear and occur in antiformal as well as synformal structures. The Kalgoorlie region is a wedge-shaped region which lies between the Southern Cross and Laverton regions. It is characterized by higher metamorphic grades, structural complexity, at least two ages of greenstone belts, an abundance of nickel sulfide occurrences, and a near absence of banded iron formation. Boundaries between these regions are, at least in part, shear zones or faults.

The Murchison subprovince is similar to the Eastern Goldfields subprovince in many respects (De la Hunty, 1975). It differs chiefly by its ENE structural trend. The Southwestern subprovince (also known as the Wheat Belt) is underlain chiefly by a high-grade Archean terrane (I.R. Williams, 1975b). It contains supracrustal remnants metamorphosed up to the granulite facies surrounded by a complex of tonalitic gneisses. It also contains at least some granitic rocks older than the other subprovinces of the Yilgarn. The Southwestern subprovince may represent a more deeply eroded level of the greenstone-granite terranes (Glikson and Sheraton, 1972).

Rb-Sr isochron dates and model lead dates suggest that the oldest rocks in the Yilgarn Province are tonalite-trondhjemite gneisses and related granulites approximately 3.3—3.4 b.y. in age (Arriens, 1971; Oversby, 1975). The first well-preserved terranes, however, are 2.9—3.0-b.y.-old gneisses and granitic rocks found in the Southwestern subprovince. The most widespread period of greenstone volcanism, plutonism, deformation, and metamorphism occurred between 2.6 and 2.75 b.y. (Compston and Arriens, 1968; Arriens, 1971; Cooper et al., 1978). Detailed work in the Agnew area (Eastern Goldfields subprovince) indicates that lower greenstones in this area are about 2.72 b.y. in age (Cooper et al., 1978). A minimum age for upper greenstones is given by the ages of granitic intrusive bodies (the Lawlers tonalite and Mt. Keith granodiorite) of 2.63—2.65 b.y. (Roddick et al., 1976; Cooper et al., 1978). The Jones Creek and Lawlers Conglomerates were deposited and the first major deformation occurred between 2.6 and 2.63 b.y. as bracketed by intrusive granites. A second deformation is bracketed between 2.48 and 2.56 b.y. also by intrusive bodies. Late-stage, mostly post-tectonic granites and quartz monzonites were emplaced between 2.45 and 2.55 b.y. and mafic dikes were intruded at about 2.35 b.y. A second period of chiefly post-tectonic plutonism is recorded in some areas at 2.1—2.25 b.y.

Pilbara Province

The Pilbara Province is a terrane elongated in an east-west direction underlying perhaps an area of 100,000 km^2 in Western Australia (Fig. 1-1). Structural trends in the province are complex reflecting up to five periods of deformation (Hickman, 1975). The province is overlain in the south and east by Proterozoic sediments and on the north by Phanerozoic sediments

(Rutland, 1976). It is comprised of a typical granite-greenstone terrane and recent data suggest that at least two ages of greenstone belts are present (Hickman, 1975; Blockley, 1975). The lower greenstone succession, known as the Warrawoona Group, is composed chiefly of mafic to ultramafic volcanic rocks and some sediments. In the eastern part of the province, this succession is unconformably overlain by the Gorge Creek Group which consists chiefly of clastic sediments, chert, and iron formation.

The oldest date from the Pilbara Province is a U-Pb zircon date of 3.45 b.y. from a dacite in the lower part of the Warrawoona Group near Marble Bar (Pidgeon, 1978). Although granitic rocks of this age have not as yet been reported, it is likely that they exist. Most ages from tonalite-trondhjemite gneisses and granitic plutons in the province, however, fall in the range of 2.9—3.0 b.y. suggesting that this period was the major period of plutonism (Compston and Arriens, 1968; De Laeter and Blockley, 1972; Oversby, 1976). Although some granitic rocks appear to have been emplaced between 2.6 and 2.7 b.y. (Trendall, 1975; De Laeter et al., 1975), magmatism of this age is less widespread than in other Archean provinces. The age of the sediments of the Gorge Creek Group, which unconformably overlie the Warrawoona Group is not known, but may be 2.6—2.7 b.y. As in the Yilgarn Province, the last period of granitic plutonism occurred at 2.1—2.25 b.y.

The existence of Archean inliers and evidence from diabase dike swarms indicate that the Pilbara and Yilgarn Provinces were continuous by late Archean time (Horwitz and Smith, 1978).

Asia

Indian Province

Most of India (including Ceylon) appears to be underlain by Archean and early Proterozoic rocks (Fig. 1-1) (Pichamuthu, 1967). In general, the Indian Province can be divided into six subprovinces (Fig. 1-16) (Naqvi et al., 1974). Each subprovince is characterized by its structural trend, metamorphic grade distribution, and lithologic proportions. Boundaries between the subprovinces are poorly known. The Peninsular Gneiss subprovince underlies the southern tip of India and Ceylon. It is characterized by complex structural trends reflecting polyphase deformation. It is composed chiefly of gneissic complexes containing interlayered granulite-facies rocks including charnockites and khondalites (Subramanian, 1967; Pichamuthu, 1976). Metamorphic grade ranges from amphibolite to granulite facies. Remnants of supracrustal rocks in gneiss terranes, as represented by the Sargur schist belt near Mysore (Viswanatha and Ramakrishnan, 1975), contain quartzite, mica schist, carbonates, calc-silicate rocks, and anorthosite. North of the Peninsular Gneiss subprovince is the Karnataka subprovince which is composed of a typical granite-greenstone terrane (Pichamuthu, 1974). The contact between the Karnataka and Peninsular Gneiss subprovinces is a subject of considerable

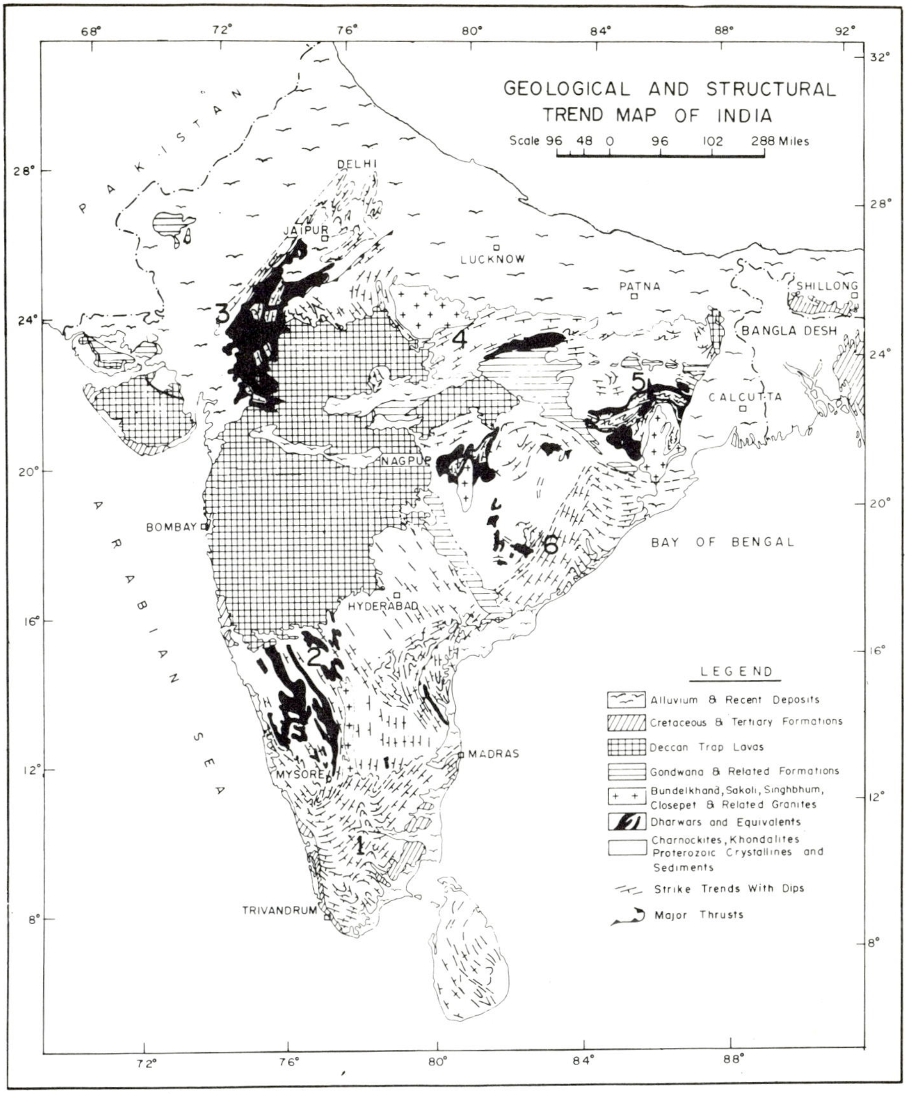

Fig. 1-16. Geologic map of the Indian Province (after Naqvi et al., 1974). Subprovinces: *1* = Peninsular Gneiss, *2* = Karnataka, *3* = Aravalli, *4* = Bundelkhand, *5* = Gangpur, *6* = Eastern Ghats.

discussion and disagreement (Nautiyal, 1965; Naqvi et al., 1974, 1978a; Pichamuthu, 1976). Although some of the Peninsular Gneisses are clearly intrusive into greenstone belts of the Karnataka subprovince, in most areas contacts are either concordant or poorly exposed, or both. Two types of greenstone successions are recognized in the Karnataka subprovince (Radhakrishna and Vasudev, 1977; Naqvi et al., 1978a): an older succession, the Bababudan

Group, contains an abundance of mafic volcanics, and a younger succession, the Chitradurga Group, containing an abundance of immature clastic sediments and cherts and intermediate to felsic pyroclastic rocks. Structural trends within the Karnataka subprovince are dominantly NNW to NW. The Aravalli subprovince in Rajasthan contains greenstone belts rich in argillaceous sediments with only minor amounts of volcanic rock (Pasco, 1950). This subprovince is characterized by a northeastern trend (Fig. 1-16). The Bundelkhand, Gangpur, and Eastern Ghats subprovinces in east-central India are characterized by greenstone successions rich in quartzite, mica schist, and dolomitic marble with notable occurrences of iron and manganese formation (Pichamuthu, 1967). Structural range from ENE to E-W in the Bundelkhand-Gangpur subprovinces to variable in the Eastern Ghats subprovince.

Although many radiometric dates are available from the Archean terranes in the Indian Province, few are of high quality. The oldest reliable dates fall in the range of 3.0—3.4 b.y. and come chiefly from the Peninsular Gneiss Complex (Crawford, 1969; Radhakrishna and Vasudev, 1977). A recent Rb-Sr isochron date of 3.36 b.y. has also been obtained from these gneisses west of Hassan in Karnataka State (Beckinsale et al., 1980). Uncertain dates suggest that at least three greenstone belts in the Karnataka subprovince (the Kolar, Holenarasipur, and Nuggihalli) are $\gtrsim$ 3.4 b.y. in age (Naqvi et al., 1978a). Granitic plutons have been emplaced in the Peninsular Gneiss and Karnataka subprovinces at 2.9—3.0, 2.6—2.7, and 2.2—2.4 b.y. (Venkatasubramanian, 1974; Radhakrishna and Vasudev, 1977). Recent data suggest that the Bababudan- and Chitradurga-type greenstone belts formed prior to 3.0 b.y., the age of intrusive granites (S. Moorbath, unpublished data, 1980). The major period of granulite-facies metamorphism in the Peninsular Gneiss subprovince appears to have occurred between 2.6 and 2.7 b.y. (Ramiengar et al., 1978).

Europe

Kola Province

The Kola Province underlies an area of about 600,000 km^2 in the northeastern part of the Baltic Shield (Fig. 1-17). This province is composed of both high-grade and granite-greenstone terranes and has been partially overprinted by Proterozoic orogeny (Simonen, 1971; Bowes, 1976, Gaál et al., 1978). Despite the large amount of geologic mapping in this province, only recently have Archean greenstone belts been recognized as such (Gaál et al., 1978; Blais et al., 1978). The contacts of the province are not well known. The western boundary appears to be gradational with the Svecofennian mobile belt on the west. The southwestern contact in southern Finland is marked by a major shear zone, the Ladoga-Raahe shear (Salop and Scheinmann, 1969; Gaál et al., 1978). In general, the province is characterized by a NNW structural grain (Fig. 1-17). The granite-greenstone terrane comprises a broad belt

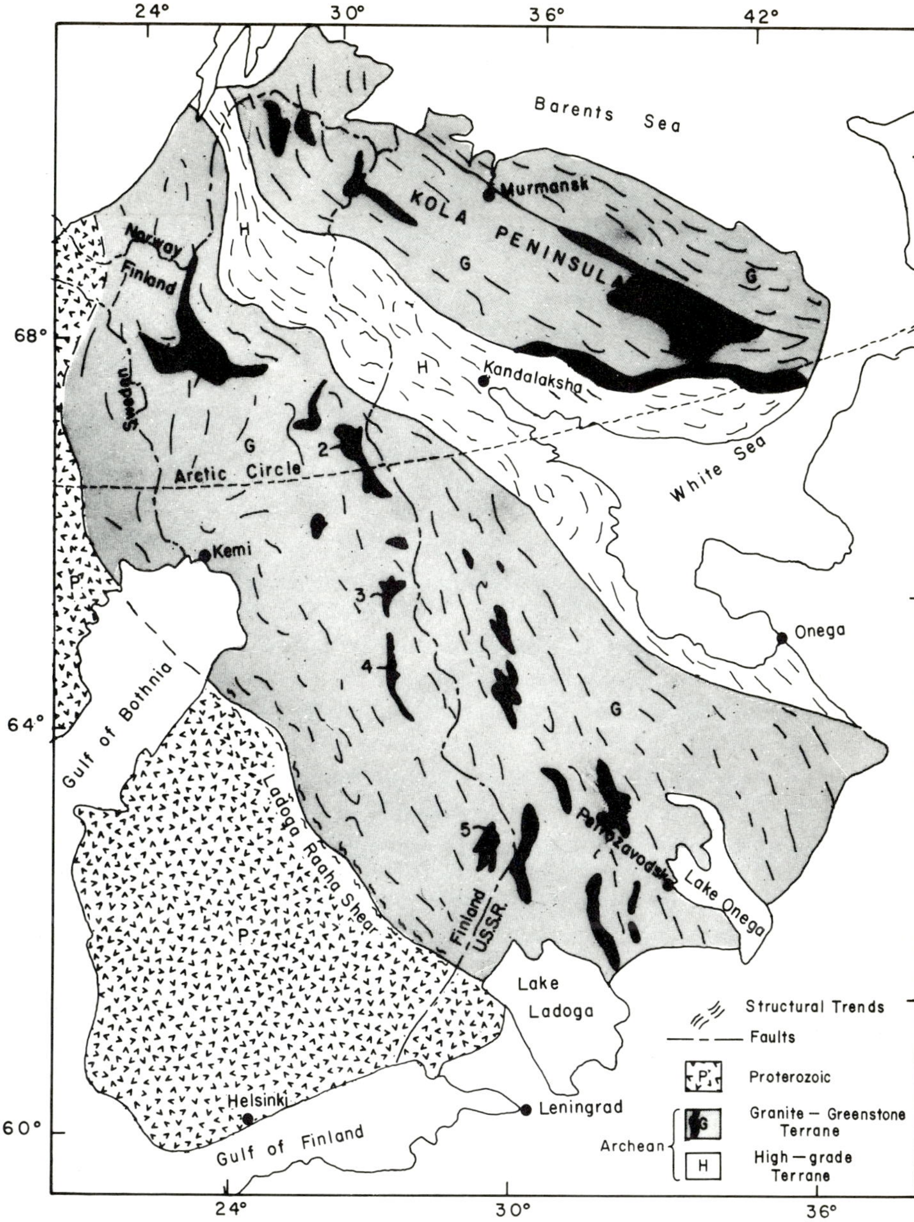

Fig. 1-17. Geologic map of the Kola Province, northern Europe (after Salop and Scheinmann, 1969; Gaál et al., 1978 and miscellaneous sources). Major greenstone belts: *1* = Kittilä, *2* = Salla, *3* = Suomussalmi, *4* = Kuhmo, *5* = Ilomantsi. Proterozoic cover rocks are not shown.

extending from Lake Ladoga to the Arctic Ocean in Norway. Metamorphic grade in the greenstone belts is somewhat higher than usual ranging from upper greenschist to amphibolite facies. Some of the major greenstone belts, have recently been the object of detailed geochemical studies (Blais et al., 1978; Jahn et al., 1979). Granite-greenstone terrane also appears to underlie much of the central part of the Kola Peninsula. High-grade Archean rocks, in part reworked during the Proterozoic, comprise a northwesterly trending belt separating the two granite-greenstone terranes.

The oldest rocks from the Kola Province occur in the high-grade terranes of the Kola Peninsula. Mafic rocks of the Chuna-Moncha-Volch'i tundras yield a lead isochron age of about 3.7 b.y. and are interpreted as representing the basement for younger Archean greenstone belts (Turgarinov and Bibikova, 1975). The dominant periods of granitic plutonism, metamorphism, and deformation occur at 2.9—3.0 and 2.6—2.7 b.y. and most of the greenstone belts appear to have formed between 2.6 and 2.7 b.y. (Gerling et al., 1968; Kratts et al., 1974; Lobach-Zhuckenko et al., 1976; Gaál et al., 1978).

Ukrainian Province

The Ukrainian Province is exposed over an area of about 150,000 km^2 in the Soviet Union north of the Black Sea. This province contains a great variety of rocks, metamorphic grades, and has been in part or entirely reworked during Proterozoic orogenic events (Semenenko et al., 1968). Both high-grade and granite-greenstone Archean terranes are known with the latter occurring in the eastern part of the province (Fig. 1-18). Structural trends are complex and variable with the overall trend ranging from north to northwest. The oldest rocks in the Ukrainian Province are greenstone belts of the Konkian Series and associated granitic complexes. These rocks fall in the age range of 3.2—3.5 b.y. (Semenenko et al., 1968). A second granite-greenstone cycle, called the Aulian orogenic cycle, appears to range from about 2.8 to 3.0 b.y. During this time the Aulian, Romankoya, Volnyanka, and Tarom greenstone successions were formed. A third greenstone-granite cycle, the Bazavlukian cycle, is recorded between 2.3 and 2.7 b.y. (probably chiefly at 2.6—2.7 b.y.). The high-grade rocks represented by such series as the Bug, Teterev, and Gnilopyat and associated granitic rocks were last deformed at 2.0—2.3 b.t. An Archean high-grade terrane (2.0—2.7 b.y.) dominates in the western part of the province. A belt of deformed Proterozoic rocks separates the high-grade terrane from the granite-greenstone terrane.

South America

São Francisco Province

The São Francisco Province is exposed in two areas in eastern Brazil amounting to approximately 350,000 km^2 (Fig. 1-1). Later sediments cover the province between the two exposures. The province is comprised of both

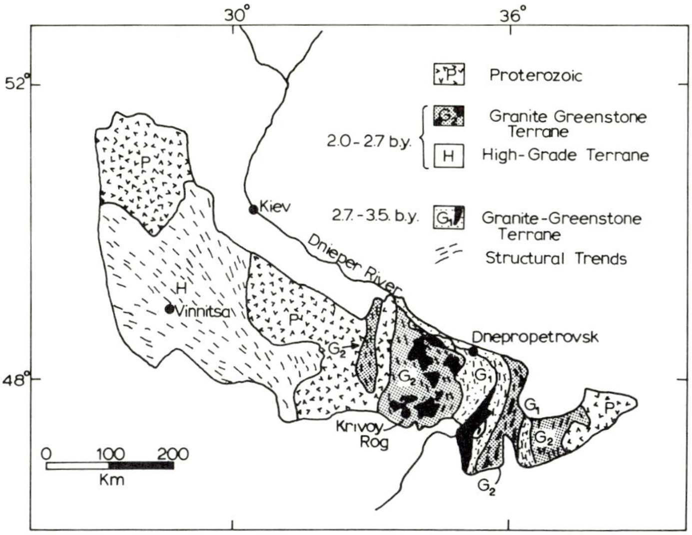

Fig. 1-18. Geologic map of the Ukrainian Province, U.S.S.R. (after Semenenko et al., 1968; reproduced by permission of the National Research Council of Canada).

high-grade and granite-greenstone terranes, although the latter has only been recently recognized as such (Mascarenhas, 1973; Wernick, 1979, 1980). Nine greenstone belts of Archean or probable Archean age have been recognized in this province (Fig. 1-19). Although structural trends in the province are complex due to polyphase deformation and later reworking during the Proterozoic, northerly trends are common. The greenstone belts range from 2000 to 7000 km^2 in area and exhibit other characteristic features of Archean greenstone belts (Wernick, 1980). They are surrounded by gneisses and granites. The boundaries of the São Francisco Province are poorly known, but it appears to represent an Archean "island" within a belt of Proterozoic orogeny (Cordani et al., 1973).

Detailed radiometric age studies, until recently, were available only for the coastal area of Brazil. Existing data suggest that the Archean rocks in the São Francisco Province formed between 2.6 and 3.0 b.y. (Herz, 1970; De Almeida et al., 1974; Wernick, 1980).

Discussion

The geochronologic results from Archean granite-greenstone provinces are summarized in Fig. 1-20. Results from the high-grade North Atlantic Province are included for comparison. The episodic nature of Archean magmatism is clear from the data. Major periods of magmatism (± orogeny) occur at 3.1—3.8, 2.8—3.0, and 2.6—2.7 b.y. Two or more periods appear to be represented in the 3.1—3.8-b.y. group at 3.5—3.8, 3.3—3.4 and 3.1—3.2 b.y.

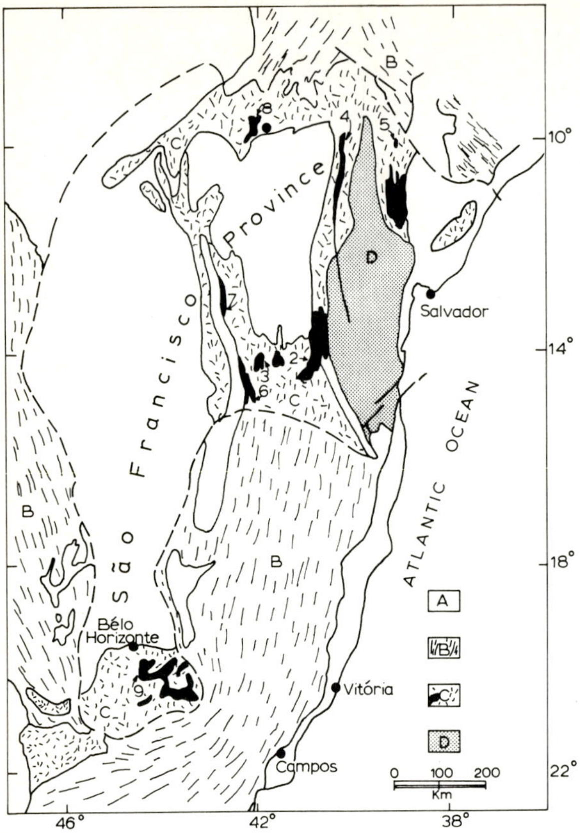

Fig. 1-19. Geologic map of the Sao Francisco Province, Brazil (after Wernick, 1979a). Symbols: A = Post-Archean sediment cover, B = Proterozoic mobile belts, C = Archean granite-greenstone terrane (greenstones in black), D = Archean high-grade terrane. Greenstone belts: *1* = Serrinhas, *2* = Contendas-Mirante, *3* = Brumado, *4* = Jacobina, *5* = Capim, *6* = Urandi, *7* = Boquira, *8* = Colomi, *9* = Rio das Velhas.

From areas in which accurate radiometric dates are available, it appears as if a typical greenstone-granite deformation cycle lasts 50—100 m.y. The oldest known rocks are tonalite-trondhjemite gneisses from the Superior, Kola, Rhodesian, and North Atlantic Provinces. These gneisses, however, contain inclusions of mafic and ultramafic rock and of metasediments indicating the existence of still older crustal rocks. Dating of the Isua greenstone belt in Greenland, which occurs as a remnant in the Amitsoq gneisses, indicates an age indistinguishable from the gneisses (~ 3.8 b.y.; Moorbath et al., 1973). The Barberton greenstone belt and perhaps other greenstone belts in the Kaapvaal Province appear to be somewhat younger (~ 3.5 b.y.). The first episode of significant granite emplacement is recorded at 3.1—3.2 b.y. in the Kaapvaal, Central African, and Liberian Provinces. In

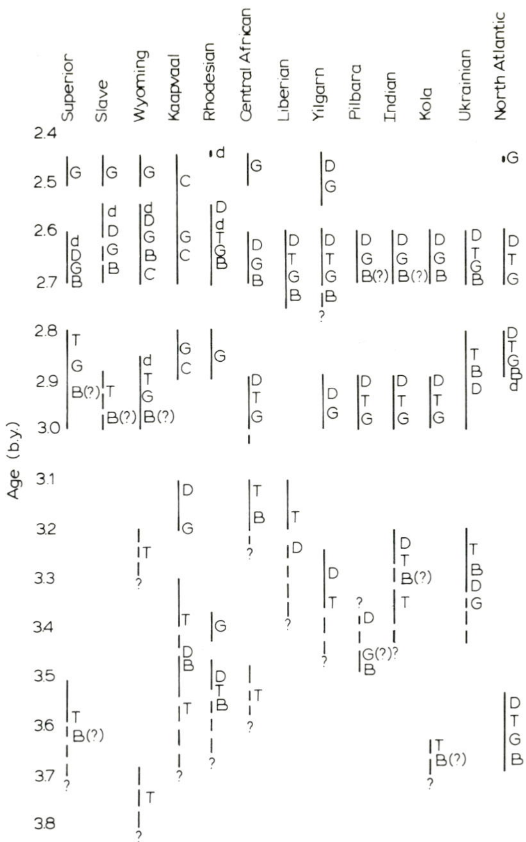

Fig. 1-20. Geochronologic summary of Archean granite-greenstone provinces. Key: B = greenstone belt development, T = intrusion of tonalite-trondhjemite, D = deformation and regional metamorphism, G = intrusion of granites and pegmatites, d = intrusion of mafic dikes, C = cratonic sedimentation ± volcanism.

the period prior to 3 b.y., data suggest diachronous evolution in that various events occur at different places at different times. During the remainder of the Archean, however, major events appear to be concentrated in two time intervals at 2.8—3.0 and 2.6—2.7 b.y. It is noteworthy that the granite-greenstone terranes which formed between 3.1 and 3.8 b.y. were contemporary with maria basaltic magmatism on the moon.

The period of time between 2.8 and 3.0 b.y. is characterized chiefly by granitic plutonism and is recorded in all granite-greenstone provinces and in the North Atlantic Province (Fig. 1-20). Clearly defined supracrustal rocks of this age have not been described and supracrustal remnants in 2.8—3.0-b.y. granites may be > 3.0 b.y. in age. By far, the major period of Archean volcanism and plutonism and perhaps the major period in all earth history is at 2.6—2.7 b.y. It is recorded in all Archean provinces. At 2.45—2.5 b.y., post-tectonic granitic plutonism of minor importance is recorded in many provinces. The Great Dyke in Rhodesia was also emplaced at this time. In any given Archean province, up to three ages of greenstone belts are preserved. The most widespread belts, however, were formed between 2.6 and 2.7 b.y. In the Rhodesian and Yilgarn Provinces, two ages of greenstone belt development are recorded within this 100-m.y. period. The younger greenstone belts in the Indian Province (2.2—2.4 b.y.) are Proterozoic in age and are unusual in terms of Archean standards in that they contain an abundance of clastic sediments (Chapter 2). Well-preserved cratonic sediments of Archean age occur only in the Kaapvaal basin. Existing evidence indicates that Archean cratons were uplifted and eroded to their approximate present-day levels of exposure in 300—400 m.y. after their stabilization (Watson, 1976b).

Orogenic episodicity has long been recognized on the earth (Gastil, 1960). Two causes for such episodicity have been discussed: changing convecting patterns in the earth with time, and secular changes in the spin axis of the earth. Runcorn (1962, 1965) suggests that orogenic episodes are related to changes in convection patterns in the mantle accompanying a growing core. Dearnley (1966), employing structural trends, supports Runcorn's model, although also requiring an expanding earth. Runcorn's model has two serious problems: (1) recent data indicate that the earth's core formed during planetary accretion or within a few hundred million years after accretion, and (2) the model does not account for pre-2.7-b.y. orogenies nor for minor orogenies of other ages. Sutton (1963, 1973) proposes a model based on four major chelogenic cycles (2.7—3.6, 1.9—2.7, 1.1—1.9, and $<$0.6 b.y.). Each cycle lasts 750—1250 m.y. and involves an evolving convection system beginning with many small cells which increase in size and eventually fuse giving rise to one earth-wide cell. Widespread orogeny occurs at the beginning of each cycle and is caused by the existence of many small convection cells. Why such a cycle should repeat itself four times during earth history is not specified in the model. R. St. J. Lambert (1978) has recently proposed that the 2.6—2.7-b.y. event may be related to the onset of deep convection in the earth.

G.E. Williams (1973) suggests that the earth's spin axis changes attitude with respect to the ecliptic plane with a period of about 2.5 b.y. while the moon stays in an orbit close to the ecliptic plane at all times. The result is two independent lunar torques applied to the outer parts of the spinning earth which alternately peak in intensity about every 620 m.y. A tidal torque which is caused by the pull of the moon on the earth's tidal bulge peaks when the moon is in its equatorial orbit and a precessional torque due to the pull of the moon on the earth's equatorial bulge peaks when the moon is in polar orbit. The alternate peaks in torque intensity occur at 3.85, 3.25, 2.6, 2.0, 1.35, 0.75, and 0.13 b.y. thus corresponding closely to many orogenic peaks.

Chapter 2

GREENSTONE BELT STRATIGRAPHY

INTRODUCTION

Stratigraphic studies of Archean greenstone successions are faced with several major difficulties (Ayres, 1977): (1) sections are often isoclinally folded and difficult to reconstruct; (2) formations may be thinned, thickened, or entirely removed by deformation and plutonism; (3) rock lithologies are similar and distinctive marker horizons uncommon, making it difficult to correlate across structural or igneous discontinuities; and (4) outcrops are often poor. A basic prerequisite of unraveling the stratigraphy of greenstone belts is an understanding of the structure. In very few belts is the structure simple enough to reconstruct the stratigraphy with a great deal of confidence. Estimated preserved thicknesses of most greenstone successions range from 10 to 20 km (Table 2-1). Such estimates are minimal in that neither the base nor the top of most successions is exposed. The thicknesses may be overestimated due to tectonic thickening and to migration of volcanic centers and sedimentary basins with time. Although the lower contacts of most greenstone successions are faults or intrusive contacts, some sections unconformably overlie older gneisses or greenstone belts.

Based chiefly on the abundances of rock types in greenstone successions, two types of greenstones can be recognized. The *bimodal type* contains an abundance of mafic and ultramafic rocks with minor felsic volcanics and cherts. Andesites are rare or absent. The *calc-alkaline type* contains a range of volcanic lithologies from ultramafic through andesitic to felsic and derivative graywacke assemblages.

To illustrate the diversity in stratigraphy of greenstone belts, several well-known successions are described below in detail.

STRATIGRAPHIC SECTIONS

The Swaziland Supergroup

Perhaps the best known Archean greenstone belt is the Barberton belt in South Africa (Fig. 1-11) (Viljoen and Viljoen, 1969a; Anhaeusser, 1971a). Radiometric ages indicate an age for volcanics in the belt of about 3.5 b.y. (Jahn and Shih, 1974). The Barberton greenstone belt is exceptionally well

STRATIGRAPHIC COLUMN OF THE SWAZILAND SEQUENCE

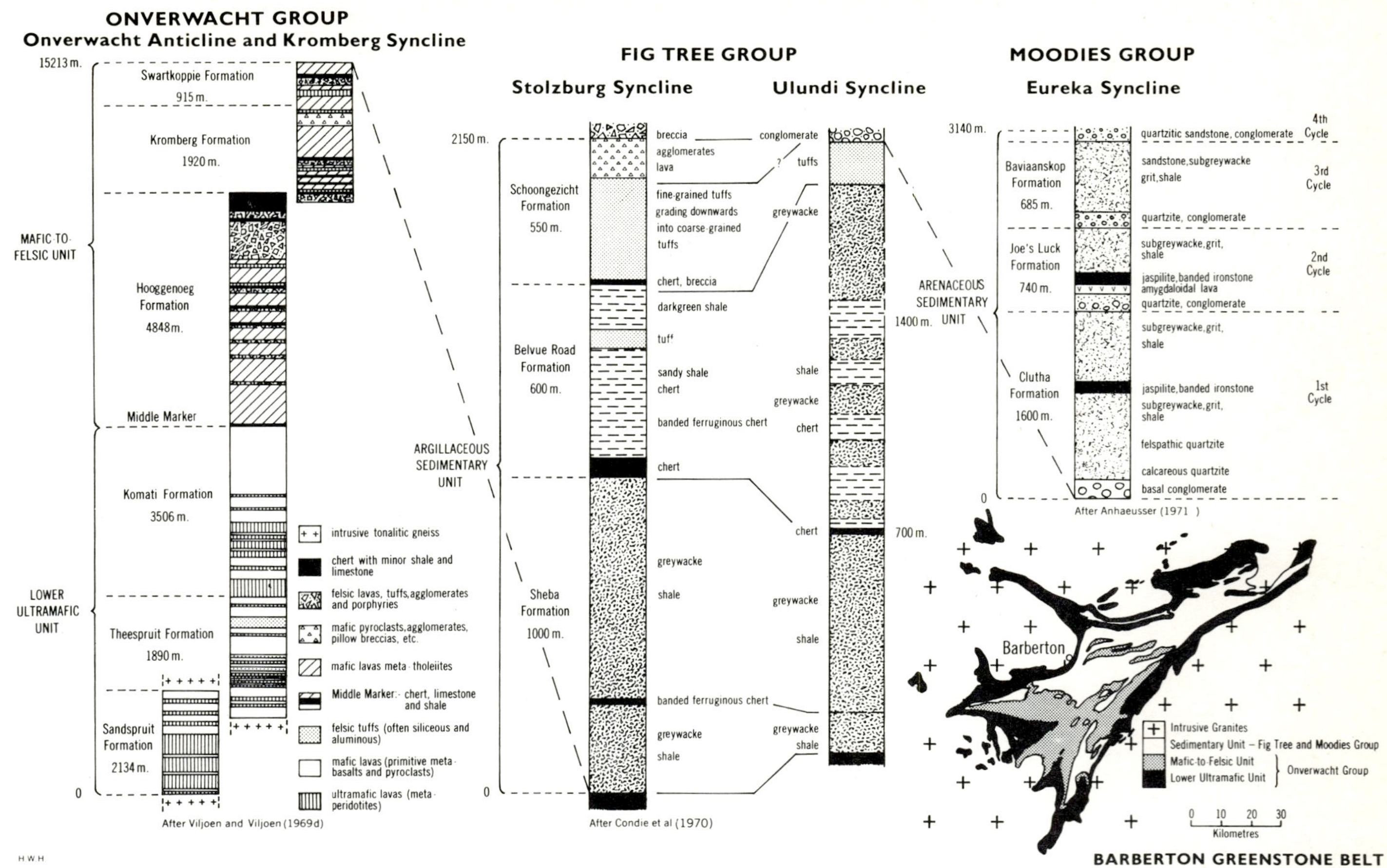

TABLE 2-1

Estimated thicknesses of Archean greenstone successions

Reference[1]	Belt	Thickness (km)
1	Abitibi belt at Noranda, Quebec	18
2	Setting Net Lake, Ontario	9.5
3	Shebandowan, Ontario	22
4	Yellowknife, N.W.T., Canada	18
5	Michipicoten, Ontario	11
6	Vermilion, Minnesota	16
7	Barberton, South Africa	20
8	Nimini Hills, Sierra Leone	5
9	Tati, Botswana	20
10	Midlands, Rhodesia	17
11	Coolgardie-Kurrawang, Australia	19.5
12	West Lake Lefroy, Australia	13

[1]References: 1 = Goodwin et al. (1972); 2 = Ayres (1977); 3 = Wilson and Morrice (1977); 4 = Baragar (1966), Henderson and Brown (1966); 5 = Goodwin (1962); 6 = Sims (1976); 7 = Anhaeusser (1971b); 8 = Williams (1978); 9 = Key et al. (1976); 10 = Harrison (1970); 11 = Glikson (1970); 12 = McCall (1969).

preserved and well exposed. Although folded, the belt is not as deformed as most greenstone belts. The stratigraphic succession, as summarized in Fig. 2-1, is probably the most accurately known greenstone section. Details of the succession are described in Viljoen and Viljoen (1969a,b,g), Anhaeusser (1973a, 1974) and Reimer (1975a).

The succession, known as the Swaziland Sequence or Swaziland Supergroup, is divided into three groups, from oldest to youngest, the Onverwacht, Fig Tree, and Moodies Groups (Fig. 2-1). The Onverwacht Group comprises most of the section attaining a thickness of over 15 km. It is composed chiefly of mafic to ultramafic volcanics with minor felsic volcanics and chert. Six formations are recognized in the Onverwacht Group as follows (after Viljoen and Viljoen, 1969a):

(1) The lowest exposed formation is the Sandspruit Formation consisting of 60—70% of ultramafic rocks and the remainder of mafic volcanics with minor calc-silicate horizons of probable sedimentary origin.

(2) The overlying Theespruit Formation is composed dominantly of

Fig. 2-1. Stratigraphic column of the Swaziland Supergroup in the Barberton greenstone belt (after Anhaeusser, 1973a).

mixed mafic and ultramafic volcanics with minor interbeds of water-layed felsic tuff.

(3) The Komati Formation is composed of mafic (70%) and ultramafic (30%) volcanics intruded by quartz-feldspar porphyries.

(4) The Hooggenoeg Formation consists of mafic-to-felsic volcanic cycles with individual cycles often ending with chert horizons.

(5) The Kromberg Formation is composed chiefly of mafic and felsic lavas and felsic pyroclastic rocks with minor amounts of chert, carbonate, and calc-silicate rock.

(6) At the top, the Swartkoppie Formation is composed of mafic-to-felsic volcanics and cherts and has been extensively sheared.

The lower three formations are referred to as the Lower Ultramafic Unit and are characterized by a relative abundance of ultramafic and mafic flows and sills. The upper three formations, which are characterized by mafic-to-felsic volcanic cycles, are known as the Mafic-to-Felsic Unit. It appears, however, that andesites are rare or absent in these cycles and hence they may more appropriately be referred to as mafic-felsic cycles. Separating the two units is a persistent black and white chert unit, the Middle Marker. This unit, which averages about 6 m thick, is composed of banded cherts with minor pyroclastic volcanic layers. The cherts also contain carbonate and black shaly layers rich in carbonaceous matter and, locally, sulfides.

Conformably overlying the Onverwacht Group is about 2 km of mixed graywackes, shales, and chert, the Fig Tree Group (Condie et al., 1970; Reimer, 1975a). The Fig Tree Group is divided into three formations. The Sheba Formation at the base is composed chiefly of graywackes and associated shales with minor chert horizons. The Belvue Road Formation, which begins with a massive chert unit, consists chiefly of graywacke siltstones, shales, and locally, felsic tuff. Ferruginous chert bands are also common. The Schoongezicht Formation is composed of felsic tuffs, breccias, and agglomerates.

The Fig Tree Group is overlain unconformably to conformably by the Moodies Group which attains a thickness of over 3 km. The Moodies is also divided into three formations (Anhaeusser, 1971a, 1974). The Clutha Formation begins with a basal conglomerate composed of clasts representing a great variety of rock types. It is composed chiefly of quartzites and subgraywackes with lesser amounts of shale, chert, and banded iron formation. The Joe's Luck Formation is similar and contains one amygdaloidal basalt flow. The Baviaanskop Formation consists chiefly of quartzites and arkoses with lesser amounts of shale, subgraywacke, and conglomerate. The topmost unit preserved in the Swaziland Supergroup is the Bickenhall Member composed of buff-colored sandstone and conglomerate.

Two general trends are observed in the igneous rocks in the Swaziland Supergroup as a function of time: (1) a decrease in the amount of ultramafic rock, becoming negligible above the Middle Marker, and (2) the ratio of pyroclastics to flows increases with stratigraphic height. Within the Fig Tree and Moodies Groups, a trend from immature to mature sediments occurs in going upwards in the section.

The Coolgardie-Kurrawang succession

Approximately 19.5 km of greenstone volcanics and sediments are exposed on the western limb of the Kurrawang syncline east of Coolgardie in the Yilgarn Province of Western Australia (Glikson, 1970, 1971a, 1972a). The lower part of the sequence is intruded by granitic rocks and hence the base is not exposed. The lowest unit, the Coolgardie greenstones[1], are composed of about 6.2 km of mixed ultramafic and mafic flows and associated sills with minor argillaceous units and intrusive quartz porphyries (Fig. 2-2A). Pillows are common in the flows. Conformably overlying the Coolgardie greenstones are the Gunga meta-argillites composed of interbedded argillite and slate (in part graphitic) and minor banded iron formation. The sequence is interlayered with up to 50% of mafic volcanic rocks. Overlying the Gunga meta-argillites are the Mt. Robinson greenstones composed chiefly of mixed mafic and ultramafic flows and quartz-feldspar porphyry intrusives. The overlying Brown Lake metasediments are composed of siltstones, phyllites, and graywackes and are unconformably overlain by the Red Lake greenstones. These greenstones are composed of mafic, intermediate (andesitic), and felsic volcanics with some intrusive ultramafics and felsic porphyries. A thin basal conglomerate is also present. The Black Flag metasediments, which conformably overlie the Red Lake greenstones, are composed chiefly of interbedded graywackes, siltstones, and minor conglomerates and mafic to felsic volcanic units. The Black Flag metasediments grade upwards into the Kurrawang Beds which are characterized by a basal graywacke unit, a middle polymict conglomerate, and a thick upper graywacke unit.

Compared to many greenstone successions, the Coolgardie-Kurrawang sequence is characterized by a large proportion of clastic sediments. The lower and largest part of the volcanic succession (below the Red Lake Greenstones) is bimodal, whereas the upper part is calc-alkaline. With increasing stratigraphic height the ratio of clastic sediments to volcanics also increases. The average grain size of clastic sediments also tends to increase upwards in the succession with the Kurrawang Beds containing an appreciable amount of coarse graywacke and conglomerate. At least one unconformity exists in the section beneath the Red Lake greenstones.

[1] The term ophiolite, originally used by Glikson (1972a), is herein replaced with the non-genetic term, greenstone.

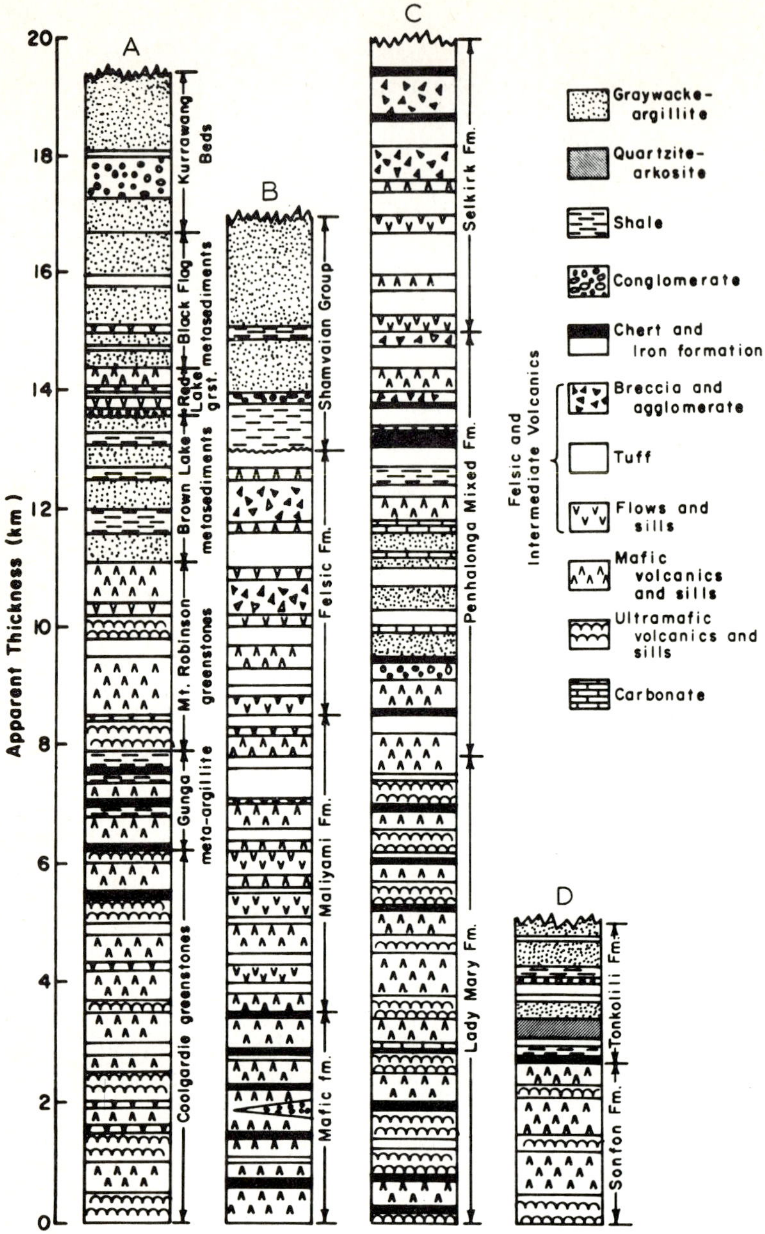

Fig. 2-2. Generalized stratigraphic sections of Archean greenstone belts in Africa and Australia. A. Coolgardie-Kurrawang succession, Western Australia (after Glikson, 1971a). B. Bulawayan-Shamvaian Groups, near Que Que, Rhodesia (after Harrison, 1970). C. Tati succession, northern Botswana (after Key et al., 1976). D. Nimini Hills succession, Sierra Leone (after Williams, 1978).

The Bulawayan-Shamvaian Groups

The Bulawayan Group is a lithostratigraphic group exposed in most of the Archean greenstone belts in Rhodesia (Fig. 1-12). One of the best exposed sections is in the Midlands belt west of Que Que (Harrison, 1970; Bliss, 1970) where a minimum of 17 km of the combined Bulawayan and Shamvaian Groups is exposed (Fig. 2-2B). The lowest exposed formation, the Mafic Formation, is intruded by the Rhodesdale batholith or ultramafic bodies. This formation is composed chiefly of pillowed mafic flows interlayered with chert and minor felsic tuffs and conglomerate. Granitic clasts in a conglomerate horizon cropping out in the Sebakwe River indicate the presence of earlier pre-greenstone sialic crust. Overlying the Mafic Formation is the Maliyami Formation which consists of a thick sequence of intercalated mafic and andesitic flows and andesitic to dacitic pyroclastics. Porphyritic and amygdaloidal volcanics are common. Conformably overlying the Maliyami Formation is the Felsic Formation composed principally of andesitic to dacitic flows, breccias, and tuffs with lesser amounts of mafic volcanics. Numerous small plugs and sills of quartz-felsite porphyry intrude the Felsic Formation. The upper contact of the Bulawayan Group is an erosional unconformity. The Bulawayan rocks were folded, uplifted, and eroded prior to deposition of overlying graywackes, phyllites, and conglomerates of the Shamvaian Group.

The Bulawayan Group in the Midlands greenstone belt is characterized by a lower bimodal unit (the Mafic Formation) overlain by a thick calc-alkaline unit. Sediments are notably uncommon in the sequence. There is also an increase in the ratio of pyroclastics to flows with stratigraphic height with coarse pyroclastics being most frequent in the Felsic Formation. Significant detrital sediments occur only in the Shamvaian Group at the top of the section.

The Tati succession

A thick sequence of volcanics and sediments are exposed in the Tati greenstone belt in northeastern Botswana (Key et al., 1976). The base of the succession is poorly exposed, but apparently unconformably underlying it is a sequence of highly deformed meta-arkoses interlayered with amphibolite, schist, and ultramafic rocks. The Lady Mary Formation, which comprises the lower one-third of the Tati succession, is composed chiefly of mafic and ultramafic flows and sills with minor felsic tuff, arkose, and chert, and very minor carbonate (Fig. 2-2C). Overlying the Lady Mary Formation (perhaps unconformably) is the Penhalonga Mixed Formation which is composed of mixed graywacke, graphitic phyllite, mafic flows, and some andesitic and felsic volcanics. Minor banded iron formation, carbonate and conglomerate also occur in this unit. The upper part of the formation is composed of andesitic breccias and tuffs. The uppermost formation, the Selkirk Formation, is composed dominantly of andesitic to felsic pyroclastics with minor basalts and cherts. The formation is also intruded with an unknown number of gabbroic sills.

The lower part of the Tati succession (the Lady Mary Formation) is a typical bimodal sequence and the upper two formations represent a calc-alkaline sequence. Coarse pyroclastic rocks increase in abundance in the upper one-third of the section. Sediments, as a whole, are infrequent although two sediments, arkose and carbonate, which are rare or absent in most Archean greenstone belts are found in minor amounts in the succession.

The Nimini Hills succession

The Nimini Hills greenstone belt is one of many Archean greenstone belts in the Liberian Province in Sierra Leone in West Africa (Fig 1-14). A minimum of about 5 km of section has been described from this belt (Williams, 1978; Rollinson, 1978). Two formations are recognized in the Nimini Hills succession (Fig. 2-2D). The Sonfon Formation consists of a lower mixed ultramafic-mafic unit overlain by a thick succession of mafic flows and sills with minor ultramafic intrusives. Overlying the Sonfon Formation is the Tonkolili Formation comprised of basal banded iron formation overlain by mixed felsic tuffs, graywackes, phyllites, and quartzites and minor conglomerate.

The Nimini Hills succession may represent only part of a much thicker greenstone succession. The break between the two formations is suggestive of a change from a bimodal- to a calc-alkaline-type succession.

The Michipicoten Group

A minimum of 11 km of greenstone succession is exposed in the Michipicoten area along the northeast shore of Lake Superior in the Superior Province (Goodwin, 1962) (Fig. 2-3A). The Michipicoten Group is characterized by a lower volcanic unit (base not exposed) composed chiefly of mafic to andesitic flows with andesitic to rhyolitic pyroclastic rocks increasing in abundance towards the top of the unit. In most places, the volcanic unit is overlain by banded iron formation which ranges up to 300 m thick. Because the iron formation is discontinuous, the lower volcanics are in some places overlain by the middle volcanic unit. This unit is composed chiefly of andesitic flows and related pyroclastic rocks. It interfingers and grades upward into the Dore sedimentary rocks. These sediments are composed chiefly of graywackes, argillite, and shale, with minor conglomerate and quartzite. The uppermost unit in the Michipicoten Group is again a volcanic unit composed of mafic to felsic volcanics with mafic, pillowed flows dominating. The unit appears to represent the resumption of a new volcanic cycle.

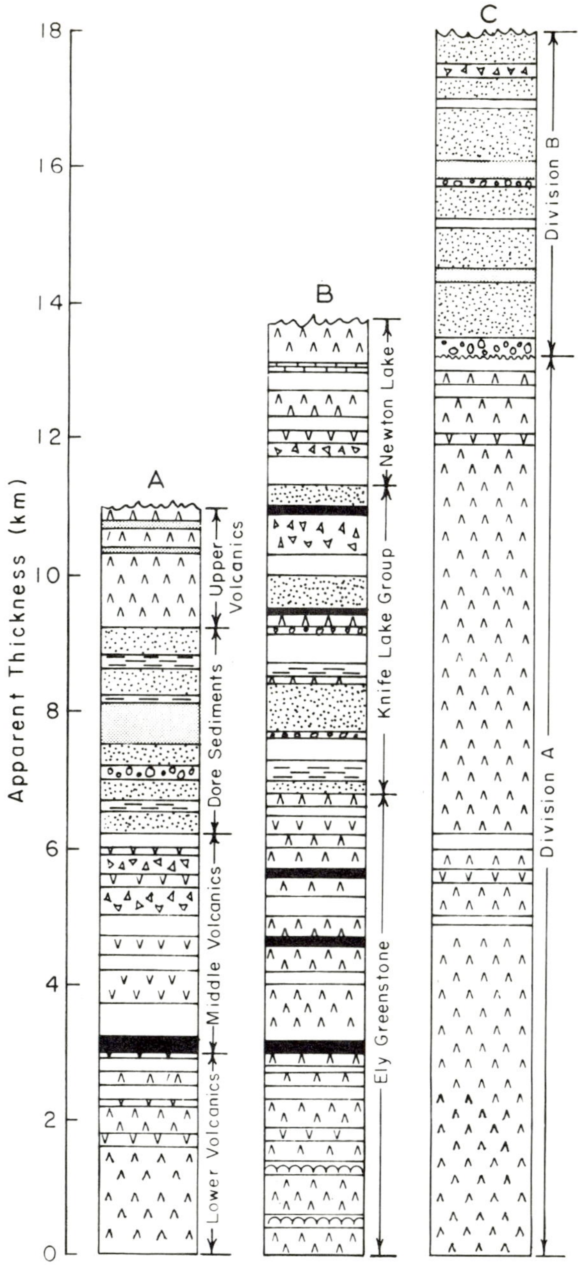

Fig. 2-3. Generalized stratigraphic sections of Archean greenstone belts in North America. A. Michipicoten Group, Ontario (Goodwin, 1962). B. Vermilion succession, Minnesota (Sims, 1972a, 1976). C. Yellowknife Supergroup, N.W.T., Canada (Henderson and Brown, 1966; Baragar, 1966). Symbols given in Fig. 2-2.

The lower and middle volcanic units and the Dore sedimentary unit constitute a three-fold stratigraphic division characterizing many greenstone belts in the Superior Province (Goodwin, 1968b; Goodwin et al., 1972). This succession is characterized by a decrease in the mafic to calc-alkaline ratio, in the pyroclastic to flow ratio, and in the amount of sediments with stratigraphic height. The Michipicoten Group, as with many other successions in the Superior, Slave, and Wyoming Provinces in North America, is a calc-alkaline greenstone succession with no evidence of an early bimodal phase. Ultramafic flows appear to be absent in the Michipicoten section. Goodwin (1968a,b) suggests that this three-fold sequence in development so common in North American greenstone belts reflects the following evolutionary stages: (1) widespread, fluid eruptions of submarine basalts and andesites; (2) explosive eruptions of andesitic to rhyolitic magma at more localized volcanic centers and associated fumarolic activity; and (3) uplift and erosion of volcanic piles (and associated granitic rocks) producing thick sections of graywacke and related sediments.

The Vermilion succession

Approximately 14 km of a greenstone succession has been described from the Vermilion district in northeastern Minnesota (Morey et al., 1970; Sims, 1972a, 1976) (Fig. 2-3B). The oldest exposed formation, the Ely greenstone, is composed of three members: a lower member comprised dominantly of mafic and minor ultramafic flows with a small amount of felsic pyroclastic rocks; a middle member, the Soudan Iron Formation, composed of about 150 m of banded iron formation and associated mafic volcanics; and an upper member of mafic lava, pyroclastics, and minor iron formation. Intrusive gabbro bodies occur throughout the Ely greenstone. The Ely greenstone is overlain by the Knife Lake Group in the central part of the district and by the Lake Vermilion Formation in the western part. Both formations are composed largely of felsic pyroclastic rocks and associated graywackes. Minor conglomerate, graphitic slate, banded iron formation, and mafic flows also occur in these formations. The Newton Lake Formation overlies the Knife Lake Group and is divided into two members. The lower member is composed of felsic to andesitic pyroclastic rocks and flows. The upper member is composed dominantly of mafic lavas, andesitic to mafic pyroclastics, and gabbroic intrusives and contains minor siliceous marble.

The Vermilion succession differs from the typical Superior Province sequence described above in that most of the succession is bimodal; only in the Newton Lake Formation do andesites appear. This bimodality in the lower parts of greenstone successions also characterizes greenstone successions in the Rice Lake area in nearby parts of Ontario (Church and Wilson, 1971) as well as the lower part of the Abitibi succession in the eastern Superior Province (Baragar, 1966; Jolly, 1975).

The Yellowknife Supergroup

The Yellowknife Supergroup is the name applied to greenstone successions in the Slave Province of northwestern Canada (Fig. 1-9). A minimum thickness of about 18 km is exposed in the Yellowknife greenstone belt at the southern margin of the province (Jolliffe, 1938; Henderson and Brown, 1966; Baragar, 1966) (Fig. 2-3C). The exposed base of the succession is an intrusive contact. The lower two-thirds of the section (Division A) is composed almost exclusively of mafic flows. Two volcanic cycles are represented, each terminating with andesitic to felsic flows and pyroclastics. Many mafic dikes and sills also occur in Division A. Division B, which may be separated from Division A by an unconformity, is composed largely of graywackes and associated argillites. A coarse conglomerate occurs at the base at some localities. Arkosic quartzite and andesitic to felsic pyroclastics occur interbedded in the graywacke sequence.

The Yellowknife Supergroup is an example of a calc-alkaline greenstone succession. Andesitic to felsic pyroclastic rocks increase in abundance at higher stratigraphic levels and two volcanic cycles are clearly defined in the lower part of the succession. Ultramafic and komatiitic rocks are unknown from this succession and appear also to be rare or absent in greenstone belts in other parts of the Slave Province.

CYCLICITY

Stratigraphic cyclicity occurs in Archean greenstone successions both in igneous and sedimentary rocks. It also occurs on varying scales (Anhaeusser, 1971b). *Mini-cycles*, which are most common in clastic sediments, occur on scales of millimeters to centimeters. Graded bedding in graywacke-argillite units is perhaps the most widespread type of mini-cycle. It is generally interpreted to reflect turbidity current deposition. Fine lamination in cherts is another example of mini-cyclicity.

Minor cycles are measured in terms of meters to tens of meters and are best developed in volcanic sections (Anhaeusser et al., 1968). *Major cycles* may include many minor cycles and range from hundreds to several thousands of meters in thickness. Minor and major cycles occur in both volcanic and sedimentary parts of greenstone successions and include the same types of stratigraphic changes. Examples of such cycles are well known in the Barberton belt (Anhaeusser et al., 1968) and in greenstone belts of the Superior Province (Hubregtse, 1976; Ayres, 1977). Many greenstone belts contain 5 to 10 major cycles, each being comprised of many minor cycles. Typical volcanic cycles in the Barberton belt begin with ultramafic or mafic flows; these are overlain chiefly by mafic flows and the cycle ends with felsic volcanic rocks, often capped with chert (Fig. 2-4). These cycles are

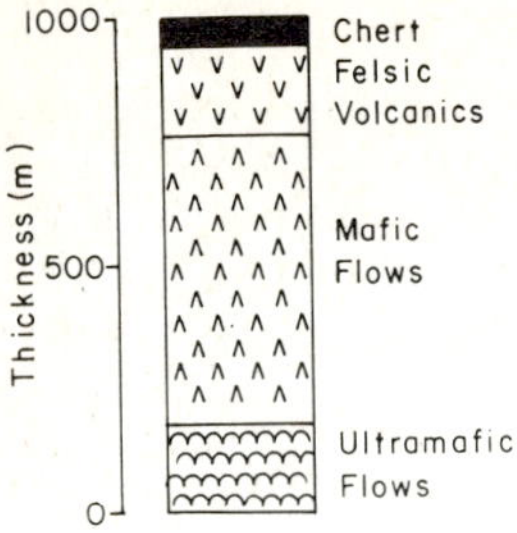

Fig. 2-4. An idealized volcanic cycle from the Swaziland Supergroup in the Barberton greenstone belt.

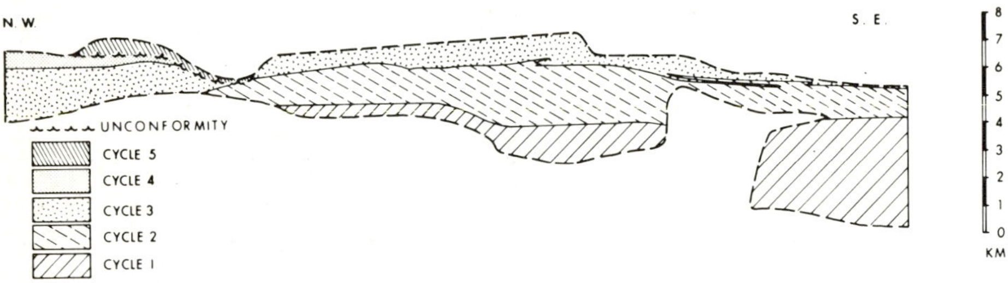

Fig. 2-5. Reconstructed (pre-folding) cross-section of volcanic cycles in the Favourable Lake greenstone succession, northwestern Ontario (from Ayres, 1977).

repeated many times with the proportion of ultramafic rock decreasing at the expense of felsic volcanics and chert in going upwards in the section.

Ayres (1977) has described five volcanic cycles in the Favourable Lake greenstone belt in northwestern Ontario which contains a minimum of 7.5 km of section. Fig. 2-5 shows a reconstructed cross-section of this area showing the time-space relationships of the cycles. The lowermost cycle (4 km thick) is a sequence of subaerial, dacitic to andesitic flows and pyroclastics that are interpreted to represent a portion of a stratovolcano. Cycles 2, 3, and 4 represent subaqueous basalt eruptions that pass upwards into andesite-dacite pyroclastic cones which provided detrital input for graywackes deposited on the flanks of the volcanoes. Cycle 5 (570 m thick) unconformably overlies the other cycles and is composed of subaerial andesitic to dacitic flows and pyroclastics. At least six volcanic cycles have been recognized in the Knee Lake—Oxford Lake greenstone belt in nearby areas in Manitoba (Hubregtse, 1976). Each cycle is characterized by a thick basal basaltic portion and a thinner overlying section of felsic volcanics and graywackes. Previously noted, two volcanic cycles are also preserved in the Yellowknife Supergroup (Fig. 2-3C), each ending with calc-alkaline pyroclastics.

Sedimentary cycles have been described by Anhaeusser et al., (1968) and Anhaeusser (1971b) in the Fig Tree and Moodies Groups in the Barberton area. In the Fig Tree Group, major and minor cycles are characterized by a

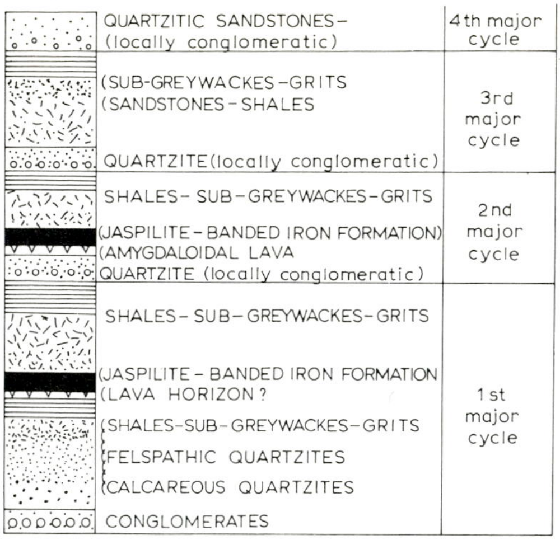

Fig. 2-6. Cyclic sedimentation in the Moodies Group, Barberton greenstone belt, South Africa (from Anhaeusser, 1971b). Vertical section represents 3.5 km.

decrease in the grain size of graywackes and an increase in the amount of shale, chert, and banded iron formation towards the top of the cycle. An example of the cyclicity in the Moodies Group is shown diagrammatically in Fig. 2-6. Three major cycles and the beginning of a fourth cycle are recognized (Anhaeusser, 1971b). The first cycle begins with a conglomerate and is followed sequentially by quartzites, shales, and banded chert. The second cycle begins with a conglomeratic quartzite and is followed by a mafic flow, a chert bed, and a sequence of shales and subgraywackes. The third cycle includes a basal quartzite overlain by shales and subgraywackes. In Western Australia, similar cycles are characterized by basal conglomerate overlain sequentially by quartzite, shale, lava, and banded iron formation. Small-scale (few meters) sedimentary cycles have also been described from graywacke-argillite successions in Western Australia (Dunbar and McCall, 1971).

The final type of cycle recognized in Archean greenstone successions is the *super-cycle* which embraces the entire stratigraphic succession. The Onverwacht, Fig Tree, and Moodies Groups (Fig. 2-1) each represent a super-cycle.

RELATIONSHIPS BETWEEN GREENSTONE BELTS

Interbelt correlation

Of fundamental importance to an understanding of greenstone belt origin is a knowledge of the relationships between greenstone belts within a crustal province. Correlation of stratigraphic sections between belts is difficult due to complex structure and to lateral facies changes both within and between belts. Similar lithostratigraphic successions may occur in greenstone belts over a broad area within a crustal province as illustrated for instance by the Sebakwian, Bulawayan, Shamvaian Groups in Rhodesian greenstone belts. Such similar successions suggest similar geologic histories for these greenstone belts. Some investigators feel that greenstone belts are infolded remnants of widespread volcano-sedimentary terranes. Glikson (1976a, 1977a) has suggested that the distribution of xenoliths in granitic gneiss terranes between greenstone belts can be used to delineate the original extent of the greenstone cover. Detailed mapping of greenstone belts, indeed, indicates that greenstone successions can be traced into gneissic terranes by trains of inclusions. Some inclusion trains link one greenstone belt to another. Some xenoliths, however, may be fragments of earlier greenstone terranes not now preserved as greenstone belts. The question of which xenoliths are fragments of preserved greenstone belts and which are fragments of earlier, not-preserved belts is a subject of considerable controversy at the present time (Glikson, 1977a; Bickle and Nisbet, 1977). In any case, enough evidence now exists from granite-greenstone provinces to indicate that greenstone belts did not develop in isolation from one another, but they are remnants of a once widespread greenstone terrane (or terranes) in these provinces.

Lateral facies changes both within and between greenstone belts have now been documented in several areas. Examples are described in Chapters 3 and 4. From studies of greenstone successions in the Superior and Rhodesian Provinces, it has been possible to define greenstone basins (Goodwin, 1973; Goodwin and Ridler, 1970; Coward et al., 1976a; Key et al., 1976; Wilson, 1978). Similar basins appear to have existed in the Yilgarn, Pilbara, and Slave Provinces (Ryan, 1965; I.R. Williams, 1973; Goodwin, 1973). Such basins are reconstructed from greenstone belt successions which represent remnants of the basin fill (Chapter 4).

Lower and upper greenstone successions

General features

Glikson (1976a,b, 1977a) was one of the first to point out that more than one age of greenstone belt may occur in the same area. This has been best documented in the Shabani belt in Rhodesia, in the Lawlers-Mt. Keith belt in Western Australia, and in the Dharwar greenstone belts in India. In these areas an unconformity separates an older greenstone-granite terrane from a younger greenstone succession. Basal unconformities have also been described with underlying granitic gneiss terranes in some areas as described in Chapter 1.

Rhodesian successions

The upper greenstone succession in the Shabani belt in Rhodesia

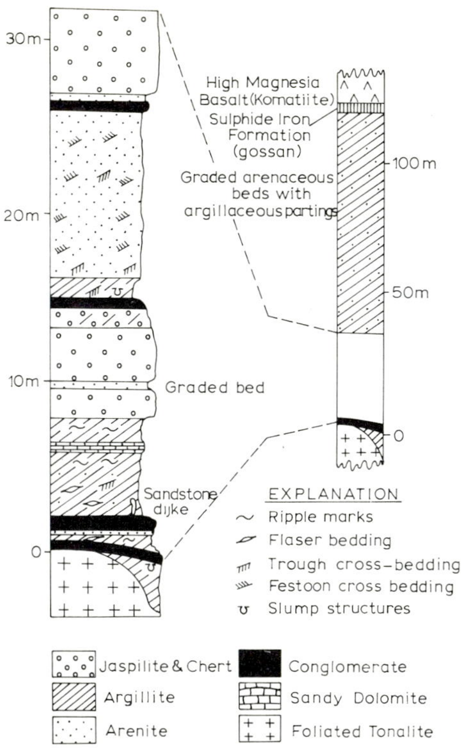

Fig. 2-7. Measured section through the basal unconformity of the upper greenstone succession near Shabani, Rhodesia (from Bickle et al., 1975).

unconformably overlies both an older greenstone succession and an older tonalitic gneiss terrane (Nisbet et al., 1977). A section through the unconformity at the gneissic contact is shown in Fig. 2-7. A course basal conglomerate containing clasts of the underlying tonalite occurs at the base of the greenstone succession and can be traced along strike for more than 30 km. Above the basal conglomerate is a sequence of shallow-water siltstones and a thin dolomite unit overlain by cherts and graywackes. A thin unit of iron formation followed by pillowed mafic and ultramafic flows overlies the graywackes. The clastic sediments, cherts, carbonate, and iron formation are collectively referred to as the Manjeri Formation. Near Belingwe, the Manjeri Formation unconformably overlies a greenstone succession. Wilson et al., (1978) suggest that this unconformity extends into other greenstone belts in Rhodesia as shown in Fig. 2-8. In general, the lower greenstone successions are characterized by pillowed mafic to ultramafic lavas, banded iron formation, and minor felsic volcanics, conglomerates, shales and quartzites. The fact that clasts of tonalite are present in the conglomerate indicates the presence of nearby, pre-lower greenstone sialic source areas. The upper greenstone succession is composed chiefly of mafic and ultramafic

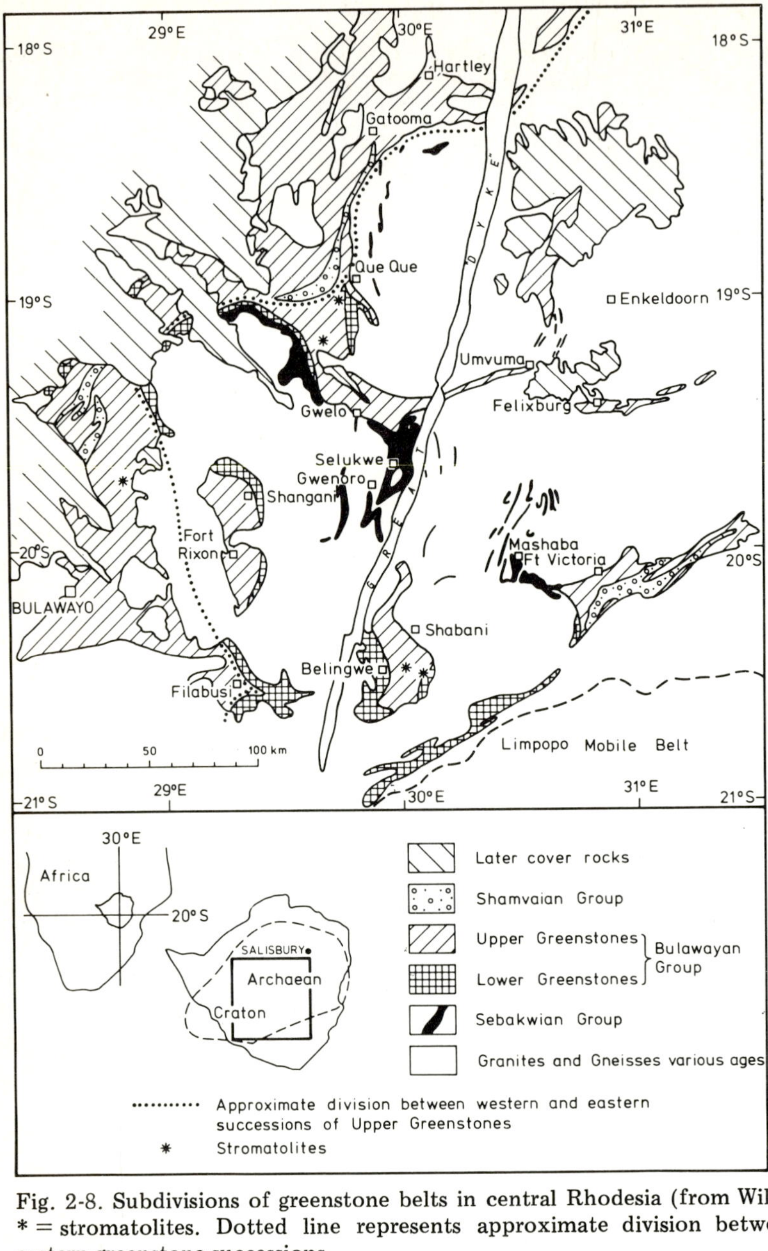

Fig. 2-8. Subdivisions of greenstone belts in central Rhodesia (from Wilson et al., 1978). * = stromatolites. Dotted line represents approximate division between western and eastern greenstone successions.

flows and calc-alkaline volcanics with smaller amounts of graywacke, phyllite and banded iron formation and minor stromatolitic dolomite. Existing radiometric ages indicate that both greenstone successions were formed between 2.6 and 2.7 b.y. (Hawkesworth et al., 1975; Jahn and Condie, 1976).

TABLE 2-2

Generalized western and eastern successions of upper Rhodesian greenstones (after Wilson et al., 1978).

Western succession	*Eastern succession*
4. Graywacke, phyllite, conglomerate	
3. Mafic and calc-alkaline volcanics	3. Not present
2. Mafic and calc-alkaline volcanics; minor graywacke, conglomerate, iron formation, and very minor carbonate	2. Mafic and minor ultramafic flows; phyllites, iron formation, minor conglomerate and carbonate
1. Mafic and ultramafic flows; minor felsic tuffs; iron formation, minor graywacke, conglomerate, and carbonate	

Remnants of a still older greenstone succession ($\gtrsim$ 3.4 b.y.) are preserved as the Sebakwian Group and associated inclusions in the 3.5 b.y. gneiss terrane in south-central Rhodesia (Fig. 2-8). The best documented section of this old greenstone succession occurs in the Selukwe belt (Stowe, 1968b, 1974). At this locality, the succession is composed chiefly of mafic and ultramafic flows and intrusive bodies and of one clastic sedimentary unit, the Wanderer Formation. Conglomerates in this formation contain granitic clasts indicating the presence of still older ($>$ 3.5 b.y.) sialic crust in Rhodesia.

A Manjeri-type marker is found in many of the greenstone belts in Rhodesia and can be used to distinguish the upper and lower greenstone successions (Wilson et al., 1978). The upper greenstones can be further subdivided into eastern and western successions based on lithologic associations (Fig. 2-8; Table 2-2). The lower portion of both successions is similar. The upper part of the eastern succession is similar to the lower part although containing more clastic sediments while the upper part of the western succession contains an appreciable amount of calc-alkaline volcanic rocks (as illustrated by the Maliyami and Felsic Formations in Fig. 2-2B). In terms of igneous components, the western succession passes upwards from a bimodal into a calc-alkaline sequence while the eastern succession remains bimodal.

The Lawlers-Mt. Keith succession

Two, and possibly the beginning of a third, greenstone successions occur in the Lawlers-Mt. Keith belt in western Australia (Fig. 2-9) (Naldrett and Turner, 1977). The lower greenstone succession, of which the base is not exposed, is composed of mafic and ultramafic flows and intrusives and minor cherts. Unconformably overlying this sequence is an upper greenstone se-

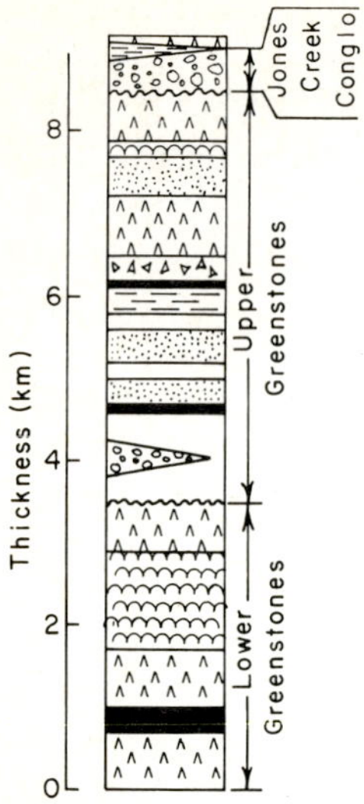

Fig. 2-9. Generalized stratigraphic section of greenstone successions in the Lawlers-Mt. Keith belt in Western Australia (after Naldrett and Turner, 1977). Symbols given in Fig. 2-2.

quence of which the lower half is composed of felsic pyroclastic rocks, chert, and shale and one conglomerate unit. These are overlain by mixed mafic and ultramafic flows, felsic pyroclastics, intrusive mafic and ultramafic rocks, and minor banded iron formation. A possible second unconformity in the Yakabindie area separates the upper greenstones from the Jones Creek Conglomerate (Durney, 1972). It is not possible, however, to verify an unconformable relationship in this area due to structural complexities (Marston and Travis, 1976). The Jones Creek Conglomerate, which ranges up to 1 km thick and can be traced along strike for over 90 km, interfingers upwards with finer sediments and possibly mafic and ultramafic volcanic rocks which may represent the beginning of a still younger greenstone succession which has been removed by faulting.

Karnataka greenstone successions

In the Western Karnataka subprovince in India, at least three ages of greenstone belts are known (Naqvi et al., 1978a). The oldest greenstone successions ($\gtrsim$ 3.0 b.y.) are the Sargur, Holenarasipur, Nuggihalli, Kolar, and related successions (Ramakrishnan et al., 1976). The stratigraphy of these rocks is at present poorly known and lower contacts are not exposed. Also, metamorphic grade is high ranging from the amphibolite to the granulite facies and most primary textures and structures are not preserved. A generalized stratigraphic column in the Sargur belt from oldest to youngest is as follows (Viswanatha and Ramakrishnan, 1975): Quartzites and quartz mica schists probably representing mature clastic sediments; mica schists and paragneisses representing shales and related rocks; limestones and calc-silicate rocks; more mica schists and paragneiss; amphibolites (mafic volcanics); and iron formation. The section is also intruded by numerous ultramafic and mafic bodies. Sargur-type greenstone successions differ from most other Archean greenstones by the relatively large amount of mature sediments which they contain. Successions in the Holenarasipur and Nuggihalli greenstone belts contain mixtures of ultramafic rocks, anorthosite, amphibolite, kyanite-staurolite schist, and quartzite (Naqvi et al., 1978a).

Unconformably overlying Sargur-type belts and the Peninsular Gneiss Complex are the Dharwar greenstone belts in the western part of the Karnataka subprovince (Fig. 1-16). Rocks in these greenstone belts are included in the Dharwar Supergroup which is divided into the lower Bababudan Group and the overlying Chitradurga Group (Nath et al., 1976). Generally, the Bababudan Group begins with a basal oligomictic conglomerate composed of quartz and quartzite pebbles which overlies older rocks with an angular unconformity. In the type locality (the Bababudan belt), this is overlain by mixed amygdaloidal basalts, shallow-water quartzites, schists (originally shales, etc.) and iron formation. This sequence is in turn overlain successively by mafic to felsic volcanics and ultramafic rocks, argillite, and iron formation.

Unconformably overlying the Bababudan Group is the Chitradurga Group which is best known in the Shimoga and Chitradurga belts. More than 7 km of section is exposed at the type section in the Chitradurga belt. The lower 1 km is composed of a platform suite with polymictic conglomerates (Naqvi et al., 1978b), quartzite, phyllite, limestone and dolomite, and ferruginous chert. The remainder of the sequence is composed chiefly of graywacke-argillite and mafic to felsic (calc-alkaline) volcanic rocks. Minor chert and conglomerate are also present.

Provinciality

Enough data are now available from Archean greenstone belts to quantitatively compare greenstone belts based on rock-type abundances. Results suggest a provinciality of greenstone belts (Condie, 1976b). Average abundances of ultramafic rocks, mafic to felsic volcanics, and sediments in greenstone belts from several Archean provinces are summarized in Table 2-3. These estimates are made from stratigraphic sections. Existing data from greenstone belts in the Slave and Kaapvaal Provinces suggest that the Yellowknife and Barberton successions are representative of each of these provinces, respectively. Noteworthy is the sparsity of ultramafic and related mafic rocks in the Superior and Slave Provinces. Geographic variations also occur in the relative abundances of volcanic rock types in greenstone belts (Condie, 1976b; Goodwin, 1977a). Examples are summarized in Table 2-4. On the whole, the greenstone belts from North America are similar containing

TABLE 2-3

Percentages of rock types in Archean greenstone belts

	Superior Province		Rhodesian Province (western successions)	Kaapvaal Province (Barberton)	Slave Province (Yellowknife)	Yilgarn Province (Kalgoorlie system)
	Abitibi	Wabigoon				
Ultramafic-mafic[1]	4	3	10	16	⩽1	23
Mafic to felsic	72	81	75	51	73	48
Sediments	24	16	15	33	26	29

[1] Mafic rocks in this category are tholeiite and basaltic komatiite flows closely associated with ultramafic flows.

References: Superior Province, Goodwin et al., (1972); Rhodesian Province, K.C. Condie (unpublished data); Kaapvaal Province, Anhaeusser (1974); Slave Province, Henderson and Brown (1966); Yilgarn Province, Glikson (1976b).

TABLE 2-4

Proportions of volcanic rock types in Archean greenstone belts (in percent)

	Ultramafic-mafic[1]	Mafic	Andesite	Felsic
North America				
1. Birch-Uchi	4	54	29	13
2. Wabigoon	4	58	26	12
3. Abitibi	5	50	37	8
4. Yellowknife	⩽1	65	20	14
Africa				
1. Midlands-Bulawayo	10	42	40	8
2. Fort Victoria	10	75	5	10
3. Shabani	12	75	10	3
4. Barberton	24	72	—	4
5. Western Kenya	—	10	75	15
Australia				
1. Coolgardie-Norseman	20	62	5	13

[1] See footnote to Table 2-3.

References: Condie (1976b); Baragar and Goodwin (1969); Goodwin et al. (1972); Anhaeusser (1974); Glikson (1972a, 1976b); Goodwin (1977a).

TABLE 2-5

Characteristics of upper and lower greenstones (modified after Glikson, 1977a)

	Lower greenstones	Upper greenstones
Volcanic rocks	ultramafic and mafic flows and intrusives common; felsic tuffs minor, but ubiquitous	mafic flows and sills dominate; ultramafic rocks of variable importance; calc-alkaline volcanic rocks relatively abundant especially in upper parts of successions, pyroclastic rocks important
Volcanic compositions	bimodal association characteristic	bimodal and/or calc-alkaline associations characteristic
Sedimentary rocks	generally minor compared to volcanics; cherts, phyllites, and iron formation most common	graywacke-argillite and associated felsic to intermediate pyroclastics common in upper parts of successions; chert, phyllite, iron formation minor components throughout sections

about 50% mafic rocks, 30% andesite, 10% felsic volcanics, and $\lesssim$ 5% of ultramafic and related mafic rocks. Ultramafic and related mafic rocks, however, typically comprise 10—25% of greenstone belts in the Yilgarn, Kaapvaal, and Rhodesian Provinces. Western Kenya is unique in containing a large amount of andesite. The Australian and most of the African belts appear to define an ultramafic-rich province while the North American belts define a major calc-alkaline province.

Glikson (1977a) has further suggested that there are fundamental differences in rock abundances between lower and upper greenstone successions as summarized in Table 2-5. In general, lower greenstones appear to be characterized by an abundance of ultramafic to mafic rocks and chert with minor felsic pyroclastics and thus are bimodal in character. Upper greenstone belts contain a greater variety of volcanic and sedimentary rock types and may be bimodal, calc-alkaline, or both. As pointed out by Bickle and Nisbet (1977), ultramafic rocks may occur in both lower and upper greenstones. Also, these authors indicate that, at least in part, the relatively greater abundance of clastic sediments in the upper greenstones than in the lower, may be related to erosion level. The lower greenstones may have been uplifted and the sediment-rich upper parts of the sections removed by erosion prior to burial beneath the upper greenstones. Although upper greenstone belts may be bimodal or calc-alkaline in nature, lower greenstones are not observed to be calc-alkaline. In terms of available radiometric ages, it appears that bimodal greenstones formed throughout the Archean (2.6—3.7 + b.y.) while calc-alkaline types were chiefly formed between 2.6 and 2.7 b.y.

GENERAL STRATIGRAPHIC FEATURES

Despite the fact that a good deal of diversity exists between greenstone successions, certain overall features are widespread as summarized below.

(1) There is a decrease in the amount of both ultramafic and mafic rocks with stratigraphic height.

(2) There is an increase in the ratio of pyroclastics to flows with stratigraphic height and a parallel increase in the amount of felsic, and in some cases andesitic, volcanics.

(3) Immature clastic sediments often dominate in the upper parts of greenstone successions.

(4) Two basic types of greenstone successions are recognized based on volcanic abundances. The bimodal type is composed chiefly of ultramafic and related mafic rocks with minor amounts of chert and felsic volcanics. The calc-alkaline type is composed chiefly of mafic to felsic volcanics and derivative clastic sediments.

(5) Bimodal successions may unconformably underlie or evolve into calc-alkaline successions. Calc-alkaline successions are not observed to underlie bimodal successions.

(6) Greenstone belts exhibit lithologic provinciality. Greenstone belts in southern Africa and Australia are typically bimodal and contain relatively large amounts of ultramafic and associated mafic rocks. Many greenstone belts in North America are typically calc-alkaline.

(7) Cyclicity occurs in both sediments and volcanics in greenstone successions and ranges from 10^{-3} to 10^4 m per cycle. Individual greenstone belts may contain 5 to 10 major volcanic cycles.

(8) Greenstone belts may represent remnants of large Archean volcano-sedimentary basins.

(9) Mature clastic sediments, carbonates, and alkali-rich volcanic rocks are uncommon in Archean greenstone successions.

Chapter 3

VOLCANIC AND HYPABYSSAL ROCKS

INTRODUCTION

Volcanic rocks and associated hypabyssal intrusive rocks (dikes, sills, plugs, etc.) are the dominant rocks in Archean greenstone belts. Many detailed descriptions of these rocks are available in the literature and suggest that magmatic eruptions were largely subaqueous. Available chemical analyses indicate a wide range in composition from ultramafic to felsic and alkaline with mafic rocks greatly dominating. Estimated stratigraphic thicknesses of greenstone volcanics range up to 20 km (Goodwin et al., 1972).

Several methods have been employed in the classification of Archean volcanic and hypabyssal rocks. Rittman (1952), Irvine and Baragar (1971), and Church (1975) propose methods based on major element compositions. The method of Irvine and Baragar (1971) has been widely adopted by Canadian investigators. It employs several chemical variation diagrams which broadly classify rocks into subalkaline (tholeiite and calc-alkaline), alkaline, and peralkaline categories and then makes use of normative color index and plagioclase composition to assign individual rock names. More simplified methods based chiefly on SiO_2-K_2O relationships have been employed by Taylor et al. (1969) and Condie and Moore (1977). Classification schemes of Archean volcanic rocks based on major elements are faced with the ever-present problem of element mobility during alteration and metamorphism. Recently Winchester and Floyd (1977) have proposed a classfication based on relatively immobile elements (Ti, Zr, Y, Nb, etc.), the chief disadvantage of which is having to know the concentrations of such elements before one can classify a rock.

Recent studies in the Canadian Shield have been instructive in better understanding the nature of Archean volcanism. Goodwin and Ridler (1970) have defined volcanic complexes in the Abitibi belt which are now deformed, but originally ranged from 100 to 175 km in diameter (Fig. 3-1). Each volcanic complex is characterized by its own mafic to felsic volcanic sequence (including hypabyssal rocks) and associated sediments and granitic plutons. Intermediate and felsic volcanic rocks are abundant at volcanic centers within the complexes. Each volcanic complex contains many volcanic centers. Some studies focus on reconstructing individual volcanic centers and determining the relationships of volcanics to sediments (Dimroth, 1976; Hallberg et al., 1976; Page and Clifford, 1977; Tasse et al., 1978; M. B. Lambert, 1978). Studies of the facies distribution of volcanic rocks, employing such features as vesicularity, the proportion of massive, pillowed, and brecciated rock types, and the types of sedimentary structures in

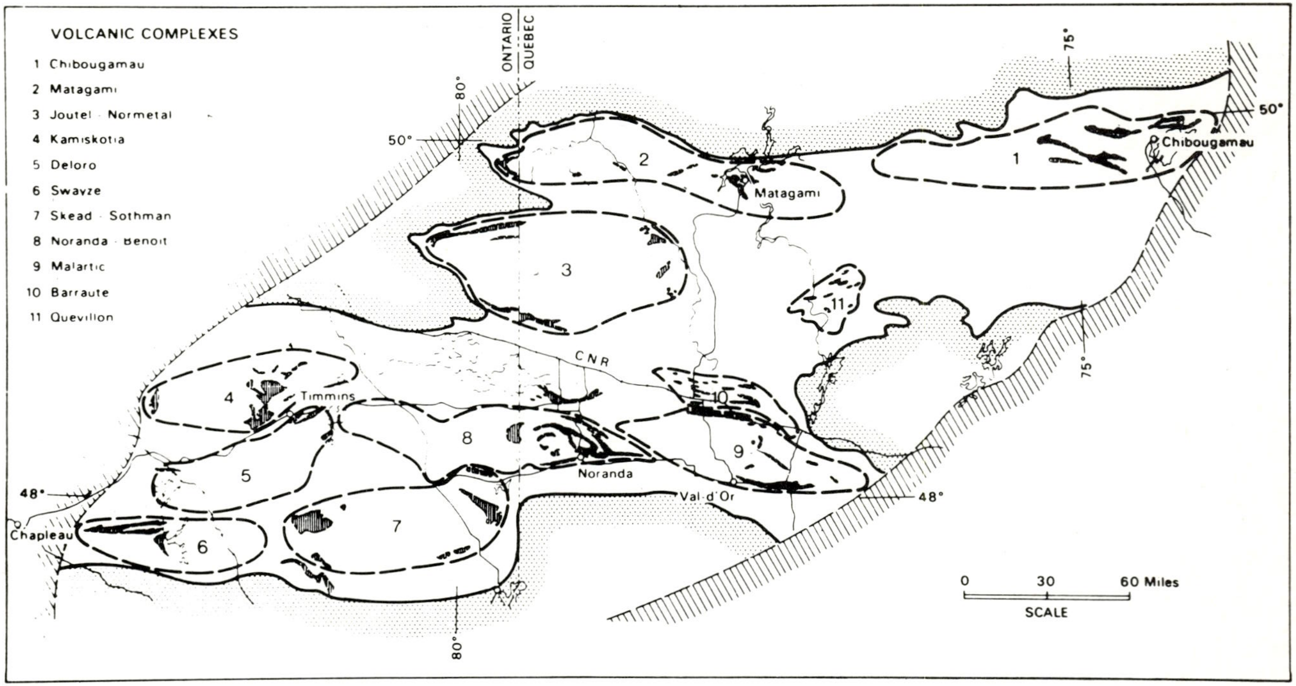

Fig. 3-1. Distribution of volcanic complexes in the Abitibi greenstone belt, Canada (from Goodwin and Ridler, 1970). Felsic volcanic rocks shown by vertically ruled pattern.

pyroclastic-hyaloclastic rocks, have been informative in terms of mechanism of emplacement and bathymetry at the sites of emplacement (Dimroth, 1976; Dimroth et al., 1979). The Blake River Group in the Abitibi belt is characterized by the subaqueous eruption of a vast, thick basalt sheet upon which overlapping shield volcanoes grew. Some of these volcanoes emerged during the late stages of Blake River volcanism and shed fans of conglomeratic turbidites composed of volcanic detritus. Emergence of these volcanoes appears to have formed a relatively large island. Some emergent volcanic centers erupted large volumes of felsic subaerial and subaqueous ash flows, such as found in the Back River Complex in the Slave Province (M. B. Lambert, 1978). At this location, the eruptions culminated in cauldron subsidence and emplacement of rhyolite domes in ring fractures beneath shallow seas marginal to the emergent volcanoes (Fig. 3-2). Buck (1975) has also described a major caldera in the Favourable Lake greenstone belt in northwestern Ontario.

ALTERATION

General features

Alteration is widespread in volcanic terranes in Archean greenstone belts (Viljoen and Viljoen, 1969b; Williams, 1971). The most common types are cabonization, chloritization, silicification, epidotization, and serpentinization. Mineral assemblages formed during alteration are similar to greenschist-facies mineral assemblages and in some instances they have formed in response to the same processes. In other cases, secondary minerals such as carbonate, quartz and sericite may replace greenschist-facies assemblages and cross-cut foliation indicating post-metamorphic alteration (Condie et al., 1977). Olivine and orthopyroxene commonly exhibit varying degrees of serpentinization beginning at grain boundaries and in cracks. Glass and groundmass minerals in volcanic rocks are recrystallized to assemblages of actinolite, chlorite, epidote, talc, and other secondary minerals as described in the following sections.

The effect of various types of progressive alteration on the distribution of major and trace elements in Archean volcanics is important in distinguishing primary from secondary compositions. Very few quantitative studies of alteration in greenstone belts are available. In terms of the results of studies of alteration in younger volcanic rocks, however, it would appear that alteration processes can significantly affect both major and trace element distributions (Christensen et al., 1973; Hart et al., 1974; Winchester and Floyd, 1976; Scott and Hajash, 1976; Humphris and Thompson, 1978; Ludden and Thompson, 1978). The results of recent studies are summarized in Table 3-1. Hart et al. (1970a) have shown that the order of mobility of

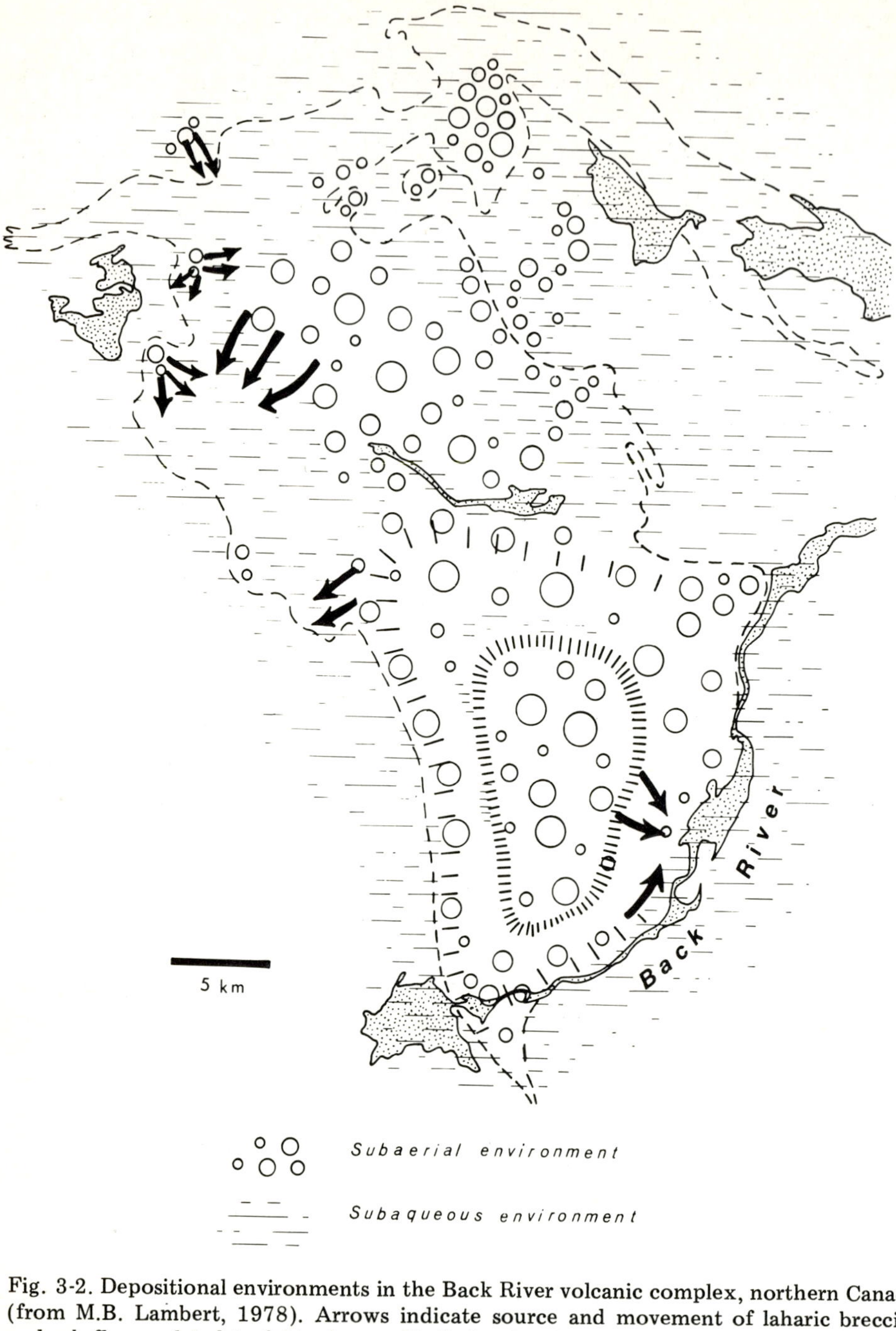

Fig. 3-2. Depositional environments in the Back River volcanic complex, northern Canada (from M.B. Lambert, 1978). Arrows indicate source and movement of laharic breccias and ash flows related to felsic domes. Radiating hatched pattern represents ring-fracture system.

TABLE 3-1

Summary of changes in element concentrations during progressive alteration and metamorphism[1] (after Condie, 1979b)

Little or no change	Significant change
TiO_2, Na_2O, Y, REE, Zr, Zn, V, Sc, Hf, Nb, Ta, Co, (total Fe), (Cr), (Sr), (Ni), (Cu)	Fe^{3+}/Fe^{2+}, K, Cs, Rb, H_2O, SiO, CaO, Al_2O_3, MgO, F, Cl, CO_2, Th, U, (Ba)

[1] ( ) indicates element sometimes falls in other category.

alkali elements in Archean and younger volcanic rocks is Cs > Rb > K $\simeq$ Sr. Beswick and Soucie (1978) have proposed a graphical method to correct for losses and gains during secondary processes in Archean volcanics. To employ the method it is necessary to make two assumptions: (1) the altered rocks originally had a composition which conformed to well-defined igneous trends, and (2) Al_2O_3 remained immobile during alteration. Results of this approach on greenstone volcanics from the Timagami belt in Ontario indicate Na, K, Ca, Mg, and Fe were mobilized in varying amounts in different parts of the belt. Rimsaite (1974) has shown that progressive alteration of greenstone volcanic rocks in Quebec results in significant changes in mineral compositions. Progressive chloritization of biotite, for instance, results in losses of Si, Al^{VI}, Ti, K, F, Cl and gains in Al^{IV}, Mg and OH. Studies of the effects of alteration on rare earth element (REE) distributions in Archean volcanics suggest that light-REE concentrations and perhaps Eu anomalies may be modified in small amounts by secondary processes (Condie et al., 1977; Sun and Nesbitt, 1978). Let us now examine some of the major kinds of alteration in more detail.

Serpentinization

All Archean ultramafic rocks exhibit the effects of serpentinization in varying degrees. Viljoen and Viljoen (1969b) found that in tracing relatively unserpentinized peridotites along strike, they become progressively more serpentinized. The fresh peridotites are characterized by incipient serpentine development around olivine grains. Along strike, serpentinization increases until the entire rock is serpentinized with original minerals completely destroyed. However, as illustrated in Fig. 3-3A, original olivine grains may be outlined by magnetite producing a mesh texture. Massive fine-grained serpentinites often exhibit spinifex texture indicating a volcanic origin for these rocks. Progressive serpentinization (as measured by increasing H_2O^+) is accompanied by a decrease in specific gravity and also in Al_2O_3, CaO, $Na_2O + K_2O$, SiO_2, and FeO (total Fe). Some of the depleted major elements may have gone into forming calc-silicate and carbonate rocks

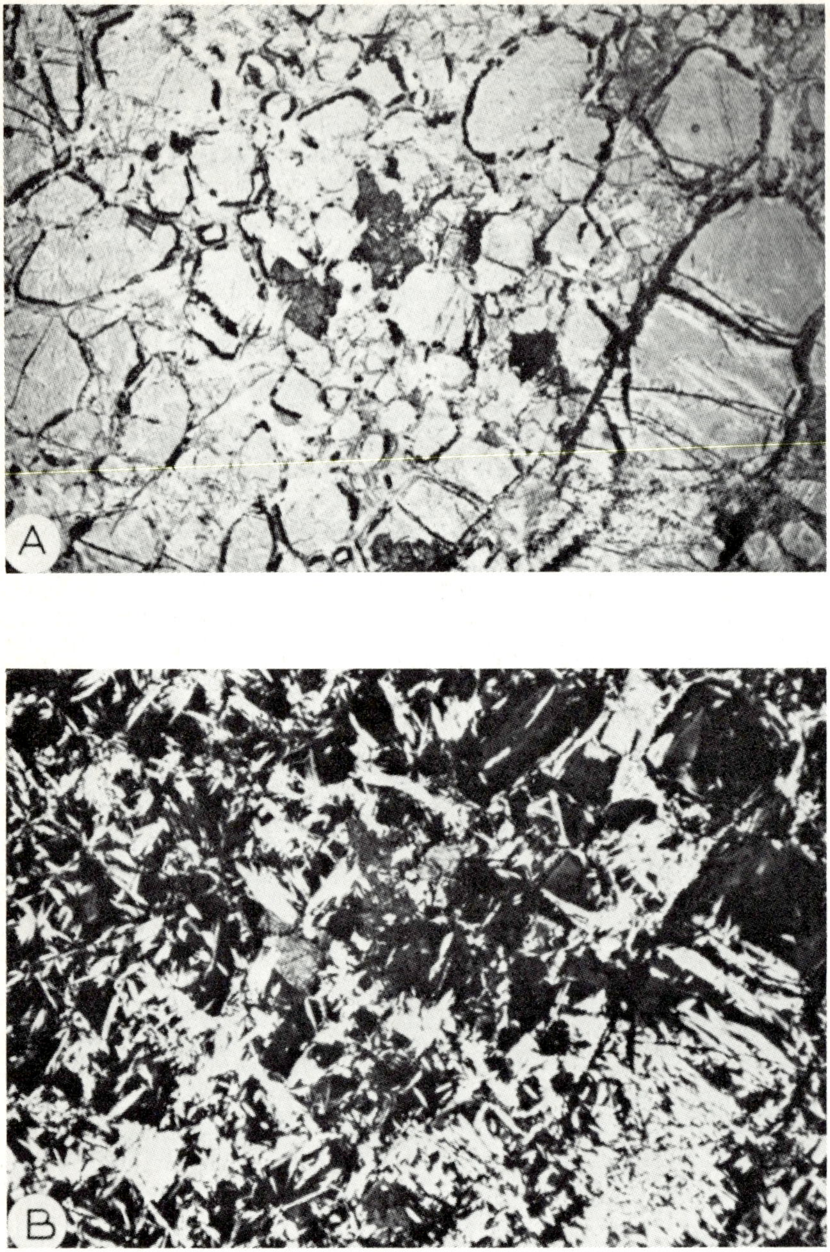

Fig. 3-3. Photomicrograph (× 20) of a completely serpentinized peridotite for the Komati Formation, South Africa (from Viljoen and Viljoen, 1969f). A Serpentinized olivine grains outlined with fine magnetite (black). Plane light. B. Same view under crossed polars showing two varieties of serpentine (fibrous variety white; dense varity black).

which are common within and adjacent to highly serpentinized rocks in the Barberton belt. Studies of Archean ultramafic rocks from Australia suggest that light-REE abundances and Eu anomalies may also vary during serpentinization (Sun and Nesbitt, 1978).

Carbonization

Carbonization is widespread in many Archean greenstone belts. Carbonate occurs disseminated in the matrices of volcanic rocks, as lenses up to a few meters long, often within shear zones, and as anastomosing veinlets. Pillowed lavas are typically more carbonated than massive flows or sills (Viljoen and Viljoen, 1969b). Common carbonate minerals are calcite and ankerite and they may be associated with quartz, epidote, amphibole, sphene and sometimes, sulfides. Often, carbonate can be observed to replace plagioclase and amphibole in volcanic rocks (Condie et al., 1977); calcite pseudomorphs of these minerals are present in some samples. Such textures, together with cross-cutting veins, indicate that much carbonization is post-metamorphic. Hence, it should not be equated with deuteric alteration, seawater interaction (halmyrolysis), or low grades of metamorphism (Condie et al., 1977). SiO_2, Al_2O_3, MgO, CaO, and K_2O decrease and FeO, TiO_2, and H_2O increase during carbonization of tholeiites from the Hooggenoeg Formation in the Barberton belt. Wilson et al., (1965) report similar changes accompanying carbonization of tholeiites in Canadian greenstone belts. Recent geochemical studies of a Barberton tholeiite flow progressively carbonated to 10% indicate small enrichments in Fe^{2+}, Ti, H_2O, Ga Zn, Y, Ta, Nb, and light REE, major losses of Sr, Ba, Cr, and Cs, and small losses of Na, Fe^{3+}, Mn, Sb, Au, and U (Condie et al., 1977). Elements least affected are Hf, Ni, Co, Zr, Th, and heavy REE. A summary of losses and gains of trace elements is given in Fig. 3-4. Light REE appear to be very slightly enriched during carbonization (Fig. 3-5). Carbonization of the flow is accompanied also by some chloritization. Brooks et al., (1969) have shown that the $^{87}Sr/^{86}Sr$ ratio in disseminated carbonate in Archean volcanics from the Superior Province is similar to that of the host rock while carbonate veins may have considerably greater isotopic ratios.

The origin of carbonate in Archean greenstone volcanics is one of the major problems in greenstone belt development. Possible sources are as follows: (1) late magmatic or deuteric fluids associated with volcanism; (2) volatiles liberated by intrusive granites; (3) reaction with seawater during or soon after eruption; and (4) mobilization and concentration of volatiles already present in the volcanics. Although late magmatic or deuteric sources may be important for some carbonate (such as that in alteration zones of massive sulfide deposits, Chapter 7), the fact that carbonates clearly replace metamorphic minerals in many rocks indicates a post-metamorphic source for much of the carbonate. Because carbonization does not increase towards

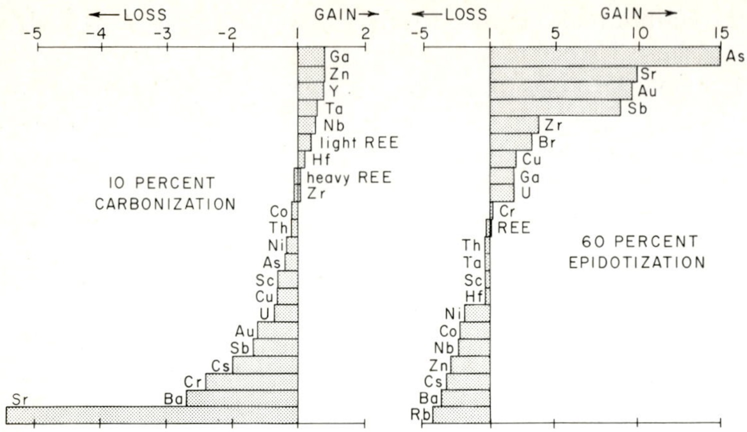

Fig. 3-4. Summary of losses and gains of trace elements accompanying carbonization and epidotization of Barberton tholeiites (from Condie et al., 1977).

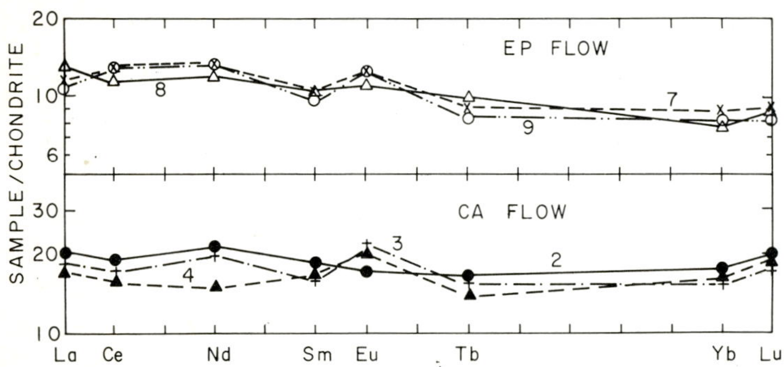

Fig. 3-5. Chondrite-normalized REE patterns in progressively altered tholeiite flows from the Barberton belt (from Condie et al., 1977). CA = carbonated flow [carbonate percentages: 4(< 1%), 3(7.9%), 2(9.0%)] EP = epidotized flow [epidote percentages: 7(7%), 8(24%), 9(59%)].

intrusive contacts, a plutonic source is not appealing (Boyle, 1959). Textural relations also do not favor a seawater reaction source and recent studies (field and experimental) of seawater volcanic interactions do not show additions of carbonate during such interactions (Scott and Hajash, 1976). Boyle (1959) and Viljoen and Viljoen (1969b) favor the fourth source. Boyle points out that carbonated Archean volcanic rocks are most widespread in greenschist and prehnite-pumpellyite facies terranes and they are uncommon in higher-grade terranes near granitic plutons. He suggests that volatiles in volcanic rocks are mobilized and migrate down thermal gradients away from intrusive plutons into low-grade terranes and here are deposited as carbonates. Vein carbonates with high $^{87}Sr/^{86}Sr$ ratios, however, cannot be of the same origin as disseminated carbonate.

Epidotization

Besides occurring as a primary metamorphic phase, epidote occurs in much the same way as carbonate. It often is found in fractures. As with carbonization, epidotization appears to be primarily a post-metamorphic process. Condie et al. (1977) describe compositional changes accompanying up to 60% epidotization of a Barberton tholeiite, Fe^{3+} and Ca increase and Fe^{2+}, Mg, Si, and H_2O decrease during the epidotization as dictated by mineral stoichiometry. A summary of losses and gains of trace elements is given in Fig. 3-4 and corresponding REE patterns in Fig. 3-5. It is notable that such a large amount of epidotization does not appreciably affect REE distributions. The remarkable increase in As, Au, and Sb during the Barberton epidotization appears to reflect accompanying increases in sulfide phases. Those elements least affected by epidotization are Hf, Ta, Sc, Cr, Th, and the REE.

ULTRAMAFIC AND MAFIC IGNEOUS ROCKS

The komatiite problem

The use of the term komatiite is a topic of considerable controversy and discussion. It was originally defined by Viljoen and Viljoen (1969c) to describe a suite of Mg-rich ultramafic and basaltic lavas from the Barberton greenstone belt. They are characterized by MgO contents greater than 9%, high CaO/Al_2O_3 ratios (> 1), and low alkali contents (< 1% K_2O). In addition they are often characterized by spinifex (quench) textures. Based on MgO and other major element contents komatiites are subdivided into peridotitic komatiite (PK) and basaltic komatiite (BK). The BK are further broken down into three subgroups (Fig. 3-6). Brooks and Hart (1974) indicate that many cumulus ultramafic and mafic rocks have the geochemical features of PK and BK, yet young quench-textured lavas with these characteristics, although not absent (Upadhyay, 1978), are uncommon. They propose the following definition for komatiites after surveying the compositions of many thousands of Phanerozoic ultramafic and mafic igneous rocks: komatiites are *non-cumulate* rocks with $CaO/Al_2O_3 > 1$, $MgO > 9\%$, $K_2O < 0.9\%$, and $TiO_2 < 0.9\%$.

Mg-rich lavas from the Abitibi belt in Canada (Arndt et al., 1977), and from greenstone belts in India (Viswanathan, 1974), Rhodesia (Bickle et al., 1975; Nisbet et al., 1977), and Western Australia (Nesbitt and Sun, 1976) share most of the geochemical characters of the Barberton komatiites but, in general, lack the high CaO/Al_2O_3 ratios (Tables 3-3 and 3-4). Nesbitt and Sun (1976) have suggested that the CaO/Al_2O_3 ratio can be affected by metamorphism and hence, should not be included in a komatiite definition.

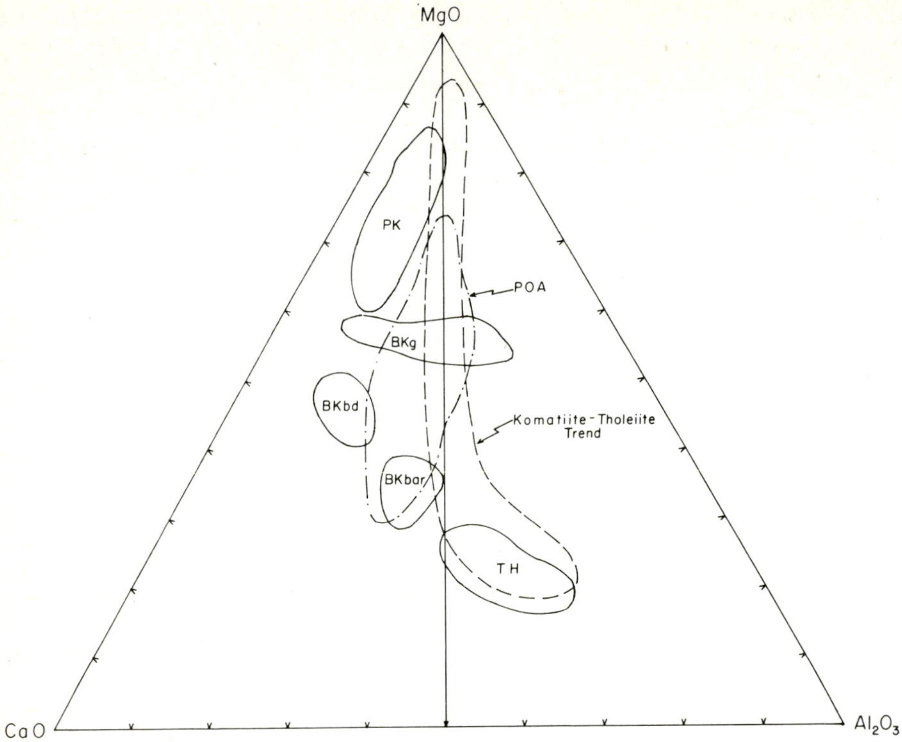

Fig. 3-6. MgO-CaO-Al_2O_3 diagram showing the distribution of komatiitic and tholeiitic rocks (Viljoen and Viljoen, 1969c; Arndt et al., 1977). PK = peridotitic komatiites; BK = basaltic komatiites from the Barberton area (g = Geluk type; bd = Badplaas type; bar = Barberton type); TH = tholeiites; POA = field of picrite-oceanite-ankaramite association (Brooks and Hart, 1974). Samples from greenstone belts in Canada, Rhodesia, and Australia define the komatiite-tholeiite trend.

A modification of the Barberton definition for komatiite has been proposed by Arndt et al., (1977). On an MgO-CaO-Al_2O_3 diagram (Fig. 3-6), ultramafic and mafic rocks from the Abitibi belt in Canada, from several greenstone belts in Western Australia, and probably from the Roodekrans belt in South Africa define a continuous trend herein called the komatiite-tholeiite trend (Arndt et al., 1977; Naldrett and Turner, 1977; Anhaeusser, 1976b). Komatiitic rocks from the Munro Township in the Abitibi belt fall into three categories (Arndt et al., 1977): peridotitic (PK), pyroxenitic (PYK), and basaltic komatiites (BK). They include both cumulus and spinifex-textured flows. Arndt et al. (1977) suggest that these rocks define a komatiite series which can be distinguished from the tholeiite series on an Al_2O_3 vs. $FeO_T/FeO_T + MgO$ diagram. The rocks have $< 16.5\%$ Al_2O_3, $> 8.5\%$ MgO, and $< 1\%$ TiO_2. They also have high Ni and Cr contents. It is noteworthy that on the Al_2O_3 vs. $FeO_T/FeO_T + MgO$ diagram, the

Barberton komatiites fall into the tholeiite series and many mid-ocean ridge tholeiites fall into the komatiite series, thus confusing the nomenclature still further.

Although a rigorous definition of the term komatiite has not been agreed upon, any definition to be of widespread application to high-Mg Archean volcanic rocks must encompass the following features: (1) it must be restricted to lavas with clearly defined quench textures (i.e., spinifex texture) in order to avoid including a great variety of cumulus rocks; and (2) it must not be restricted to rocks with CaO/Al_2O_3 ratios greater than one or most Archean high-Mg ultramafic and mafic lavas will be excluded from the definition. Nisbet et al. (1977) suggest a tentative definition for komatiites which includes spinifex-textured lavas that satisfy the chemical definition of Brooks and Hart (1974) except that rocks with CaO/Al_2O_3 ratios as low as 0.8 are included. This definition is herein adopted.

Ultramafic volcanic rocks

Occurrence

In Archean greenstone belts, ultramafic rocks occur as flows and pyroclastic units, as portions stratiform intrusions, and as miscellaneous sills, dikes, and related intrusive bodies. The intrusive occurrences are considered in later sections. Ultramafic rocks range from partially to entirely serpentinized and primary textures may be well preserved or entirely obliterated by deformation and recrystallization. Most extrusive ultramafic rocks are flows which range up to 30 m thick. Rocks comprising flows range in color from black to dark green or gray in the more highly serpentinized varieties. They range from fine to medium grained to rarely coarse grained. Fine-grained and fractured flow tops are sometimes preserved (Willett et al., 1978). Pillows occur in some flows and range in size from 20 to 50 cm across (Nisbet et al., 1977). Vesicles, amygdules, and variolites are generally lacking within ultramafic pillows (Viljoen and Viljoen, 1969d). Chilled and cracked margins (up to 2 cm thick) and small radial joints characterized some pillows. Pillowed flows are generally interlayered with massive, non-pillowed flows and boundaries are often gradational between the two types.

Ultramafic flows are interbedded with basaltic komatiite and tholeiite flows and sills as illustrated by a stratigraphic section in the Munro Township (Fig. 3-7). Cyclic variation is common as monitored by the MgO content. The lower two cycles begin with thin ultramafic flows and end with thicker massive mafic flows. The succession in Munro Township is characterized by 45% PK, 35% PYK, and 20% BK (Arndt et al., 1977). Approximately 60 flow units are present in an outcrop width of 135 m (Pyke et al., 1973). Individual flows range in thickness from 0.5 to 15 m and some can be traced for up to 180 m along strike. It is noteworthy that the lower contacts of individual flows conform to the topographic surface of underlying flows

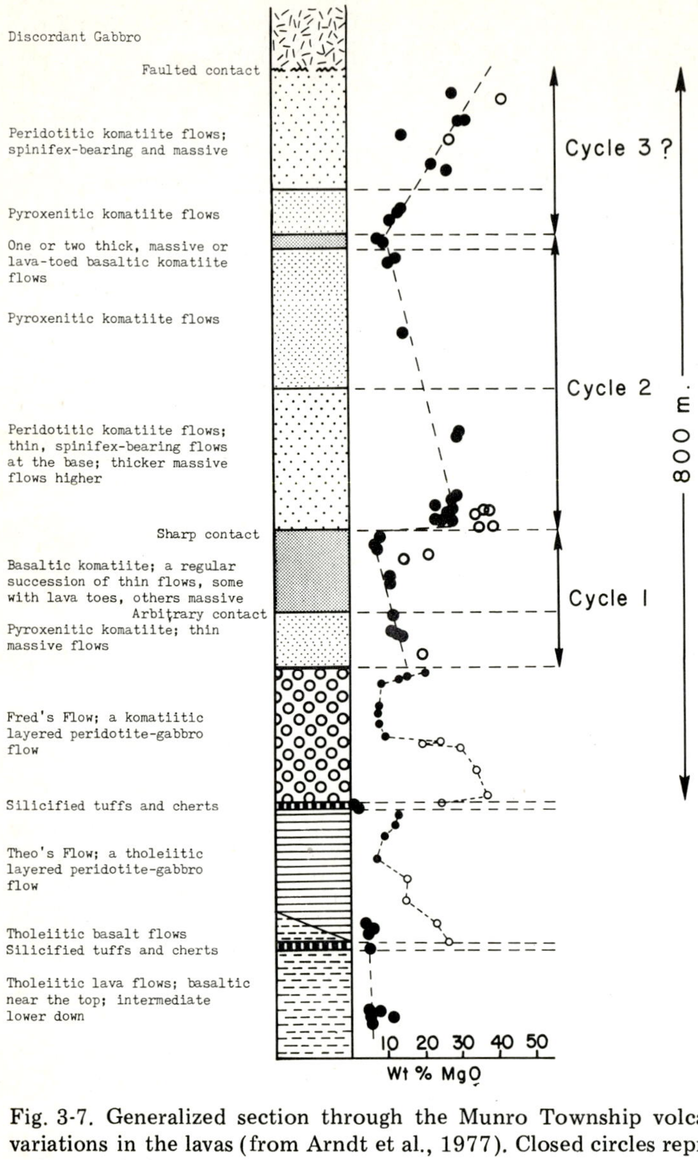

Fig. 3-7. Generalized section through the Munro Township volcanic pile showing MgO variations in the lavas (from Arndt et al., 1977). Closed circles represent quenched liquids and open circles, cumulates.

(Fig. 3-8). Also, no flow unit or part thereof cross-cuts another unit. Felsic tuffs are locally interlayered with ultramafic and mafic flows as illustrated by the lower Onverwacht Group in South Africa (Viljoen and Viljoen, 1969c).

Spinifex texture is commonly preserved in the upper portions of ultramafic and basaltic komatiite flows. The texture is characterized by randomly

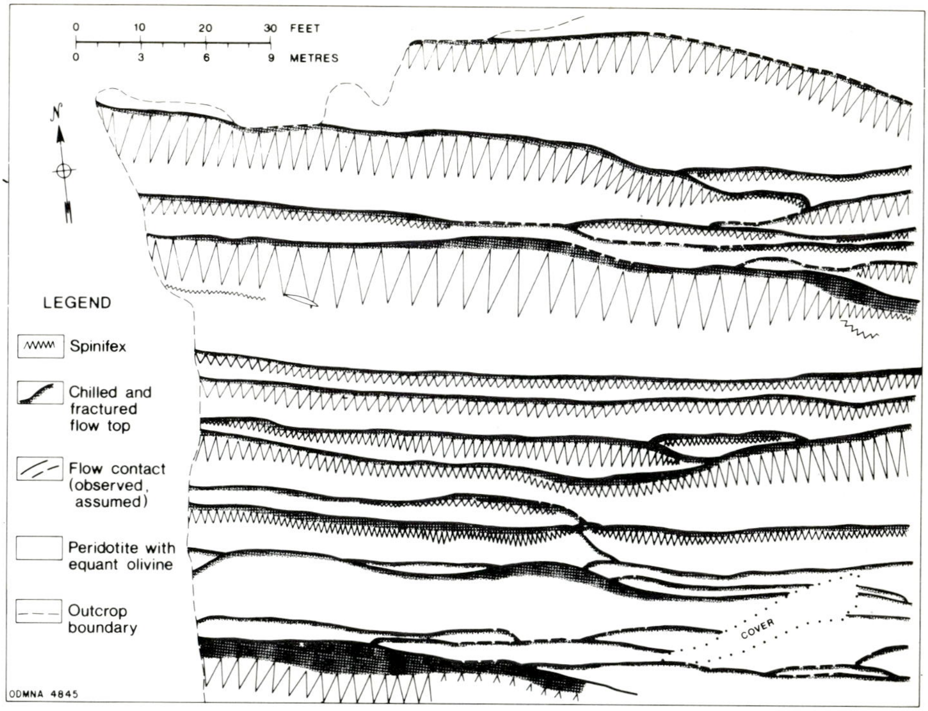

Fig. 3-8. Map showing ultramafic flow units in part of the Munro Township section (from Pyke et al., 1973).

oriented skeletal crystals of olivine or pyroxene similar in appearance to spinifex grass in Western Australia (hence the name)(Fig. 3-9A). It appears to result from the comparatively rapid cooling of Mg-rich magmas. Spinifex has not been reported from rocks with less than 9% MgO. Spinifex textures have long been recognized in olivine slags. Studies of Donaldson (1974) suggest that skeletal spinifex crystals are indicative of extreme supersaturation and may actually form over a range of cooling rates. It is necessary, however, that sufficient crystal nuclei for normal growth are not available. Nesbitt (1971) recognizes four types of spinifex in Archean volcanics based on the form of the olivine: plate, radiating, porphyritic, and harrisitic. Plate spinifex is characterized by subparallel plates of skeletal crystals; radiating spinifex, which is most common, shows randomly oriented plates of skeletal crystals (Fig. 3-9B). Individual plates range up to 0.5 mm thick and up to 7 cm long, averaging 3—4 cm long. Cores of olivine or clinopyroxene are rarely preserved in the skeletal crystals. Porphyritic spinifex is characterized by equant skeletal olivine crystals in a pyroxene-chlorite matrix and harrisitic spinifex by groups of equant crystals that maintain a common optical orientation over distances up to 15 cm.

The most detailed studies of ultramafic flows are those at Pyke Hill in the

Fig. 3-9. Spinifex-textured peridotite. A. Sample from the Barberton belt showing randomly oriented bundles of skeletal olivine crystals (Viljoen and Viljoen, 1969d). B. Photomicrograph of radiating spinifex in rock from the Barberton belt in South Africa (× 6). Contains serpentine pseudomorphs (light-colored) after skeletal olivine (from Naldrett, 1970).

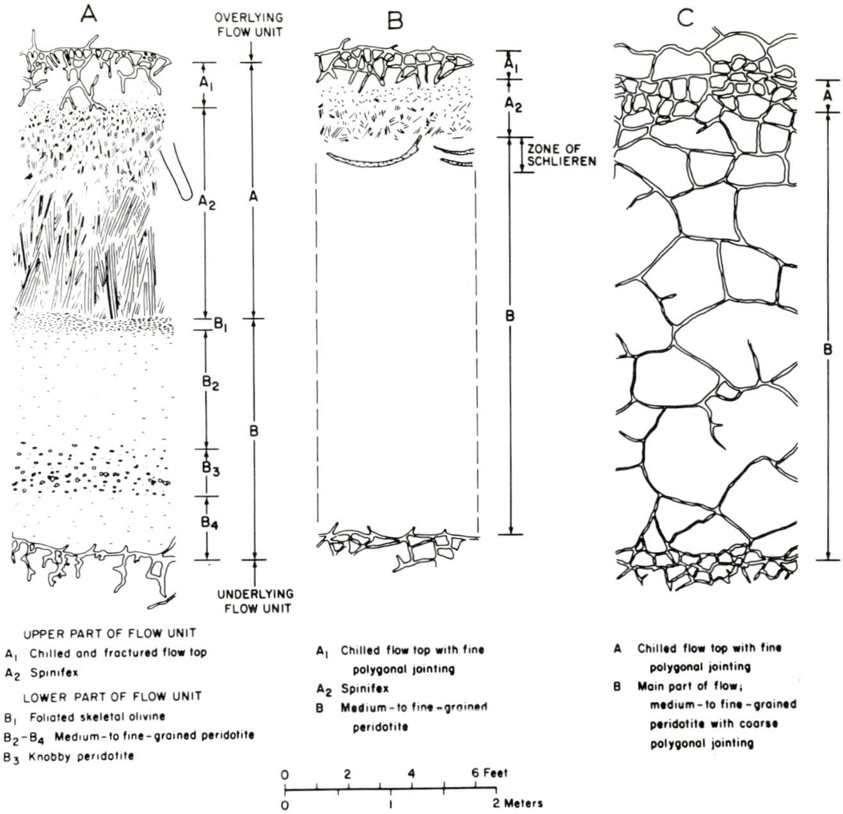

Fig. 3-10. Diagrammatic sections through three types of peridotitic komatiite flows from Munro Township, Canada (from Arndt et al., 1977).

Munro Township (Pyke et al., 1973; Arndt et al., 1977). Diagrammatic sections of three representative flows from this area are given in Fig. 3-10. A continuum of types exist between these three end members. Flows without spinifex texture are thicker, but shorter than flows with spinifex texture. In flow type A, the size of bladed olivine crystals increases downward and the orientation changes from random in the upper part of the spinifex zone (A_2) to blades or sheaths that are approximately at right angles to the flow surface at greater depths. Zone A_2 is in sharp contact with B and in some cases, such as in flows at Spinifex Ridge, LaMotte Township, Quebec, appears to represent an erosional surface (Lajoie and Gelinas, 1978). Skeletal crystals in B_1 grade downwards into more equant and progressively smaller grains in B_2 to B_4. At least half of the flows at Pyke Hill do not have spinifex texture (type C). This type of flow is composed chiefly of equant olivine crystals with skeletal overgrowths. Flow type B apears to represent a case intermediate between A and C, exhibiting only limited development of the spinifex zone. Flow types B and C commonly compose lava toes or lobes from the main flows. Systematic changes in olivine composition either within a flow or within the entire sequence of flows is not observed. Chemical analyses indicate an abrupt compositional change between A and B zones with MgO enriched and CaO and Al_2O_3 depleted in the B zone. There is also a suggestion of a slight progressive decrease in MgO and increase in CaO and Al_2O_3 downwards within zones A and B.

The existence of both spinifex and equant olivine in ultramafic flows indicates different conditions of crystallization probably related to some combination of differing cooling rates, the availability of nuclei, and the degree of supersaturation (Donaldson, 1974; Walker et al., 1976). Skeletal crystals appear to have formed in a nuclei-free lava by relatively rapid cooling and equant grains probably grew on available nuclei during slower cooling. Arndt et al. (1977) suggest the following cooling model. Rapid eruption with the formation of a chilled surface (A_1) followed by rapid settling of olivine phenocrysts to form a basal cumulus zone B, leaving the upper part of the flow devoid of crystal nuclei. Rapid cooling and the absence of crystal nuclei in the upper part of the flow leads to supersaturation and the production of skeletal olivine blades to form the spinifex texture. The average composition of the original magma is probably best represented by the average composition of zone A_2. Lajoie and Gelinas (1978) suggest that the erosional contact between zones A and B in some flows may result from continued motion of the B zone after formation of the spinifex zone A. Differences in the flow types A, B, and C appear to reflect differences in degree of gravitational separation of olivine phenocrysts. Hence type A results from complete settling, type B from partial settling, and type C from no settling.

Ultramafic pyroclastic rocks are uncommon in greenstone belts. They have been found interlayered with ultramafic lavas at Spinifex Ridge in Quebec where two types are described (Gelinas et al., 1977b). The first type is tuffaceous material deposited between ultramafic lava tubes. Tuff layers are well bedded and often exhibit graded bedding. They contain abundant pseudomorphs of shards, globules, and skeletal crystals and may be either hyaloclastic or pyroclastic in origin. The second type of deposits are volcanic breccias underlain by erosional surfaces. They contain angular blocks and, for the most part, are devoid of sedimentary structures.

Petrography

Original minerals in Archean ultramafic rocks vary in degree of preservation. Olivine, pyroxenes, and chromite are the most common primary minerals. They are replaced in varying amounts by such secondary minerals as serpentine, chlorite, talc, magnetite, carbonate, and tremolite (Worst, 1956; Viljoen and Viljoen, 1969c, d; Harrison, 1969, 1970; Oliver and Ward, 1971; Hancock et al., 1971; Glikson, 1972a). Olivine occurs as partially serpentinized skeletal crystals in spinifex-textured rocks (up to 4 cm long)

TABLE 3-2

Summary of typical sequence of alteration in Archean ultramafic rocks (after Nisbet et al., 1977)

Increasing degree of alteration →				
Olivine	→	serpentine and magnetite ± chlorite	→	serpentine, magnetite, tremolite, talc and/or chlorite
Clinopyroxene	→	clinopyroxene ± tremolite ± magnetite	→	tremolite, magnetite, chlorite ± talc ± serpentine
Matrix	→	serpentine, chlorite, magnetite	→	serpentine, tremolite, talc, chlorite, magnetite
		igneous textures often preserved	→	igneous textures rarely preserved

and as remnants of equant grains (0.5—5 mm in size). Its former presence is also attested to by serpentine or talc-tremolite pseudomorphs. Olivine composition ranges from Fo_{85} to Fo_{95}, often within the same flow (Arndt et al., 1977). Clinopyroxene occurs both as skeletal crystals in PYK and as small remnants of equant grains. It is commonly partially altered to chlorite and tremolite. Orthopyroxene is rare in ultramafic volcanics. Chromite and some sulfide minerals occur as minute euhedral grains between skeletal olivine crystals and in fine-grained matrices. Matrix material is generally completely recrystallized to a mixture of chlorite, serpentine, tremolite, talc, and magnetite. A summary of the mineralogical changes that occur with increasing degree of alteration in Archean ultramafic flows is given in Table 3-2.

Of the secondary minerals, serpentine is the most widespread. It occurs in fibrous and platey varieties and in veins, generally as crysotile. Tremolite (± actinolite) occurs as fine to coarse interlocking mats of prismatic crystals up to 3 cm in length. Magnetite occurs as veinlets and as fine, dusty grains associated with serpentine. Talc is white to buff in color in hand specimen and occurs as fine-grained aggregates replacing amphiboles, olivine, or serpentine. It is commonly associated with patches of fine-grained carbonate. Brown to green chlorite is an alteration product of clinopyroxene and amphiboles and occurs both in veinlets and in irregular patches. In the pyroclastic rocks described by Gelinas et al. (1977b), chlorite replaces shards and other volcanic fragments.

Composition

As discussed above, the spinifex-textured portions of ultramafic flows are thought to be representative of original magma composition. Average compositions of spinifex-textured peridotitic komatiites (STPK) from Archean

TABLE 3-3

Average compositions (oxides in wt.%, trace elements in ppm) of Archean spinifex-textured peridotitic komatiite (STPK) lavas compared to peridotite (values in parentheses calculated water-free)

	1	2	3	4	5
	Average garnet peridotite nodule from kimberlite	Average peridotite from Phanerozoic ophiolite	STPK Komati Formation, South Africa	STPK Abitibi belt, Canada	STPK Western Australia
SiO_2	44.70	43.25 (45.8)	42.52 (45.2)	42.9 (46.8)	41.9 (46.0)
TiO_2	0.20	0.10 (0.11)	0.18 (0.20)	0.36 (0.39)	0.23 (0.25)
Al_2O_3	3.23	2.66 (2.76)	3.44 (3.66)	7.46 (8.14)	5.22 (5.73)
Fe_2O_3	1.66	2.63 (2.73)	4.92 (5.23)	2.9 (3.2)	3.62 (3.97)
FeO	7.58	5.55 (5.75)	5.87 (6.24)	6.50 (7.08)	5.21 (5.72)
MgO	39.71	39.67 (41.1)	30.27 (32.2)	24.0 (26.2)	29.86 (32.8)
CaO	2.38	2.43 (2.52)	4.96 (5.28)	7.21 (7.86)	4.69 (5.14)
Na_2O	0.27	0.05 (0.05)	0.41 (0.44)	0.13 (0.14)	0.22 (0.24)
K_2O	0.07	0.01 (0.01)	0.16 (0.17)	0.06 (0.07)	0.02 (0.02)
P_2O_5	0.03	0.03 (0.03)	0.02 (0.02)	0.02 (0.02)	0.02 (0.02)
MnO	0.13	0.13 (0.14)	0.19 (0.22)	0.22 (0.24)	0.18 (0.20)
H_2O	0.10	4.0	7.1	6.0	8.39
CaO/Al_2O_3	0.74	0.91	1.4	0.97	0.90
FeO/Fe_2O_3	4.6	2.1	1.2	2.2	1.4
MgO/FeO	5.2	7.1	5.2	3.7	5.7
Al_2O_3/TiO_2	16	27	19	21	23
Cr	2500	3000	2200	2700	3000
Ni	2500	2000	2000	1300	1600
V	50	100	90	170	120
Co	100	100	60	110	105
Cu	30	10	45	89	95
Zn	33	40	102	90	130

TABLE 3-3 (continued)

	1	2	3	4	5
	Average garnet peridotite nodule from kimberlite	Average peridotite from Phanerozoic ophiolite	STPK Komati Formation, South Africa	STPK Abitibi belt, Canada	STPK Western Australia
Zr	40	30	34	35	15
La	1.3		0.80	0.45	0.55
Ce	2.9		2.1	1.18	1.65
Nd	1.6		1.7	1.33	1.53
Sm	0.32		0.51	0.45	0.56
Eu	0.10		0.18	0.21	0.22
Gd	0.29		0.65	0.50	0.75
Tb	0.04		0.15	0.09	0.15
Er	0.11		0.48	0.41	0.64
Yb	0.07		0.47	0.88	0.63
Lu	0.01		0.10	0.08	0.11
Y	1	2	3	10	6
Ni/Co	25	20	33	12	15
Ti/Zr	30	20	31	62	111
Zr/Y	40	15	11	3.5	2.4
Ti/V	24	6	12	13	12
$(La/Sm)_N$	2.2		0.86	0.55	0.54
Eu/Eu^*	1.0		1.0	1.3	1.1
$(Yb/Gd)_N$	0.30		0.90	2.2	1.0

N = chondrite-normalized ratio.

References: 1 and 2, various sources; 3—5, Viljoen and Viljoen (1969c); Villaume and Rose (1977); Hermann et al. (1976); Nesbitt and Sun (1976); Sun and Nesbitt (1978); Arndt et al. (1977); Arth et al. (1977).

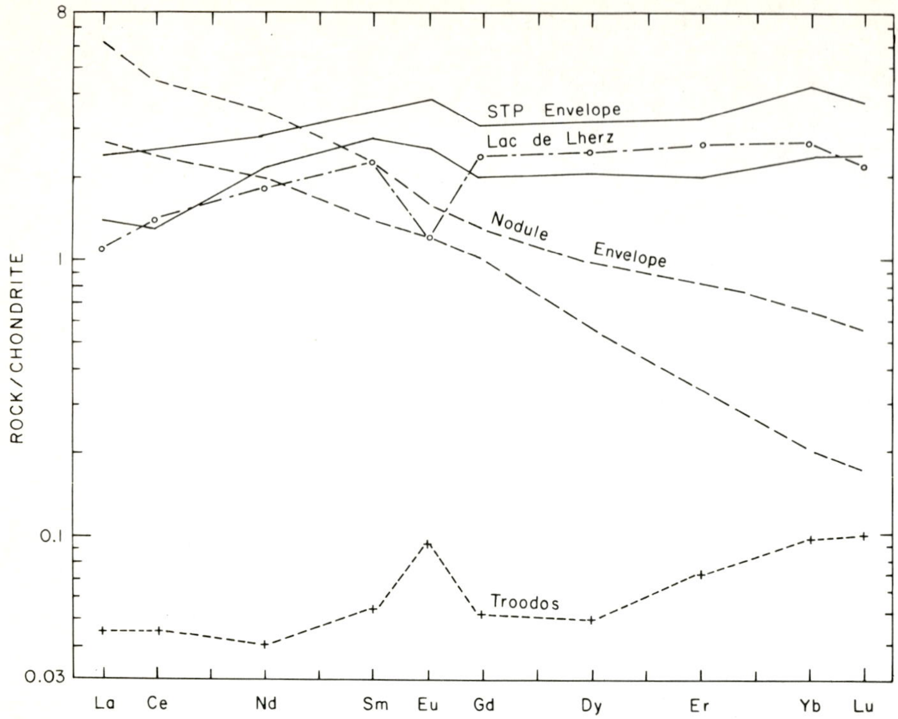

Fig. 3-11. Envelopes of variation of chondrite-normalized REE contents in Archean spinifex-textured peridotitic komatiite (STPK) and garnet peridotite nodules from kimberlite. Also shown are two alpine peridotites (Lac de Lherz and Troodos). References: Frey et al. (1971), Kay and Senechal (1976) and as given in Table 3-3.

greenstone belts in Canada, Africa, and Australia are given in Table 3-3. There is considerable variability between the three averages in Al_2O_3, CaO, Ti/Zr, Zr/Y, and in many of the transition metal contents. The Barberton average, as previously mentioned, is distinct in its high CaO/Al_2O_3 ratio. These differences probably reflect different amounts of fractionation of such minerals as olivine, spinel, ilmenite, clinopyroxene, and perhaps garnet (Nesbitt and Sun, 1976). When compared to an average garnet peridotite or alpine peridotite (Table 3-3), the STPK are high in TiO_2, CaO, MnO, Zn, Cr and Y, and in some cases Al_2O_3, V, and Ti/Zr; on the other hand, they are low in MgO, Zr/Y, and in some cases Ni. These differences are apparent for both hydrous and H_2O-free analyses. Compared to ultramafic rocks of mid-ocean ridges, STPK are high in Zr and low in Ni (Villaume and Rose, 1977).

REE in STPK are depleted in light REE, may show some heavy-REE enrichment, and have no Eu anomalies or small positive anomalies (Fig. 3-11). These patterns are similar to alpine peridotites although alpine peridotites exhibit a much greater range in absolute REE abundances, light-REE depletion, and in magnitude and direction of Eu anomaly as

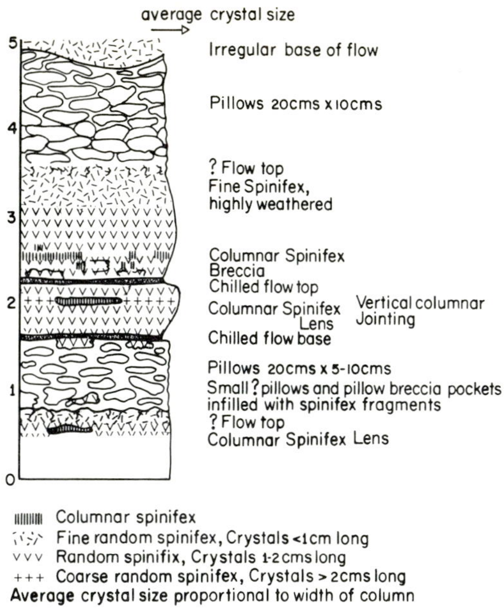

Fig. 3-12. Characteristic textures in several basaltic komatiite flows in the Belingwe greenstone belt, Rhodesia (from Nisbet et al., 1977).

indicated by two alpine peridotites in Fig. 3-11. REE distributions in STPK are unlike those of garnet peridotite nodules in kimberlite some of which may be representative of relatively fertile upper mantle (Table 3-3; Fig. 3-11).

Mafic volcanic rocks

Occurrence

Mafic igneous rocks have the same four occurrences in greenstone belts as ultramafic rocks. Most occur as flows and sills. Individual flows range from about 2 to 250 m in thickness, averaging less than 10 m (Nisbet et al., 1977; Goodwin et al., 1972), are commonly pillowed, and may, or may not have basal breccias and brecciated flow tops. Grain size ranges from very fine to medium and flows may change along strike from massive to pillowed. Pillows vary from 0.5 to 2 m in length attaining up to 5 m in deformed rocks (Viljoen and Viljoen 1969e). They are often vesicular, amygdaloidal, or variolitic. Spinifex texture may be present in BK flows. A section of several BK flows from the Belingwe greenstone belt in Rhodesia illustrates the variation in flow textures (Fig. 3-12). Flow tops and bottoms are very fine grained and contain skeletal pseudomorphs of equant olivine and quenched clinopyroxene needles in a devitrified glassy groundmass. Below flow tops, clinopyroxene spinifex textures are visible in hand specimen with skeletal clinopyroxene rosettes up to 5 cm in diameter. Columnar spinifex zones up to 10 cm thick and 30—50 cm long occur in the interiors of some flows.

BK flows in the Barberton greenstone belt exhibit a great variety of pillow shapes and sizes (Viljoen and Viljoen, 1969d). True mafic pyroclastic rocks are minor in most greenstone belts. When found, they occur as well-layered tuffs (often partially chertified) and minor volcanic breccias interbedded with flows and hyaloclastic rocks (Viljoen and Viljoen, 1969e).

Some of the most thorough descriptions of pillowed tholeiite flows and hyaloclastites are given by Hargreaves (1975) and Dimroth et al. (1978). Typical flows in the Rouyn-Noranda area in Quebec consist of, from base to top: massive lava, pillowed lava, pillow breccia, and hyalotuff (Dimroth et al., 1978). Variations in this succession occur due to absence of one or more of the divisions. Individual flows may be traced laterally for up to 15 km and marker units, composed of several flows, up to 70 km. Individual flows within marker units are often not present over the entire strike length. Massive lavas range from 2 to >100 m thick and have chilled margins a few centimeters thick. They may exhibit flow layering, varioles, vesicles, and porphyritic to glomeroporphyritic textures. Vesicularity may reach 30% with vesicles ranging from 1 mm to several centimeters across. Columnar jointing is also locally preserved. The pillowed divisions are composed of closely packed pillows of varying sizes, shapes, and degrees of deformation. Some exhibit chilled crusts with concentric cooling cracks. Material between pillows is hyaloclastic and sometimes cherty. The pillow breccia zone is comprised of both whole and fragmented pillows set in a hyaloclastic matrix. Pillows typically break-up along cooling fractures. Some pillow fragments are imbricated in the direction of flow. The contact between pillowed lava and pillow breccia is typically gradational. Hyalotuffs are layered units composed of hyaloclastic shards and pillow fragments. They may be graded over distances of centimeters to a few meters.

Dimroth et al. (1978) propose the following model for subaqueous eruptions in the Rouyn-Noranda area of the Abitibi greenstone belt. They interpret the massive lava facies as representing near-vent eruptions on broad shield volcanoes. These flows become pillowed at greater distances by budding and branching of the subaqueous flows. Large flow lobes or tubes occupy an intermediate position and are represented by megapillows (several meters across). Field relations between massive and pillowed lavas suggest that massive flows formed by surging advances of large volumes of hot lava of low viscosity. In response to falling temperature and viscosity, pillows formed at distal flow fronts. Pillow breccias form in the waning stages of eruption by the strong interaction of lava and water. The hyalotuffs are interpreted as lava-fountain deposits reworked in shallow water during the final stages of eruption and carried into deeper water by turbidity currents.

Layered mafic flows have been described from the Munro Township in Canada (Arndt, 1977a; Arndt et al., 1977). An example of a komatiitic flow is Fred's flow which has a maximum thickness of 120 m and extends laterally for over 3 km. It is characterized from top to bottom (relative

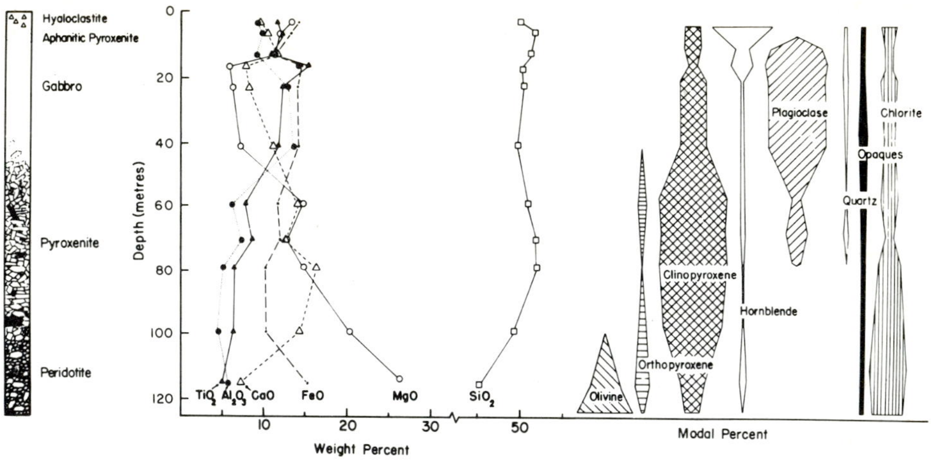

Fig. 3-13. Variation in major element and mineral contents in Theo's flow, a layered tholeiitic flow from Munro Township (from Arndt et al., 1977). TiO_2 scale enlarged ten times.

percentages in parentheses) by a MgO-rich flow-top breccia (5%); an olivine spinifex zone (< 1%); a clinopyroxene spinifex zone (5%); a gabbro zone (32%); a clinopyroxene cumulate (6%); an olivine cumulate (47%); and a lower ultramafic border zone (5%). Chemical and modal variation of a layered tholeiite flow from Munro Township is illustrated in Fig. 3-13. It differs from the komatiitic flow in that it does not exhibit spinifex texture, has a thin cumulus olivine zone (11%) and is dominated by a cumulus clinopyroxene zone (40%).

Fine-grained anorthosites and anorthositic basalts occur as sheets and lenses up to 20 m wide within mafic and ultramafic volcanics in the Holenarasipur greenstone belt in India (Drury et al., 1978). Grain size and mafic mineral content are quite variable in these units which share some geochemical properties in common with lunar anorthositic gabbros.

Variolites

Variolites are volcanic rocks that contain varioles within a fine matrix (Carstens, 1963). Such rocks are common among Archean mafic volcanics and have received attention in regards to their origin (Ferguson and Currie, 1972; Gelinas et al., 1976). Varioles are spherical bodies that are lighter colored than the host rock (Fig. 3-14) and range in diameter from 0.05 mm to over 5 cm. They occur in both tholeiite and BK flows and in massive as well as pillowed units. In pillows, they are often concentrated in the marginal zones. In some massive flows they are confined to horizons that can be traced discontinuously for many tens of kilometers (Dimroth et al., 1973).

Varioles typically exhibit sharply defined contacts with a narrow outer zone composed of iron oxides and chlorite. A relict spherulitic to dendritic

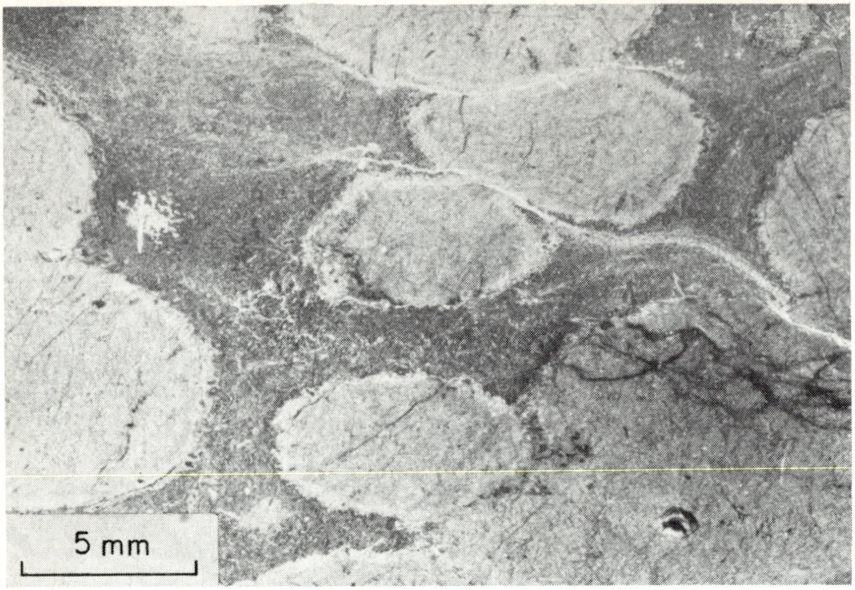

Fig. 3-14. Archean variolite showing partially coalesced varioles in sharp contact with surrounding matrix (Gelinas et al., 1976).

texture is often preserved within varioles which are composed principally of mixtures of secondary quartz, plagioclase, tremolite-actinolite, chlorite, carbonate, and epidote. Compositional data indicate that Archean variolites range from mafic to felsic in bulk composition with the varioles and matrices representing extremes in composition in any given flow.

Textural studies of Gelinas et al. (1976) indicate that varioles and matrix represent quenched fractions of two immiscible magmas and that the two magmas were in contact prior to eruption. In most cases, one magma corresponds to a low-K rhyolite (variole) and the other to tholeiite (matrix). Consistent with this interpretation is the fact that both variole and matrix compositions fall within or close to the field of known immiscibility in lunar glasses. Hughes (1977) suggests that Archean varioles may have formed in response to secondary processes. However, the fact that varioles are stratigraphically controlled and often preserve relict spherulitic textures does not favor such an origin (Gelinas et al., 1977a).

Petrography

Archean mafic volcanic rocks are composed chiefly of secondary minerals (McCall, 1958; Harrison, 1970; Hallberg, 1972; Sims, 1972b; Glikson, 1972a; Arndt et al., 1977). Relict phenocrysts occur in some volcanics. Plagioclase occurs both as phenocrysts (An_{65}—An_{80}) and as a common matrix constituent. Groundmass plagioclase may be randomly oriented microlites (An_{45}—An_{60}) of primary origin or granoblastic aggregates

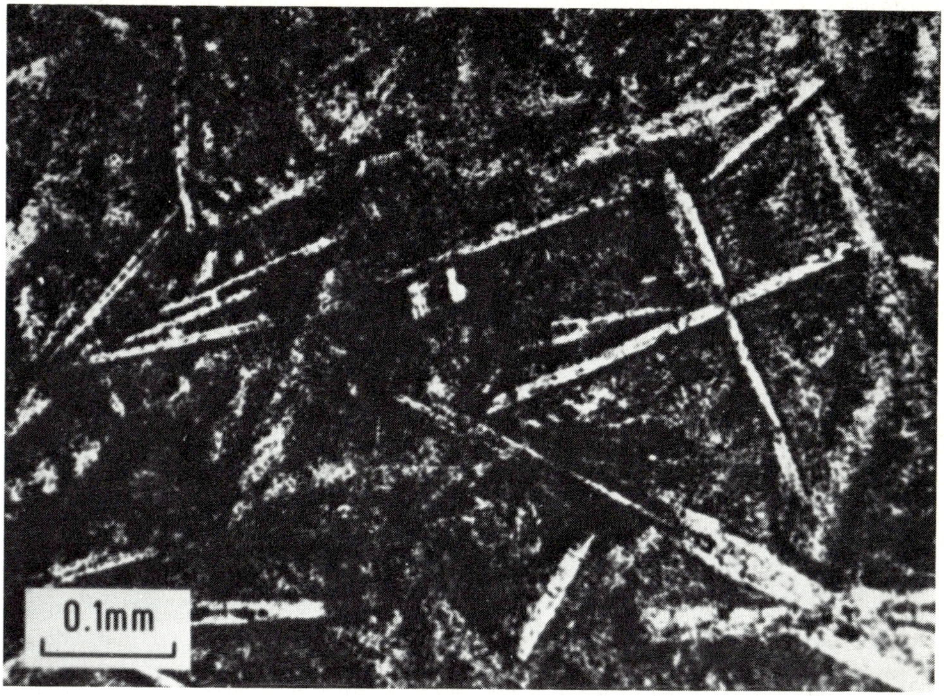

Fig. 3-15. Photomicrograph of Archean tholeiite showing calcic quench plagioclase (rosettes) and a fine matrix of intergrown plagioclase, clinopyroxene, and secondary minerals (from Gelinas and Brooks, 1974).

(An_{20}—An_{40}) of probably secondary origin. Distinctive flows with glomeroporphyritic aggregates of plagioclase up to 18 cm across are reported in some Archean greenstone successions (N. L. Green, 1975). Some flows in the Barberton belt contain up to 50% of short, stubby, primary clinopyroxene crystals (Viljoen and Viljoen, 1969e). Orthopyroxene (bastite) and olivine are rare, both as phenocryst and groundmass phases, in tholeiites but occur in most BK flows. Olivine is usually partly to completely replaced with serpentine. Fine-grained ilmenite and magnetite may be partly of primary origin. Matrices of mafic volcanic rocks are almost entirely recrystallized to various combinations of tremolite-actinolite, chlorite, sodic plagioclase, quartz, epidote-clinozoisite, prehnite-pumpellyite, iron oxides, sphene (± leucoxene), carbonate, and sulfides. Sulfides (primarily pyrite) are generally minor and widely dispersed. At higher metamorphic grades, biotite, hornblende, and garnet become important constituents (Satyanarayana et al., 1974). Amygdules and veinlets are commonly filled with quartz, epidote, chlorite, and carbonate.

Porphyritic to glomeroporphyritic textures occur in some tholeiites. Delicate groundmass crystals preserved in Archean tholeiites are interpreted by Gelinas and Brooks (1974) to represent quench textures (Fig. 3-15).

TABLE 3-4

Average compositions (oxides in wt.%, trace elements in ppm) of Archean basaltic komatiites (BK)

	BK1	BK2	BK3
SiO_2	50.5	48.8	47.3
TiO_2	0.60	0.73	0.50
Al_2O_3	11.0	13.0	9.08
Fe_2O_3	1.53	1.94	2.98
FeO	9.23	9.68	7.80
MgO	10.2	11.8	21.0
CaO	11.8	8.24	7.81
Na_2O	1.87	1.48	0.87
K_2O	0.17	0.15	0.16
P_2O_5	0.06	0.09	0.09
MnO	0.20	0.21	0.19
H_2O	2.4	3.0	2.5
CaO/Al_2O_3	1.1	0.63	0.86
FeO/Fe_2O_3	6.0	5.0	2.6
MgO/FeO	1.1	1.2	2.7
Cr	920	900	1750
Ni	360	390	640
V	250	270	235
Co	50	69	80
Zr	33	37	30
Ba	20		
Sr	100	96	60
La	3.0	1.9	0.86
Ce	7.9	5.9	2.9
Nd	5.2	4.8	2.7
Sm	1.6	1.5	1.0
Eu	0.55	0.57	0.39
Gd	2.0	2.2	1.4
Dy	2.5	2.8	1.8
Er	1.5	1.8	1.1
Yb	1.5	1.8	1.1
Lu	0.23	0.29	0.18
Y	17	22	16
Ni/Co	7.2	5.7	8.0
Ti/Zr	109	118	100
Zr/Y	1.9	1.7	1.9
Ti/V	14	16	13
$(La/Sm)_N$	1.0	0.70	0.47
Eu/Eu^*	0.95	0.97	1.0
$(Yb/Gd)_N$	0.93	1.0	0.99

N = chondrite-normalized ratio.

Chief references: Viljoen and Viljoen (1969d); Sun and Nesbitt (1978); Hawkesworth and O'Nions (1977); Arndt et al. (1977); Jahn et al. (1979).

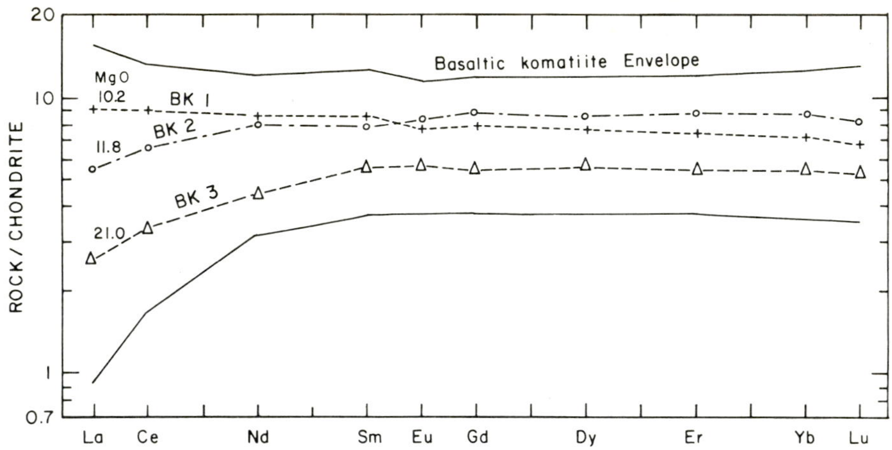

Fig. 3-16. Chondrite-normalized REE distributions in Archean basaltic komatiites (BK). Shown is the envelope of BK variation and three BK averages from Table 3-4. Average MgO values (in percent) for each BK group are also given.

In thin-section, these textures are similar to those observed in modern ocean-ridge basalts.

Composition

Basaltic komatiites (BK), as defined previously, can be subdivided into three groups (BK1, BK2, BK3) based on REE distributions (Table 3-4, Fig. 3-16). A continuum probably exists between the groups as indicated by the REE envelope. Each of the three types may occur interbedded in the same greenstone succession as for instance found in the Tipasjarvi belt in Finland (Jahn et al., 1979). All three groups are characterized by high MgO, low TiO_2, low large-ion-lithophile (LIL) element contents, flat heavy-REE patterns, and generally negligible Eu anomalies (Nesbitt and Sun, 1976; Arth et al., 1977; Hawkesworth and O'Nions, 1977; Sun and Nesbitt, 1978; Jahn et al., 1979). BK1 exhibits flat to very slightly enriched light REE, BK2 slightly depleted light REE, and BK3 strongly depleted light REE. Two BK from Finland have been reported with sloping heavy-REE patterns (Jahn et al., 1979). With regard to most other elements, BK1 and BK2 are similar. BK3, however, shows significant enrichment in MgO, Cr, Ni, and Co. As pointed out by several investigators (Hawkesworth and O'Nions, 1977; Arth et al., 1977), the degree of light-REE depletion increases with MgO (Fig. 3-16) (and presumably Cr, Ni, and Co) and is inversely correlated with the CaO and Sr contents and with the FeO/Fe_2O_3 ratio. The high CaO/Al_2O_3 ratio in BK1 (Table 3-4) reflects this ratio in Barberton BK and in some BK from Finland greenstone belts. Transition metal contents within each group are quite variable although the averages are high. Ratios of these metals and of other transition elements are much less variable.

TABLE 3-5

Element concentrations in Archean basaltic komatiites compared to Phanerozoic high-Mg lavas normalized to 20% MgO (after Cox, 1978)

	1	2	3	4	5
K_2O (%)	0.1	0.07	0.2	0.5	2.5
TiO_2 (%)	0.5	0.78	1.25	1.5	2.9
P_2O_5 (%)	0.03	0.09	0.15	0.2	0.4
Ba (ppm)	10	49	69	200	1000
Rb (ppm)	3	1.3	3.6	10	40
Sr (ppm)	20	126	215	200	1000
Zr (ppm)	20	55	82	100	375

1 = average Archean BK (Nesbitt and Sun, 1976); 2 = average of 24 Tertiary olivine basalts from Baffin Island; 3 = average of 24 Tertiary basalts from West Greenland; 4 = Deccan picritic basalts, western India; 5 = Nuanetsi olivine-rich basalts.

It is of interest to compare Archean BK with Phanerozoic high-Mg lavas. Although this can be accomplished by comparing lavas exhibiting the same degree of fractionation, it should be re-emphasized that, with rare exceptions, Phanerozoic lavas do not exhibit spinifex textures and unlike many Archean BK flows, some may contain cumulus material. In Table 3-5, the concentrations of several incompatible elements in Archean BK are compared to Phanerozoic high-Mg lavas, normalized to 20% MgO (after Cox, 1978). Most of the elements in the Archean BK are lower and some significantly lower (Sr, Zr) than in the Phanerozoic high-Mg lavas.

Many hundreds of Archean basaltic volcanic rocks have been chemically analyzed (see, for instance, Hallberg, 1972; Goodwin, 1977a; Naldrett et al., 1978). Almost all are quartz and/or hypersthene-normative tholeiites. Many investigators have pointed out the gross similarity in composition of Archean tholeiites and modern mid-ocean ridge basalts (MORB) (Glikson, 1971b, Naqvi and Hussain, 1973a, b). An example of the variation in composition of Archean tholeiites from the Eastern Goldfields subprovince in Western Australia is given in Table 3-6. Flows and intrusive diabase and gabbro are grouped separately. SiO_2, Al_2O_3, FeO_t, MgO, CaO, and Na_2O all have relative deviations $\lesssim$ 26% and many trace elements also exhibit relatively low dispersions. The very high dispersion for some LIL elements (K_2O, Rb) probably reflects mobility of these elements during secondary processes and the moderate dispersion of Cu may reflect variable sulfide distribution in the rocks (Hallberg, 1972).

Archean basalts have been classified according to both major elements as previously discussed (Irvine and Baragar, 1971; Naldrett and Goodwin, 1977), and according to trace elements (Condie, 1976c). The trace element groups appear to be most definitive in terms of basalt genesis (Condie and

TABLE 3-6

Compositional variations (oxides in wt.%, trace elements in ppm) in Archean mafic igneous rocks from Western Australia (from Hallberg, 1972)

	123 flows				84 diabases and gabbros			
	x	*m*	*s*	*C* (%)	*x*	*m*	*s*	*C* (%)
SiO_2	51.4	51.0	1.7	3.3	50.8	50.9	2.0	3.9
Al_2O_3	14.8	15.0	1.0	6.7	14.5	14.7	1.3	8.9
FeO_T[1]	10.4	10.3	1.7	16.3	11.7	11.5	1.9	16.2
MgO	6.7	6.5	0.9	13.4	6.9	7.0	1.5	21.7
CaO	10.7	10.7	0.9	8.4	9.9	10.1	1.7	17.1
Na_2O	2.7	2.7	0.4	14.8	2.7	2.7	0.7	25.9
K_2O	0.18	0.17	0.06	33.3	0.25	0.23	0.19	76.0
TiO_2	0.92	0.93	0.19	20.6	1.16	1.01	0.45	38.7
Co	59	57	6	10.6	57	59	6	10.5
Cr	395	410	105	26.5	314	275	82	26.1
Cu	98	89	44	44.8	111	120	35	31.5
Li	5	5	1	22.0	6	7	2	29.5
Ni	161	159	26	16.2	145	134	29	20.0
Rb	9	4	10	110.0	9	13	8	88.8
Sr	105	96	23	22.3	91	96	19	20.8
V	320	311	38	11.8	307	295	39	12.7
Y	22	21	3	15.0	22	23	3.5	15.9
Zn	112	117	19	17.4	107	112	12.5	11.6
Zr	60	67	10	16.6	54	56	10	18.5

[1] Total iron as FeO.

Mean (*x*), median (*m*), standard deviation (*s*), and relative standard deviation (*C*).

Baragar, 1974; Condie and Harrison, 1976; Condie, 1976c). Condie (1976c) proposed a two-fold classification of Archean tholeiites based on REE patterns. Even with the larger number of analyses now available, these two categories still emerge (Table 3-7; Fig. 3-17). TH1 (originally referred to as DAT) is characterized by flat REE patterns ($\sim 10\times$ chondrites) with or without small Eu anomalies. TH2 (originally referred to as EAT) is characterized by enriched light REE and a sloping REE pattern. Both groups differ from BK by their higher TiO_2, Al_2O_3, Na_2O, K_2O, Ba, Sr, Zr, Y, Zr/Y, and Ti/V and by their lower MgO/FeO (< 1), Cr, Ni, Co, and Ni/Co. In addition to REE patterns, TH1 differs from TH2 by its higher TiO_2, K_2O, P_2O_5, V, Zr, Sr, Y, and Zr/Y and by its lower CaO, Cr, FeO/Fe_2O_3, and Ti/Zr. Some Archean tholeiite samples exhibit rather unique REE patterns not falling in either the TH1 or TH2 categories. Several samples are reported to have slightly depleted light-REE and unfractionated to sloping heavy-REE patterns (Sun and Nesbitt, 1978; Jahn et al., 1979). These may constitute one or more additional groups of Archean tholeiite. TH1 is the common volcanic rock type in most greenstone belts composing 50—80% of a typical section. TH2 becomes more abundant at higher stratigraphic levels in many belts although it appears to be absent in some.

REE patterns in both TH1 and TH2 are characterized by either the absence of Eu anomalies or the existence of small negative anomalies ($1 > Eu/Eu^* > 0.9$). In only a few cases have positive anomalies been reported (Jahn et al., 1979). The Eu anomalies may be produced by plagioclase fractionation at shallow depths, alteration, or Eu depletion in the mantle source. Each has been suggested for various rocks (Condie and Baragar, 1974; Sun and Nesbitt, 1978; Jahn et al., 1979). In the Abitibi greenstone belt, negative Eu anomalies are characteristic of both basalts and andesites (Figs. 3-17 and 3-25) (Condie and Baragar, 1974; Smith, 1977). These anomalies do not appear to be related either to alteration or to plagioclase removal as discussed later. The Abitibi tholeiites (TH1a) appear to constitute a rather unique subgroup of TH1. In addition to the Eu anomalies, they have lower MgO, CaO, FeO/Fe_2O_3, Cr, and higher Na_2O, K_2O, Sr, Zr, and overall REE contents than TH1.

Compositional variation within Archean basaltic pillows has been examined by several investigators (Glikson, 1972a; Hallberg, 1972). As with studies of younger pillows (Scott and Hajash, 1976), Archean pillows exhibit inconsistent changes in the concentration of many elements (and especially LIL elements) from pillow margin to core.

The anorthositic basalts described from India exhibit slight REE enrichment with significant positive Eu anomalies (Drury et al., 1978; Naqvi and Hussain, 1979) and are similar in this respect to other terrestrial and lunar anorthosites.

The average composition of each of the Archean basalt groups is compared to the average compositions of modern basalts in Table 3-7. REE patterns of TH1 are compared to those of MORB and arc tholeiites and TH2 to calc-alkaline tholeiites in Fig. 3-17. Although an overall similarity in composition exists between TH1 and MORB and modern immature arc tholeiites and between TH2 and modern calc-alkaline (and oceanic island) tholeiites (Condie, 1976c), several significant differences also stand out. All Archean tholeiites differ from modern groups by their high FeO and other transition metal contents, their high FeO/Fe_2O_3, and their low Al_2O_3. Also, as pointed out by Gill (1979), most Archean tholeiites have Mg/Mg + Fe ratios larger than those characteristic of MORB. Although there is considerable scatter when Archean tholeiites are plotted in the Ti-Zr-Y-Sr figures of Pearce and Cann (1973), most averages fall in the plate-margin fields on the Ti-Zr-Y plot and in the MORB or calc-alkaline fields on the Ti-Zr-Sr plot (Fig. 3-18). On the F_1-F_2-F_3 major element discriminant diagrams of Pearce (1976), most fall in the low-K tholeiite or MORB fields (Yellur and Nair, 1978). On the FeO_T-MgO-Al_2O_3 diagram of Pearce et al. (1977), most samples plot in ocean-ridge, oceanic island, and continental categories. Caution should be

TABLE 3-7

Average compositions (oxides in wt.%, trace elements in ppm) of Archean and modern tholeiites (after Condie, 1976c)

	Archean		Modern			
	TH1 (DAT)	TH2 (EAT)	MORB	arc	calc-alkaline	continental rift
SiO_2	50.2	49.5	49.8	51.1	50.2	50.3
TiO_2	0.94	1.49	1.5	0.83	1.0	2.2
Al_2O_3	15.5	15.2	16.0	16.1	17.7	14.3
Fe_2O_3	1.63	2.80	2.0	3.0	3.9	3.5
FeO	9.26	9.17	7.5	7.3	6.3	9.3
MgO	7.53	6.82	7.5	5.1	5.4	5.9
CaO	11.6	8.79	11.2	10.8	9.8	9.7
Na_2O	2.15	2.70	2.8	2.0	2.7	2.5
K_2O	0.22	0.69	0.14	0.30	0.9	0.8
P_2O_5	0.10	0.17	0.20	0.15	0.2	0.16
MnO	0.22	0.18	0.17	0.17	0.2	0.2
H_2O	1.62	2.04	1.3	0.50	0.70	0.65
CaO/Al_2O_3	0.75	0.58	0.70	0.67	0.55	0.68
FeO/Fe_2O_3	5.7	3.3	3.8	2.4	1.6	2.7
MgO/FeO	0.81	0.74	1.0	0.70	0.86	0.63
Cr	490	250	300	50	50	100
Ni	140	125	100	25	50	100
V	260	365	300	270	150	300
Co	52	55	32	20	40	40
Cu	110	100	70	80	80	90
Zn	80	120	75	80	80	90
Zr	53	135	100	60	100	200
Ba	80	90	11	60	100	200
Sr	100	190	135	225	300	350
La	3.6	13	3.5	3.9	9.2	27
Ce	9.2	30	12	7	25	140
Nd	6.6	17	11	6	15	61
Sm	2.0	4.0	3.9	2.2	3.8	8.2
Eu	0.73	1.3	1.5	0.9	1.3	2.0
Gd	2.6	3.8	6.2	2.5	4.5	6.5
Dy	3.1	4.2	7.0	2.7	4.8	6.1
Er	2.0	2.3	3.6	1.8	2.6	3.0
Yb	1.9	2.2	3.0	2.0	2.5	2.5
Lu	0.31	0.38	0.3	0.3	0.5	0.4
Y	20	30	30	20	23	30
Ni/Co	2.7	2.3	3	1.3	1.3	2.5
Ti/Zr	106	66	90	83	60	66
Zr/Y	2.7	4.5	3.3	3.0	4.3	6.7
Ti/V	22	24	30	18	40	44
$(La/Sm)_N$	0.99	1.8	0.49	0.97	1.3	1.8
Eu/Eu^*	0.96	1.0	0.92	1.2	1.0	0.85
$(Yb/Gd)_N$	0.91	0.73	0.60	1.0	0.69	0.48

N = chondrite-normalized ratio.

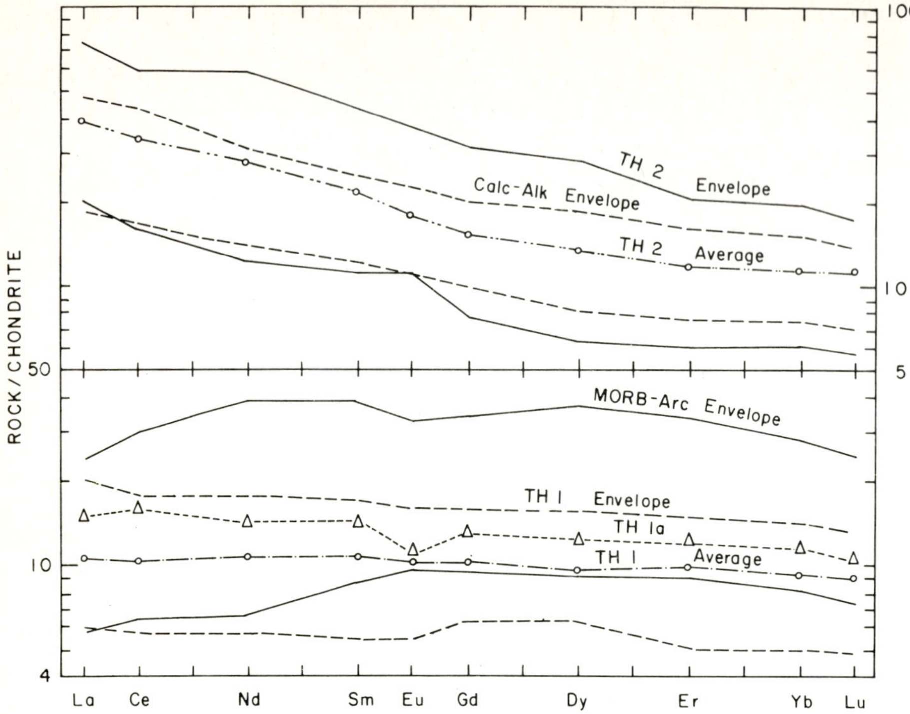

Fig. 3-17. Chondrite-normalized envelopes of variation of Archean tholeiite groups TH1 and TH2 compared to envelopes of variation of modern calc-alkaline tholeiite and oceanic rise and immature arc tholeiites (MORB-arc) (after Condie, 1976c). Also shown are average TH1, TH1a, and TH2.

exercised in deducing tectonic settings in the Archean from geochemical diagrams as discussed in Chapter 10.

Dikes and sills

Occurrence

Mafic and, to a smaller extent, ultramafic dikes and sills are a minor yet widespread component in Archean granite-greenstone terranes. Individual dikes, which may cross-cut both greenstone volcanics and granitic gneiss terranes, range in width from <1 m to >100 m and can be followed along strike for up to 50 km (Fahrig and Wanless, 1963; Prinz, 1964). Dikes commonly occur in swarms and such swarms often have a consistent strike over hundreds of kilometers. The most extensive dike swarms in Archean terranes of the Canadian Shield are, however, Proterozoic in age (McGlynn and Henderson, 1972; Fahrig and Wanless, 1963). Among the Archean dikes and associated sills, ages of emplacement range from pre-metamorphic (i.e., those associated with volcanism) to post-metamorphic (Heimlich et al.,

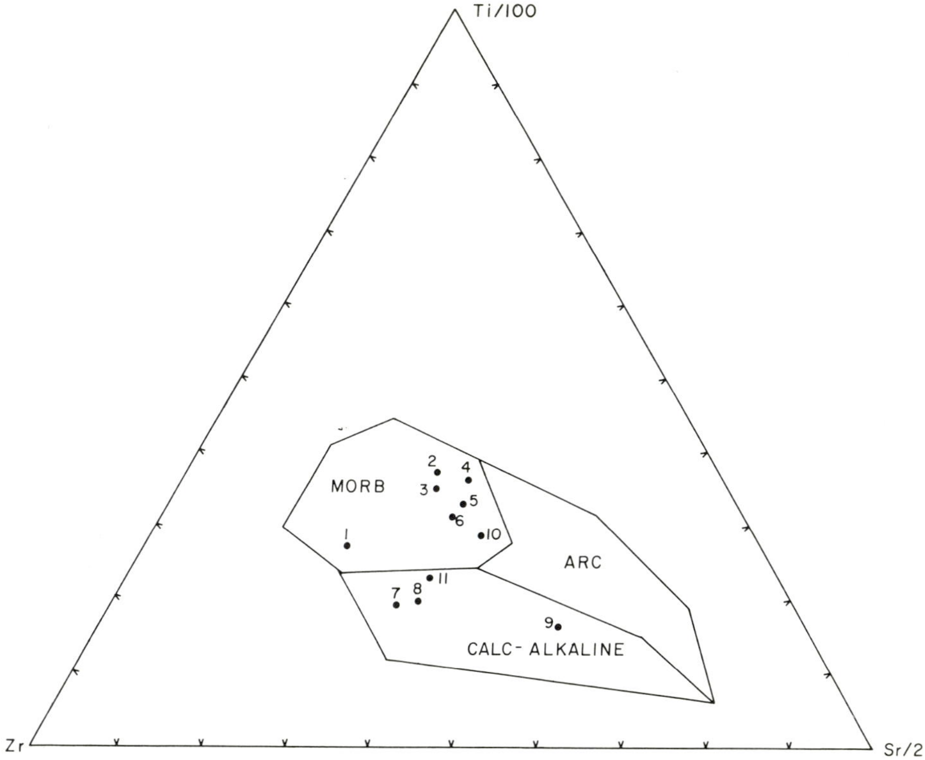

Fig. 3-18. Ti-Zr-Sr plot of average Archean tholeiite compositions. Fields for modern rise (MORB), arc, and calc-alkaline tholeiites from Pearce and Cann (1973). Key to greenstone belts: *1* = Abitibi; *2* = Coolgardie; *3* = Norseman; *4* = Yellowknife; *5* = South Pass; *6* = Mafic Formation, Midlands belt; *7* = Birch-Uchi; *8* = Lake-of-the-Woods; *9* = Maliyami Formation, Midlands belt; *10* = Nyanzian, western Kenya; *11* = Sturgeon Lake.

1974). Mafic sills in greenstone successions vary in abundance and in some belts may comprise up to 50% of the section (McCall, 1973; Hallberg, 1972). They range in thickness from < 5 to > 100 m and are often closely associated with and often difficult to distinguish from penecontemporary flows (Viljoen and Viljoen, 1969e). Rarely are contact metamorphic aureoles found in rocks adjacent to mafic or ultramafic sills and dikes. Dike and sill margins may exhibit chill zones up to several centimeters thick which often contain fragments of country rock. Often, ophitic to subophitic textures are observed in the field. Some mafic dikes are porphyritic with plagioclase, the dominant phenocryst. Rarely such phenocrysts attain sizes up to 5 cm and may be glomeroporphyritic (Prinz, 1964; Manzer and Heimlich, 1974).

Petrography

In thin section, mafic dikes and sills often exhibit well-preserved ophitic to subophitic textures (Fig. 3-19) even when original minerals are almost

Fig. 3-19. Photomicrograph of an Archean diabase dike from the Bighorn Mountains, Wyoming (from Heimlich et al., 1974). Note the subophitic texture of relatively fresh plagioclase and augite. Vertical dimension $\simeq$ 0.5 mm.

entirely replaced by actinolite, epidote, chlorite, sodic plagioclase, and magnetite of secondary origin. Some are porphyritic containing 10—30% of plagioclase and clinopyroxene phenocrysts. Many Archean diabase dikes from Wyoming have been studied in detail (Prinz, 1964; Condie et al., 1969; Mueller and Rogers, 1973; Heimlich et al., 1974; Armbrustmacher, 1977). These dikes exhibit considerable variability in mineral percentages from area to area. Plagioclase occurs both as a phenocryst and groundmass constituent. Phenocrysts may vary in composition from An_{65} to An_{80} and groundmass microlites are typically in the range of An_{40}—An_{60}. Phenocrysts are commonly zoned. Clinopyroxene occurs as subrounded phenocrysts and is often composed of cores of pigeonite surrounded by augite. Orthopyroxene is a rare microphenocryst phase. Traces of olivine occur in some dikes and magnetite is ubiquitous. Hornblende varies considerably in abundance and often partially replaces primary plagioclase and clinopyroxene (uralite). Secondary minerals such as chlorite, magnetite, quartz (in part primary), epidote, sphene, carbonate, and sericite comprise much of the matrices of these rocks.

Studies by Ross and Heimlich (1972) indicate that with increasing distance from contacts to dike centers in several dikes from the Bighorn Mountains in Wyoming mineral percentages vary in different manners in different dikes. Plagioclase becomes less calcic towards the center of all dikes, however.

Composition

Average compositions of Archean dikes and sills from three different continents are given in Table 3-8. In general, the compositions are similar, the most notable exception being the high MgO in the Nuggihalli dikes. The dikes share compositional characteristics in common with Archean tholeiites (Table 3-7). The Wyoming dikes, which are post-tectonic and not directly associated with greenstone volcanics are similar in composition to TH2 although they contain less Zr and more Ba and light REE. The Nuggihalli and Western Australia dikes and sills, which are more closely associated with greenstone volcanics, share more features in common with TH1 (although REE data are not available to classify them clearly as such). The Sr depletion trend in the Wyoming dikes, originally proposed by Condie et al. (1969), is perhaps best interpreted to reflect Sr loss from the dikes by secondary processes rather than removal of plagioclase during crystallization as originally suggested. The absence of an increasing negative Eu anomaly with decreasing Sr does not allow removal of significant amounts of plagioclase.

The following compositional changes are reported by Manzer et al. (1971) in going from the margin to the center of the Archean mafic dikes from the Bighorn Mountains in Wyoming: enrichment in SiO_2, Na_2O, K_2O, and FeO_T and depletion in CaO. Jolly (1977) reports that mafic lavas and associated intrusive bodies from the Abitibi belt in Canada lie on the same differentiation paths when major-element contents are plotted on conventional variation diagrams (Fig. 3-33). These relations are interpreted as indicating that the intrusive bodies represent fractionating holding chambers from which the lavas are erupted. Some stratiform mafic ultramafic complexes (like the Dore Lake Complex in Canada) may represent crystal cumulates remaining after removal of mafic magmas which formed lavas and shallow sills.

Stratiform igneous complexes

Occurrence

Stratiform or layered igneous complexes are well represented in Archean granite-greenstone terranes where they occur both as conformable units within greenstone successions and as discordant units in granitic gneiss terranes. They show a large range in size and shape and may range in bulk composition from ultramafic (PK) to BK and tholeiitic. The largest known Archean body is the Great Dyke in Rhodesia (2.46 b.y.) which has a strike length of 500 km and an average width of 6 km (Tyndale-Biscoe, 1949; Worst, 1958) (Fig. 1-12). The Stillwater Complex in Montana (2.7 b.y.) crops out along a belt for 50 km with a maximum width of 8 km (Jackson, 1961). Typical bodies in greenstone belts have thicknesses of 0.5—1 km and extend laterally for up to 20 km. Bodies range from sill-like or lensoid in shape to irregular. They may be intruded into volcanic or sedimentary

TABLE 3-8

Average compositions (oxides in wt.%, trace elements in ppm) of Archean mafic dikes

	1 ($n = 27$)	2 ($n = 6$)	3 ($n = 84$)
SiO_2	49.0	48.1	50.8
TiO_2	1.44	0.78	1.16
Al_2O_3	13.7	11.5	14.5
Fe_2O_3	2.76	2.33	
FeO	11.0	8.45	11.7[1]
MgO	7.07	15.7	6.9
CaO	9.43	8.55	9.9
Na_2O	2.18	2.22	2.7
K_2O	0.95	0.42	0.25
P_2O_5	0.10	0.20	
MnO	0.18	0.18	
CaO/Al_2O_3	0.69	0.74	0.68
FeO/Fe_2O_3	4.0	3.6	
MgO/FeO	0.64	1.85	
Cr	300		314
Ni	143	270	145
V	175	218	307
Co	45	83	57
Cu	150	214	111
Zn		256	107
Zr	75	47	54
Ba	320		
Sr	186		91
La	19		
Ce	49		
Nd	29		
Sm	6.0		
Eu	1.7		
Gd	6.3		
Dy	6.2		
Er	3.4		
Yb	3.2		
Lu	0.45		
Y	28		22
Ni/Co	3.2	3.3	2.5
Zr/Y	2.7		2.5
$(La/Sm)_N$	1.7		
Eu/Eu^*	0.90		
$(Yb/Gd)_N$	0.63		

[1] Total Fe as FeO.

n = number of samples; N = chondrite-normalized ratio.

1 = average Archean diabase from Wyoming (Condie et al., 1969; Armbrustmacher, 1977); 2 = average Archean diabase from Nuggihalli greenston belt, India (Satyanarayana et al., 1973); 3 = average Archean diabase-gabbro, Western Australia (Hallberg, 1972).

sections of greenstone belts (Viljoen and Viljoen, 1969c; McCall, 1971, 1973; Williams and Hallberg, 1973) or in the case of larger bodies, such as the Great Dyke, the Stillwater Complex, and the Mashaba Complex, into granitic gneiss or both granitic gneiss and greenstone terranes. Contacts with surrounding rocks are generally sharp and may be either concordant or discordant to surrounding structures. Small complexes are generally concordant to surrounding volcanic-sedimentary rocks and appear to be closely related to contemporary volcanism (Raudsepp, 1975). The larger bodies mentioned above are all post-tectonic because they cross-cut gneissic foliation and/or greenstone stratigraphy. Some bodies contain xenoliths of surrounding country rock as exemplified by those in the Kaapmuiden area in the Barberton greenstone belt (Viljoen and Viljoen, 1970). Contact metamorphism is generally minor and irregularly distributed. Aureoles representing pyroxene hornfels facies range from a few to a few tens of meters thick and are generally discontinuous along strike (Wilson, 1968). Most bodies and especially the smaller ones, have been affected by varying degrees of serpentinization and regional metamorphism. Original layers and textures may be completely replaced with assemblages of serpentine, talc, tremolite, and chlorite (Williams, 1971). Depending on the degree of deformation, such original features may or may not be preserved as pseudomorphs. Many of the stratiform bodies in Rhodesian greenstone belts have been highly serpentinized and deformed such as to destroy all evidence of primary layering or cumulus textures (Harrison, 1969, 1970; Worst, 1956).

Stratigraphic thicknesses of Archean stratiform bodies are quite variable and, in the case of the larger bodies, are not completely exposed. The thickest reported sections are for the Windimurra Complex in Western Australia (5500 m; McCall, 1971), the Stillwater Complex (4900 m; Hess, 1969), and the Great Dyke (3000 m; Worst, 1958). The uppermost contacts of the Stillwater and Great Dyke Complexes are not exposed and the original thickness of the Stillwater may have been as much as 8200 m (Hess, 1960). Stratiform bodies often have chilled contact zones up to tens of meters thick. In general, they are characterized by lower ultramafic zones followed by anorthositic and gabbroic zones. Contacts between major compositional zones are generally sharp. Some complexes, as for example many of those found in Western Australia, appear to represent closely spaced sills representing individual magma injections (McCall and Doepel, 1969; McCall, 1971). To illustrate the overall zoning in Archean stratiform complexes, several bodies of varying size will be considered in detail.

The Stillwater Complex is characterized by a lower border zone (*BZ*) composed of fine- to coarse-grained mafic and ultramafic rocks overlain by a layered ultramafic zone (*UmZ*) composed chiefly of cumulates of harzburgite (*P*), chromitite, and bronzitite (*B*) (Fig. 3-20). This is in turn overlain by a norite zone (*NZ*) in which the only cumulus phases are plagioclase and orthopyroxene; clinopyroxene and quartz are intercumulus phases.

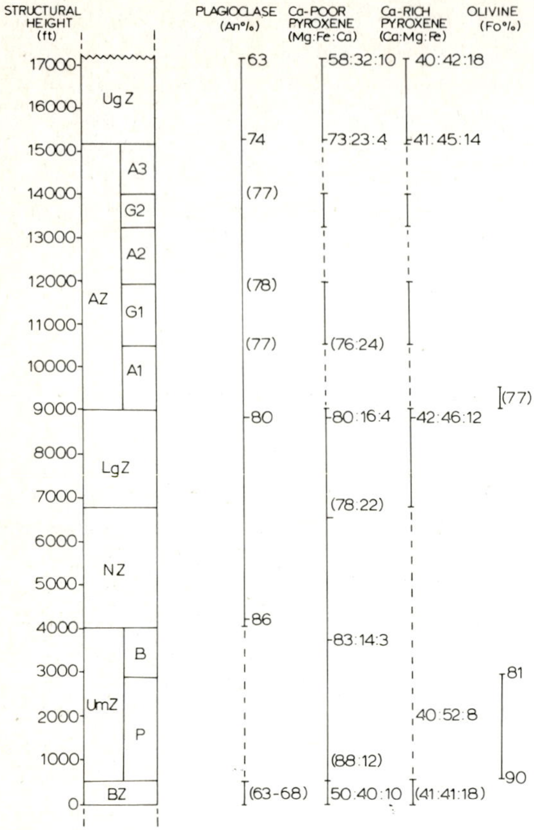

Fig. 3-20. Zones of the Stillwater Complex, Montana, showing cryptic layering of minerals (from L. R. Wager and G. M. Brown, *Layered Igneous Rocks*, W. H. Freeman and Co., Copyright © 1967).

The lower gabbro zone (*LgZ*) is marked by the appearance of cumulus augite. In the thick anorthosite zone (*AZ*), plagioclase is the major cumulus phase and traces of quartz and ilmenite-magnetite occur as intercumulus phases. Finally, the upper gabbro zone (*UgZ*) is marked by the return of cumulus pyroxenes.

The Mt. Thirsty sill complex in Western Australia (Fig. 3-21) is characterized by a basal zone of harzburgite-dunite which is partially serpentinized and overlain by a noritic gabbro zone (McCall, 1971). A layer of engulfed metasediments separates this lower part of the complex from overlying bronzitites, norites, and gabbros. Granophyric gabbro occurs in irregularly distributed patches near the top of the gabbroic sequence. The rocks in the sill complex are chiefly cumulates exhibiting rhythmic and cryptic layering characteristic of larger stratiform bodies.

Viljoen and Viljoen (1969c; 1970) have recognized three types of layered

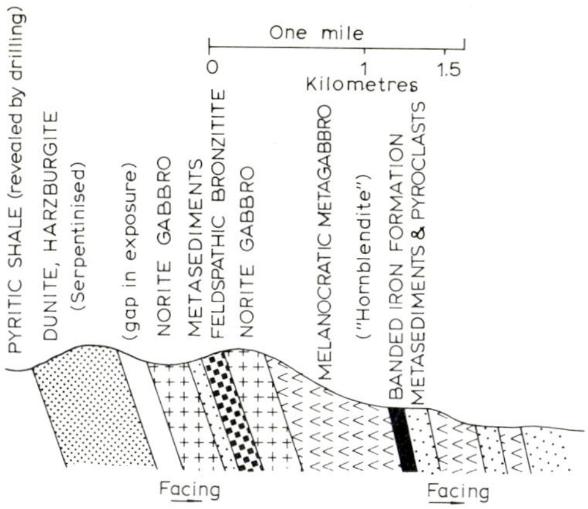

Fig. 3-21. Diagrammatic section of the Mt. Thirsty sill complex, Western Australia (after McCall, 1971).

igneous complexes in the Barberton belt which differ from each other in the proportion of cumulate rock types and in the presence or absence of cyclic units. The Kaapmuiden type has been studied in most detail and is represented by three bodies which are interpreted as equivalents of each other. The bodies are characterized by a basal peridotite chill zone 15-30 m thick followed by cumulus zones of dunite-peridotite, orthopyroxenite, websterite, anorthositic gabbro-norite, and dunite-peridotite. In addition, intrusive, late-stage gabbroic pegmatites are found.

The Dundonald Sill near Chibougamau, Ontario, has been studied by Naldrett and Mason (1968). Other sills similar to the Dundonald Sill have been described from greenstone belts in the Superior Province (Irvine and Smith, 1967; MacRae, 1969; Goodwin et al., 1972). Some, like the Garner Lake Body in Manitoba are composed entirely of ultramafic layers (Scoates, 1971). A diagrammatic cross-section of the Dundonald Sill showing the major units and minerals is given in Fig. 3-22. This body is characterized by a basal peridotite with cumulus olivine and chromite overlain by a clinopyroxenite with cumulus augite. The upper part contains three gabbroic layers with cumulus plagioclase, augite, and magnetite and intercumulus granophyric intergrowths.

Archean stratiform complexes are characterized by rhythmic layering on both coarse and fine scales. Such layers range from a few centimeters up to 100 m thick in the Stillwater and Great Dyke Complexes (Jackson, 1961; Worst, 1958). Graded and inversely graded beds occur in many bodies. Igneous lamination is pronounced in the gabbroic and anorthositic zones of some bodies and slump structures are described from the Stillwater Complex (Hess, 1960).

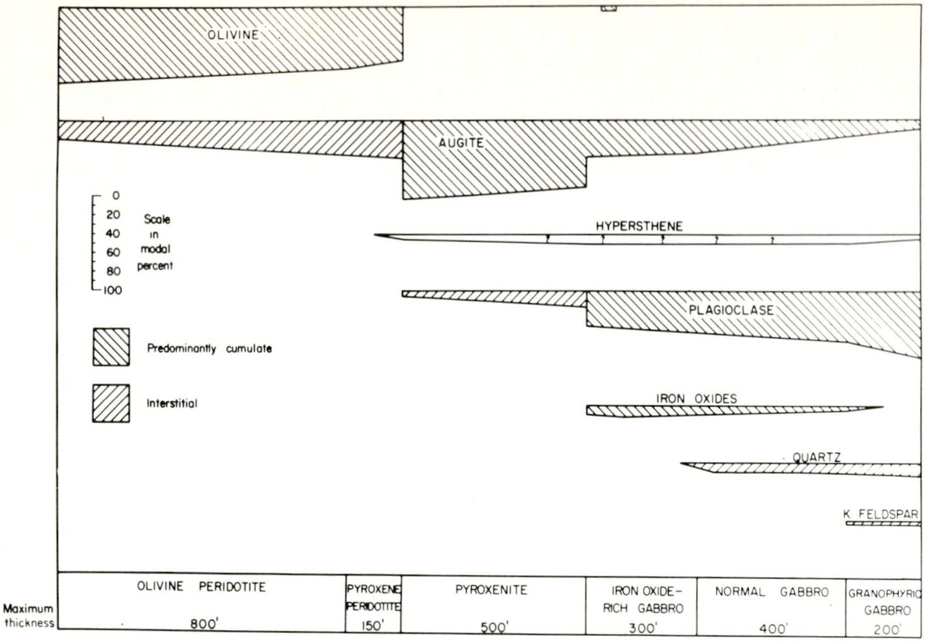

Fig. 3-22. Diagrammatic cross-section of the Dundonald Sill, Ontario (after Naldrett and Mason, 1968).

Petrography

Cumulus textures are well preserved in many Archean stratiform complexes. Even when completely altered to secondary minerals, such textures are often preserved as pseudomorphs in portions of the bodies.

Olivine is a major cumulus phase in many bodies and may range up to 1 cm in size. Cryptic compositional variation is from Fo_{80} to Fo_{100} (Wager and Brown, 1967). Olivine is often partly to completely replaced with serpentine or tremolite. Orthopyroxene occurs as a major cumulus phase in the lower parts of most bodies and as an intercumulus phase in some (Hess, 1960). It may range in composition from the base upwards from En_{95} to En_{60} (Hess, 1950; Worst, 1958) and often has exsolved clinopyroxene lamellae. Clinopyroxene occurs as both cumulus and intercumulus material and may range significantly in Fe/Mg ratio (Fig. 3-20). It often contains orthopyroxene exsolution lamellae. Cumulus plagioclase, which appears in anorthosite and gabbro zones, may vary in composition upwards with stratigraphic height from An_{80} to An_{60}. Plagioclase with An contents of An_{92}—An_{94} has been reported from anorthosites in the Kaapmuiden bodies in South Africa (Viljoen and Viljoen, 1970). Intercumulus plagioclase occurs in the ultramafic zones of some bodies. Cumulus chromite is common in the ultramafic zones of the Stillwater, Great Dyke, and Mashaba Complexes. Late-stage intercumulus or trapped liquid phases are quartz, apatite, magnetite-ilmenite, biotite and rarely K-feldspar. Only rarely is primary hornblende found in stratiform complexes in greenstone terranes although it is common in stratiform complexes in Archean high-grade terranes (Windley et al., 1973). Hornblende found in ultramafic intrusive bodies in the Quetico area in western Ontario has been interpreted as magmatic by Watkinson and Irvine (1964).

Composition and origin

Stratiform complexes in Archean greenstone-granite terranes, like most post-Archean stratiform complexes, exhibit cryptic layering with olivine and pyroxene becoming more Fe-rich and plagioclase more Na-rich with stratigraphic height (Wager and Brown, 1967). Typical cryptic layering in the Stillwater Complex is shown in Fig. 3-20. Similar changes are reported in the Kaapmuiden bodies in the Barberton area (Viljoen and Viljoen, 1970). Changes in magma composition in these bodies follow the komatiite or tholeiite series (Fig. 3-32) which reflect rather dry conditions and late crystallization of magnetite. Estimates of bulk composition of the original magmas are often made from chilled-border facies or from weighted analyses of cumulate rocks (Hess, 1960). Although both approaches are faced with problems, estimated compositions of original magmas tend to converge on tholeiite or BK (Hess, 1960; Naldrett and Mason, 1968). Some appear to have been ultramafic in composition (Viljoen and Viljoen. 1970; Scoates, 1971).

Models for the origin of Archean stratiform complexes are diverse and it is clear from existing data that no single model can explain all complexes. The rhythmic layering, cumulus textures, and cryptic composition changes, however, suggest that fractional crystallization has played the major role in the formation of stratiform complexes (Wager and Brown, 1967). Some models propose that a single injection of magma from the mantle filled a crustal reservoir and that cooling and crystallization of this magma occurred under approximately closed-system conditions (to prevent oxygen replenishment and early crystallization of magnetite) (Hess, 1960). Assuming such a model, Hess (1960) calculates a cooling rate for the Stillwater Complex of 10 cm/yr; thus the exposed thickness would take 49,000 years to crystallize. Convection currents in such magmas have been called upon to explain slump structures, igneous lamination, and graded bedding (Wager and Brown, 1967). Other investigators have emphasized the need for more than one injection of magma (Jackson, 1961; Viljoen and Viljoen, 1970; McCall, 1971). In particular, the sill complexes in Western Australia seem to necessitate several to many injections of magma from deeper chambers where fractional crystallization occurs (McCall, 1971).

ANDESITES

Occurrence

Andesites are an important rock type in many Archean greenstone belts (Condie, 1979b). They occur in calc-alkaline volcanic centers that are tens of kilometers across (Goodwin et al., 1972; Hallberg et al., 1976). These centers appear to have been, at least in part, subareal in character. Archean andesites occur as tuffs, breccias, agglomerates, flows, and as shallow intrusive bodies in order of decreasing abundance (Condie, 1979b). The ratio of pyroclastics to flows increases both with increasing stratigraphic height and with decreasing distance to eruptive centers (Goodwin et al., 1972). Breccias and agglomerates are abundant near volcanic vents. These rocks are typically gray to green and contain volcanic fragments ranging from a few millimeters to over 30 cm in size (Shackleton, 1946; Henderson and Brown, 1966; Harrison, 1970). Fragments comprise from 2 to 80% of some units (Tasse et al., 1978) and are generally similar in lithology to enclosing matrix; they range from angular to rounded. Breccia units are typically poorly sorted and vary from <1 to >100 m in thickness. Individual units can be traced laterally over distances up to a few kilometers where they grade into or interfinger with tuffs of similar composition. Andesitic tuffs are well-bedded with beds ranging from a few centimeters to tens of meters thick (Fig. 3-23). Some thick beds can be traced for great distances and provide distinctive marker units (Henderson and Brown, 1966). Graded-bedding, and less commonly, cross-bedding are locally preserved within tuff units. Two types of calc-alkaline pyroclastic units have been recognized in the Noranda region of the Abitibi greenstone belt (Tasse et al., 1978; Dimroth and Demarcke, 1978). One type is characterized by thick beds, coarse fragments, and reverse grading and is interpreted as a debris flow and turbidity current deposit. The second type of deposit is finer-grained and exhibits typical turbidite features indicative of turbidity current deposition.

Andesite flows range from homogeneous to amygdaloidal and porphyritic. Pillows are less frequent than in associated mafic flows and when found are often poorly developed (Harrison, 1970). They are commonly small (<20 cm across) and closely packed. Amygdules are filled with some combination of quartz, epidote, carbonate, and prehnite. Streaky to lenticular flow banding is preserved in some flows (McCall, 1958). Andesitic dikes and sills, which appear to be penecontemporary with eruptive units, occur in some greenstone successions. Textures of these bodies range from aphanitic or porphyritic to ophitic or subophitic.

Petrography

Primary textures and minerals are often preserved in Archean andesites

Fig. 3-23. Parallel layering in Archean andesitic tuffs from the Noranda region of the Abitibi greenstone belt (from Tasse et al., 1978).

(Shackleton, 1946; Huddleston, 1951; McCall, 1958; Goodwin, 1962; Harrison, 1970; Goodwin et al., 1972; Hallberg et al., 1976). Moorehouse (1970) presents an excellent series of photomicrographs of Archean andesites and modern counterparts which illustrates how well Archean textures can be preserved. Many Archean andesites are porphyritic (Fig. 3-24). Plagioclase (An_{25}—An_{35}) is the most widespread phenocryst phase comprising from 10 to 40% of some rocks. It ranges from 1 to 5 mm in length and is partially sericitized or saussuritized. Zoned crystals are common in some terranes

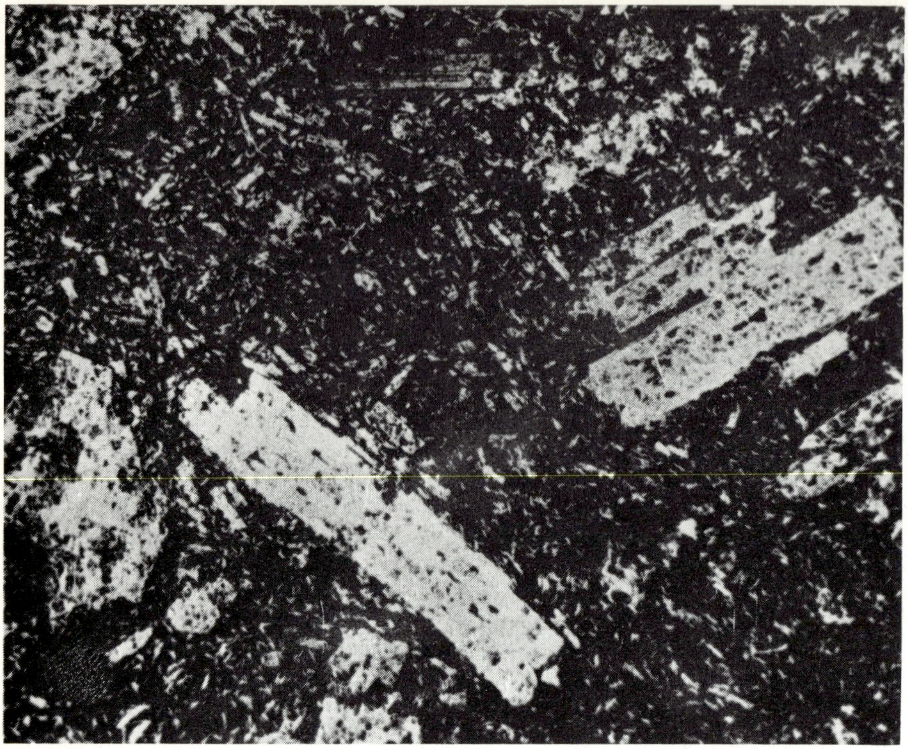

Fig. 3-24. Photomicrograph of porphyritic Archean andesite from Lake Timiskaming, Ontario (from Moorehouse, 1970). Altered plagioclase phenocrysts in a matrix of plagioclase and secondary minerals. Plane light, ×50.

(Harrison, 1970; Hallberg et al., 1976). Smaller, blue-green hornblende occurs as phenocrysts in some andesites. Less frequent phenocryst phases are quartz, pyroxene, and magnetite. Quartz occurs as small equidimensional phenocrysts sometimes embayed by surrounding matrix. Clinopyroxene (augite) is the most common pyroxene and ranges from 1 to 2 mm in length. Remnants of brown orthopyroxene occur in some andesites. Most orthopyroxene is partly to completely replaced by chlorite, iron oxides, and actinolite. Small magnetite phenocrysts partially replaced with secondary iron oxides, sphene, and leucoxene occur in some andesites.

Aphyric andesites and the groundmass of porphyritic varieties are composed of a fine-grained intergrowth of plagioclase microlites, clinopyroxene and a variety of secondary minerals including some combination of chlorite, actinolite, carbonate, epidote, zoisite, iron oxides, sphene, quartz, prehnite, zeolites, and pyrite (Fig. 3-24). The plagioclase is generally similar in composition to phenocrysts and may exhibit a pilotaxitic or trachytic texture. Augite is generally highly chloritized and orthopyroxene occurs only as pseudomorphs. Pyrite, carbonate, and quartz often occur in veinlets indicating a post-metamorphic origin. Pseudomorphs of perlitic cracks have been reported in some andesitic rocks that were originally glassy (Harrison, 1970).

Composition

Average compositions of andesites from six Archean greenstone belts and an average for the Superior Province are given in Table 3-9. The variation between the averages ranges by a factor of 2 to 3 for most elements. Light REE and especially the La/Yb ratio are even more variable. Although the Na_2O/K_2O ratio ranges from 2 to 7, most values are 3 to 4. SiO_2, Al_2O_3, Zn, Cu, and Co are similar in all averages. Employing REE, which as previously discussed are examples of elements least susceptible to mobilization during secondary process, it is possible to classify Archean andesites into three types, I, II, and III (Table 3-9) (Condie, 1979b). Envelopes of variation of REE patterns for each type are given in Fig. 3-25. Type I shows slightly enriched light REE (~ 50x chondrites) and negligible Eu anomalies. It also has higher FeO, MgO, Ni, Cr, and Zn and lower K_2O, Rb, and Ba than the other types. Type II andesites are notably enriched in light REE (~ 200x chondrites) and also exhibit negligible Eu anomalies. Some greenstone belts, such as the Yellowknife belt in Canada and the Marda complex in Western Australia contain only one type of andesite while others such as the Midlands in Rhodesia and the Nyanzian belts in Kenya contain both types I and II. In Kenya these types appear to be mixed stratigraphically, although the stratigraphy is not well known in this area (Davis and Condie, 1976). In the Midlands belt, on the other hand, type I andesites occur only in the Maliyami Formation and type II only in the overlying Felsic Formation. Type III andesite, which thus far has been described only from the Abitibi belt in Canada (Condie and Baragar, 1974), is characterized by flat REE patterns (30—40x chondrites) and negative Eu anomalies. They are closely associated with tholeiites with similar REE patterns although lower REE concentrations. Compared to types I and II, these rocks are also low in Sr and high in Y. The only igneous rocks reported to have similar REE patterns and negative Eu anomalies are lunar basalts (Gast, 1972).

Modern andesites can also be divided into three categories based on composition and tectonic setting (Jakeš and White, 1972; Condie, 1967a) (Table 3-9). Arc andesites (AA) occur in immature, oceanic island arcs (such as the Marianas) and near the trench side of mature arcs. Calc-alkaline andesites (CA) are most widespread in modern arc systems and high-K calc-alkaline andesites (HKA) occur in some continental margin arc systems (such as the Andes) which are underlain by thick lithosphere. Although in terms of many major elements, it is tempting to equate each of the Archean andesite types I, II, and III with modern andesites CA, HKA, and AA,

TABLE 3-9

Average compositions (oxides in wt.%, trace elements in ppm) of Archean andesite groups compared to modern andesites (after Condie, 1976c, 1979b)

	Archean			Modern		
	I	II	III	arc	calc-alkaline	high-K calc-alkaline
SiO_2	56.7	58.9	55.1	57.3	59.5	60.2
TiO_2	0.92	0.65	0.95	0.58	0.70	0.95
Al_2O_3	14.0	15.5	15.9	17.4	17.2	16.9
Fe_2O_3	2.3	1.5	1.99	2.5	2.5	2.6
FeO	7.0	4.5	5.86	2.7	5.0	2.8
MgO	5.4	4.5	4.3	3.5	3.4	2.2
CaO	6.6	5.1	5.9	8.7	7.0	5.5
Na_2O	3.4	4.0	3.9	2.6	3.7	3.7
K_2O	0.67	1.9	1.1	0.7	1.6	2.8
H_2O	3.0	3.0	2.8	1.0	1.0	1.0
FeO/Fe_2O_3	3.0	3.0	2.9	1.1	2.0	1.2
Na_2O/K_2O	5.1	2.1	3.4	3.7	2.3	1.2
Cr	125	88	105	40	90	90
Zn	97	81	77	60	65	
Cu	60	36	64	70	100	40
Ni	70	60	55	20	25	40
Co	25	23	29	20	25	20
Sr	278	580	210	240	475	700
Rb	22	75	30	20	40	80
Ba	230	547	361	150	300	700
Zr	150	190	104	90	110	200
La	13	34	12	3	12	43
Ce	31	70	30	6.8	25	84
Nd	17	35	22	6	14	37
Sm	3.6	6.7	7.3	2.3	3.0	5.1
Eu	1.1	1.9	2.0	0.9	1.0	1.4
Gd	3.6	6.2	8.5	3.5	3.6	4.0
Dy	3.8	5.8	11	4.5	4.5	3.5
Er	2.0	3.0	6.4	2.6	2.0	1.8
Yb	1.8	2.4	6.1	2.3	1.9	1.6
Lu	0.3	0.3	1.1	0.4	0.4	0.27
Y	25	35	40	25	20	10
K/Rb	253	210	315	291	332	208
Ni/Co	2.8	2.6	1.9	1.0	1.0	2.0
La/Yb	7.2	14	2.0	1.1	6.3	2.7
Eu/Eu^*	0.96	0.92	0.78	1.0	0.94	1.0
$(La/Sm)_N$	2.0	2.8	0.90	0.72	2.2	4.6
$(Yb/Gd)_N$	0.62	0.48	0.89	0.82	0.66	0.50

N = chondrite-normalized ratio.

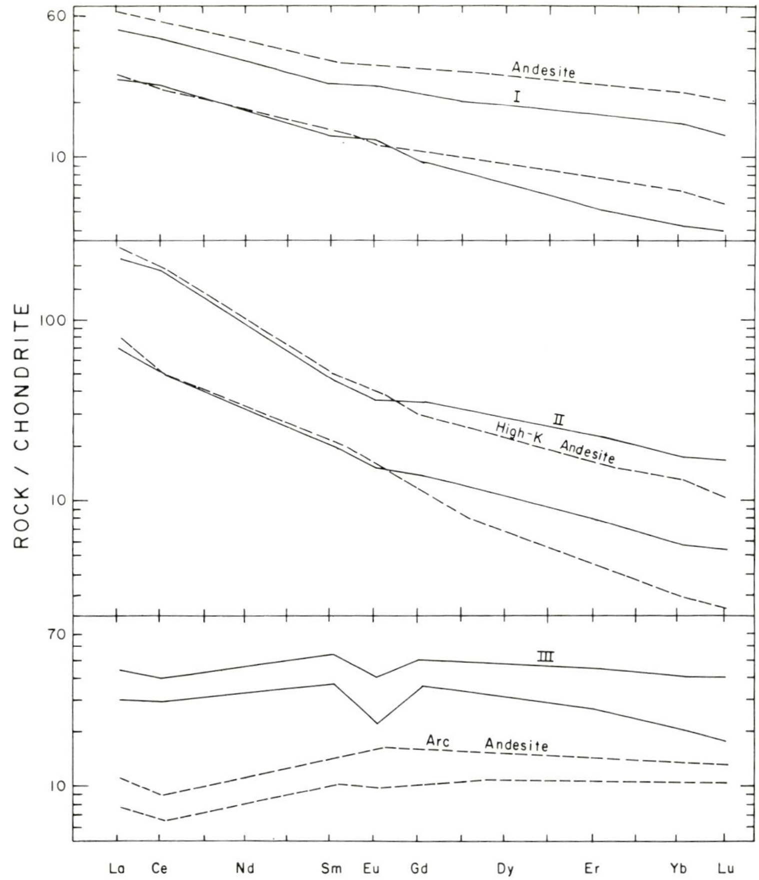

Fig. 3-25. Envelopes of variation of chondrite-normalized REE distributions in Archean andesite groups I, II, and III compared to envelopes of modern andesite groups (from Condie, 1979b).

respectively, several important differences render such correlations improbable. First of all, all Archean andesites differ from modern andesites in terms of their low Al_2O_3 contents and their high FeO, MgO, Y, and FeO/Fe_2O_3, and Ni/Co ratios. Among the transition trace elements, Ni, Cr, Co, and Zn are also enriched in Archean andesites. In addition to the overall differences, most arc andesites differ from type III andesites in having lower concentrations of REE and no Eu anomalies (Fig. 3-25). CA and HKA are also somewhat higher in K_2O, Rb, Sr, and Ba than most type I or II andesites, respectively. Their REE patterns are, however, strikingly similar to the modern groups.

Occurrence

Felsic igneous rocks in Archean greenstone terranes occur as pyroclastic-hyaloclastic-epiclastic rocks, as flows, and as intrusive porphyries. Some of the most extensive descriptions of these rocks are given in Wilson (1964), Goodwin (1962), Henderson and Brown (1966), Viljoen and Viljoen (1969e), Harrison (1970), and Sims (1972b). The term felsic is generally used in a broad sense to include dacite, rhyodacite, quartz latite and rhyolite compositions, which in most greenstone belts, decrease in relative abundances in the order listed. Hyaloclastic and pyroclastic rocks are most common. A typical section of breccias and tuffs in the Hooggenoeg Formation in the Barberton belt is given in Fig. 3-26. The section can be divided into three major units. The lowest is a mixed breccia and tuff unit which is comprised of several cycles (each 10—20 m thick) each beginning with a coarse breccia and grading upwards into progressively finer tuffs. Breccia units in the upper part of each cycle are lensoid in shape. The middle unit consists chiefly of water-worked felsic tuffs becoming finer grained with stratigraphic height. In addition, these tuffs contain many sedimentary structures such as cross-bedding, slump structures, and load casts suggesting an epiclastic origin. The upper unit in the section is composed of finely banded tuffs which grade upwards into laminated cherty tuffs. Large scour and fill channels filled chiefly with angular black chert clasts in a highly carbonated matrix are found in this unit. Textures and structures preserved in the Hooggenoeg section are interpreted to reflect subaqueous volcanic deposition with the rocks representing mixed hyaloclastites and epiclastites (Viljoen and Viljoen, 1969e).

In general, felsic breccias in Archean greenstone successions are characterized by units with broadly lensoid shapes which may range up to 300 m thick and can be traced for up to several kilometers along strike. Coarse units may grade laterally into fine units over distances as short as 5 km (Page and Clifford, 1977). Fragments in breccias are chiefly felsic volcanics and range up to 3 m across although generally averaging 10—30 cm. Units are poorly sorted and fragments are generally angular although some units are composed of well-rounded fragments (agglomerates). Goodwin (1962) describes felsic breccia domes from the Michipicoten area of the Superior Province. The domes are broadly lensoid shaped and may be up to 25 km across and 3 km thick. They are characterized by rhyolitic breccia cores that grade upwards into breccias of mixed calc-alkaline compositions.

Felsic tuff units vary from coarse to fine and are generally well bedded. Individual beds range from 1 cm to several meters thick. The color of these rocks is highly variable depending, in part, on degree of alteration. Spherulites (1—15 cm in diameter) are common in some units. Vitric, crystal, and

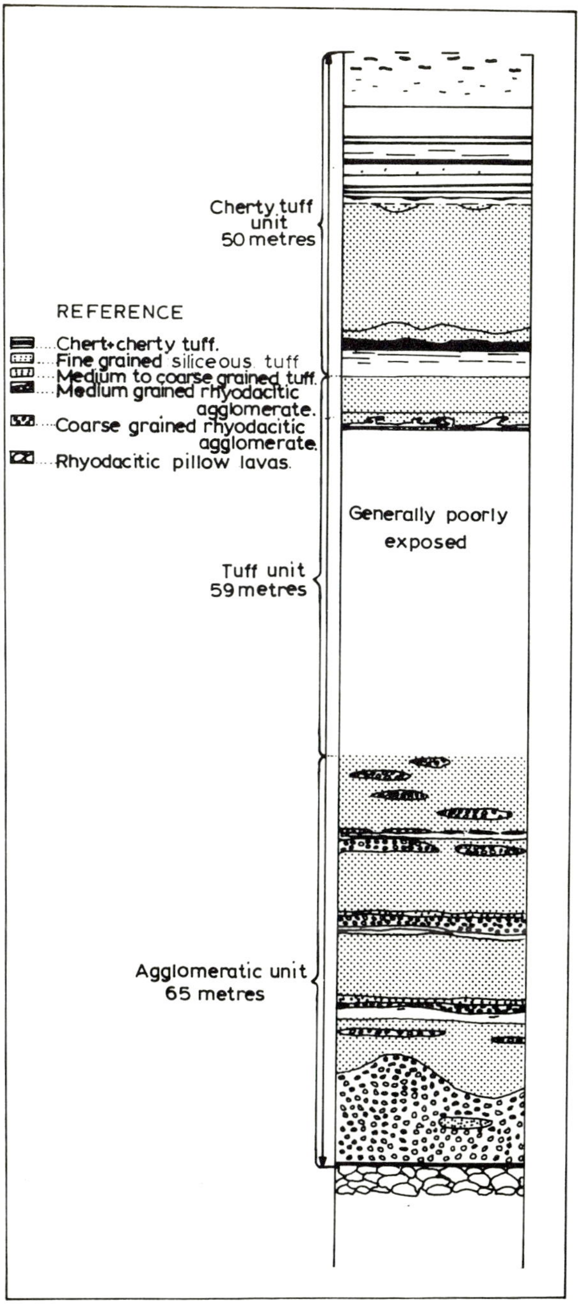

Fig. 3-26. The upper felsic volcanic zone in the Hooggenoeg Formation, Barberton belt, South Africa (from Viljoen and Viljoen, 1969e).

Fig. 3-27. Photomicrograph of an Archean ash-flow tuff from the Marda Complex in Western Australia (from Hallberg et al., 1976). Note the well-preserved eutaxitic texture.

lithic tuffs are all represented and primary structures such as graded-bedding, cross-bedding, and scour channels are common in many tuffs. Ash-flow tuffs have been described from the Michipicoten area and from the Marda Complex in Australia (Goodwin, 1962; Hallberg et al., 1979). These units contain flattened pumice fragments (now recrystallized) and often exhibit eutaxitic textures (Fig. 3-27). Individual flows up to 300 m thick have been traced for 3 km in the Michipicoten area.

Felsic flows are uncommon in most greenstone successions. An exception is the Nyanzian System in western Kenya, where felsic flows appear to comprise most of the greenstone successions (Huddleston, 1951; Saggerson, 1952; McCall, 1958). Locally, flows may be abundant in other belts such as in the Newton Lake Formation in northeastern Minnesota (J.C. Green, 1972). Flows are characterized by short lateral extent, bulbous flow tops and streaky, irregular flow banding. Vesicles and spherulites (some up to 60 cm in diameter) are common. Some flows contain pillows (Viljoen and Viljoen, 1969e) which are usually smaller than those found in mafic and andesitic flows. In the upper Onverwacht section, flows grade upwards into bedded white cherts which terminate volcanic cycles (Fig. 2-4).

Felsic porphyries occur in all greenstone belts and may be of intrusive or extrusive origin. The intrusive nature of most of them is attested to by field relationships (Henderson and Brown, 1966; Viljoen and Viljoen, 1969d;

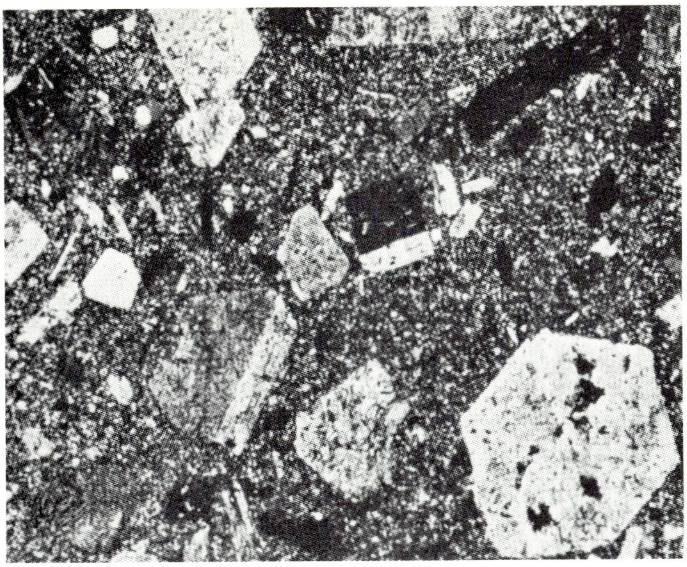

Fig. 3-28. Photomicrograph of an Archean dacite porphyry from Kakagi Lake, Ontario (from Moorehouse, 1970). Phenocrysts of quartz, albite, and chloritized biotite in a quartz-feldspar-chlorite matrix. Crossed polars, × 49.

Harrison, 1969, 1970; Glikson, 1972a). Extrusive porphyry flows which contain flow banding and spherulites have been recognized in some areas (Wilson, 1964; O'Beirne, 1968). Intrusive bodies occur as sills, dikes, plugs, ring dikes, and irregular-shaped bodies with contacts ranging from concordant to discordant. They may be injected before or after the major period of deformation, but almost always exhibit evidences of regional metamorphism (foliation, etc.). Such bodies occur almost entirely within greenstone belts and generally do not possess contact metamorphic aureoles. Individual dikes and sills may range up to 200 m thick (or rarely 1000 m) and can be traced laterally for distances of 1—3 km. The rocks range from white to gray to buff or brown in color and contain large phenocrysts of plagioclase and sometimes quartz.

Petrography

Petrographic descriptions of felsic volcanic and hypabyssal rocks are given in Huddleston (1951), Saggerson (1952), McCall (1958), Henderson and Brown (1966), Viljoen and Viljoen (1969c, e), Harrison (1970), and in Glikson (1972a). Porphyritic varieties are common and contain chiefly plagioclase phenocrysts ranging up to several millimeters in size (up to 10 mm in intrusive porphyries) (Fig. 3-28). Porphyries contain 10—40% of such phenocrysts. Crystals are generally short and stubby, range in composition from An_{10} to An_{30}, and may be zoned. They vary from slightly to strongly altered with mixtures of sericite, epidote, chlorite, and iron oxides as the common alteration products. Quartz phenocrysts are found in some rocks where they comprise up to 15% of the rock, range from 1 to 5 mm in size, and are usually partially resorbed by the matrix. K-feldspar phenocrysts are rare. Hornblende phenocrysts occur in some dacitic units and range up to 1 mm in length. They are generally partially altered to epidote, chlorite, and carbonate. Common accessory minerals are magnetite, apatite, ilmenite (± leucoxene), and rarely sphene.

Most matrix minerals are secondary in origin although micrographic intergrowths are rarely preserved. Felted to trachytic textures are often present even in highly altered rocks. Common secondary minerals are sericite, carbonate, quartz, epidote, chlorite, and iron oxides. Some units contain up to 60% carbonate. Others may be highly silicified containing up to 80% fine grained quartz. Metamorphic minerals such as andalusite, pyrophyllite, and chloritoid are reported from tuffs (Viljoen and Viljoen, 1969e). Shard pseudomorphs are present in some tuffs and ash-flow tuffs.

Composition

In terms of chemical composition, felsic volcanic and hypabyssal rocks will be grouped into two categories: rhyolite (including quartz latite) (> 69% SiO_2) and dacite (including rhyodacite) (63—69% SiO_2). Using REE distributions, it is possible to subdivide felsic volcanics into two groups FI and FII, originally referred to as DSV and USV, respectively (Condie, 1976c). FI is characterized by strong depletion in heavy REE (down to 1× chondrites) while FII is not (Fig. 3-29). Available data suggest that one type or the other dominates or is the only type represented in a given greenstone belt. FI felsic volcanics only are reported in the Midlands belt in Rhodesia (Condie and Harrison, 1976), the Vermilion belt in northeastern Minnesota (Arth and Hanson, 1975), and in the Suomussalmi belt in Finland (Jahn et al., 1979). FII volcanics only are reported from the Nyanzian belts in western Kenya (Davis and Condie, 1976), the Marda Complex in Western Australia (Taylor and Hallberg, 1977), and the Yellowknife belt in Canada (Condie and Baragar, 1974). Both types are reported in the Barberton belt in South Africa (Glikson, 1976c) and in the Prince Albert Group in northern Canada (Fryer and Jenner, 1978). In addition to exhibiting heavy-REE depletion, FI is characterized by high contents of Al_2O_3, Na_2O, Na_2O/K_2O, Ti/Zr, Zr/Y, and relatively large amounts of many transition metals and low Zr, Ba, Y, and Ti/V compared to FII (Table 3-10). Eu anomalies are also absent or negligible in FI while negative Eu anomalies characterize FII (Fig. 3-29).

As shown in Table 3-10, FII dacites and rhyolites are grossly similar to modern calc-alkaline dacite and rhyolite. They differ, however, in containing greater concentrations of transition trace metals and high Ni/Co ratios.

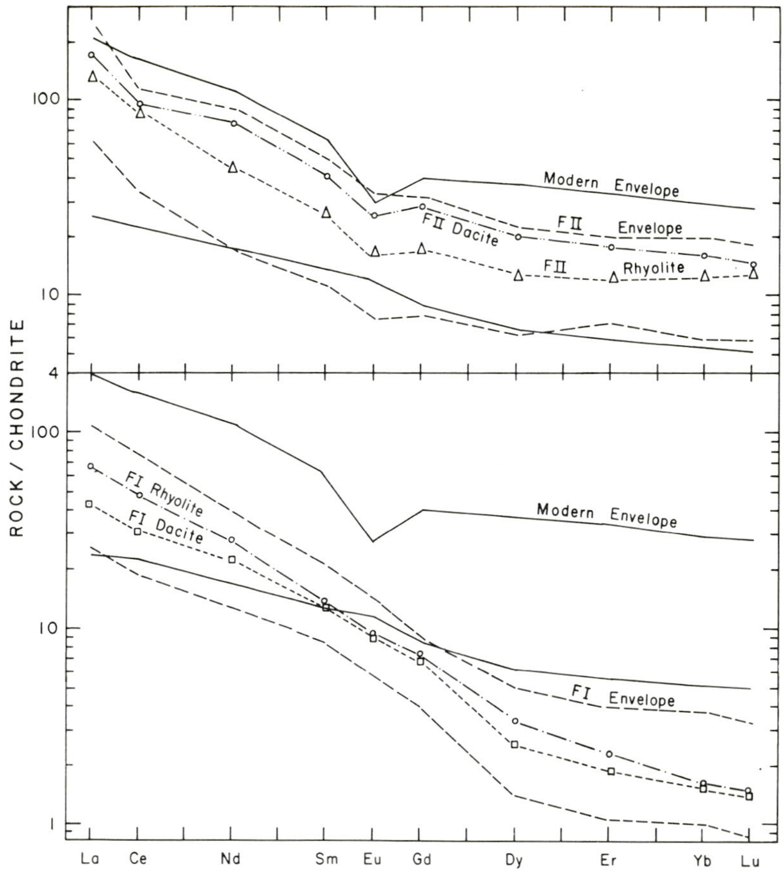

Fig. 3-29. Envelopes of variation of chondrite-normalized REE distributions in Archean felsic volcanic rock groups FI and FII compared to envelopes of modern felsic volcanic rocks (after Condie, 1976c). Also shown are average REE patterns for Archean rhyolite and dacite (including rhyodacite) for each group from Table 3-10.

Although most modern felsic volcanics have REE patterns similar to FII (Fig. 3-29), some have been reported which exhibit heavy-REE depletion like FI (Pecerillo and Taylor, 1976). A depletion in heavy REE also characterizes many plutonic rocks of the tonalite-trondhjemite suite of various ages (Barker et al., 1976a; Frey et al., 1978).

ROCKS WITH ALKALINE AFFINITIES

Occurrence

Volcanic and hypabyssal rocks in Archean greenstone belts with alkaline affinities are uncommon. They comprise up to a few percent of some belts

TABLE 3-10

Average compositions (oxides in wt.%, trace elements in ppm) of Archean and modern felsic volcanic rocks (after Condie, 1976c)

	Archean				Modern		
	FI (DSV)		FII (USV)				
	dacite-rhyodacite	rhyolite	dacite-rhyodacite	rhyolite	arc dacite	dacite	rhyolite
SiO_2	67.1	70.9	68.4	76.0	66.8	64.9	74.0
TiO_2	0.28	0.23	0.25	0.11	0.20	0.60	0.25
Al_2O_3	16.5	15.8	14.8	12.1	18.2	16.0	13.3
Fe_2O_3	0.94	0.64	1.20	0.57	1.30	3.2	1.3
FeO	1.02	1.49	2.85	0.58	1.0	1.0	0.5
MgO	1.60	0.90	1.58	0.63	1.5	1.7	0.30
CaO	3.90	1.10	3.20	0.93	3.2	4.7	1.5
Na_2O	5.23	5.58	4.00	3.83	5.0	4.2	4.0
K_2O	1.72	1.72	1.65	4.12	1.0	1.8	3.5
P_2O_5	0.10	0.14	0.25	0.03	0.05	0.06	0.04
MnO	0.04	0.02	0.08	0.04	0.10	0.10	0.03
H_2O	0.65	1.55	1.25	0.74	0.6	0.7	0.5
Na_2O/K_2O	0.3	3.2	2.4	0.93	5.0	2.3	1.1
FeO/Fe_2O_3	1.1	2.3	2.3	1.0	0.77	0.31	0.39
Cr	70	12	40	11	5	10	2
Ni	15	10	20	12	1	8	1
V	35	31	20	11	20	50	20
Co	20	8	13	6	8	15	3
Cu	32	11	15	10	7	20	5
Zn	70	60	55	28	60	70	50
Zr	160	150	260	275	80	100	160
Ba	650	440	1000	1080	250	400	900
Sr	500	221	320	42	200	500	150
La	14	23	65	43	6	15	30
Ce	30	42	87	77	15	26	70
Nd	14	17	47	27	8.4	14	33
Sm	2.4	2.5	7.6	4.8	2.0	2.9	5.5
Eu	0.67	0.66	1.8	1.1	0.7	1.0	1.5
Gd	1.7	1.8	7.0	4.3	2.7	2.7	5.7
Dy	0.85	1.1	6.7	4.1	3.5	2.9	6.7
Er	0.38	0.48	3.7	2.4	2.0	1.6	3.8
Yb	0.32	0.34	3.2	2.5	2.0	1.4	3.5
Lu	0.05	0.05	0.50	0.44	0.40	0.20	0.50
Y	12	10	32	26	25	30	10
Ni/Co	0.75	1.3	1.5	2.0	0.13	0.5	0.2
Ti/Zr	11	9.2	5.8	2.4	15	36	9.4
Ti/V	34	45	75	60	60	72	75
$(La/Sm)_N$	3.2	5.0	4.7	4.9	1.6	2.8	3.0
Eu/Eu^*	1.0	0.95	0.75	0.74	0.91	1.1	0.83
$(Yb/Gd)_N$	0.23	0.24	0.57	0.72	0.92	0.65	0.76

N = chondrite-normalized ratio.

in the Canadian Shield (Goodwin, 1977a) but are absent, or at least not preserved, in most belts. The Kirkland Lake area of the Abitibi belt is unique in that about 13% of the volcanic rocks are alkaline (Cooke and Moorehouse, 1968). Archean rocks with alkaline or shoshonitic affinities have also been described from the Oxford Lake Group in northeastern Manitoba (Hubregtse, 1976) and in the Schoongesicht Formation in the upper part of the Swaziland Supergroup in South Africa (Visser, 1956; Condie et al., 1970; Anhaeusser, 1974), and in the Suomussalmi belt in Finland (Jahn et al., 1979). At each locality they are interbedded with calc-alkaline volcanic rocks.

Alkaline igneous rocks occur as both volcanic and intrusive varieties with the former usually being more widespread. In all occurrences, the alkaline volcanics are intimately mixed with calc-alkaline volcanics. Pyroclastics usually exceed flows in abundance. Alkaline breccias in the Kirkland Lake area are lensoid in shape and extend for up to 200 m along strike where they interfinger with tuffs of similar composition. Most alkaline pyroclastic rocks are porphyritic and many have amygdules and spherulites. Trachytes and trachyandesites appear to be the most abundant compositional types present. In the Kirkland Lake area, trachyte, leucite trachyte, mafic trachyte, and quartz trachyte (in order of decreasing abundance) are interlayered with tholeiites and andesites (Cooke and Moorehouse, 1968). They are associated with small syenite intrusive bodies which may have served as feeders for the volcanics. Flows, when found, exhibit flow banding and, in some cases, pillows.

Petrography

Alkaline volcanics are commonly porphyritic with sodic plagioclase and augite being the two principal phenocryst phases. Trachytes from the Kirkland Lake area are composed of 25—60% of olivine, augite, plagioclase, and biotite. Plagioclase phenocrysts may be zoned and range in composition from An_5 to An_{15}. Small phenocrysts of K-feldspar and hornblende occur in some rocks. Pseudomorphs of pseudoleucite phenocrysts ranging from 0.5 mm to 2 cm across occur in some trachytes from Kirkland Lake (Fig. 3-30). These pseudomorphs are composed of K-feldspar, sericite, sodic plagioclase, carbonate, and chlorite. The matrices of alkaline volcanics are composed almost entirely of secondary assemblages of such minerals as sodic plagioclase, sericite, chlorite, iron oxides, and carbonate. In some rocks as much as 80% of the groundmass is altered to carbonate. In the less altered varieties, fluidal and trachytic textures are often preserved.

Composition

Few analyses are available of alkaline Archean volcanics. The trachytes at Kirkland Lake are typical trachytes with $Na_2O > K_2O$ and may or may

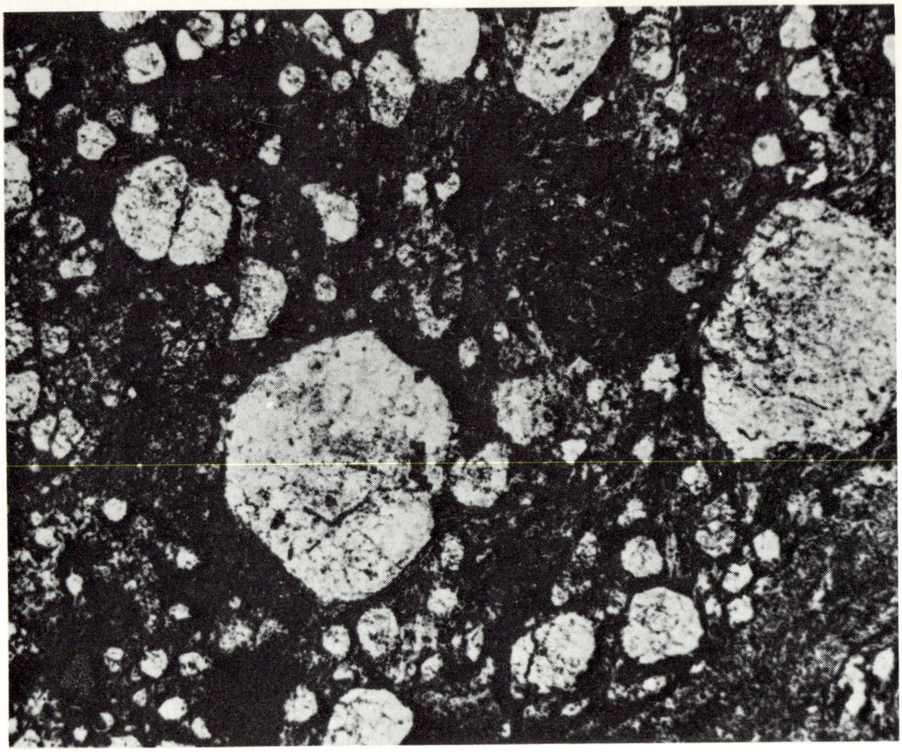

Fig. 3-30. Photomicrograph of altered leucite tuff from the Kirkland Lake area, Ontario (from Moorehouse, 1970). Pseudoleucite phenocrysts in altered matrix. Plane light, × 50.

not be nepheline normative (Cooke and Moorehouse, 1968). The leucite-bearing volcanics are high in total $Na_2O + K_2O$ (9—11%) and Ba and $K_2O > Na_2O$. In terms of major element composition, these rocks are similar to young leucite-bearing volcanics in Italy and Indonesia. Associated syenite plutons have similar compositions and appear to represent intrusive phases of the same magma. Analyses of volcanic rocks with alkaline or shoshonitic affinities have also been reported from several other greenstone belts in the Superior Province (Hubregtse, 1976; Goodwin, 1977a), from the Schoongezicht Formation in the upper part of the Barberton section in South Africa (trachytes and trachyandesites) (Visser, 1956), and from the Suomussalmi belt in Finland (Jahn et al., 1979). Major and trace elements contents of alkaline rocks from the Oxford Lake Group in Manitoba are strikingly similar to those of young shoshonites from Papua (Hubregtse, 1976). Two alkali basalts from the Finland occurrence are similar in composition, including REE distributions (light REE = 200 × chondrites, heavy REE = 10× chrondrites), to modern alkali basalts.

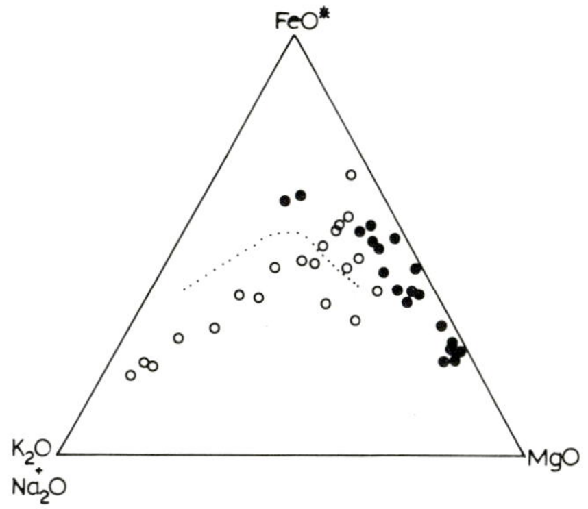

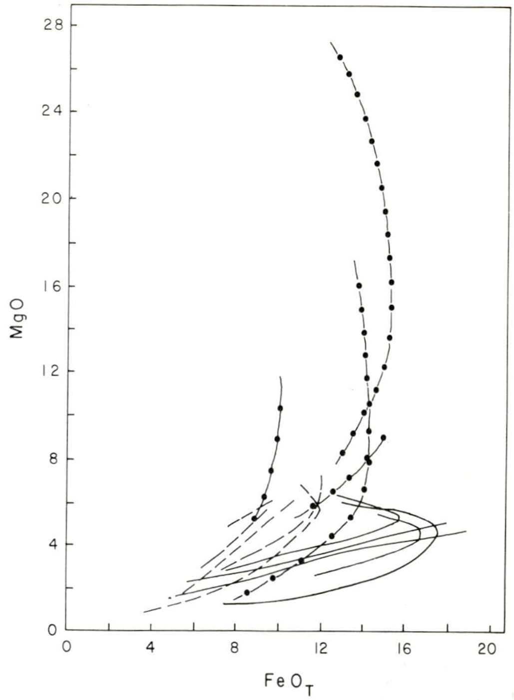

Fig. 3-31. AFM diagram showing Bulawayan volcanic rocks from Rhodesia (after Hawkesworth and O'Nions, 1977). Filled circles: the combined tholeiite-komatiite series. Open circles: the calc-alkaline series. Dashed line (from Irvine and Baragar, 1971) separates calc-alkaline and tholeiite fields.

Fig. 3-32. MgO-FeO$_T$ diagram for various traverses across the Abitibi greenstone belt (from Jolly, 1975). —•—•— = komatiite series; ——— = calc-alkaline series; ——— = tholeiite series. Each line represents a separate traverse.

IGNEOUS ROCK SERIES

Each of the three well-established igneous rock series, the tholeiite, calc-alkaline, and alkaline, are recognized in Archean greenstone successions. The alkaline series, however, is of very limited extent. In addition, a fourth series referred to as the komatiite series (Arndt et al., 1977; Blais et al., 1978) or the high-magnesian series (Jolly, 1975, 1977) is important in some belts. Each series contains rocks ranging in composition from mafic or ultramafic to intermediate or felsic. Although volcanic and hypabyssal rocks of two or more series are commonly in close association stratigraphically, there is a clear decrease in importance of the komatiite and tholeiite series at the expense of the calc-alkaline series with stratigraphic height.

The tholeiite, komatiite, and calc-alkaline series are illustrated on chemical variation diagrams for Bulawayan volcanics in Rhodesia and for several traverses across the Abitibi belt in Canada in Figs. 3-31 and 3-32. The

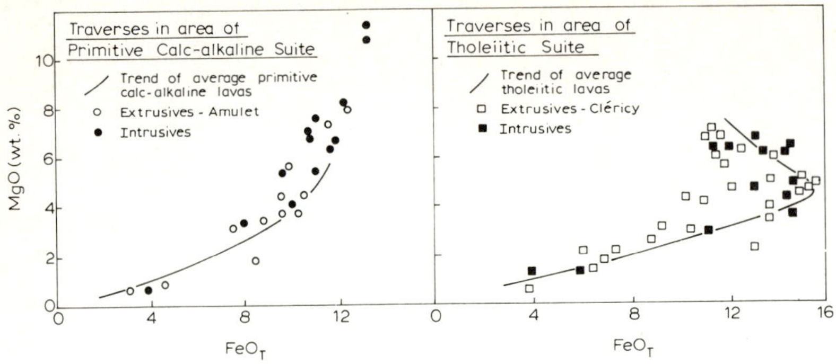

Fig. 3-33. MgO-FeO_T diagrams for traverses across the Abitibi belt showing relations of intrusive to extrusive rocks (from Jolly, 1977).

tholeiite and komatiite series, which are indistinguishable on an AFM diagram (Fig. 3-31), are characterized by rapid iron enrichment. In addition, the komatiite series exhibits rapid changes in MgO for small changes in FeO_T (Fig. 3-32). The calc-alkaline series is characterized by an almost constant Fe/Mg ratio and increasing alkalies. The komatiite series comprises volcanic rocks ranging in composition from ultramafic to andesitic and cumulate rocks ranging from ultramafic to mafic (Arndt et al., 1977). All members have high MgO, Ni, and Cr contents and low TiO_2 contents ($\leqslant 1\%$). On an MgO-CaO-Al_2O_3 diagram (Fig. 3-6), the komatiite series leads into the tholeiite series; the constancy of the CaO/Al_2O_3 ratio in the komatiite series favors a dominant olivine control. The similarity in composition of closely associated hypabyssal and volcanic rocks is illustrated for several traverses across the Abitibi belt in Fig. 3-33. In the Clericy traverses, both groups of rocks show strong iron enrichment, whereas in the Amulet traverses both groups of rocks exhibit a calc-alkaline trend (Jolly, 1977). Jolly has suggested that the rocks in each traverse represent intrusive and extrusive phases of the same magmas. Chemical trends observed in Archean stratiform complexes are also indicative of the komatiite or tholeiite series (Hess, 1960; Arndt et al., 1977).

Naldrett and Goodwin (1977) have shown that the average sulfur content increases rapidly with average FeO in volcanic rocks of the Blake River Group in the Abitibi belt (Fig. 3-34). This relationship has also been observed in other Canadian greenstone belts (Naldrett et al., 1978). Unlike Archean mafic volcanics, MORB appear to have lost large amounts of sulfur through seawater reaction. Naldrett et al. suggest the reason for retention of sulfur in Archean volcanics may be due to a rapid accumulation rate such that they are exposed to direct interaction with seawater for a much shorter time than MORB.

Some Archean greenstone belts, as discussed in Chapter 2, are bimodal in that intermediate volcanic compositions are rare. Examples are the greenstone

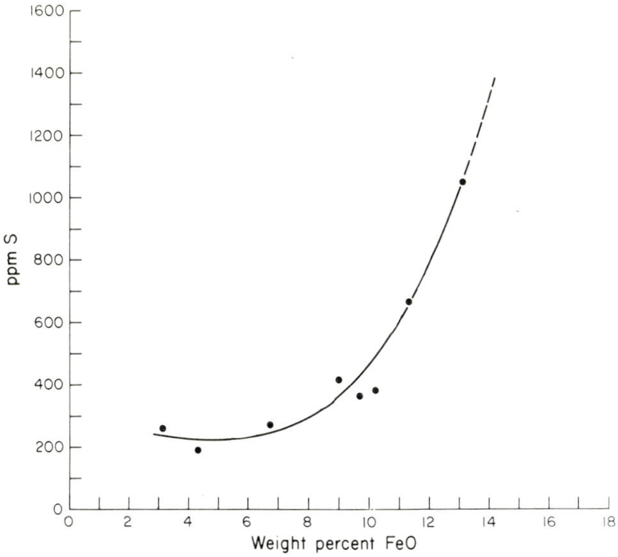

Fig. 3-34. Plot of the mean sulfur content versus mean FeO in volcanic rocks of the Blake River Group, Abitibi greenstone belt (after Naldrett and Goodwin, 1977). Dots represent mean values.

belts in the Eastern Goldfields subprovince in Western Australia (Hallberg, 1972), the Vermilion greenstone belt in northeastern Minnesota (Arth and Hanson, 1975), and the Sturgeon Lake belt in Ontario (Franklin, 1978). In Western Australia, Hallberg (1972) reports that in over 400 available analyses, not one lies in the range of 55—60% SiO_2 and only nine lie between 55 and 65% SiO_2. Total iron, MgO, and CaO also reflect a sparsity of intermediate values (Fig. 3-35). The bimodal distribution in the Sturgeon Lake belt is clearly evident on a contoured TiO_2-SiO_2 plot (Fig. 3-36).

STRATIGRAPHIC VARIATIONS IN COMPOSITION

The major stratigraphic changes found in Archean greenstone belts were discussed in Chapter 2. It is of interest to examine compositional changes as a function of stratigraphic height more closely in successions that are well known. The proportion of rock types in three stratigraphic sections in each of two belts in the Superior Province is summarized in Fig. 3-37. The sections are divided into upper and lower portions and the distribution of the dominant igneous rock series is also shown on the figure. Goodwin (1977a) makes the following conclusions with regard to these sections:

(1) Each greenstone succession displays a compositional change from dominantly tholeiite in the lower parts, through increasing proportions of

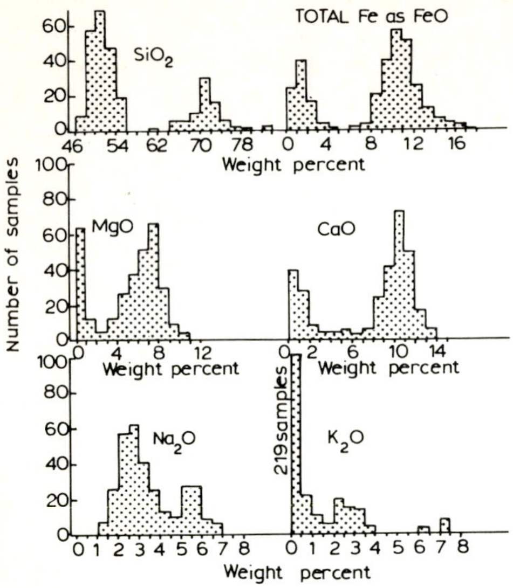

Fig. 3-35. Frequency distribution of six major oxides in Archean volcanic and related rocks from the Eastern Goldfields subprovince, Western Australia (from Hallberg, 1972).

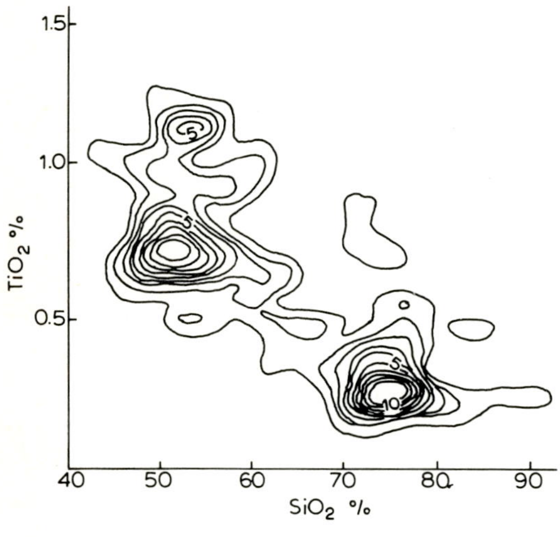

Fig. 3-36. Contoured TiO_2-SiO_2 diagrams for volcanic rocks from the Sturgeon Lake belt, Ontario (from Franklin, 1978).

andesite in the middle and upper parts, to dominantly dacite and rhyolite in the upper parts.

(2) Members of the tholeiite (± komatiite) series dominate in both belts (57%) followed by the calc-alkaline series (38%); the alkaline series comprises about 5%.

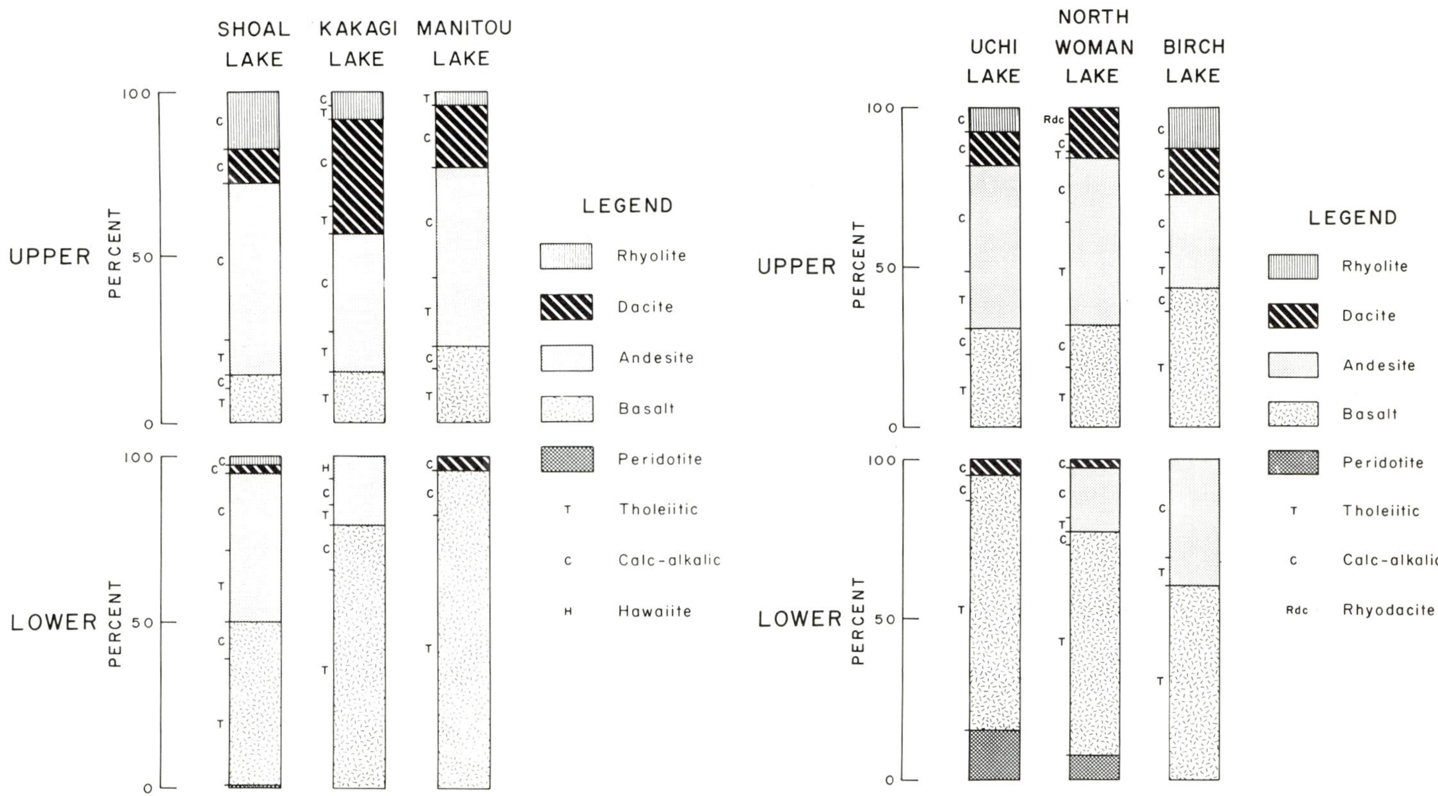

Fig. 3-37. Weighted mean abundances of volcanic classes in the Lake of the Woods and Birch-Uchi greenstone belts, Canada (from Goodwin, 1977a). Each column represents a separate stratigraphic section divided into an upper and lower division.

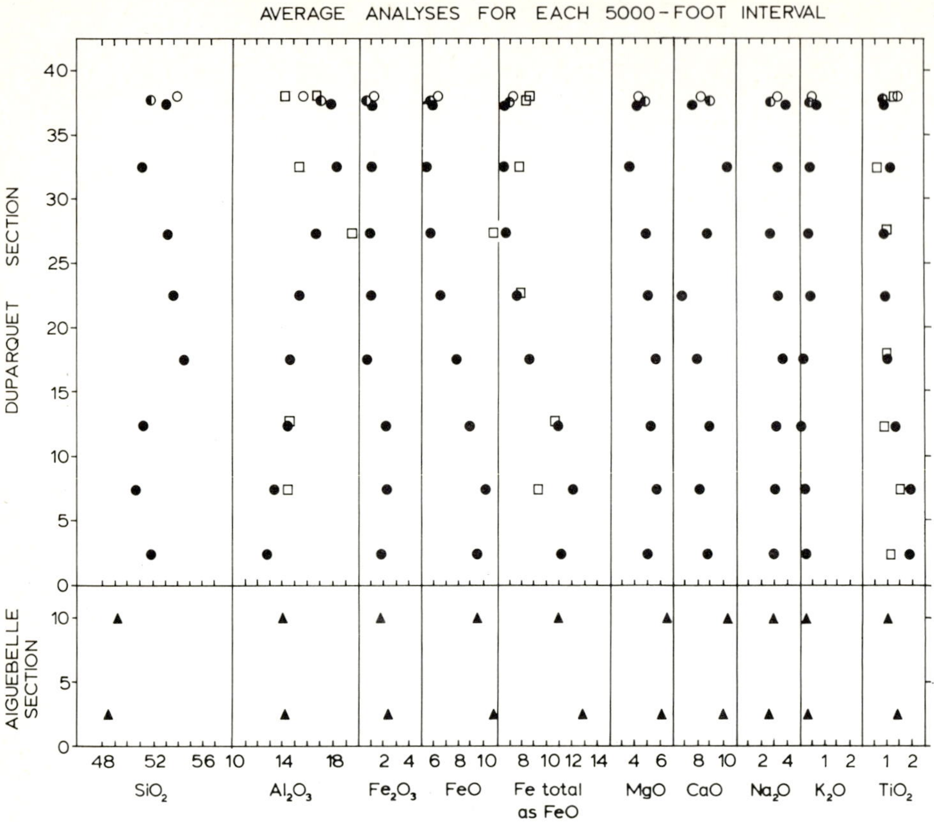

Fig. 3-38. Major element contents averaged over for 5000-ft (~ 1500 m) intervals in the Duparquet section of the Abitibi greenstone succession (after Baragar, 1968).

(3) Tholeiite components dominate in the lower parts of the successions (76%) and calc-alkaline components in the upper parts (62%). Alkaline components have very limited geographic and stratigraphic distributions.

(4) The lower parts of two of the sections (Uchi and Manitou Lake) are bimodal, lacking andesite.

The most extensive studies of stratigraphic changes in composition of greenstone volcanic successions are those in the Abitibi belt in Canada (Baragar, 1968, 1972; Jolly, 1975; Gelinas et al., 1977b; Goodwin, 1979). The average major element compositions of a 12-km-thick section of volcanic rocks near Duparquet is summarized in Fig. 3-38 as a function of stratigraphic height. Several trends are evident in the diagram (Baragar, 1968). Al_2O_3 and K_2O increase steadily with stratigraphic height and FeO, total Fe as FeO, MgO, and TiO_2 decrease. Farther to the east and over a stratigraphic thickness of about 4.5 km, Gelinas et al. (1977b) recognize two volcanic cycles in the Deguisier tholeiitic series. Geochemical trends within these

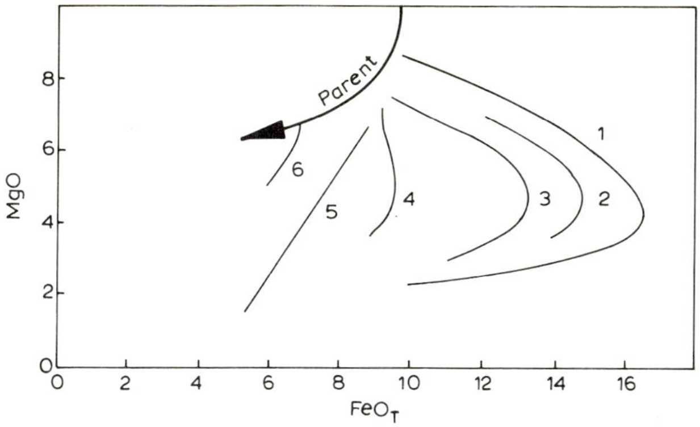

Fig. 3-39. MgO-FeO_T diagram for samples from the Duparquet section of the Abitibi belt (after Jolly, 1977). Each line represents a suite of samples numbered in order of increasing stratigraphic height.

cycles are not as clearly defined as those reported by Baragar (1968). There is a tendency, however, for the lower cycle (~ 2 km thick) to show, with increasing stratigraphic height, increasing total Fe and decreasing MgO and SiO_2. When the samples from Baragar's traverse are considered on an MgO-FeO_T diagram, a strong iron enrichment is observed in the lowest volcanics (Fig. 3-39). This enrichment decreases with stratigraphic height and an abrupt shift to Fe depletion occurs between trends 3 and 4 with the trends above this being more calc-alkaline in nature. The distribution of samples indicate, however, that lavas associated with any given trend are side-by-side with lavas from other trends indicating that magmas exhibiting various degrees of fractionation were erupted in close succession at least partly without mixing with each other. The possible compositions of the parent magmas for each of the trends is also noted in the figure. Analyses of REE in samples from the Duparquet traverse indicate an increase in overall REE content with stratigraphic height, but no appreciable change in REE patterns (Condie and Baragar, 1974). Considering the entire volcanic sequence in the Abitibi belt in the Noranda-Kirkland Lake area, Jolly (1975) has proposed a three-fold stratigraphic division. Rocks of the lowest level are dominated by volcanic and hypabyssal rocks of the komatiitic (high-magnesian) series and very rich in MgO, Ni, and Cr (Fig. 3-32). The middle and upper divisions are characterized by an abundance of the tholeiite and calc-alkaline series, respectively. Existing data suggest that the centers of volcanism shifted eastwards with time in the Abitibi belt (Goodwin, 1977a).

Geochemical variations in volcanic rocks of the Yellowknife belt indicate the presence of two volcanic cycles (Baragar, 1966). Each cycle is composed chiefly of tholeiites with calc-alkaline volcanics appearing rather abruptly at the top of each cycle. All major elements except Na_2O and Al_2O_3 show this change. Smaller scale cyclical trends are also observed within each of these cycles. The degree of light-REE enrichment is greater in tholeiites of the upper cycle (20—30 × chondrites) than it is in tholeiites of the lower cycle (10—20 × chondrites) (Condie and Baragar, 1974). Hubregtse (1976) reports five volcanic cycles in the Knee Lake greenstone belt in Manitoba with each cycle showing a progression from more tholeiitic components at the base to more calc-alkaline components at the top.

Chapter 4

SEDIMENTARY ROCKS

INTRODUCTION

Sedimentary rocks comprise 15—30% of most Archean greenstone belts reaching 85% in belts in the Slave province. Although in most greenstone successions sediments become important only in the upper parts, some successions contain major sediment horizons throughout. Clastic sediments, in particular the graywacke-argillite suite, dominate and non-clastic sediments (principally chert) are minor, but widespread. Sediments are particularly important in reconstructing the tectonic history of greenstone belts. They contain information not only relevant to distance from source area and energy of the sedimentary environment, but also contain clues about the composition of their source areas which may, in part, represent significantly older crust. Employing primary textures and structures, one can learn about water depth, mechanism of deposition, and current directions. Clastic sedimentary mineral assemblages also reflect the composition of the Archean atmosphere and oceans and the nature of Archean weathering. Finally, it is through the study of sediments that one can learn more about the size and distribution of Archean basins and their relationships to each other.

CLASTIC SEDIMENTS

Graywacke-argillite

General features

Interbedded graywacke and argillite (or slate) are by far the most abundant sediments in Archean greenstone belts. Graywacke and argillite occur as "couplets", often as parts of graded beds with graywacke usually dominating (McGlynn and Henderson, 1970). Individual couplets range in thickness from about 1 cm to over 1 m. Although the contact between couplets may be sharp, the change from graywacke to argillite within a couplet is usually gradational involving an intermediate siltstone. Although it is difficult because of folding to estimate accurate thicknesses of graywacke-argillite sections, minimum thicknesses of the order of 5 km are reported in some localities (Henderson, 1972; Bayley et al., 1973). Individual beds appear to be broadly lensoid in shape and they often can be traced along strike for

TABLE 4-1

Bedding characteristics of two measured graywacke-argillite sections in the Vermilion greenstone belt, Minnesota (from Ojakangas, 1972)

	1	2
Thickness of section (m)	20.8	55.4
Number of beds in section	426	252
Graywacke beds:		
percent of total thickness	59	99.5
average bed thickness (cm)	6	24
range of bed thickness (cm)	0.5—120	1—124
number and percent of total	201, 47%	228, 90%
number and percent graded	128, 64%	143, 62%
percent with mud-chips	15	tr
percent with load casts or flames at base	tr	11
percent with convolutions or cross-lamination	0	6
percent composite beds	12	5
Siltstone beds:		
percent of total thickness	24	0.5
average bed thickness (cm)	4	1.7
range of bed thickness (cm)	0.5—35	1—4
number and percent of total	100, 24%	15, 7%
number and percent graded	9, 9%	0
percent with cross-laminations	4	0
Slate beds:		
percent of total thickness	22	0
average bed thickness (cm)	3.7	0
range of bed thickness (cm)	0.5—23	0
number and percent of total	125, 29%	0

tens of meters. The graywacke-argillite association may grade laterally or vertically into pyroclastic volcanic or conglomeratic horizons. An example of the bedding variations in two typical graywacke-argillite sections from the Vermilion greenstone in Minnesotá are given in Table 4-1. The sections reveal that graywacke is the dominant rock type and that nearly two-thirds of the beds are graded. The abundance of other primary structures varies considerably between the sections.

Graywacke and argillite range from brown to tan in color although argillite horizons may be black if carbonaceous matter is present. Generally,

Fig. 4-1. Photomicrographs of coarse-grained Archean graywacke (from Henderson, 1972). Upper: crossed polarizers; lower: plane light. Bar length 1 mm.

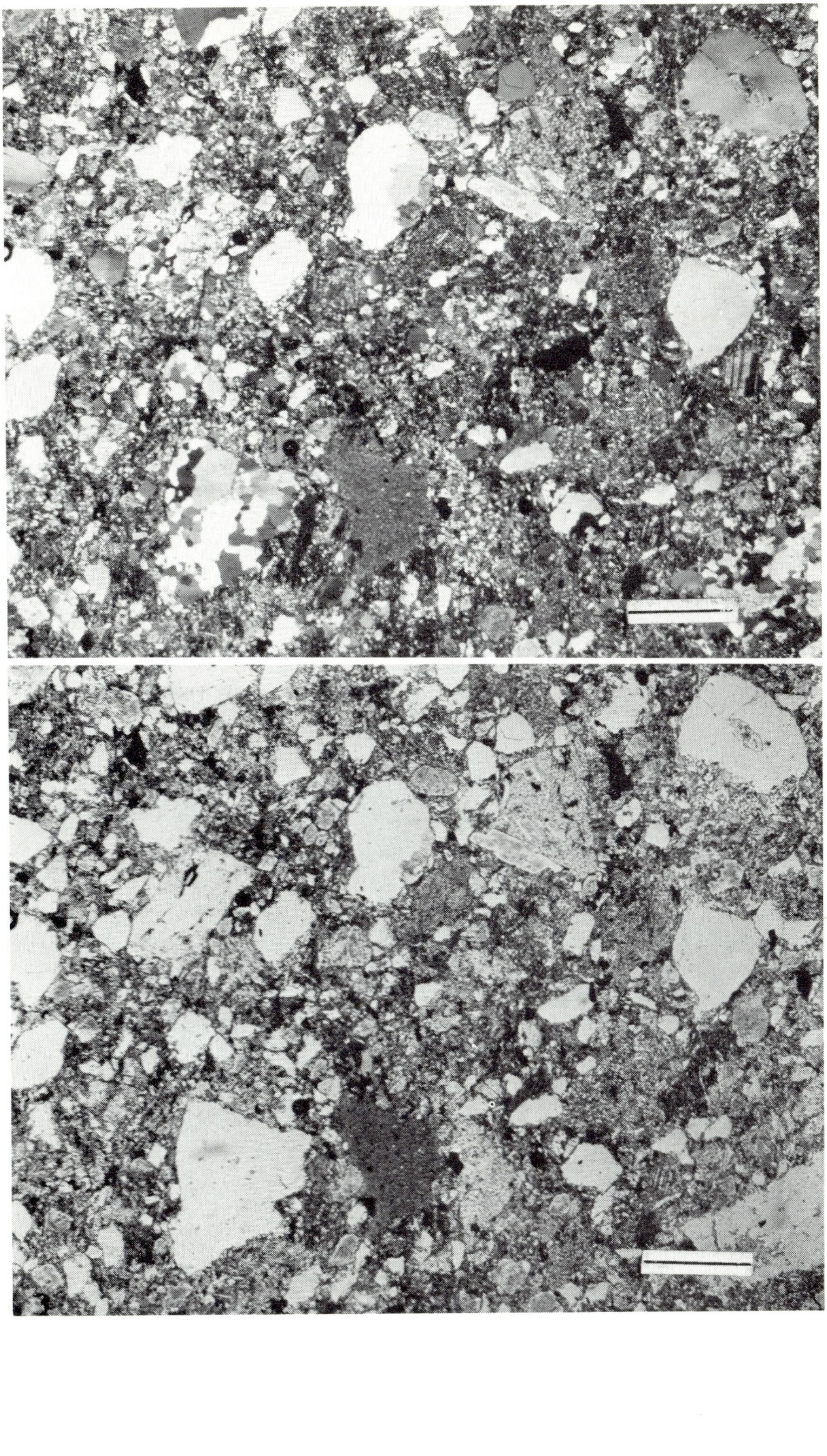

TABLE 4-2

Average modal analyses of Archean graywackes

	1 $n = 18$	2 $n = 8$	3 $n = 7$	4 $n = 9$	5 $n = 10$	6 $n = 8$	7 $n = 19$
Quartz			9			50	24
monocrystalline	24	30		16	3		
polycrystalline	15	5		9	3		
Plagioclase	2	4	10	12	25	8	9
K-feldspar	6	15	2	—	<1	2	<1
Carbonate	8	1	1	0.1	3		tr
Other[1]	4	7	3	1	3		8
Rock fragments	3	5					
felsic volcanic			20	8	22		12
mafic-intermediate volcanic			30	4	2		tr
plutonic			—	1			tr
sedimentary			5	9			5
Matrix[2]	38	33	20	40	39	40	42

[1] Principally amphibole, epidote, mica, chlorite, opaques.
[2] $\lesssim$ 0.03-mm fraction.

n = number of samples.
1 = Sheba Formation South Africa (Condie et al., 1970); 2 = Belvue Road Formation South Africa (Condie et al., 1970); 3 = Lake Timiskaming area, Ontario (Boutcher et al., 1966); 4 = Burwash Formation N.W.T. (Henderson, 1972); 5 = Vermilion district, Minnesota (Ojakangas, 1972); 6 = North Spirit Lake area, Ontario (Donaldson and Jackson, 1965); 7 = South Pass greenstone belt, Wyoming (Condie, 1976a).

Fig. 4-2. Graded graywacke bed with convoluted top (from Ojakangas, 1972).

TABLE 4-3

Summary of components in Archean graywackes

Monocrystalline quartz:	angular to subrounded grains; sharp to undulatory extinction; some crystal faces present
Polycrystalline quartz:	undulatory extinction; grains with irregular shapes and sizes; crystals range from very fine (chert) to coarse; some cataclastic textures; probably represents dominantly plutonic and chert sources
Plagioclase:	twinned to untwinned; commonly altered; some zoned crystals; composition albite or oligoclase
K-feldspar:	altered; commonly perthitic; microcline twinning common; rare in most graywackes
Volcanic rock fragments:	fine-grained, equigranular; some contain feldspar or amphibole phenocrysts; difficult to distinguish felsic fragments from chert; mafic and intermediate types merge with matrix
Sedimentary rock fragments:	chiefly fine-grained argillite; dark colored; often flattened and squeezed around other grains
Plutonic rock fragments:	coarse, composite grains of quartz and feldspar
Amphibole:	hornblende or actinolite; blue-green pleochroism; occurs as phenocrysts in rock fragments, as metamorphic mineral in the matrices, and possibly as detrital grains
Epidote:	fine, irregular masses in matrices; chiefly metamorphic in origin
Carbonate:	irregular masses in matrix and euhedral crystals; secondary origin
Opaques:	irregular to disseminated iron oxides; euhedral to irregular-shaped pyrite
Chlorite-mica:	fine-grained, metamorphic matrix constituents
Accessory minerals:	sphene, zircon, apatite, epidote, magnetite, chromite, pyrite, andalusite, cordierite, garnet, sillimanite

graywackes are chemically, mineralogically, and texturally immature. They are poorly sorted (Fig. 4-1) and contain grains ranging in size from < 0.1 to > 3 mm. Most larger grains are subangular to subrounded in shape. The general high degree of crystallinity of graywacke matrices ($\lesssim 0.02$ mm) suggests a diagenetic or metamorphic origin (Condie, 1967a; Reimer, 1975a).

As indicated by modal analyses summarized in Table 4-2, quartz and feldspars are the most abundant clasts in most graywackes (Condie, 1967a;

Fig. 4-3. Series of graded graywacke beds in the Burwash Formation near Yellowknife, Canada (from Henderson, 1972). The degree of grading is reflected by the color change from light (coarse) to dark (fine).

McCall et al., 1970; Henderson, 1972; Ojakangas, 1972). A general description of the major components in Archean graywackes is given in Table 4-3. Rock fragments comprise variable amounts with volcanic fragments usually dominating. Matrix is composed of fine-grained intergrowths of feldspars, quartz, chlorite, mica, and sometimes carbonate. All gradations exist between volcanic and argillite rock fragments and matrix suggesting a diagenetic origin for much of the matrix as suggested by Cummins (1962).

Rock fragments provide important information about provenance. They range up to > 50% in abundance and may have been even more abundant in some graywackes where they merge with the matrix rendering identification difficult. Generally, felsic volcanic (± hypabyssal) fragments are most abundant with intermediate volcanic and/or graywacke-siltstone fragments of secondary importance. The abundance of granitic fragments is variable but usually they are absent or only of minor importance.

Primary structures

Many primary textures and structures are preserved in Archean graywackes (Dunbar and McCall, 1971; Glikson, 1971a; Ojakangas, 1972; Walker and Pettijohn, 1971; Pettijohn, 1972). The most common is graded-bedding (Fig. 4-2). When considered together with associated bedding features (convolutions, flame structures, small-scale cross-bedding, etc.), a turbidity current origin is suggested for the grading. Graded and non-graded units are typically interbedded with color changes often monitoring the grading as illustrated in Fig. 4-3. Bouma (1961) describes five units within a complete turbidite consisting of: (1) a basal graded unit; (2) a lower parallel laminated unit; (3) a ripple-laminated unit; (4) an upper parallel laminated unit; capped with (5) a pelitic unit. Rarely is this complete cycle observed in Archean turbidites. Usually some of the upper units are missing (Fig. 4-4A) although in some beds, lower units may not be present. Reverse graded-bedding can be produced in a turbidite by metamorphic recrystallization with large metamorphic minerals such as biotite or andalusite developing in fine-grained argillaceous tops of graded units.

Cross-bedding is generally limited to the ripple-laminated unit of a turbidite generally occurring on scales of a few centimeters (Fig. 4-4A). Festoon cross-bedding with channels up to 1 m deep, however, may occur in non-graded graywacke beds (Donaldson and Jackson, 1965). Mud chips occur in some graywacke beds (Fig. 4-4B) and appear to have formed by the erosion of underlying muddy beds by turbidity currents. Individual chips range from < 1 cm to about 30 cm long and up to 10 cm thick. Scour channels are present, rarely ranging up to 1 m deep and 3 m across (Dunbar and McCall, 1971). Such channels are thought to have formed by high-velocity turbidity currents. Features of soft-rock deformation are also common in graywacke-argillite successions of which load casts are most common. Flame structures are found locally in some graded units (Fig. 4-4C). Convolute laminations (Fig. 4-2), slump structures, and dewatering structures have also been reported from some Archean graywacke-argillite sequences (McCall et al., 1970; Dunbar and McCall, 1971; Henderson, 1972).

The origin of turbidity currents is generally ascribed to earthquake-activated submarine slumping (Kuenen and Migliorini, 1950). It is possible that some graywacke-argillite couplets represent individual volcanic eruptions which upon entering the sea became turbidity currents. Observed and estimated sedimentation rates of modern turbidites (Hand and Emery, 1964)

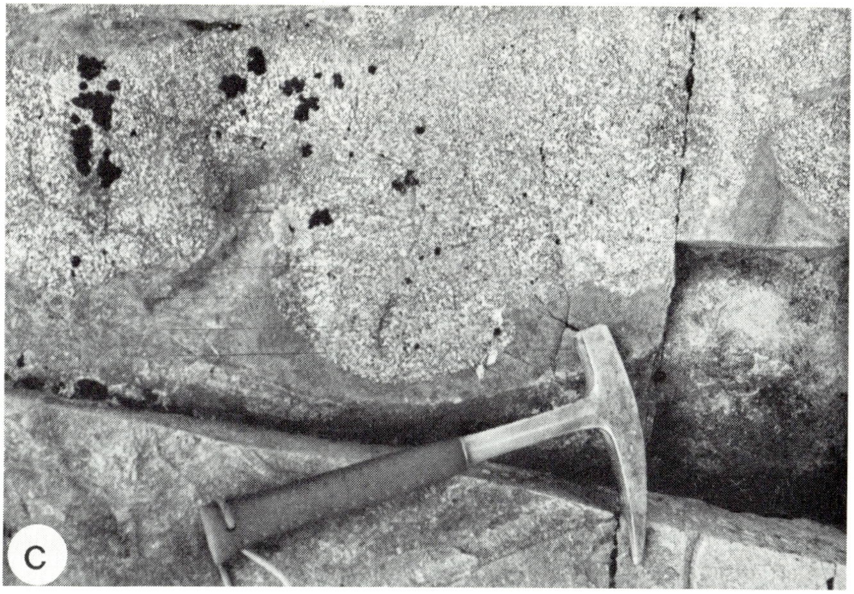

Fig. 4-4. A. Two turbidites showing ripple marks (from Henderson, 1972). Grain size grades upward from very fine sand to silt in each bed. Scale in tenths and hundredths of feet. B. Graywacke bed with mudchips oriented parallel to cleavage (from Ojakangas, 1972). C. Flame structures at the base of a graded graywacke turbidite (from Ojakangas, 1972).

suggest that each argillite layer may take hundreds or thousands of years to accumulate. Thus frequent earthquake and volcanic activity may account for the common absence of the upper pelitic layers in Archean turbidites. Walker (1967) has proposed a method using detailed structures of turbidites to identify distal and proximal (near-shore) facies. Such methods have been successful in some Archean successions in determining source directions and estimating basin size. Campbell (1971), for instance, has shown from studies of Archean graywackes in eastern Manitoba that a change from proximal to distal facies occurs over a distance of 8—10 km. Other studies indicate the presence of turbidites with "distal" and "proximal" features interbedded with each other (Ojakangas, 1972; Henderson, 1972).

Composition

Average major element compositions of Archean graywackes and argillites (from graywacke-argillite couplets) are compared to other compositions in Tables 4-4 and 4-5. The graywackes are similar in composition to Phanerozoic graywackes (column 5), to high-Ca granitic rocks (columns 6 and 7) (especially granodiorite), and to an estimated average composition of the Precambrian continental crust (column 8). They differ from most Phanero-

TABLE 4-4

Average compositions (wt.%) of Archean graywackes

	1	2	3	4	5	6	7	8
SiO_2	63.7	66.2	63.3	64.4	69.2	66.9	70.5	65.2
TiO_2	0.57	0.52	0.56	0.62	0.53	0.57	0.3	0.57
Al_2O_3	14.9	10.2	13.3	15.5	13.7	15.7	14.6	15.8
Fe_2O_3	1.01	1.63	1.0	1.05	1.14	1.33	0.77	1.2
FeO	4.67	5.38	4.9	4.94	3.05	2.59	1.50	3.4
MgO	2.99	4.50	3.7	3.12	1.6	1.57	1.44	2.2
CaO	2.63	1.97	3.4	2.22	1.8	3.56	3.55	3.3
Na_2O	3.14	1.80	2.9	3.74	3.1	3.84	4.45	3.7
K_2O	2.30	1.58	2.1	2.44	2.0	3.07	1.32	3.23
H_2O	2.17	2.76	2.0		2.4	0.65	0.75	0.8
CO_2	1.49	2.59	1.0		0.3			0.2
P_2O_5	0.14	0.08	0.15		0.12	0.21	0.12	0.17
MnO	0.11	0.10	0.1		0.10	0.07	0.05	0.08
Na_2O/K_2O	1.4	1.1	1.4	1.5	1.6	1.3	3.4	1.1
Al_2O_3/Na_2O	4.8	5.7	4.6	4.2	4.4	4.1	3.3	4.3
FeO/Fe_2O_3	4.6	3.3	4.9	4.7	2.7	2.0	2.0	2.8

1 = average of 20 Archean graywackes (Henderson, 1972); 2 = average of 17 Archean graywackes from the Sheba Formation, SouthAfrica (Condie et al., 1970); 3 = composite Archean graywacke (Condie, 1976c); 4 = average of 23 Archean graywackes, South Pass greenstone belt, Wyoming (Condie, 1967a); 5 = average Phanerozoic graywacke (Condie et al., 1970); 6 = average granodiorite (Nockolds, 1954); 7 = average Archean tonalite (Hunter, 1973, and other sources); 8 = average Precambrian continental crust (Eade and Fahrig, 1971).

zoic graywackes only in having greater amounts of Fe, Mg, and Ca (and of transition trace metals) (Condie, 1976c) and in their larger FeO/Fe_2O_3 ratios. The Al_2O_3/Na_2O chemical maturity index, originally proposed by Pettijohn (1957), ranges from about 4 to 6. Such values are generally interpreted to reflect composition of source materials and diagenetic processes (Na mobilization) rather than degree of weathering and erosion (Condie et al., 1970). The data in Table 4-4 suggest that, with exception of source areas of Archean graywackes being somewhat more mafic and less oxidized than source areas of Phanerozoic graywackes, graywacke source areas have not significantly changed in composition with time. If the source-area composition of greywacke is equated with average Precambrian crust (column 8, Table 4-4), then the average composition of continental crust has not changed with time (Condie, 1967b).

On the whole, Archean argillites are also similar in composition to their modern counterparts (Table 4-5). With exception of the Sheba Formation argillites which are anomalous, Archean argillites differ from Phanerozoic

TABLE 4-5

Average compositions (wt.%) of Archean argillites from graywacke-argillite couplets

	1	2	3	4
SiO_2	57.8	59.2	60.6	63.0
TiO_2	0.70	0.69	0.67	0.8
Al_2O_3	18.4	20.2	12.6	18.2
Fe_2O_3	1.67	1.15	2.35	1.3
FeO	6.21	4.85	8.24	4.5
MgO	3.93	3.34	4.71	2.5
CaO	1.89	1.38	0.68	1.0
Na_2O	2.19	2.67	0.84	1.3
K_2O	3.26	2.49	2.28	3.5
H_2O	3.11	3.60	4.0	4.0
CO_2	0.17	0.05	2.6	0.2
P_2O_5	0.19	0.15	0.10	0.2
MnO	0.09	0.06	0.08	0.1
Na_2O/K_2O	0.67	1.1	0.37	0.37
Al_2O_3/Na_2O	8.4	7.6	15	14
FeO/Fe_2O_3	3.7	4.2	3.5	3.5

1 = average of 20 Archean slates (Henderson, 1972); 2 = average Archean slate from graywacke-argillite couplet (Pettijohn, 1972); 3 = average pelite from the Sheba Formation, South Africa (Reimer, 1975a); 4 = average Phanerozoic argillite from greywacke-argillite couplet (Schwab, 1971, and other sources).

argillites chiefly by their higher contents of Na_2O, CaO, and MgO and lower K_2O content. The higher Na_2O in the Archean samples appears to reflect a greater content of sodic plagioclase than is found in most younger argillites.

Chemical analyses of many graywackes from the South Pass greenstone belt in Wyoming indicate significant interbed compositional variability (Condie, 1967a). With exception of SiO_2 and Al_2O_3, all oxides have relative standard deviations from the mean of $\gtrsim$ 10%. Such large interbed variations may be due to one or a combination of the following mechanisms:

(1) Slumping and turbidity current generation at different sites on a submarine slope. The difference in turbidity current composition would result from original differences in the sediments deposited on the slope. Such original differences may result from the segregation of minerals as a function of transport distance from the shoreline due to differing grain sizes and settling velocities.

(2) Turbidity current generation on different sides of a partially enclosed basin (bay or lagoon). The composition of the sediment arriving at the basin's edge would vary from one point to another along the shoreline depending on the composition of the immediate source area.

Conglomerate

Conglomerates are minor but important sediments in greenstone belts in that they provide direct evidence of sediment provenance. They occur as broadly lensoid units ranging from < 1 m to over 1 km thick and have been traced along strike in some sections for over 15 km (McGlynn and Henderson, 1970; Naqvi et al., 1978b). The Jones Creek Conglomerate in Western Australia has been traced along strike for over 90 km (Durney, 1972; Marston, 1978). Conglomerates occur throughout greenstone successions and are not always found at local or major unconformities (Pettijohn, 1972). Some grade laterally into graywackes and others may taper out between volcanic units. Boulders in conglomerates range up 1 m across although generally averaging between 5 and 10 cm. Most conglomerates are poorly sorted, contain clasts ranging from subangular to rounded, and exhibit a wide range in compositions both in matrix and in clast lithologies (Boutcher et al., 1966; McCall et al., 1970). Some fragments of reworked argillite are squeezed into irregular shapes. Most Archean conglomerates are polymictic although some oligomictic varieties have been described (Nath et al., 1976). Matrices can range from quartzite to graywacke or arkose in composition. Some conglomerates are crudely graded, some exhibit scour channels, and a few show imbricate structure.

Archean conglomerates can be classified into two broad groups (Naqvi et al., 1978b): pyroclastic (discussed in Chapter 3) and sedimentary. The sedimentary group can be further subdivided into contact (pebbles touching) or disrupted framework types. Disrupted framework types are generally

TABLE 4-6

Abundances of clasts in Archean conglomerates from the Superior Province (in percent)

	Knife Lake Group		Lake Timiskaming area	North Spirit Lake area
	2A	4		
Felsic volcanic clasts	24	4	34	39
Mafic-Intermediate volcanic clasts	36	58	26	25
Amphibolite, gabbro	14	9	5	tr
Quartz-porphyry, aplite	2	1	14	5
Chert (including quartz and quartzite)	2	2	13	28
Plutonic rocks	16	18	3	3
Graywacke-argillite	6	8	5	3

References: Knife Lake Group, McLimans (1972); Lake Timiskaming area, Boutcher et al. (1966); North Spirit Lake area, Donaldson and Jackson (1965).

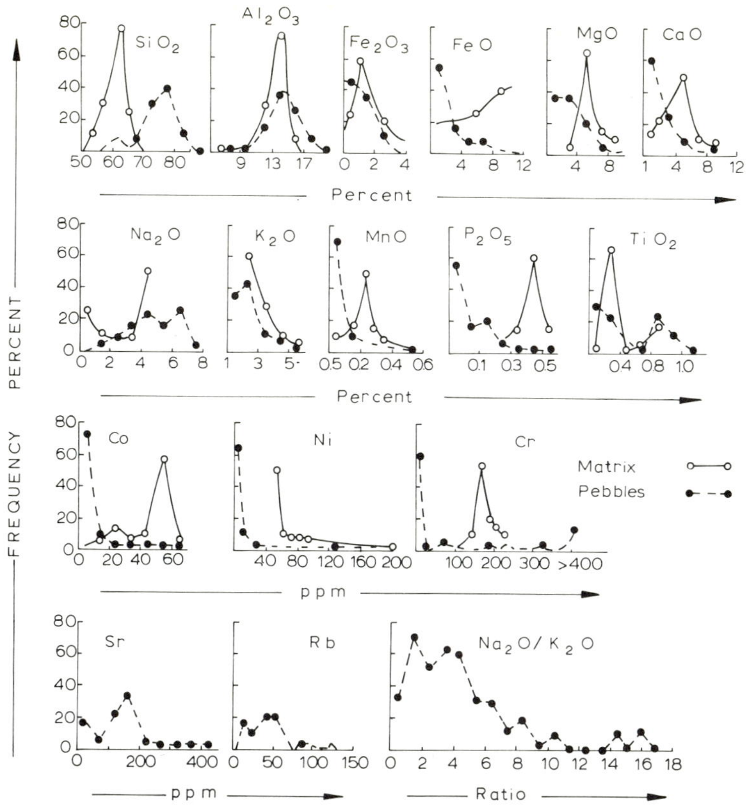

Fig. 4-5. Frequency distribution of compositional constituents in pebbles and coexisting matrix of Archean conglomerates from India (from Naqvi et al., 1978b).

associated with graywacke turbidites and are thought to have formed during subaqueous slumping (Walker, 1978). Most contact types, on the other hand, are thought to represent alluvial-fan deposits formed near rapidly uplifted source areas; in part, they may be subaerial (Gordanier, 1976).

Pebble lithologies in Archean conglomerates, although varied, are often dominated by felsic volcanic fragments (McCall et al., 1970; Boutcher et al., 1966; Goodwin, 1962). Measured modes of clast lithologies in Archean conglomerates are summarized in Table 4-6. The results show considerable diversity with volcanic clasts exceeding plutonic clasts. Locally, however, tonalite and quartzite are the major clasts as in some of the Indian greenstone belts (Naqvi et al., 1978b).

Recent geochemical studies of conglomerates from Archean terranes in India (Naqvi et al., 1978b) show that mafic components in the source (Fe, Mg, Ca, Mn, Ti, Co, Ni, Cr) are reflected in conglomerate matrices (Fig 4-5) and not by the dominant trondhjemite-tonalite pebbles. This observation was first described by Naqvi and Hussain (1972). Matrix compositions in Fig. 4-5 are often intermediate between granitic and mafic pebbles. The fact that the matrix is more mafic than the dominant pebbles probably reflects the ease with which mafic and ultramafic source materials break down during weathering. The high Na_2O content of the matrices appears to reflect the presence of fine-grained Na-rich plagioclase derived from trondhjemite-tonalite sources.

Quartzite and arkose

Quartzite and arkose are not common sediments in most greenstone belts. The most abundant quartzites are reported in Indian greenstone belts (Nath et al., 1976), in the Moodies Group in the Barberton area (Anhaeusser et al., 1968), in the Prince Albert Group in northern Canada (Schau, 1977), and in some greenstone belts in Sierra Leone (Rollinson, 1978). Quartz-rich graywackes (subgraywackes) and quartzite pebbles have also been described from the Spirit Lake area in northwestern Ontario (Donaldson and Jackson, 1965; Donaldson and Ojakangas, 1977). Arkoses are described from the Moodies Group in the Barberton area (Anhaeusser et al., 1968), greenstone belts in northeastern Botswana (Key et al., 1976), and from the Minnitaki Basin in northwestern Ontario (Walker and Pettijohn, 1971).

Fig. 4-6. Cross-bedded Archean quartzite from the Prince Albert Group, Canada (from Schau, 1977).

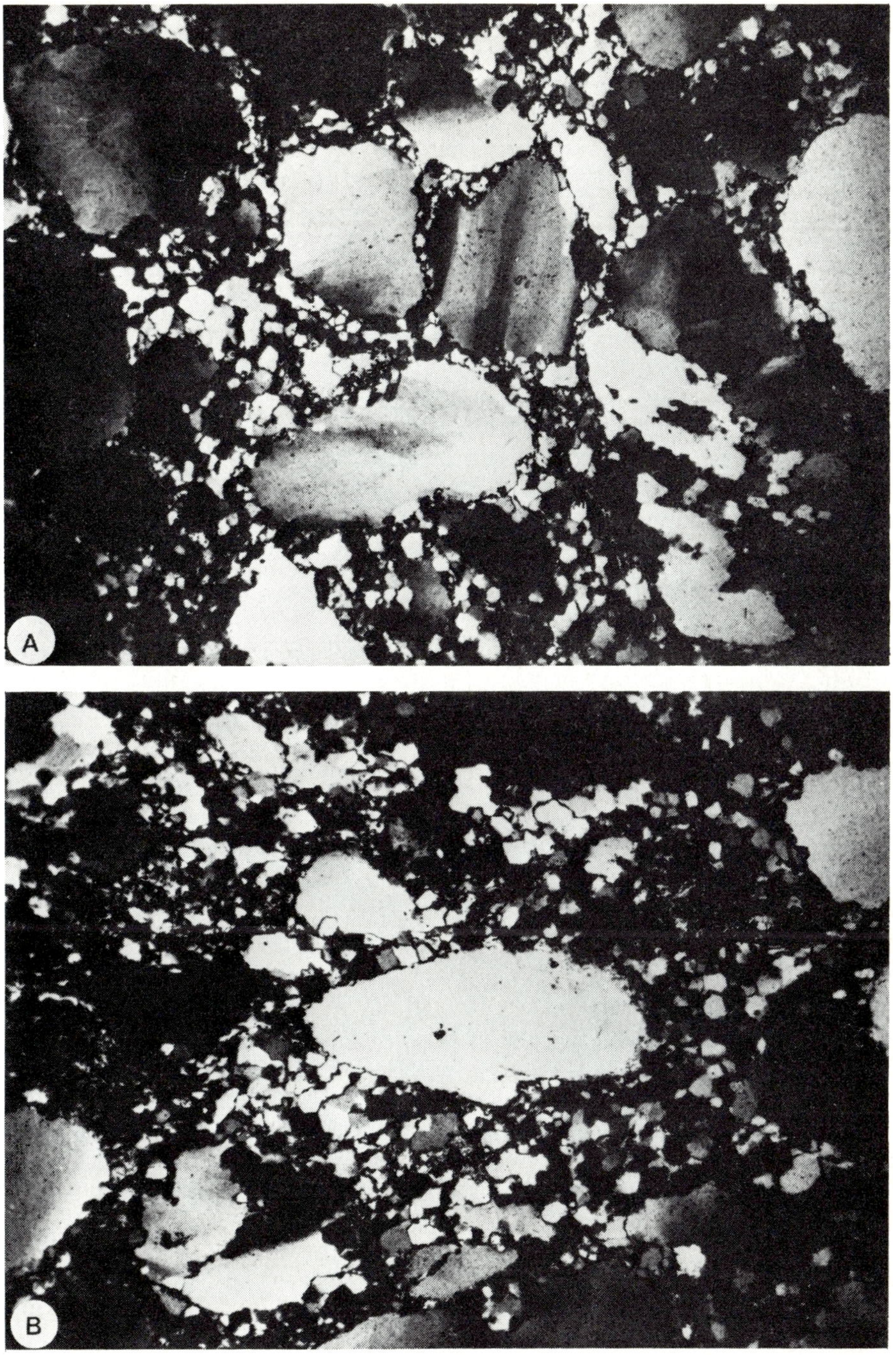

Fig. 4-7. Photomicrographs of quartz grains in two Archean quartzite pebbles (from Donaldson and Ojakangas, 1977). Both fields are 2.5 mm wide.

Several major quartzite, feldspathic quartzite, and arkose units occur in the Moodies Group (Anhaeusser, 1974). The thickest, in the Joe's Luck Formation, attains 300 m. Most quartzite beds are massive, thick-bedded, and range from white to brown in color. Local conglomerate and shale horizons occur in some units and cross-bedding and graded-bedding are locally abundant. Cross-bedding is common in quartzites of the Prince Albert Group in Canada (Fig. 4-6) where individual cross-beds up to 30 m long and 10 m high are reported (Schau, 1977). Scour channels and current markings are reported from some quartzites in the Moodies Group (Anhaeusser, 1974). Mineralogically, quartzites range from almost entirely quartz to mixtures of quartz, feldspar, and mica. Cr-bearing muscovite is common in some quartzites (Naqvi and Hussain, 1972) and may have developed during metamorphism from detrital chromite. Common accessory phases in quartzites are magnetite, zircon, tourmaline, apatite, and rutile. Quartz grains are usually welded together and sometimes original grain shapes are preserved. Sorting is variable in both quartzites and arkoses and sand grains may range from well-rounded to subangular, although not in the same bed. Donaldson and Ojakangas (1977) describe quartzite pebbles from an Archean conglomerate in the Spirit Lake area of Ontario with a bimodal texture of well-rounded quartz grains set in a fine quartz mosaic (Fig. 4-7). Although many quartzites appear to be of clastic origin as evidenced by primary textures and structures, some are devoid of such features and may represent recrystallized chert.

Shale

Thick successions of shale and related rocks (argillite, mudstone, phyllite, slate) are rare in most Archean greenstone belts. Some shale horizons occur as distinct members in the Moodies Group in South Africa (Anhaeusser, 1974). The most extensive sections of fine-grained Archean clastic rocks occur in the Coolgardie-Kurrawang sequence in western Australia (Fig. 2-2a) (McCall, 1969; Glikson, 1971a). Pelitic rocks in this section range from brown to black in color, the black varieties being carbonaceous. Fissility is developed in varying degrees. Fine-scale cross-bedding, grading, and structures developed from soft-sediment deformation are common. Locally, pelitic rocks are interlayered with subgraywackes. Mineralogically, Archean shales are composed chiefly of micas, chlorite, quartz, and feldspars, with trace amounts of such minerals as magnetite, graphite, pyrite (usually diagenetic), and hematite. Magnetite-rich shales have been reported in part of the Moodies Group.

Compared to most Phanerozoic shales, shales from the Coolgardie-Kurrawang succession are high in SiO_2 and Na_2O, and low in TiO_2, FeO, MgO, and MnO.

PROVENANCE

Introduction

The major constraint on provenance of Archean clastic sediments is provided by pebbles in conglomerates and rock fragments in graywackes. Many studies of such clasts reveal the importance of felsic volcanics in the source areas of these clastic sediments (Boutcher et al., 1966; Ayres, 1969; Condie et al., 1970; Glikson, 1971a; Ojakangas, 1972; Henderson, 1972, 1975a). As previously discussed, major element concentrations in Archean graywackes indicate that the average bulk composition of their source terranes ranges between tonalite and granodiorite (Pettijohn, 1957; Condie, 1967b). Together with rock fragment distributions, Condie et al. (1970) have shown that the K_2O-Na_2O distributions monitor graywacke provenance. This is illustrated for four groups of Archean graywacke in Fig. 4-8. The increasing trend in both K_2O and Na_2O in going from the Sheba Formation to the Wyoming graywackes reflects increasing amounts of granitic (K_2O-rich) detritus at the expense of mafic and intermediate volcanic detritus. The low-K_2O values in the Kalgoorlie graywackes reflects a dominant tonalitic (or dacitic) source as also indicated by the abundance of sodic porphyry rock fragments in these rocks (Glikson, 1971a).

Although it is possible, in part, to reconstruct source area compositions from the study of detrital clasts in graywackes, both in the field and in thin section, it is difficult to study the fine-grained matrices. Such matrices may compose up to 50% of graywacke samples and hence may be of major importance in deducing source area composition. Trace element distributions in graywackes can enhance our understanding of the provenance of such fine-grained matrices (Condie, 1976c). An example of provenance studies for

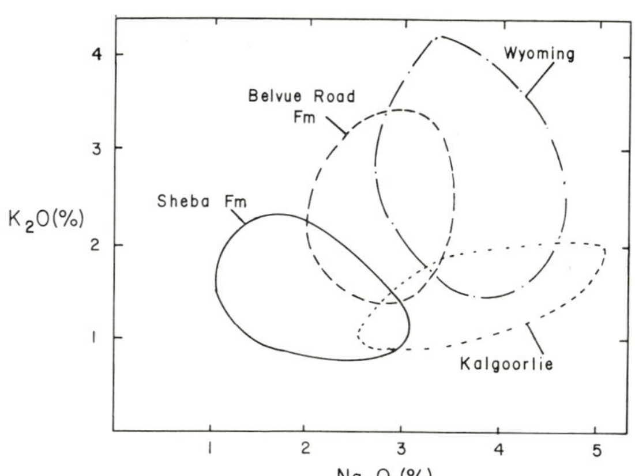

Fig. 4-8. K_2O-Na_2O distribution in Archean graywackes (after Condie et al., 1970; Glikson, 1971a)

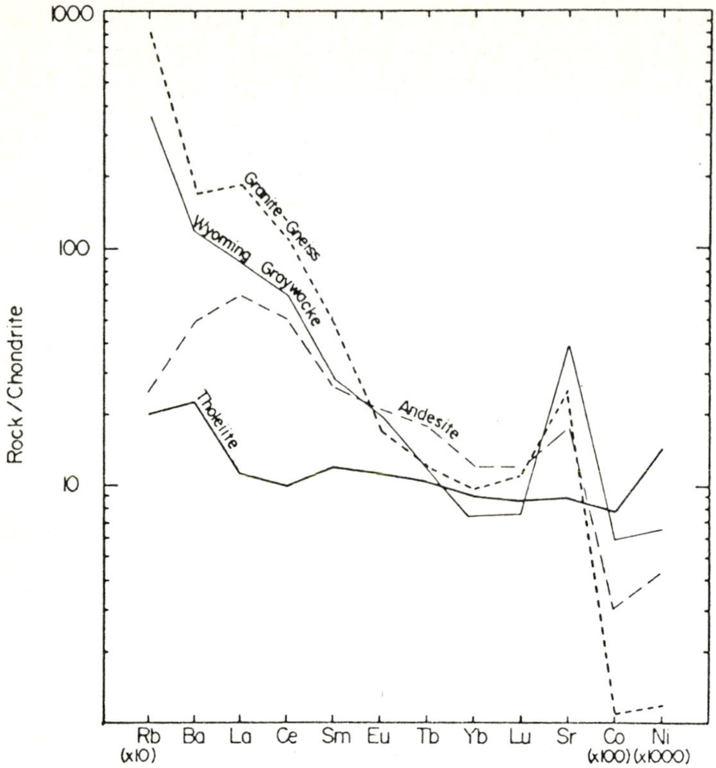

Fig. 4-9. Chrondrite-normalized trace element distributions in Wyoming Archean graywacke and in average Archean granite-gneiss, andesite, and tholeiite (from Condie, 1976c).

average Archean graywacke from the South Pass greenstone belt is summarized in Fig. 4-9. Shown in this figure are chondrite-normalized trace element concentrations for Wyoming graywacke and three possible source terranes. The trace element data indicate that with few exceptions, the graywacke can be derived by the weathering and erosion of an average Archean granitic-gneiss terrane (± felsic volcanics of similar composition). As indicated in the figure, only minor intermediate to mafic volcanic input is allowed. Such a conclusion is in harmony with the lithologies of clasts found in the graywackes.

The high Ni and Co contents of the Wyoming graywackes cannot readily be explained by reworking of granitic gneiss and volcanic rocks. It is noteworthy in this respect, that Archean graywackes are enriched in transition metals and Mg compared to Phanerozoic graywackes (Condie et al., 1970; Condie, 1976c). Although preferential absorption of transition metals by Archean clays may have contributed to this difference, the relative abundance of detrital oxides, sulfides, and mafic silicates in the matrices of Archean graywackes probably account for most of the enrichment. The abundances of these minerals can be explained by a proportionally larger fraction of mafic and ultramafic rocks in the Archean source terranes.

Combined petrographic and geochemical results for many greenstone belts seem to suggest that volcanic sources dominated during most of the greenstone belt evolution with granitic sources becoming important locally or during the late stages of development. In greenstone belts with basal conglomerates, erosion of older gneissic rocks is recorded at the onset of greenstone belt development. It is of interest to review now some specific studies dealing with the provenance of clastic Archean sediments.

Case studies

Knife Lake Group, Minnesota

Detailed studies of graywackes and conglomerates from the Knife Lake Group in the Vermilion greenstone belt in northeastern Minnesota provide data bearing on their provenance (Ojakangas, 1972; McLimans, 1972). As indicated by the clast modes in Table 4-6, pebbles of felsic volcanics dominate in most conglomerates in this area. However, as the Saganaga tonalite is approached on the east, more and more fragments of this body are recognized in the conglomerates and graywackes. The primary source of the Knife Lake sediments appears to have been from erosion of a calc-alkaline volcanic suite (Ojakangas, 1972). Unroofing of the Saganaga tonalite, however, led to the input of tonalite detritus on the east side of the basin. Both the maximum and average size of clasts in Knife Lake conglomerates decrease westward away from the Saganaga tonalite (Fig. 4-10). It appears that transport by turbidity currents and slumping was from the eastern side of the basin along the present strike of the Knife Lake Group.

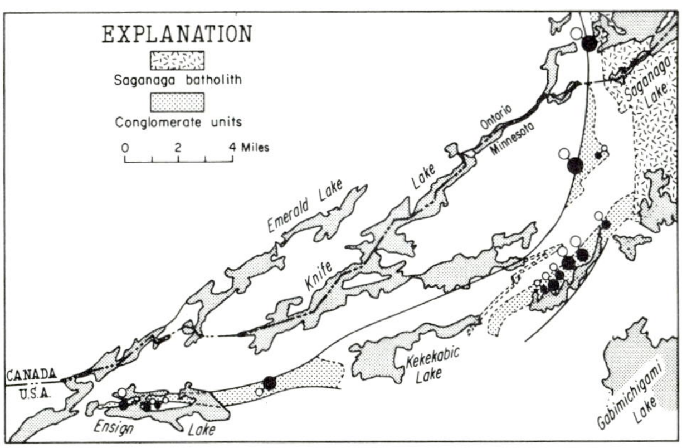

Fig. 4-10. Map showing maximum diameter (solid circles) and average diameter (open circles) of Saganaga tonalite clasts in conglomerates of the Knife Lake Group, northeastern Minnesota (from McLimans, 1972).

Fig Tree Group, South Africa

The Fig Tree Group in the Barberton greenstone belt of South Africa is composed of approximately 2 km of chiefly graywacke-argillite, tuff, and chert. Detailed petrographic and geochemical studies of the graywackes (Condie et al., 1970; Reimer, 1975a) provide some major constraints on source area composition of these rocks. Graywacke trace element distributions within both the Sheba and Belvue Road Formations (see Fig. 2-1) have been described by Condie et al., (1970). Graywackes from the Sheba Formation show a notable depletion in Sr relative to Rb compared to other graywackes and to common igneous rocks. This depletion has been interpreted in terms of Sr-depleted source rocks and diagenetic alteration of plagioclase (Reimer, 1971). A large enrichment in Ni in Sheba graywackes is interpreted to reflect an ultramafic component in the source. With increasing stratigraphic level in the Sheba Formation, volcanic rock fragments, Ti, Na, Zr, and the Na/K ratio decrease and granitic-metamorphic rock fragments, Ca, and Sr increase. A marked increase in K, Ba, and Rb also occurs near the top. Such trends suggest an over-all increase of a granitic-metamorphic component in the source material at the expense of a volcanic component. An increase in K-feldspar over this same interval also records this change in source area composition. The amount of granitic detritus continues to increase upward into the Belvue Road Formation and into the lower part of the overlying Moodies Group. Granitic pebbles compose up to 2.5% of conglomerates in the lower Moodies Group. Higher in the Moodies, however, the feldspar content of the sandstones decreases; such a decrease may record recycling of lower Moodies sediments. The overall progressive increase in sialic detritus in the upper Swaziland Supergroup appears to represent the progressive unroofing of a granitic-metamorphic terrane southeast of the Barberton greenstone belt, which was initially covered by a thick assemblage of volcanic rocks of probable Onverwacht affinities.

Results indicate the early crust that served as a source area for the Fig Tree graywackes in South Africa was composed of a diversity of igneous rock types. As evidenced by volcanic rocks of the Onverwacht Group, great quantities of volcanic rock were extruded at about 3.4 b.y. and intermittent volcanic activity continued into Fig Tree and Moodies times. It is possible that much of the silica now in the form of chert in the Fig Tree and Onverwacht sections was derived from submarine volcanic emanations or desilication of deeper rocks. The fact that granitic detritus is abundant in the Fig Tree Group, yet granitic rocks of pre-Fig Tree age have not been found associated with the underlying Onverwacht Group, indicates that the Onverwacht section or its lateral equivalents were not the sole source of Fig Tree sediments. The granitic-metamorphic source rocks for the Fig Tree sediments may be preserved in the Ancient Gneiss Complex described by Hunter (1970) in central Swaziland. The increasing amount of sialic component in the Fig Tree Group with stratigraphic level indicates that significant parts of the early crust in the area had been engulfed with granitic rocks prior to and perhaps during Fig Tree time. This early period of granite formation may be recorded only indirectly in the sediments derived from its erosion.

Minnitaki Basin, Ontario

Careful field and petrographic studies of Archean clastic sediments in the Minnitaki Basin in northwestern Ontario have proved valuable in reconstructing the provenance of these sediments (Walker and Pettijohn, 1971). Four main sedimentary facies have been recognized in the eastern part of Minnitaki Lake. Clasts in graywackes and conglomerates reflect a dominantly granitic source terrane and those in arkoses, a dominantly volcanic and granitic source. Transport direction in arkosic conglomerates is from east to west as deduced by a decrease in boulder size in this direction. A quartz-porphyry stock now exposed on the east end of the lake appears to represent a major source for the arkosic facies. This investigation shows a change in source-area with time. Uplift along the east side of the basin first provided arkosic sediments from the volcanics and quartz-porphyry stock. Removal of much of this material was followed by exposure of a dominantly tonalitic gneiss terrane which provided source material for the graywacke, slate and conglomeratic facies.

The quartz problem

Although detrital quartzite is not an abundant rock type in Archean greenstone belts, locally it is important. Significant quantities of quartzite occur in Archean rocks of the Kaapvaal Basin, the large cratonic basin in southern Africa which began by 3.0 b.y. and extended well into the Proterozoic (Chapter 1). The problem of where significant volumes of detrital quartz come from has been discussed by Donaldson and Jackson (1965). Possible source rocks for such quartz are summarized in Table 4-7. Rocks in the first category provide an inadequate source because they do not contain free quartz or the free quartz is too fine grained to form quartz sand grains. Derivation of quartz sand grains from phenocrysts in felsic volcanics (or hypabyssal rocks) requires intense weathering or selective concentration of quartz. Although some quartz grains appear to have this source (Donaldson and Ojakangas, 1977), production of large quantities of detrital quartz in this manner seems unlikely. Vein quartz is very minor in Archean terranes and is likely to serve as only a local source for quartz. Evidence for extensive volumes of silicified rocks have not been found in Archean terranes and hence such a source is not favored. Only a few polycrystalline quartz grains found in Archean quartzites represent unmetamorphosed chert as evidenced by their polygonial intergrowths. This leaves two possible major sources for Archean detrital quartz: granitic gneiss terranes and metachert (or recycled

TABLE 4-7

Possible sources of Archean detrital quartz (after Donaldson and Ojakangas, 1977)

Source rocks	Comments
1. Basalt, gabbro, non-porphyritic calc-alkaline volcanics	inadequate source
2. Quartz-phenocryst bearing felsic volcanics	requires extreme weathering and/or selective concentration of quartz
3. Vein quartz	local source only
4. Silicified source rocks	no evidence
5. Unmetamorphosed chert	requires polygonization of fine-grained quartz intergrowths
6. Granitic gneiss terrane	requires prolonged weathering
7. Metachert, recycled quartzite	requires prolonged weathering except as a local source

quartzite). If a granitic source existed, profound chemical weathering must occur to provide feldspar-free quartz sand in one cycle. Metachert and recycled quartzite may supply polycrystalline quartz sand but it is unlikely that a significant quantity of single-crystal quartz grains are produced in this manner. Also, chert and quartzite are minor rocks in Archean supracrustals and hence prolonged weathering again seems necessary. Clearly, the problem of the origin of large volumes of detrital quartz is one of the major problems in Archean sedimentation.

Rare earth elements in Archean sediments

Wildeman and Haskin (1973) and Wildeman and Condie (1973) showed that except for the usual presence of a positive Eu anomaly, Archean sediments have similar REE distributions to Phanerozoic sediments. This is illustrated in Fig. 4-11 for four Archean sediments normalized to a North American shale (NAS) composite. Similar REE patterns have been reported by Nance and Taylor (1977) for some Archean sediments from the Kalgoorlie area in Western Australia. Except for the possible oxidation of Ce, existing data suggest that the REE are not fractionated from each other by metamorphism or sedimentary processes (Wildeman and Haskin, 1973; Green et al., 1969). Hence, the relative enrichment in Eu in Archean sediments must be inherited from source materials. Jakeš and Taylor (1974) and Nance and Taylor (1976) have shown that Phanerozoic graywackes and calc-alkaline volcanic rocks have REE distributions similar to Archean sediments (Fig. 4-12). The fact that Archean graywackes and associated shales and argillites appear to have been derived largely from calc-alkaline

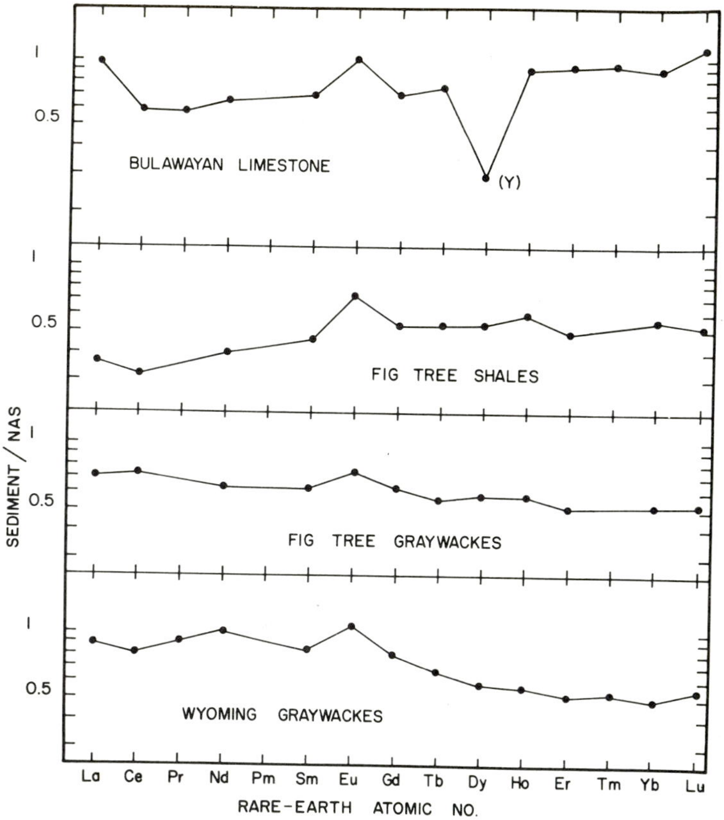

Fig. 4-11. NAS-normalized REE distributions in Archean sediments (from Wildeman and Haskins, 1973). NAS = composite North American shales of Paleozoic age.

volcanic (± plutonic) source areas is consistent with their inheriting calc-alkaline REE patterns. Similar REE patterns, however, can also be produced by mixing felsic and tholeiitic components (i.e., a bimodal suite) during erosion and sedimentation.

Jakeš and Taylor (1974) have proposed that the depletion of Eu in post-Archean sediments (relative to Archean sediments) has developed in response to continental growth. They suggest that partial melting of the lower continental crust produces granodiorite magmas which rise into the upper crust leaving a plagioclase-rich residuum in the lower crust. Plagioclase is known to preferentially retain Eu^{2+} compared to other trivalent REE and hence Eu is preferentially left behind. As this process continues throughout geologic time, the upper crust becomes depleted in Eu and such depletion is passed on to derivative sediments.

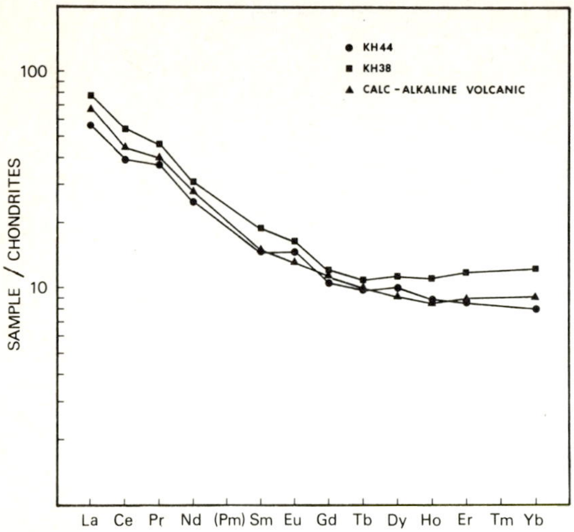

Fig. 4-12. Chondrite-normalized REE patterns for a young calc-alkaline volcanic and two Archean sediments (KH44, KH38) from the Kalgoorlie area, Australia (from Nance and Taylor, 1977).

NON-CLASTIC SEDIMENTS

Chert

Chert is a minor but widespread rock type in Archean greenstone successions. It can occur in all parts of a greenstone section, sometimes associated with volcanic rocks and sometimes with graywacke-argillite. It is also a major part of banded iron formation (see Chapter 7). It is often layered or banded on a scale of centimeters; massive beds are less prevalent. Individual chert units, which can be traced along strike for many kilometers in some greenstone belts (Harrison, 1970), range up to 50 m thick. Brecciated horizons are common in some areas (Goodwin, 1962) and are thought to have formed soon after deposition by wave action or slumping. Cherts often contain interlayers of black, siliceous phyllite and/or carbonate up to 10m thick. The phyllites contain carbonaceous matter and often, sulfides. Cherts terminate volcanic cycles in the Barberton greenstone belt (Fig. 2-4). Oolitic chert horizons have been described from the Swaziland Supergroup (Reimer, 1975b) although these may actually represent silicified accretionary-lapilli tuffs. A typical chert horizon from the Kromberg Formation in the Swaziland Supergroup is shown in Fig. 4-13. The main rock type is dense, black carbonaceous chert with minor interlayers of carbonate and shale. Microstructures of probable organic origin have been described from this chert horizon (Engel et al., 1968) (Chapter 8). Rill and scour marks have also been described and are interpreted to reflect shallow-water deposition (Viljoen and Viljeon, 1969e).

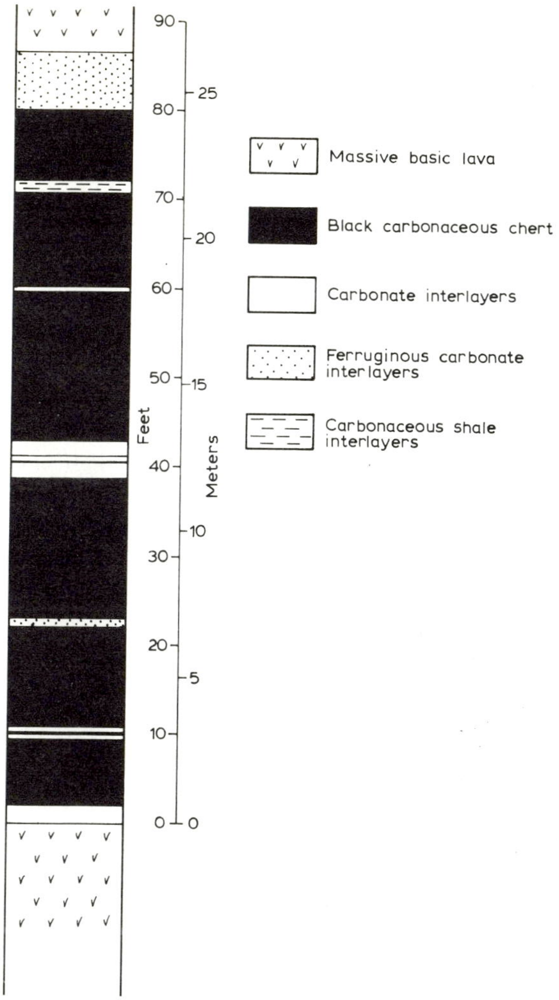

Fig. 4-13. Details of a typical chert horizon from the Kromberg Formation, Barberton greenstone belt, South Africa (from Viljoen and Viljoen, 1969e).

Archean cherts are composed chiefly of a fine-grained polygonized intergrowth of quartz with minor amounts of one or more of the following: iron oxide minerals, chlorite, amphibole, muscovite (sometimes Cr-bearing), carbonate and graphite. Grain size varies considerably due to recrystallization. Pyrite often occurs as cross-cutting cubes which appear to have developed after the crystallization of the silica (Naqvi, 1967). Siderite and calcite are the common carbonate phases present.

Most Archean cherts are thought to represent chemical or biochemical precipitates, or both. Some appear to have formed by the chertification of tuff or graywacke as evidenced by the preservation of relict volcanic or detrital textures. The close association of many cherts with volcanics suggests that the SiO_2 was derived from contemporary volcanic sources (Goodwin, 1962; Taliferro, 1933).

Carbonates

Sedimentary carbonates are rare in Archean greenstone belts (Pettijohn, 1943; Ronov, 1964). Armstrong (1960) has noted five occurrences of carbonates in the Superior and Churchill Provinces: (1) carbonate-quartzite within a dominantly sedimentary succession; (2) carbonate-quartzite within a dominantly volcanic succession; (3) carbonate-chert and/or black shale; (4) carbonate-volcanics; and (5) limey, clastic sediments. Carbonate horizons in each of these associations are minor ranging from 3 to 100 m thick and extending along strike for up to a few kilometers. Such horizons are typically thick-bedded and fine-grained unless metamorphosed to the amphibolite facies where they are recrystallized to marbles. They are usually gray to white or brown in color and composed chiefly of dolomite, or less often, calcite. Some are stromatolitic (see Chapter 8). Carbonate units associated with chert or black shale are often rich in siderite (Goodwin, 1962).

In thin section, Archean carbonates exhibit a crystalline texture and are often lithographic. Although composed chiefly of dolomite or calcite, traces of quartz and mica also occur.

The question of the rarity of Archean carbonates has recently been reviewed by Cameron and Baumann (1972). These authors showed that other Archean sediments (notably shales) could not provide a sink for Archean Ca and hence it must have remained in solution or precipitated elsewhere. Three possible causes for the scarcity of Archean carbonates are considered: (1) the pH of Archean seawater was too low for carbonate deposition; (2) carbonates were deposited on stable shelves during the Archean and later eroded; or (3) carbonates were deposited in deep ocean basins during the Archean and later destroyed by plate tectonic processes.

A higher P_{CO_2} in the Archean atmosphere has been appealed to by some to explain the sparsity of Archean carbonates (Strakhov, 1964; Cloud, 1968). The reasoning is that a high P_{CO_2} in the atmosphere results in more CO_2 being dissolved in seawater thus lowering its pH and allowing Ca^{2+} to accumulate because of its increased solubility. Although the P_{CO_2} in the Precambrian atmosphere probably was higher than at present (Holland, 1965), the CO_2 content is not the only factor controlling seawater pH. Recent evidence suggests that although the CO_2-CO_3^{2-} equilibria may have short-term control on pH, silicate equilibria have long-term control (Pytkowicz, 1967). In seawater held at a constant pH by silicate buffering reactions, the solubility of Ca^{2+} actually *decreases* with increasing P_{CO_2} (Holland, 1965) and thus the CO_2 mechanism does not seem capable of explaining the sparsity of Archean carbonates.

The second explanation was originally suggested by Pettijohn (1943). The only preserved stable-shelf association of any extent in the Archean is the Kaapvaal-Basin succession in southern Africa (Anhaeusser, 1973a). During the Archean, the Pongola and Witwatersrand Supergroups were deposited in this basin and carbonate is only a minor rock type in both Supergroups. Hence, erosion of such Archean stable-shelf associations will not solve the missing Archean carbonate problem. Cameron and Baumann (1972) favor the third explanation. They suggest that because of the near absence of stable cratons in the Archean, carbonate sedimentation was largely confined to deep ocean basins, controlled perhaps by the deposition of planktonic algae which played a role similar to that played by foraminifera (principally *Globigerina*) today. Spheroids of probable organic origin found in Archean cherts may represent remnants of such unicellular planktonic algae (Chapter 8). Carbonates would also accumulate in minor amounts in the tectonically active greenstone belts when local tectonic stability occurred. The large volumes of deep-sea carbonate, however, would be largely destroyed by later plate tectonic processes.

Barite

Archean sedimentary barite formations are described from the Swaziland Supergroup in the Barberton belt (Heinrichs and Reimer, 1977), from the Sargur Schist Complex in India (Viswanatha and Ramakrishnan, 1975), and from the Warrawoona Group in Western Australia (Dunlop et al., 1978). The Australian occurrences are associated with altered volcanic rocks and are interbedded with chert. The chert beds contain primary textures indicative of shallow-water to supratidal sedimentary environments. Primary textures are not preserved in the Indian barites which are associated with fuchsitic quartzites and quartz-mica schists.

Detailed descriptions are available for the barite beds in the Barberton belt (Heinrichs and Reimer, 1977). The barite occurs associated with chert and shale in the lower part of the Fig Tree Group and can be traced along strike for over 1000 m. The barite zone, which consists of two or three barite beds (collectively 100—250 cm thick) ranges up to about 7 m thick. Fine lamination and rare cross-bedding are occasionally preserved in the barite beds, indicating a detrital origin, at least in part. Primary barite grains are mixed with detrital chromite, quartz, zircon, chert, and pyrite. The detrital nature of the barite beds is interpreted by Heinrichs and Reimer (1977) to reflect local reworking of precipitated barite and mixing with other sediments in a near-shore environment. The barium may have been derived from hydrothermal waters associated with contemporary felsic volcanism. Sulfur isotope studies of the Barberton barites are consistent with an origin for the barite by the oxidation of volcanic-derived H_2S or S_2 to SO_4^{2-} by photosynthetic algae and precipitation as $BaSO_4$ (Perry et al., 1971). Alternately, these and the Western Australia deposits have been interpreted as evaporite deposits in which original gypsum of non-volcanic origin was later replaced with Ba^{2+} preserving the gypsum sulfur isotope ratios (Lambert et al., 1978).

SEDIMENTARY ENVIRONMENTS

General features

Sedimentary environments in Archean greenstone belts can be considered both in terms of local and regional environments. The largest proportion of greenstone belt sediments are clastics and appear to have been deposited in tectonically active basins by slumping and turbidity-current activity (Pettijohn, 1972). Local, more stable environments probably existed when chert, carbonate, and quartzite were deposited. Although the depth of the sedimentary basins is unknown, most were deep enough such that turbidites were not disrupted and broken up by wave action. Sedimentological studies in the 3.5-b.y. Warrawoona Group in the Pilbara Province of Western Australia indicate widespread shallow-water deposition and possibly the existence of evaporites in this area (Barley et al., 1979). Stromatolitic carbonates and massive, cross-bedded quartzites, however, were likely deposited on small platform areas near basin margins or around volcanic-plutonic centers. The only record preserved of a widespread platform depository is the Kaapvaal Basin in South Africa. Conglomerates of the contact framework type were also probably deposited near shore lines and, in part, may be terrestrial. A major conglomerate unit in the Favourable Lake greenstone belt in Ontario grades laterally into graywacke-argillite and is interpreted as a subaerial alluvial fan that emptied into a subaqueous turbidite basin (Gordanier, 1976). Extensive thicknesses of shale such as occur in the Kalgoorlie area in Western Australia, may represent extremely distal, turbidite depositories. Cherts and banded iron formation may have been deposited in either deep or shallow water, or both.

It is of interest now to review sedimentary environments in two greenstone belts that have been well studied: the Barberton and Yellowknife belts.

Deposition of the Fig Tree and Moodies Groups

Studies by Reimer (1975a), Anhaeusser (1974), and Eriksson (1977) in the Fig Tree and Moodies Groups in the Barberton greenstone belt in South Africa (Fig. 2-1) have been important in enhancing our understanding of greenstone-sedimentary environments. Reimer (1975a) proposes a model for the Sheba Formation in the lower part of the Fig Tree Group in which sediment is supplied from a land mass on the south (Fig. 4-14) composed of 40%

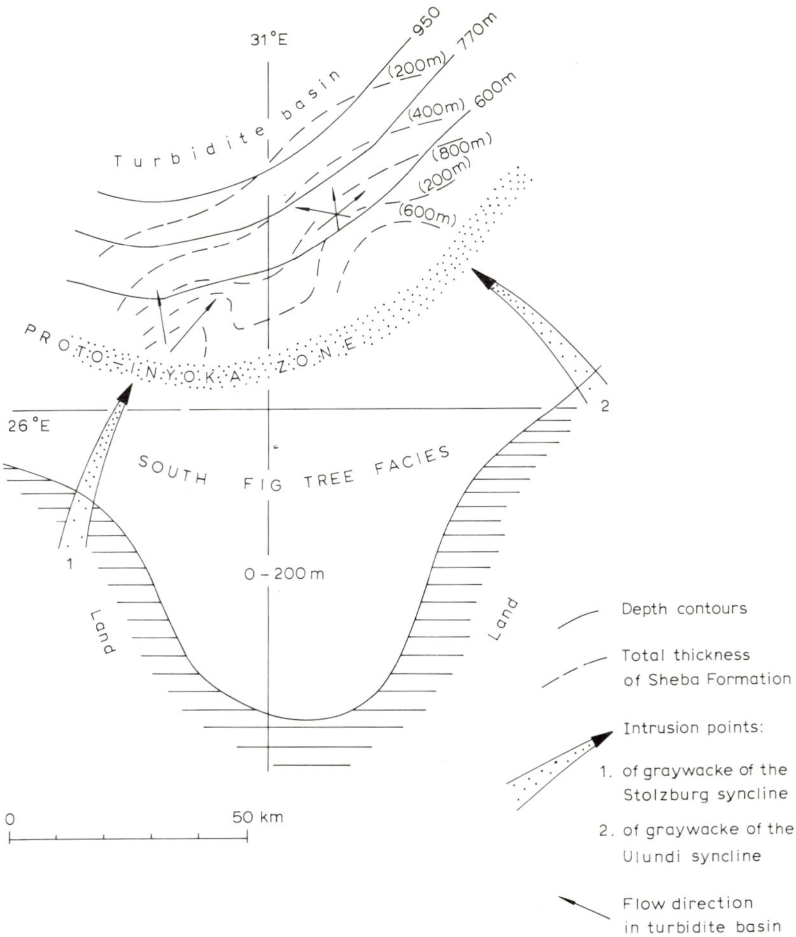

Fig. 4-14. Reconstruction of the sedimentary environment of the Sheba Formation, South Africa (from Reimer, 1975a).

mafic and ultramafic rock, 30% granitic rock, 15% chert, and 15% felsic to intermediate volcanics. This source probably represents an unroofed portion of the Onverwacht Group as previously discussed. Subsidence of the Fig Tree Basin was aided by the underlying thick succession of mafic and ultramafic rocks. Slumping along the land mass margin gave rise to turbidity currents which deposited the graywacke-argillite series in a deep basin on the north.

The overlying Moodies Group consists chiefly of quartzites and shales, in places cross-bedded and ripple-marked. Festoon cross-bedding is important in some units suggesting transportation in braided stream channels (Anhaeusser, 1974). Overall the sedimentary structures preserved suggest shallow-water deposition. The presence of high-K granitic rock fragments

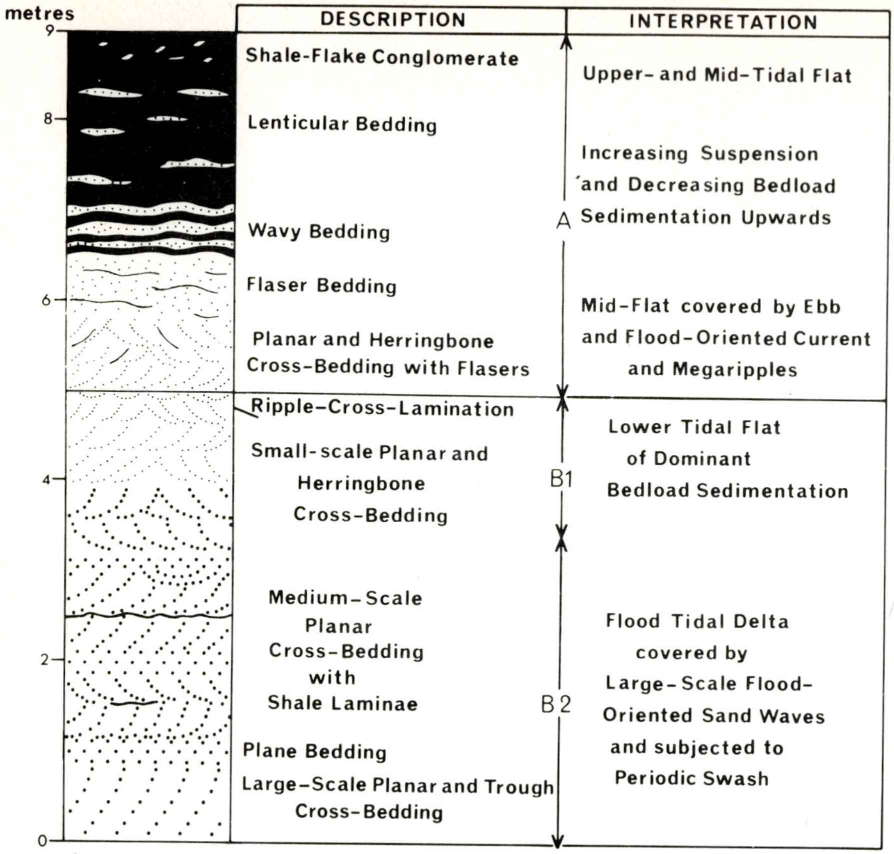
metres
9
8
6
4
2
0
DESCRIPTION
INTERPRETATION
Shale-Flake Conglomerate
Lenticular Bedding
Wavy Bedding
Flaser Bedding
Planar and Herringbone Cross-Bedding with Flasers
Ripple-Cross-Lamination
Small-scale Planar and Herringbone Cross-Bedding
Medium-Scale Planar Cross-Bedding with Shale Laminae
Plane Bedding
Large-Scale Planar and Trough Cross-Bedding
Upper- and Mid-Tidal Flat
Increasing Suspension and Decreasing Bedload Sedimentation Upwards
A
Mid-Flat covered by Ebb and Flood-Oriented Current and Megaripples
Lower Tidal Flat of Dominant Bedload Sedimentation
B1
Flood Tidal Delta covered by Large-Scale Flood-Oriented Sand Waves and subjected to Periodic Swash
B2

A

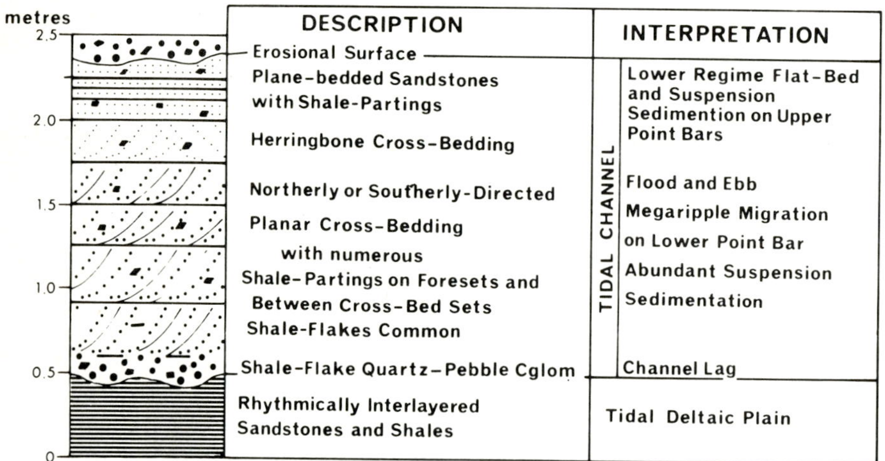
metres
2.5
2.0
1.5
1.0
0.5
0
DESCRIPTION
INTERPRETATION
Erosional Surface
Plane-bedded Sandstones with Shale-Partings
Herringbone Cross-Bedding
Northerly or Southerly-Directed Planar Cross-Bedding with numerous Shale-Partings on Foresets and Between Cross-Bed Sets
Shale-Flakes Common
Shale-Flake Quartz-Pebble Cglom
Rhythmically Interlayered Sandstones and Shales
TIDAL CHANNEL
Lower Regime Flat-Bed and Suspension Sedimention on Upper Point Bars
Flood and Ebb Megaripple Migration on Lower Point Bar
Abundant Suspension Sedimentation
Channel Lag
Tidal Deltaic Plain

B

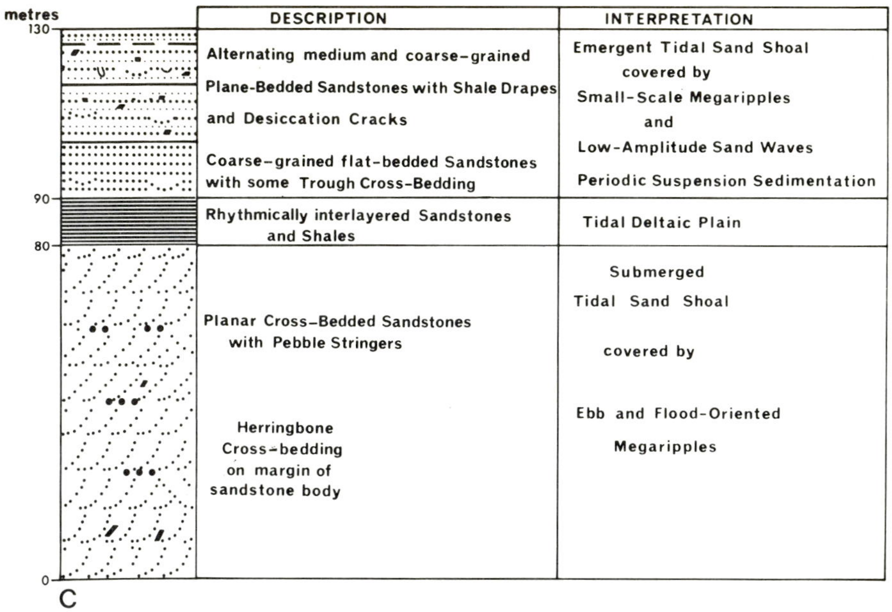

Fig. 4-15. Tidalite facies in the Moodies Group, South Africa (from Eriksson, 1977). A. Sandstone-shale facies. B. Conglomerate-sandstone facies. C. Medium- to coarse-grained sandstone facies.

and of feldspars in Moodies sediments indicates that such granitic rocks were important in the source area. Most of the Moodies sediments appear to have been deposited in a relatively high-energy, near-shore environment although localized quiescent conditions probably existed where banded magnetic shales and cherts were deposited.

Cyclic sedimentation, as previously discussed (Chapter 2), is well developed in the Moodies Group and has recently been interpreted in terms of intertidal and deltaic deposition (Eriksson, 1977, 1979). Four tidally influenced facies are recognized (Fig. 4-15). The sandstone-shale facies (*A*) with abundant flaser, lenticular, and wavy bedding is interpreted as a tidal mud-flat deposit. Two subfacies are recognized within the medium to fine-grained sandstones. The first (B_1) is characterized by small-scale herringbone cross-bedding and is thought to represent deposits formed on lower intertidal sand flats. The second (B_2) probably represents a complex of flood-tidal deltaic environments. The conglomerate-sandstone facies contains upward-fining units enclosed within rhythmically interlayered argillaceous sands and clays and is interpreted as tidal-channel deposits which meander across estuarine tidal flats. The medium- to coarse-grained sandstone facies is thought to represent subtidal and intertidal sand shoals. A diagrammatic sketch of the Moodies tidal environment (Fig. 4-16) shows a southward source area, as with the Fig Tree Group. Results also indicate that the barrier tidal flats formed on the flanks of an open estuary.

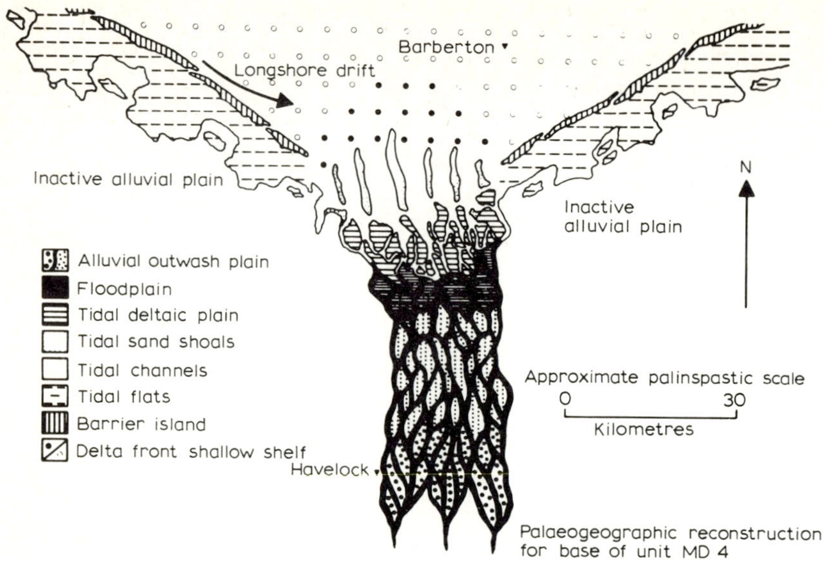

Fig. 4-16. Paleoenvironmental reconstruction of the Moodies tidal sediments (from Eriksson, 1977).

Deposition of the Yellowknife Supergroup

Continuing studies of the Yellowknife greenstone belt in the Slave Province in northern Canada have provided valuable information on the depositional environment of the graywackes and related sediments of the Yellowknife Supergroup (McGlynn and Henderson, 1970, 1972; Henderson, 1972, 1975a). Existing data suggest that volcanics and sediments of this succession were deposited on subsiding sialic crust. Paleocurrent studies indicate that sediments were derived from a western highland and poured into an adjoining basin (Fig. 4-17). Source materials were mixed volcanic and granitic rocks. The granitic component is thought to be derived from granitic and gneissic highlands which lay some distance to the west and the volcanic component, chiefly from a nearby active volcanic system along the western margin of the basin (Fig. 4-18). The basin first began to fill with felsic volcanic debris derived from erosion of the western highland. Continued uplift of this highland led to unroofing of granitic plutons which became important source material. Alluvial fans formed along the basin margin and eventually overwhelmed and buried the marginal mafic volcanic chain. Slumping along the steep marginal slope activated by volcanic explosions and/or earthquakes produced turbidity currents which deposited the graywacke-argillite of the Burwash Formation. Turbidity currents flow into the basin across a complex of large submarine fans. Most turbidity currents are restricted to valleys on the fans resulting in thick-bedded deposits with

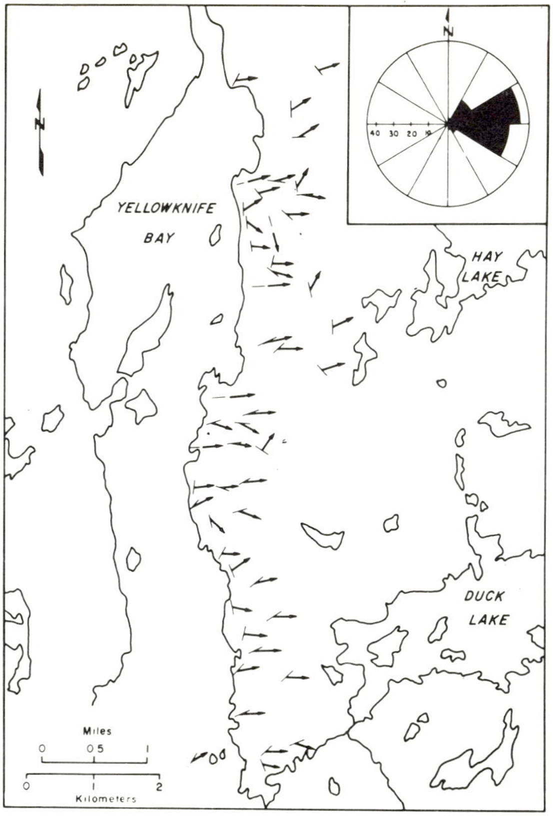

Fig. 4-17. Paleocurrent directions in graywacke turbidites of the Burwash Formation, northern Canada (from Henderson, 1972).

"proximal" characteristics within the valleys and thin-bedded, fine-grained "distal"-type deposits (analogous to overbank deposits on rivers) in the intervalley areas (Fig. 4-19). Some turbidites with "distal" characteristics are also deposited in the fan valleys from local slumping resulting in turbidites with mixed "proximal" and "distal" features in the same succession. Distal-type deposits alone characterize the inter-valley areas. Approximately 5 km of sediment accumulated in such a manner in the Yellowknife area. Pyroclastic felsic volcanic activity is recorded in the turbidite section by occasional tuffs and volcanic breccias.

It is noteworthy in terms of the numbers and sizes of sedimentary basins in the Slave Province that Ross (1962) records paleocurrent data from graywackes 300 km north of Yellowknife that suggest a northeasterly source. As at Yellowknife, this source area is now underlain dominantly by granitic rocks. These data support the interpretation of McGlynn and Henderson (1970) that the sedimentary belts in the Slave Province are not merely downfolded "keels" of large sedimentary basins that covered much greater areas,

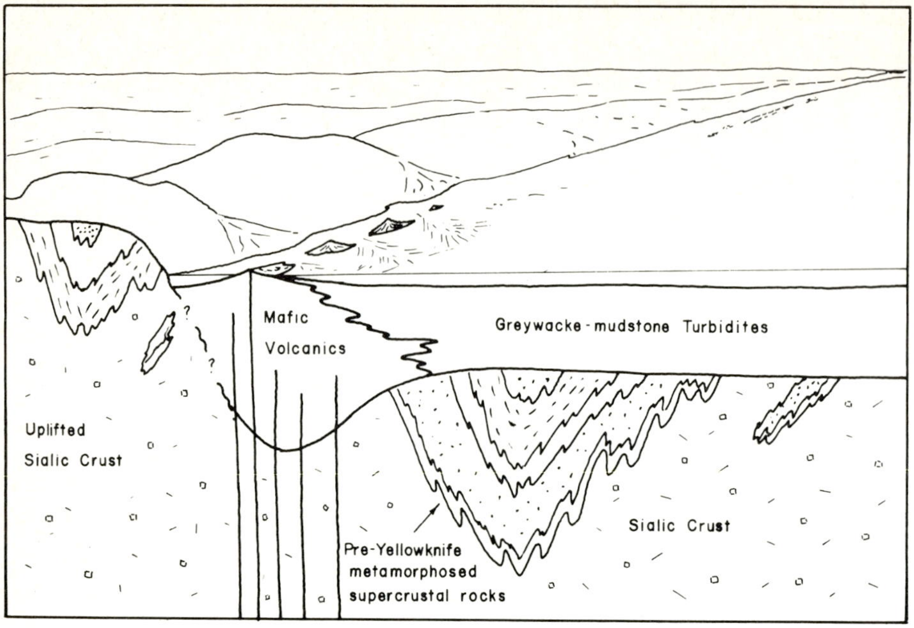

Fig. 4-18. Diagrammatic reconstruction of the tectonic sedimentary environment in the Yellowknife greenstone belt (from McGlynn and Henderson, 1970).

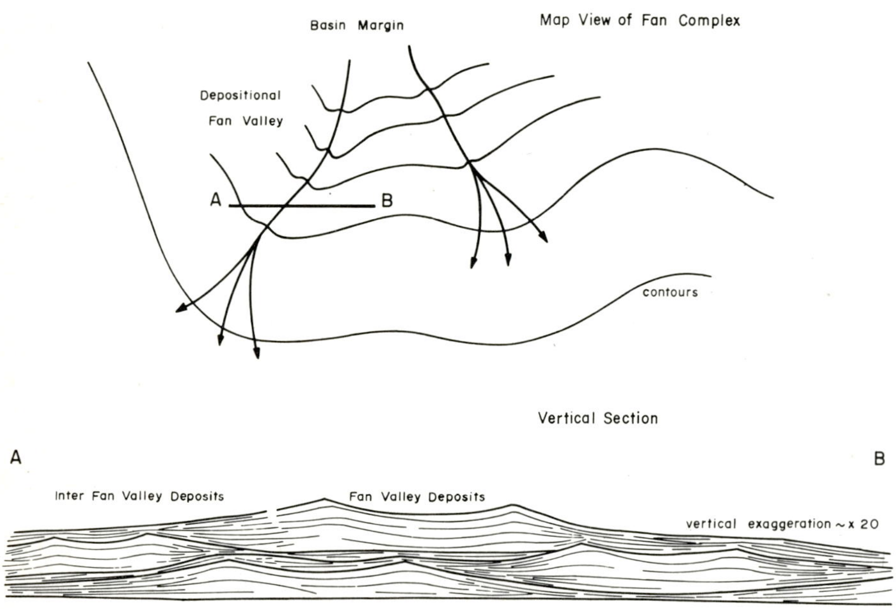

Fig. 4-19. Diagrammatic plan and cross-section of fan complex deposited by turbidites of the Burwash Formation (from Henderson, 1972).

but that the present-day borders of the basins are approximately coincident with the original margins. Such basins would have been the order of 100—200 km across, much smaller than the basins proposed in the Canadian Shield by Goodwin (1973) and discussed in the next section.

Archean basins of the Canadian Shield

Goodwin (1973) has suggested the existence of ten Archean basins in the Canadian shield of which seven are in the Superior Province (Fig. 4-20). Such basins were first suggested in the Wabigoon, Michipicoten, and Abitibi areas of the Superior Province based on data from many greenstone belts (Goodwin and Shklanka, 1967; Goodwin and Ridler, 1970; Goodwin, 1973). Since that time, they have been used as the starting point for evolutionary models of the Canadian Shield (Goodwin, 1974, 1976, 1977b). The basins in their present structurally deformed state are 600—800 km long and 200—400 km wide. All basins are deformed and fragmented and hence their reconstruction is necessarily incomplete. The basin margins are defined by a three-fold association of (1) extensive, chert-rich oxide-facies iron formation; (2) calc-alkaline volcanic piles with distinctive, commonly mineralized, felsic volcanic centers; and (3) the common presence of conglomerates and breccia. Interior parts of the basins are characterized by the four-fold association of (1) dominant tholeiitic basalt lacking well-developed felsic volcanic centers and associated mineral deposits; (2) numerous, thin layers of chert-poor, sulfide-facies iron formation; (3) distal-type graywacke-argillite; and (4) numerous granitic batholiths.

It is informative to examine two typical basins in more detail. In this regard, the Algoma basin in the southern Superior Province is perhaps best known (Goodwin and Shklanka, 1967; Goodwin, 1973) (Fig. 4-20). Details of the geology of this area are given in Goodwin (1962) and summarized in Chapter 2. The stratigraphic succession increases in thickness from about 3 km on the west to $\gtrsim$ 12 km in the east suggesting the basin deepens in this direction. The proposed basin has three divisions, shelf, margin, and core, each with a characteristic rock association as summarized in Fig. 4-21. An increase in grain size of the Dore sediments in going from east to west clearly indicates an eastward-sloping shelf. Coarse, felsic breccias and agglomerates occur only in the western marginal area grading into finer-grained pyroclastic rocks to the east. Iron formations (as discussed in Chapter 7) grade from oxide to carbonate to sulfide facies in going from west to east and are interpreted by Goodwin (1973) to reflect progressively deeper, less-oxidizing water.

The Abitibi Basin (Figs. 3-1 and 4-20) is the largest in the Canadian Shield comprising all of the Abitibi greenstone belt (Goodwin and Ridler, 1970). It is underlain by 58% volcanic, 32% granitic, and 10% sedimentary rocks. Numerous volcanic complexes are present as discussed in Chapter 3.

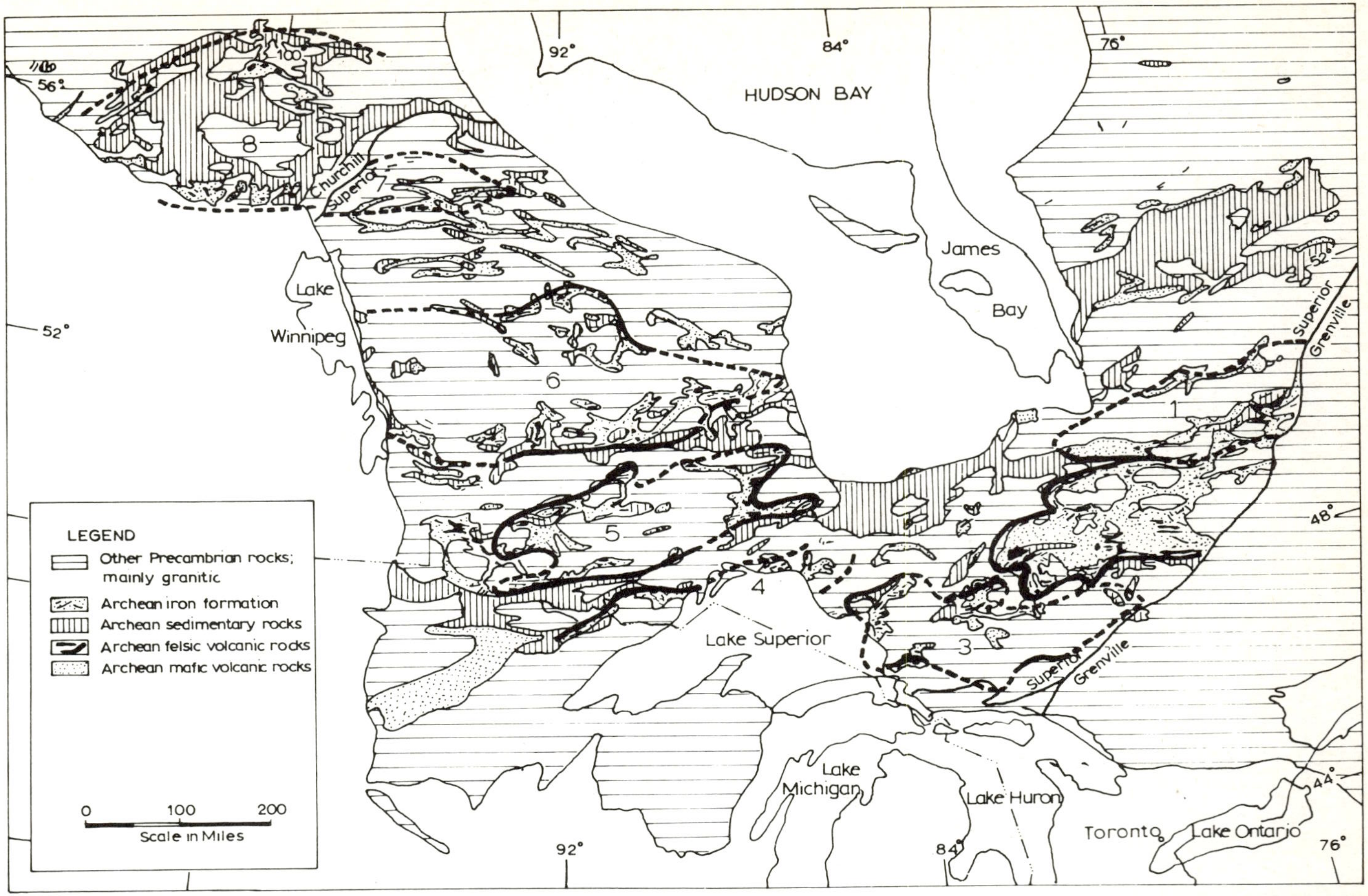

Fig. 4.20. Precambrian basins in the southern Canadian Shield (from Goodwin, 1973). Heavy lines indicate approximate basin margins; dashed where they are inferred. Basins: *1* = Matagami, *2* = Abitibi, *3* = Algoma, *4* = Superior, *5* = Keewatin, *6* = Berens, *7* = God's Lake, *8* = Kisseynew.

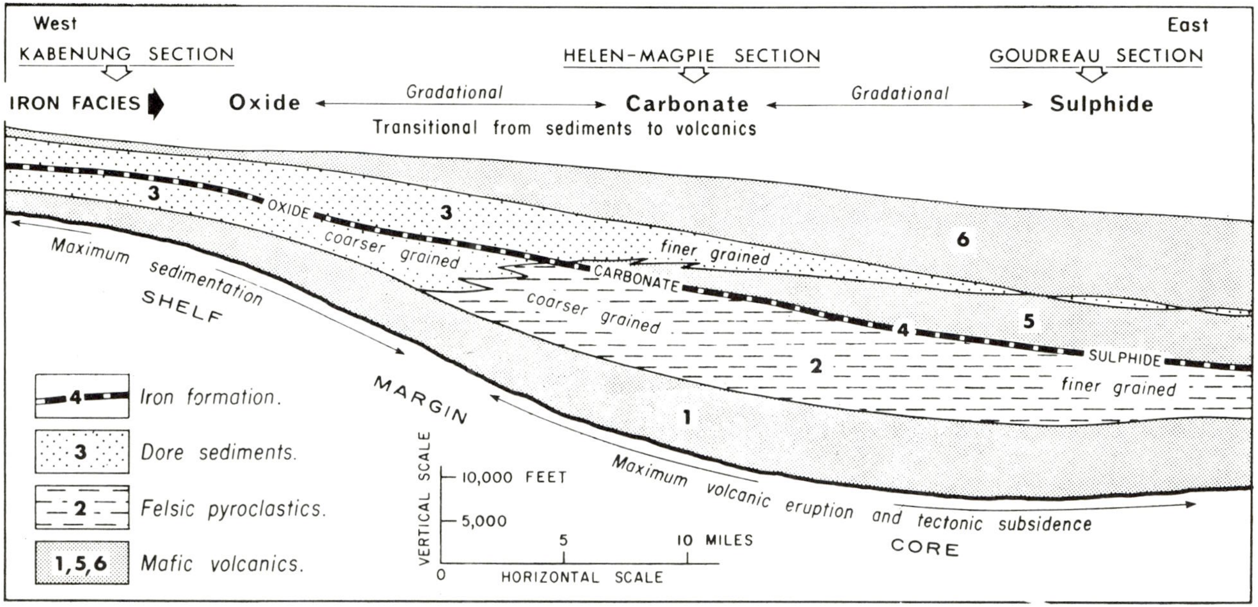

Fig. 4-21. Reconstructed cross-section of the western edge of the Algoma Basin (from Goodwin and Shklanka, 1967).

Calc-alkaline volcanic rocks with mineralized felsic volcanic centers are concentrated near the margins of the belt (Fig. 3-1) and tholeiites dominate in the center. The distribution of the three facies of iron formation is used to define paleoslopes (Goodwin, 1973). Despite local reversals, the main occurrence of oxide-facies iron formation is towards the margin of the basin whereas the chert-poor sulfide facies is more common in the interior. Goodwin (1977b) has recently pointed out that the geophysical and geologic features of the Abitibi Basin are also closely related. Although not as well defined, existing data suggest that basins of similar size exist in the Rhodesian Province (Coward et al., 1976a; Key et al., 1976) and in the Yilgarn and Pilbara Provinces in Australia (Glikson, 1970, 1971a; Ryan and Kriewaldt, 1964).

Goodwin's basinal concept has not gone unchallenged. Walker (1978) has recently questioned the basic criteria employed in defining the basins. He points out that the oxide-facies iron formation is sometimes closely associated with deep-water turbidites, that the iron formation facies distribution is controlled by diagenetic redox reactions and not water depth (Dimroth, 1975), and that iron formations used to define paleoslope within a given basin, in many instances, have not been shown to be correlative with each other. Also, the oxide facies of iron formation in the Abitibi Basin, in part, occurs along the axis of the basin and hence appears to reflect deep, rather than shallow-water, deposition. Conglomerates, a priori, cannot be used to define basin margins because some (the disrupted framework types in particular) are formed in the turbidite environment and can be carried great distances from the shoreline. Walker also questions why felsic volcanic centers should lie at basin margins and believes their reliability in defining such margins is suspect. It is notable in this regard that the felsic volcanic centers in the Abitibi (Fig. 3-1) and Keewatin Basins do not always lie near the basin margins.

Ayres (1969) has suggested that the sedimentary-plutonic superbelts in the Superior Province (Fig. 1-6) are linear sedimentary basins between active volcanic chains which represent the volcanic-plutonic superbelts. The diagrammatic reconstruction across the Wawa, Quetico, and Wabigoon Superbelts in Fig. 1-7 illustrates the proposed facies relationship in this area. Although "islands" of older sialic crust have been reported in some sedimentary superbelts (such as the English River belt), most existing data suggest that volcanism and sedimentation occurred, at least in part, simultaneously in adjacent superbelts. Clastic sediments in the Quetico belt may have been derived from contemporary volcanism in adjacent volcanic belts and from uplift and erosion of volcanics and granitic rocks. The reappearance of mafic volcanics in the upper part of the Wawa succession reflects onset of a new volcanic cycle.

Mineral assemblages in Archean sedimentary rocks can be employed to place constraints on the composition of the Archean atmosphere and oceans (Garrels and MacKenzie, 1974). Most evidence suggests that seawater approached its present composition by 3.7 b.y. and that with exception of localized areas, the composition, pH, and Eh have not greatly deviated from that of modern seawater (Garrels and MacKenzie, 1971; Holland, 1972). The absence of sepiolite in Archean sediments indicates that the silica content of seawater during the Archean did not exceed 25 ppm (at a pH of 8). The absence of marine brucite in Archean rocks places an upper limit on the pH of seawater of 10. Bedded chert, siderite, sulfide-rich sediments, carbonates, and iron oxides form in the oceans today as they did in the Archean and indicate a similarity in the properties of modern and Archean seawater. Archean cherts are, however, depleted in ^{18}O relative to Phanerozoic cherts (Perry, 1967). If these cherts reflect equilibration with Archean seawater, which seems likely, either seawater has increased in ^{18}O with time (perhaps by being recycled through the mantle) (Perry et al., 1978), or the earth's surface temperature has fallen from about 70° C at 3.4 b.y. to present-day values (Knauth and Lowe, 1978).

Most data suggest that the earth's atmosphere (except for O_2) has been produced by degassing of the earth (Rubey, 1955). The composition of the Archean atmosphere was controlled in part by the oxidation state in the crust and upper mantle where magmas are produced and in part by ocean-atmosphere interactions (Holland, 1962). If the source area of magmas did not contain free iron, as appears to be the case at least after 3.7 b.y., CO_2, H_2O, CO, and N_2 would probably be the most important gaseous species emitted from volcanic eruptions and collected in the atmosphere. Much of the water would condense in the oceans and some of the CO_2 would dissolve in seawater and be precipitated as carbonate.

Three lines of evidence indicate that oxygen was absent or minor in the Archean atmosphere.

(1) The presence of detrital uraninite and pyrite in Archean conglomerates such as described in the Witwatersrand Supergroup in South Africa (Schidlowski, 1970) and in the Bababudan Group in India (Viswanatha, 1968). These minerals would be readily oxidized if free oxygen were present.

(2) The first occurrences of major red beds and sulfate deposits which imply an oxygenated atmosphere do not appear in the geologic record until about 2.0 b.y. (Cloud, 1973).

(3) Archean weathering zones show decreases rather than increases in Fe^{3+}/Fe^{2+} ratios with depth of weathering (Rankama, 1955; Frarey and Roscoe, 1970).

Not all investigators interpret these observations to support a non-oxygen-bearing atmosphere during the Archean (Towe, 1978). Dimroth and Kimberley (1976) suggest that the distribution of carbon, sulfur, uranium, and iron in Archean sediments is similar to that observed in Phanerozoic sediments and that uraninite and pyrite in Archean sediments are of diagenetic rather than detrital origin as are Phanerozoic occurrences. The near absence of red beds and sulfates in the Archean are related to the general lack of continental platform-type sedimentation and not to a lack of oxygen. These authors also interpret iron formation to have formed by diagenetic replacement of carbonate sediments. The widespread presence of minor iron formation and of oxidized rinds on basaltic pillows in Archean greenstone belts is also suggestive that some oxygen may have been present in the Archean atmosphere (Dimroth and Lichtblau, 1978). As discussed in Chapter 7, however, precipitation of iron formation should have prevented any significant accumulation in the atmosphere.

Although some free oxygen may have existed in the Archean atmosphere (especially in late Archean time), data indicate that most oxygen has been added to the atmosphere by Phanerozoic photosynthetic processes (Berkner and Marshall, 1965).

Chapter 5

GRANITIC ROCKS

INTRODUCTION

As discussed in Chapter 1, granitic rocks, and in particular granitic rocks of tonalite or trondhjemite[1] composition, dominate in Archean granite-greenstone terranes. Macgregor (1932, 1951) was one of the first to suggest a classification for Archean granitic plutons in such terranes based on his work in Rhodesia. One group of plutons (Suess batholiths) is characterized by domal or polydomal geometry, overall concordant contacts with greenstone belts, gneissic foliation in marginal zones, and a relative abundance of greenstone xenoliths. A second group (Daly batholiths) is characterized by massive, often porphyritic textures, sharp, often discordant contacts, and a sparsity of xenoliths. Macgregor also suggested a "gregarious" habit for batholiths in the Rhodesian Province. Although it is now clear that many Archean plutons are composite in nature, it is a tribute to Macgregor that the overall features of his two-fold classification have withstood the test of time. For instance, in the Superior Province granitic plutons are broadly classified into large, synkinematic gneissic complexes and into small plutons (Goodwin et al., 1972). More detailed classifications have been proposed in some areas. Pichamuthu (1976) has suggested a three-fold classification for Archean granitic rocks in India: (1) early basement gneisses which migmatize greenstone belts; (2) massive to foliated plutons derived by local melting of basement gneisses; and (3) late-stage, post-tectonic plutons. A similar three-fold classification has recently been proposed for granitic plutons in the southern part of the English River Superbelt in Canada (Breaks et al., 1978). Viljoen and Viljoen (1969f) have proposed a four-fold classification and Hunter (1973), a six-fold classification for granitic rocks in the Barberton region in South Africa and Swaziland. The classification of granitic rocks according to Hunter is, in order of probable decreasing age, the Ancient Gneiss Complex, the Granodiorite Suite, tonalite diapirs, the Nelspruit Migmatite Complex and associated hood-type batholiths, and late granitic plutons. A generalized geologic map showing the distribution of these granite types is shown in Fig. 5-1. The Ancient Gneiss Complex is composed of gneisses and migmatites

---

[1] Trondhjemite is a leucotonalite whose plagioclase is oligoclase or albite and whose color index is <10 (Streckeisen, 1976; Barker, 1979).

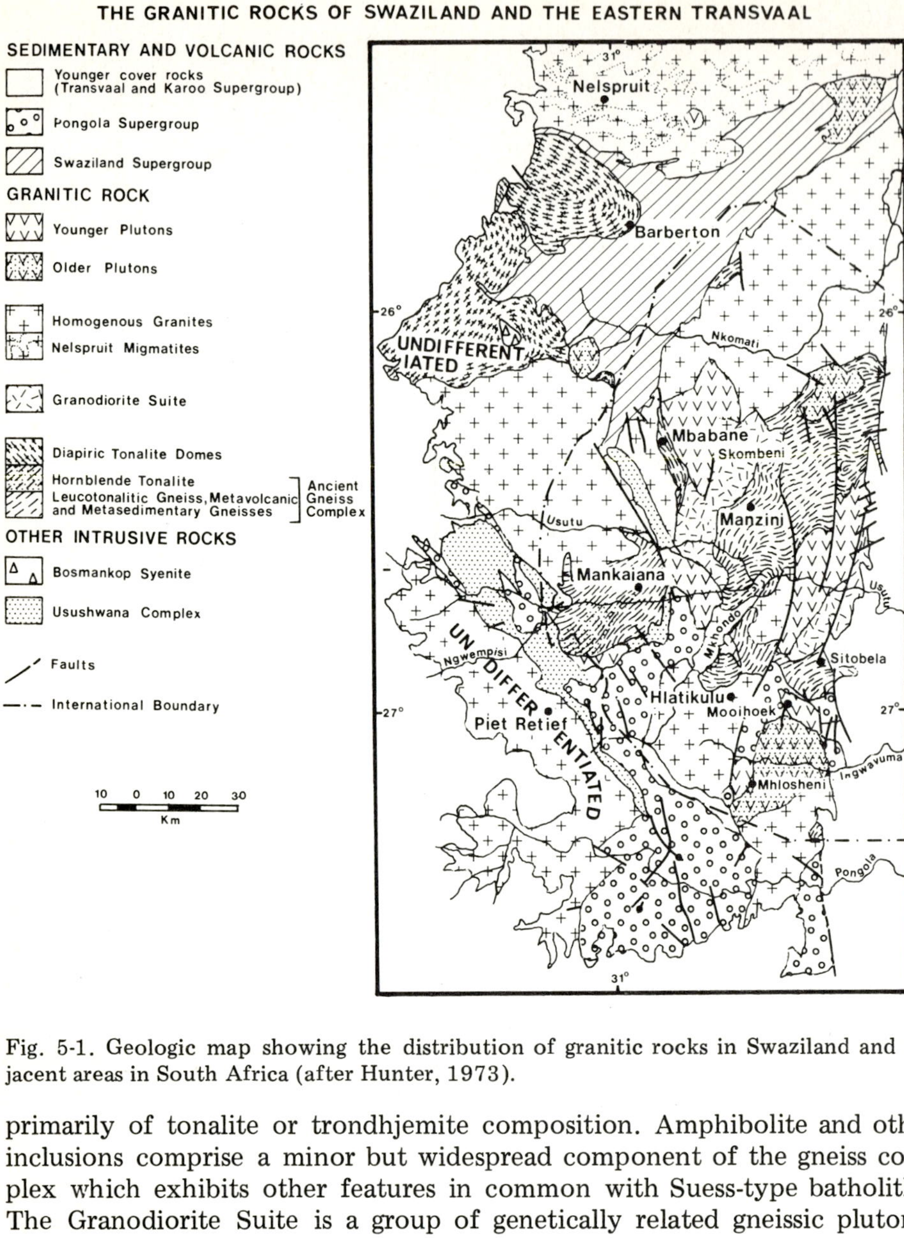

Fig. 5-1. Geologic map showing the distribution of granitic rocks in Swaziland and adjacent areas in South Africa (after Hunter, 1973).

primarily of tonalite or trondhjemite composition. Amphibolite and other inclusions comprise a minor but widespread component of the gneiss complex which exhibits other features in common with Suess-type batholiths. The Granodiorite Suite is a group of genetically related gneissic plutonic rocks ranging from tonalite through gabbro to ultramafic in composition, and again is similar to Suess-type batholiths. Tonalite diapirs are foliated plutons that are dominantly concordant with the degree and dip of foliation increasing in marginal zones. The hood-type granites are widespread, sheet-like intrusions of massive quartz monzonite which appear to grade

Fig. 5-2. Typical exposure of a young Archean pluton in the Barberton region, South Africa. The Mpageni pluton in Krokodilpoort (from Viljoen and Viljoen, 1969f).

downwards through a migmatite complex (the Nelspruit Complex) into the Ancient Gneiss Complex. The granitic plutons are massive, coarse grained, often porphyritic bodies ranging from granite to granodiorite in composition and exhibiting other features in common with Daly-type batholiths. Viljoen and Viljoen (1969f) have pointed out that a close correlation of pluton type with topography exists in the Barberton area. The Ancient Gneiss Complex and the tonalite diapirs underlie valleys and outcrops are, in general, poor. The hood-type bodies occur as plateau cappings and the late granitic plutons often form boulder-strewn koppies or large outcrops with significant relief (Fig. 5-2). The general classification of granitic rocks in the Barberton region can be likened to Buddington's (1959) depth classification with the Ancient Gneiss Complex, the Granodiorite Suite, and the tonalite diapirs classified as catazonal, the hood-type granites as mesozonal, and the late granitic plutons as epizonal.

Two generalized igneous rock trends are observed in Archean granitic terranes (Fig. 5-3). The most widespread is the tonalite-trondhjemite trend which exhibits a rather constant K/Na ratio with decreasing Ca. The other, a more typical calc-alkaline trend, exhibits an increasing K/Na with decreasing calcium. There is a broad coherence between rock composition and field occurrence in Archean granitic terranes (Hunter, 1974a,b) (Fig. 5-3). The gneissic complexes and diapiric intrusions are typically tonalitic or trondhjemitic in composition; large sheet-like batholiths are variable but average quartz monzonite in composition; and late-granitic plutons range

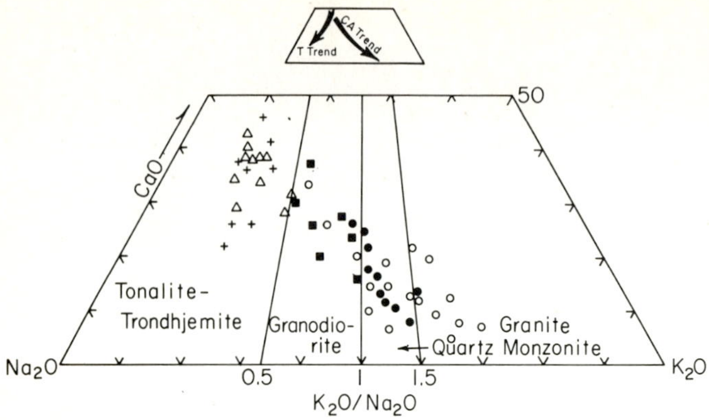

Fig. 5-3. CaO-Na_2O-K_2O diagram showing the distribution of average compositions of Archean granitic rocks from greenstone-granite terranes. △ = gneissic complexes; + = tonalite diapirs; ■ = granodiorite; ● = sheet-like batholiths; ○ = late plutons. Trends in inset: *T* = tonalite-trondhjemite; *CA* = calc-alkaline.

from tonalite to granite with granite and quartz monzonite dominating. Within the Barberton region, a clear secular trend in composition exists as evidenced by the data in Fig. 5-4. The results indicate that alkali and related elements increase rapidly in abundance in granitic rocks until about 3.0 b.y. and then increase more slowly (Hunter, 1974b). Within any given geographic locality there is a tendency for the relative abundance of tonalite to decrease while that of quartz monzonite and granite increases with time. Lateral differences in composition have been documented in the Rhodesian Province where the K/Na ratio increases towards the mobile belts on the north and south (Viewing, 1968).

FIELD ASSOCIATIONS

Gneissic complexes

Gneissic complexes comprise the dominant component in the granitic portion of Archean granite-greenstone terranes ranging in abundance from 50 to 70%. In very few areas have these complexes been mapped in detail. An example of the complexity in one area that has been mapped in Rhodesia is shown in Fig. 5-5. Also shown in the figure are associated granitic plutons and greenstone belts. Trends shown on the map are based in part on trends taken from air photographs. In general, the foliation in the gneissic complexes parallels that in greenstone belts near greenstone-gneiss contacts and becomes exceedingly variable in intervening regions. Locally the contacts are sheared. As discussed in Chapter 1, it is often difficult to ascertain whether such contacts are unconformities or intrusive contacts. Lithologic

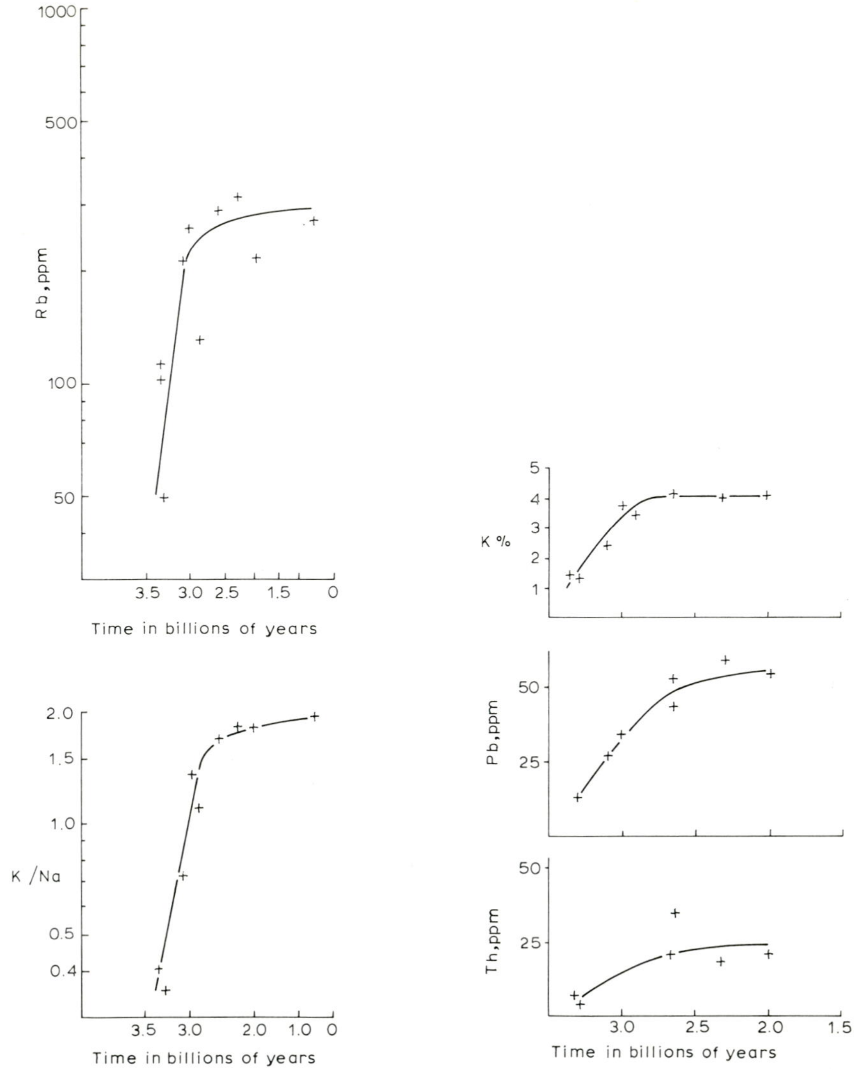

Fig. 5-4. Mean concentrations of Rb, K, Th, Pb, and K/Na in granitic rocks from the Kaapvaal Province in southern Africa versus time (from Hunter, 1974b).

variations are numerous in gneissic terranes. Rocks range from uniformly banded gneisses to faintly foliated, homogeneous gneisses. Migmatite-agmatite-nebulite terranes are also locally abundant. Contacts between gneissic complexes and granitic plutons range from sharp and discordant to gradational over hundreds of meters. Such gradational contacts have been described by Anhaeusser (1973b) in the Johannesburg-Pretoria Dome in South Africa and in the Mashaba area in Rhodesia (Wilson, 1973b). Gneissic terranes are composed chiefly of rocks of tonalite or trondhjemite composition with

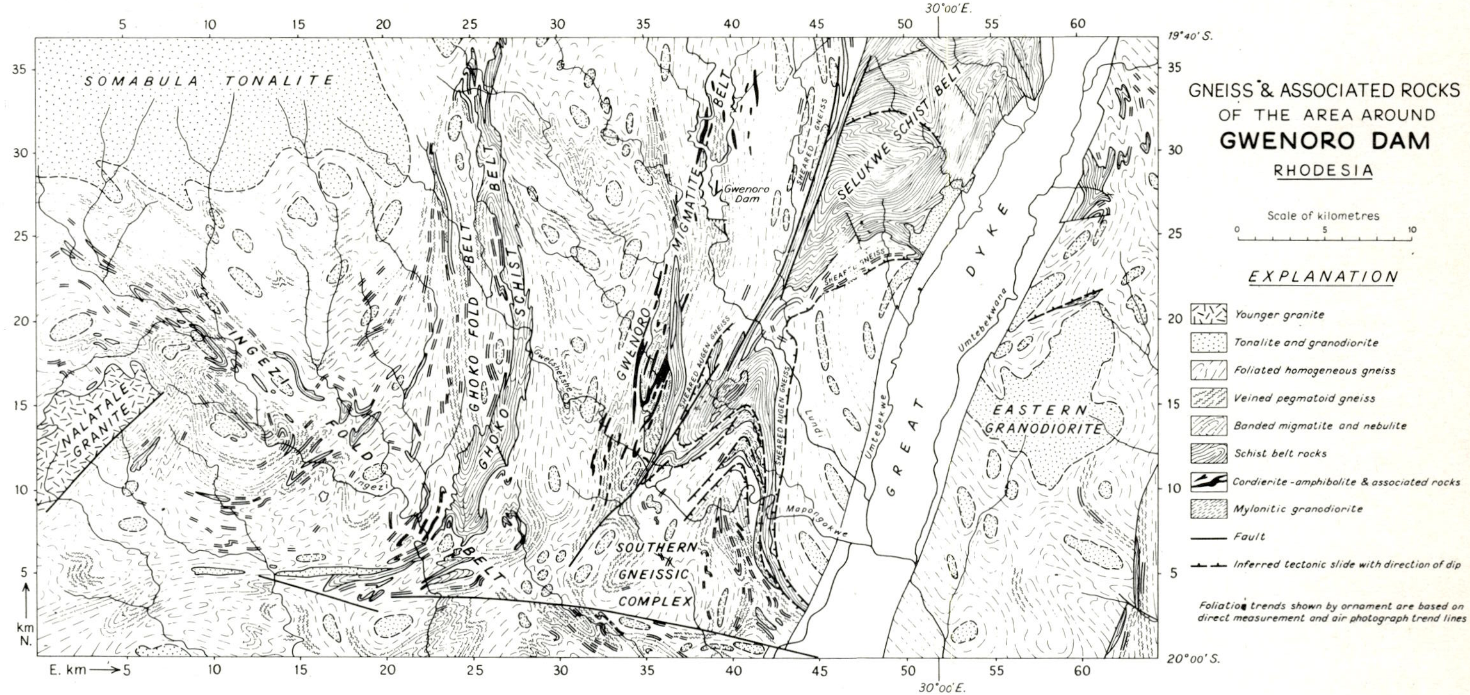

Fig. 5-5. Geologic map of the area around Gwenoro Dam, Rhodesia (from Stowe, 1973).

more K-rich granitic components being minor (Glikson, 1979a). These terranes have been referred to as bimodal (Barker and Peterman, 1974; Hunter et al., 1978) because of the association of tonalite-trondhjemite and mafic enclaves with a sparsity of rocks of intermediate composition.

Many descriptions of Archean gneissic terranes are available in the literature. Perhaps the most detailed account is presented by Hunter (1970) for the Ancient Gneiss Complex in Swaziland. Other descriptions of gneisses in the Kaapvaal Province are given in Viljoen and Viljoen (1969f, g) and Anhaeusser (1973b). Examples in Rhodesia are described by Wilson (1973b), Stowe (1973), and Phaup (1973); in North America by Heimlich (1969), Goldich et al., (1972), Harris and Goodwin (1976), Schwerdtner (1976, 1978), Breaks et al. (1978), and Peterman and Hildreth (1978); in India by Pichamuthu (1974, 1976), Rao et al. (1974), Ramakrishnan et al. (1976), and Ramiengar et al. (1978); and in Australia by Hickman (1975) and Hickman and Lipple (1975).

Faintly to prominently foliated tonalite-trondhjemite gneiss dominates in most Archean gneissic terranes. Foliation is often best developed where inclusions are abundant. Gneisses range from white to gray in color and are generally medium to coarse grained. They contain variable amounts of supracrustal inclusions showing progressive degrees of assimilation. Locally, gneisses are sheared and mylonitic and augen textures are well developed (Fig. 5-6). Pegmatitic components range from absent to abundant. Late K-feldspar megacrysts occur in some gneisses (Heimlich, 1969). Relatively homogeneous gneiss may grade into banded gneiss or migmatite. Banded gneisses, which are composed of alternating bands of quartz-feldspar-rich and biotite (± hornblende)-rich layers, greatly dominate in some areas, for example in the Teton Mountains in Wyoming (Reed, 1963). Banding occurs on two scales, 0.2—10 mm wide and 10 cm to several meters wide. It may be extremely uniform over hundreds of meters (Fig. 5-7) or it may pinch and swell producing boudins and become migmatitic. In some areas of the English River belt in Canada, such banding grades laterally into metagraywackes indicating a sedimentary precursor (Breaks et al., 1978).

Migmatite terranes are extremely heterogeneous and all of the migmatite variants described by Mehnert (1971) have been reported in Archean terranes. Small leucosome layers with mafic selvages are well developed in many metagraywacke-gneiss terranes (Fig. 5-8). With increasing amounts of migmatization, wide ranges in the ratio of leucosome to paleosome develop (Fig. 5-9). Leucosomes may be concordant, discordant, or both, and may represent several relative ages. More mafic portions of gneiss and mafic inclusions may behave as brittle solids and result in agmatite formation (Oftedahl, 1953; Parker, 1962). Agmatite blocks range from angular to subrounded and from a few centimeters to several meters in size. With increasing degree of leucosome development, both migmatite and agmatite grade into nebulite exhibiting only faint ghost-like outlines of migmatite components.

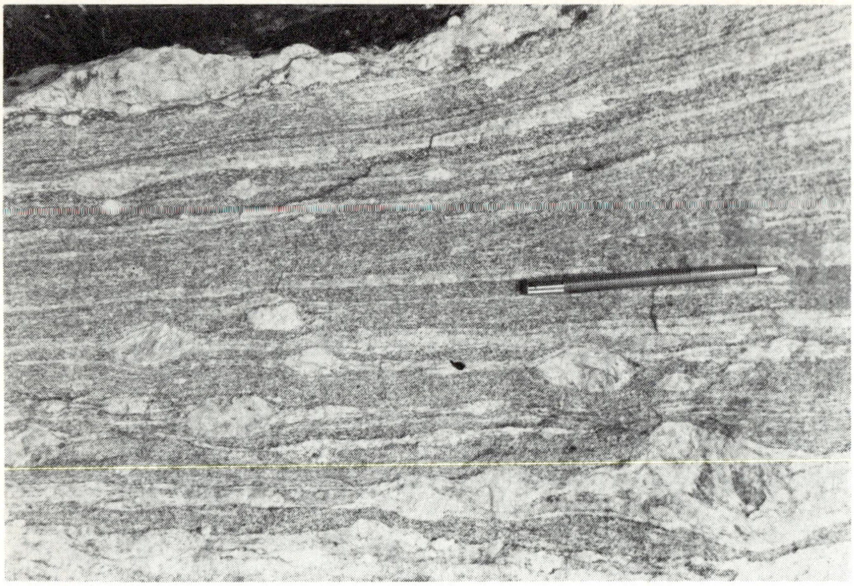

Fig. 5-6. Tonalitic augen gneiss from the Archean gneissic complex in the southern Bighorn Mountains, Wyoming (from Heimlich, 1969).

Detailed structural studies of gneissic complexes are in their infancy. Available data are very generalized (Stowe, 1968a, 1973; Condie, 1969b; Hunter, 1970, 1974a; Hepworth, 1973) yet indicate polyphase deformation involving two or more periods of isoclinal folding. Various structural domains can be defined within gneissic complexes (Condie, 1969b; Hunter, 1974a) indicating variable structural histories. Seven such domains have been defined in the gneissic complexes adjacent to the Laramie batholith in eastern Wyoming (Fig. 5-10).

A characteristic feature of Archean tonalitic gneiss complexes is the abundance of supracrustal inclusions (Hunter, 1970). Such inclusions are widespread but increase in abundance towards greenstone belt contacts (Anhaeusser et al., 1969; Viljoen and Viljoen, 1969f; Phaup, 1973). Inclusions range in size from a few centimeters to many kilometers. They also occur in various stages of digestion and fragmentation by surrounding gneisses. Trains of inclusions often connect greenstone belts as exemplified so well in Rhodesia (Phaup, 1973; Wilson, 1973a). An example of the transition from a greenstone belt into a gneissic terrane containing abundant inclusions of the belt is the transition from the Ghoko greenstone belt into the Ghoko fold belt (Fig. 5-5). Such distributions of inclusion trains strongly support an intrusive origin for the tonalitic gneisses. In order of decreasing abundance, amphibolite, ultramafic rocks, and quartzite (metachert?) are the principal inclusion lithologies. In addition, minor amounts of calc-silicate

Fig. 5-7. Uniform banded gneiss from the Archean terrane in the Teton Mountains, Wyoming (from Reed, 1963). Note the rootless isoclinal fold in the upper center of the photo.

Fig. 5-8. Well-developed metasedimentary migmatite from the English River Superbelt, Ontario (from Breaks et al., 1978). Note mafic selvages along some of the leucosome layers.

rock, mica schist, iron formation, and felsic metavolcanics occur locally. Inclusions of older gneiss and of other granitic rocks are of importance in some areas. Amphibolite inclusions, which appear to represent mafic volcanic fragments, range from small (<1 m) lense-shaped bodies to large (>1 km), linear pendants. They may be folded and fragmented into boudins. Smaller inclusions exhibit varying degrees of hybridization with surrounding gneisses (Schwerdtner, 1978). Ultramafic rocks have similar characteristics. Small inclusions of magnetite-bearing quartzite that may represent metachert, although not abundant, are widespread in many gneissic complexes. Relict bedding on a scale of a few millimeters in these rocks supports a chert precursor.

Batholiths

Batholiths are herein defined to include granitic plutonic complexes $\gtrsim 1000$ km^2 in area. Only recently have detailed field relationships of Archean batholiths become available (Glikson, 1979a). Basically such batholiths fall into two categories: simple and composite. Simple batholiths are composed of one intrusion or a series of intrusions of similar composition and composite bodies are comprised of several different plutons. Large composite batholiths, however, like the Southern California and Coast

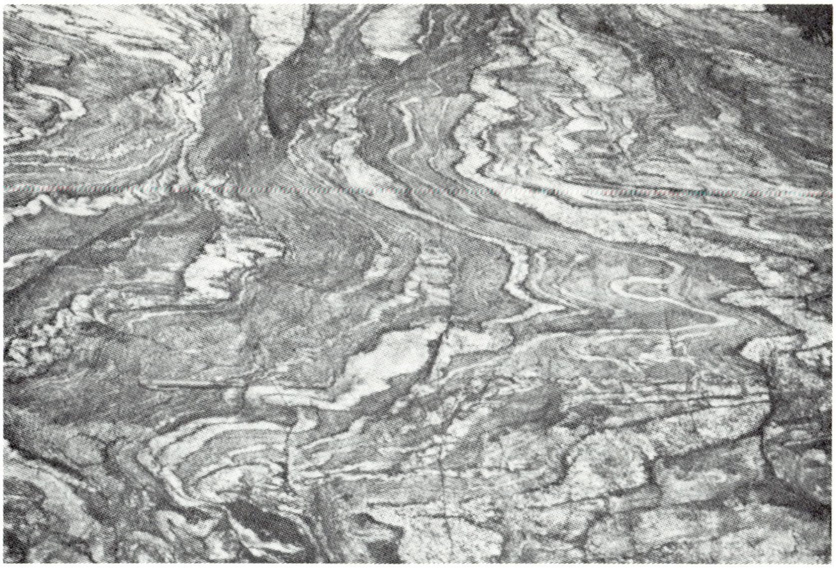

Fig. 5-9. Typical Archean migmatite complex from the Barberton region, South Africa (from Viljoen and Viljoen, 1969f).

Range batholiths of Phanerozoic age, have not been recognized in Archean terranes. Individual batholiths may range in composition from tonalite to granite such as the Vermilion and Grants Range batholiths in Minnesota (Southwick, 1972, 1978; Sims and Viswanathan, 1972), the Rainy Lake Complex in Ontario (Sutcliffe, 1978), and the North Trout Lake batholith in Ontario (Ayres, 1974). Others may exhibit only a limited range in composition such as the Lochiel batholith in South Africa (Viljoen and Viljoen, 1969f; Hunter, 1970, 1974b), the Laramie batholith in Wyoming (Condie, 1969b), and the Closepet batholith in India (Rao et al., 1972, 1974). Individual batholithic complexes may range up to 4000 km^2 in area as typified by the Closepet batholith. Contact relations with surrounding rocks are variable, even around the same batholith. They may range from sharp and discordant to concordant and gradational. Some bodies have marginal migmatite zones which grade into surrounding rocks (Eckelmann and Poldervaart, 1957; Dawson, 1966; Condie, 1969b; Casella, 1969). Marginal intrusive breccias may be present locally. The effects of contact metamorphism range from minor recrystallization accompanied by an increase in grain size of country rocks near pluton contacts, to contact aureoles up to more than 1 km wide (see Chapter 6).

Relationships between individual plutonic phases within composite batholiths are variable. In some instances, they are entirely gradational on scales ranging from meters to hundreds of meters (Condie and Lo, 1971; Hanson, 1972; Viewing and Harrison, 1973; Hunter, 1974b). In other cases, they

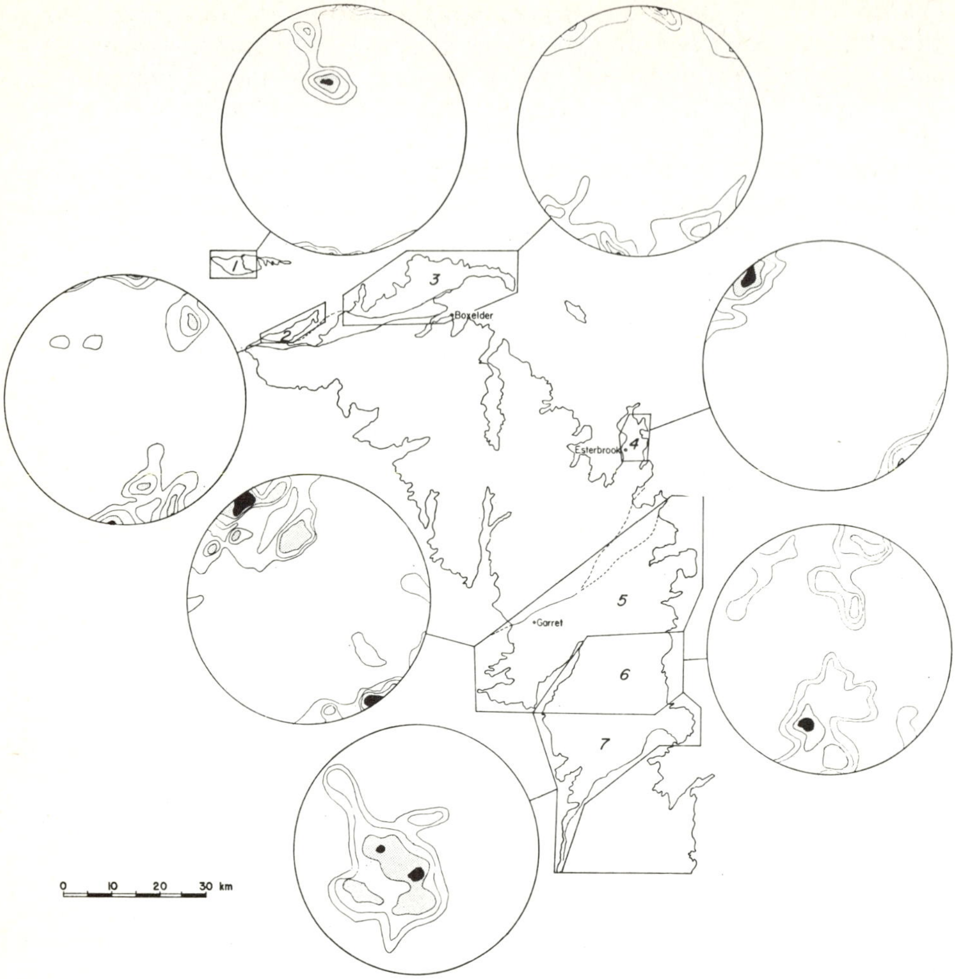

Fig. 5-10. Structural domains in Archean gneissic complexes from the Laramie Range, eastern Wyoming (from Condie, 1969b); 50—100 poles of foliation are contoured on each equal-area projection at 20, 15, 10, and 5% per 1% area.

may be sharp (Sutcliffe, 1978). Textures range from massive to foliated and from medium to coarse grained. Late, K-feldspar megacrysts produce a porphyritic texture in some bodies (Smith and Fripp, 1973). Inclusions vary in abundance generally increasing toward batholith margins. They are chiefly amphibolite, but all of the types described in the gneissic complexes have been reported. Detailed structural studies of Archean batholiths are just beginning (Hickman, 1975; Schwerdtner, 1976, 1978; Sutcliffe, 1977, 1978; Southwick, 1978). Existing data indicate complex, polyphase deformation with vertical forces dominating. Studies of Hickman (1975) and Schwerdtner

(1976, 1978) suggest that many Archean batholiths are polydomal diapirs, with typical diapiric characteristics as described in the next section. Some may represent remobilized tonalitic gneisses. Two ages of diapirism are recognized in batholithic complexes in northwestern Ontario (Schwerdtner, 1978).

Gravity studies indicate that Archean batholithic complexes, like most other Archean plutons are shallow, bottoming out between 5 and 10 km (Dawson, 1966; Brisbin, 1971; Goodwin et al., 1972; West et al., 1977). The Lochiel batholith in Swaziland and similar batholiths which have been described in northwestern Ontario (Harris and Goodwin, 1976; Goodwin, 1978) appear to represent sheet-like intrusions. Field and geochemical data suggest that the Lochiel batholith passes downwards through a migmatite zone into underlying tonalite-trondhjemite gneisses (Viljoen and Viljoen, 1969f; Hunter, 1970, 1974b). The Nelspruit Migmatites north of Barberton (Fig. 5-1) are thought to represent the exposed root zones of the Lochiel batholith or of a similar batholith.

Small plutons

All size gradations exist between batholiths and small plutons which herein include bodies < 1000 km^2 in area. Such bodies range downwards to < 100 km^2 in area and field relationships indicate that they may be pre-, syn-, or post-tectonic (Goodwin et al., 1972). Contacts are typically discordant to concordant with foliation in surrounding rocks. Intrusion breccias may be of local importance as in the Beidelman Bay pluton in Ontario (Franklin, 1978). Many small plutons have been mapped and typical detailed descriptions are given in Brownell (1941), Heimlich (1965, 1966), Viljoen and Viljoen (1969f,g), Harrison (1969, 1970), Hunter (1970, 1974b). Sims and Mudrey (1972), Sims et al. (1972), Catherall (1973), Wilson (1973b), and Stidolph (1973). Texturally, these plutons range from medium to coarse grained and are often porphyritic with feldspar megacrysts ranging up to 7 cm long (Fig. 5-11). Flow structure is reflected by aligned megacrysts in some plutons (Goldich et al., 1972) and foliation ranges from absent to well developed. Granitic rocks vary from buff to pink or orange in color and compositionally range from tonalite to syenite. Some plutons, like batholiths, are of variable composition; others are rather uniform throughout. As a whole, quartz monzonite and granite are the dominant rock types in small plutons (Ermanovics, 1971) (Fig. 5-3). Some plutons are zoned. Inclusions are generally few in number and small in size and appear to represent fragments of gneissic complexes or greenstone belts. Pegmatitic and aplitic phases of small plutons range from almost absent to quite common.

Recent detailed mapping and geophysical studies as part of the Canadian Geotraverse in western Ontario have been informative regarding small Archean plutons (Goodwin, 1978). In the traverse, 90 plutons are recognized

Fig. 5-11. K-feldspar megacrysts in the Dalmein pluton, South Africa (from Viljoen and Viljoen, 1969f).

in the volcanic-plutonic superbelts and 23 in the sedimentary-plutonic superbelts. Gravity studies (West, 1976; West et al., 1977) indicate that most plutons are shallow (2—16 km deep) and have sheet-like shapes.

Structural studies indicate that some plutons are strongly deformed and were emplaced either during or before major deformation (Schwerdtner and Sutcliffe, 1978). Deformation appears to have originated in these plutons by forceful emplacement rather than by syn- or post-tectonic regional deformation (Schwerdtner, 1976). Some plutons are virtually undeformed and appear to be post-tectonic. A group of plutons, first described in the Barberton region in South Africa, which are characterized by intense, steeply dipping, broadly concordant foliation in marginal zones have been referred to as diapirs (Viljoen and Viljoen, 1969f; Hunter, 1973, 1974a) (Fig. 5-12). Inclusions are aligned in the foliation planes along margins and become more randomly oriented in the centers of the plutons. Flattened pillows in volcanic rocks around the Bamaj-Blackstone pluton in Ontario are interpreted as indicating extension in all directions parallel to the pluton margin and compression on horizontal axes normal to the pluton margin (Clifford, 1972). These bodies, which are broadly elliptical in shape and often occupy the cores of antiforms, are similar in many respects to mantled gneiss domes (Eskola, 1948). Some diapiric plutons that are tonalite to trondhjemite in composition have been suggested as representing remobilized portions of tonalitic gneiss complexes that have risen into greenstone successions diapirically (Viljoen and Viljoen, 1969f) (Fig. 5-12). Geochemical data, however,

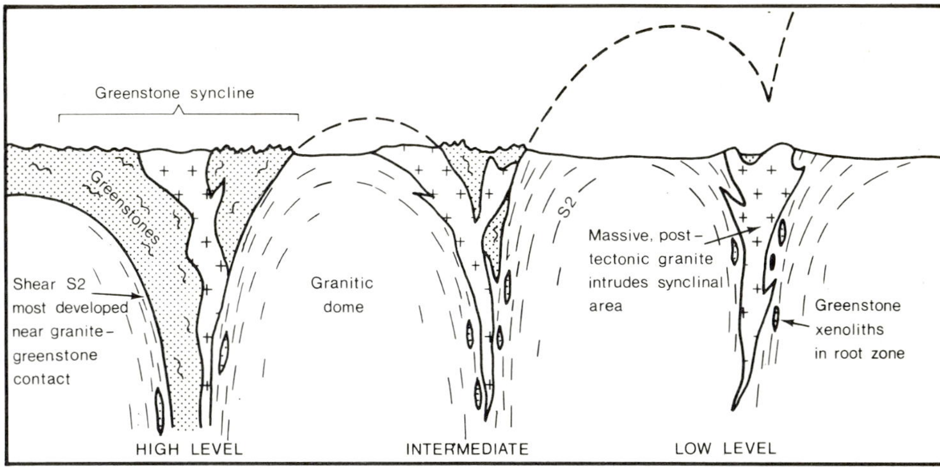

Fig. 5-12. Diagrammatic cross-section showing various structural levels of Archean diapiric plutons (granitic domes) and post-tectonic granites (after Hickman, 1975).

indicate that the tonalite diapirs in the Barberton area cannot represent remobilized samples of unchanged Ancient Gneiss Complex (Condie and Hunter, 1976). Sutcliffe (1977) has recently completed a detailed structural study of the Jackfish Lake—Weller Lake pluton which is part of the Rainy Lake batholithic complex in Ontario. This pluton is a syenodioritic pluton which has pronounced foliation and lineation. Results indicate the presence of a ubiquitous subhorizontal lineation which is best developed in the axial zone and decreases in importance towards the margins as foliation begins to dominate. In the marginal zones, the steeply dipping foliation contains the lineation which now also dips steeply. An overall decrease in the dip of both lineation and foliation is observed from the center (subhorizontal) to the edges (subvertical) of the pluton. Such a pattern has been suggested as representative of diapiric plutons in general (Anhaeusser et al., 1969) and to have developed during ascent and emplacement.

PEGMATITES AND RELATED ROCKS

Pegmatites range from locally abundant to absent in Archean granitic terranes. They occur as lensoid to irregular shaped bodies in gneissic complexes and as lensoid to dike-shaped bodies in plutons and batholiths (Dawson, 1966; Viljoen and Viljoen, 1969f; Catherall, 1973; Harris and Goodwin, 1976). They may be syn- or post-tectonic or both in a given area. They range in width up to 10 m (although typically less than 1 m) and in length up to 100 m (usually <10 m). Aplitic dikes may be associated with late-stage pegmatites and range up to 200 m long. Aplitic zones may also occur within pegmatites. Pegmatites in gneissic complexes may have mafic selvages suggesting an origin by metamorphic differentiation. Layered pegmatites have been described from the Wind River Mountains in Wyoming (Proctor and El-Etr, 1968). These bodies contain successive layers from a few centimeters to over 1 m thick and range in grain size from a few millimeters to nearly 1 m in the same pegmatite. Micrographic textures are common.

Most pegmatites are composed of one of two mineral assemblages: (1) microcline-quartz-biotite, or (2) sodic plagioclase-quartz ± muscovite. Minor amounts of garnet, tourmaline, magnetite, and allanite also occur in some pegmatites. Pegmatites with rare mineral assemblages are discussed in Chapter 7.

Granitic dikes occur in some plutonic complexes and exhibit textures and compositions similar to small granitic plutons (Anhaeusser, 1973b). They range from a few centimeters to over 1 km wide (Harris and Goodwin, 1976). Quartz veins are ubiquitous in most Archean granitic terranes and especially in plutons. They range from <1 to >100 m long, and are generally irregular to tabular shaped. In some instances, they appear to fill tension fractures (Dawson, 1966).

MINERALOGY

Tonalite-trondhjemite

Tonalite and trondhjemite are composed principally of sodic plagioclase (40—60%), quartz (25—35%), and biotite (5—10%) (Viljoen and Viljoen, 1969f; Heimlich, 1969; Hunter, 1970; Goldich et al., 1972; Glikson and Sheraton, 1972; Rao et al., 1974). In some rocks, hornblende (0—5%) and K-feldspar (0—5%) may be present. Grain size ranges from fine to coarse and foliation from well to poorly developed. The texture is typically hypautomorphic granular. Varying degrees of deformation are manifest by mortar textures and augen developed in quartz and feldspars. Sodic plagioclase (typically An_{20} to An_{35}) occurs as subhedral, partly clouded grains and may have clear overgrowths. It also may be partially sericitized and exhibit undulatory extinction. Late plagioclase megacrysts occur in some gneisses. Quartz is typically anhedral, often elongated in foliation planes, and exhibits undulatory extinction. Biotite is brown to green and partly chloritized in most rocks. K-feldspar (microcline), when present, occurs as late megacrysts (2—5 cm long) which poikilitically enclose other minerals. Common accessory phases in gneisses are some combination of epidote, magnetite, apatite, zircon, sphene, almandite, and muscovite. In some obvious paragneiss varieties, sillimanite, cordierite, or andalusite may be found.

In the amphibolite inclusions, hornblende and partially saussuritized andesine dominate with small amounts of magnetite, quartz, and sphene.

Relict clinopyroxene cores occur in some hornblende crystals (Hunter, 1970). Ultramafic inclusions are generally composed entirely of secondary minerals such as talc, serpentine, and tremolite at low grades and cummingtonite, anthophyllite, and cordierite at higher grades. Quartzite inclusions are composed chiefly of recrystallized quartz with small amounts of one or more of magnetite, clinopyroxene, grunerite, and almandite. Calc-silicate inclusions are composed of plagioclase-clinopyroxene with variable amounts of amphibole and garnet.

Other granitic rocks

Feldspar, quartz, and biotite are the principal minerals found in granodiorites, quartz monzonites, granites, and alkaline plutonic rocks. Plagioclase is usually the dominant feldspar (30—60%) and ranges in composition from An_{10} to An_{30}. It is commonly zoned, variably twinned, and partly clouded. Tapered and bent twins attest to deformation in many samples. Myrmekitic intergrowths are present in some rocks. Sericitization and saussuritization of plagioclase are found in varying degrees of development. K-feldspar (20—50%) (generally microcline) occurs chiefly as late-stage megacrysts that poikilitically enclose earlier minerals. Such crystals may reach lengths up to several centimeters. They commonly exhibit cross-hatched twinning and may be perthitic. In some plutons, such as the Kwetta in Swaziland, microcline crystals are mantled with rims of clear sodic plagioclase (Hunter, 1973). Quartz (10—30%) occurs in plutons as anhedral grains up to 3 mm in size and may show mortar textures around grain boundaries. Locally, it may occur as micrographic intergrowths in K-feldspar. Biotite (3—8%) occurs as ragged lathes and is usually, in part, altered to chlorite. Blue-green to green hornblende occurs as a minor component in some granodiorites (1—3%). Sodic pyroxenes and amphiboles occur in many alkaline plutons (Goldich et al., 1972; Sims and Mudrey, 1972; Sims et al., 1972). Accessory minerals include one or more of the following: magnetite, hematite, epidote, apatite, zircon, sphene, allanite, muscovite, and less commonly carbonate, ilmenite, sulfides, and tourmaline.

COMPOSITION

Tonalite-trondhjemite

Rocks of tonalite to trondhjemite composition (i.e., sodic granitic rocks) dominate in gneissic complexes of Archean granite-greenstone terranes and also occur in some plutons, particularly the diapiric plutons. Locally, they grade into rocks of quartz monzonite or granodiorite composition. Tonalite and trondhjemite are characterized by relatively low K_2O contents and

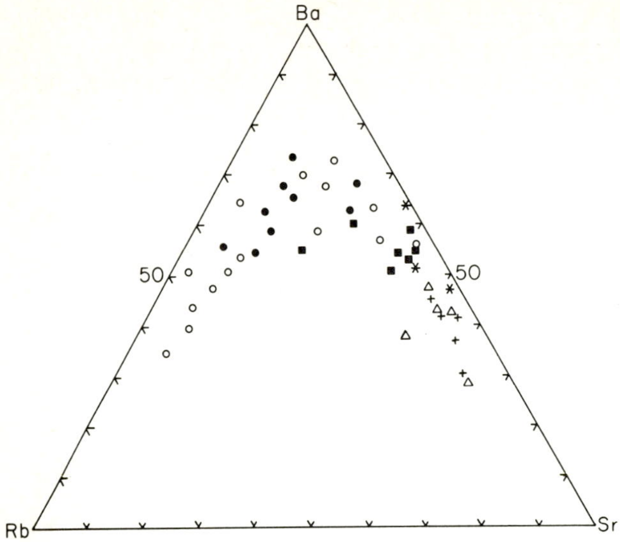

Fig. 5-13. Ba-Rb-Sr diagram showing the distribution of average Archean granitic rocks. Tonalite-trondhjemite, △ = gneisses, + = diapirs; ■ = granodiorite; quartz monzonite-granite: ● = sheet-like batholiths, ○ = late plutons; * = alkaline plutons.

exhibit a trend of increasing Na_2O with decreasing CaO (Fig. 5-3) sometimes referred to as the tonalite-trondhjemite trend. On a Ba-Rb-Sr diagram, these rocks exhibit consistently low Rb with approximately equal amounts of Ba and Sr (Fig. 5-13). In general terms, tonalite-trondhjemite can be divided into two categories based on Al_2O_3 content at 70% SiO_2 (Barker, 1979): high-Al_2O_3 ($\geqslant 15\%$) and low-Al_2O_3 ($< 15\%$) types (Table 5-1). Although a significant amount of variability exists within each group, certain geochemical features are quite distinctive. The high-Al_2O_3 type is characterized by low SiO_2, Rb, Th, and Ba/Sr and depleted heavy REE (Fig. 5-14). The degree of heavy-REE depletion and an increasingly positive Eu anomaly appears to accompany increasing SiO_2 (Hunter et al., 1978). Low-Al_2O_3 tonalite-trondhjemite is characterized by relatively lower contents of Al_2O_3, MgO, CaO, P_2O_5, Sr, Sc, Cr, Ni, and Co. It also exhibits undepleted heavy-REE patterns, negative Eu anomalies, unfractionated heavy-REE ($Yb_N/Gd_N \simeq 1$), and light-REE patterns that are less fractionated than in the high-Al_2O_3 type (Table 5-1; Fig. 5-14). Existing data suggest that high-Al_2O_3 types greatly dominate over low-Al_2O_3 type in low-grade Archean terranes. They compose most of the gneissic complexes and most or all of the tonalite diapiric plutons. An average composition of a tonalite diapir given in Table 5-1 is strikingly similar to the average high-Al_2O_3 gneiss. Low-Al_2O_3 gneisses have been reported from only a few localities such as the Ancient Gneiss Complex in Swaziland (Hunter et al., 1978), the Webb Canyon Gneiss in the Teton Mountains of Wyoming (Barker et al., 1979), and the Northern

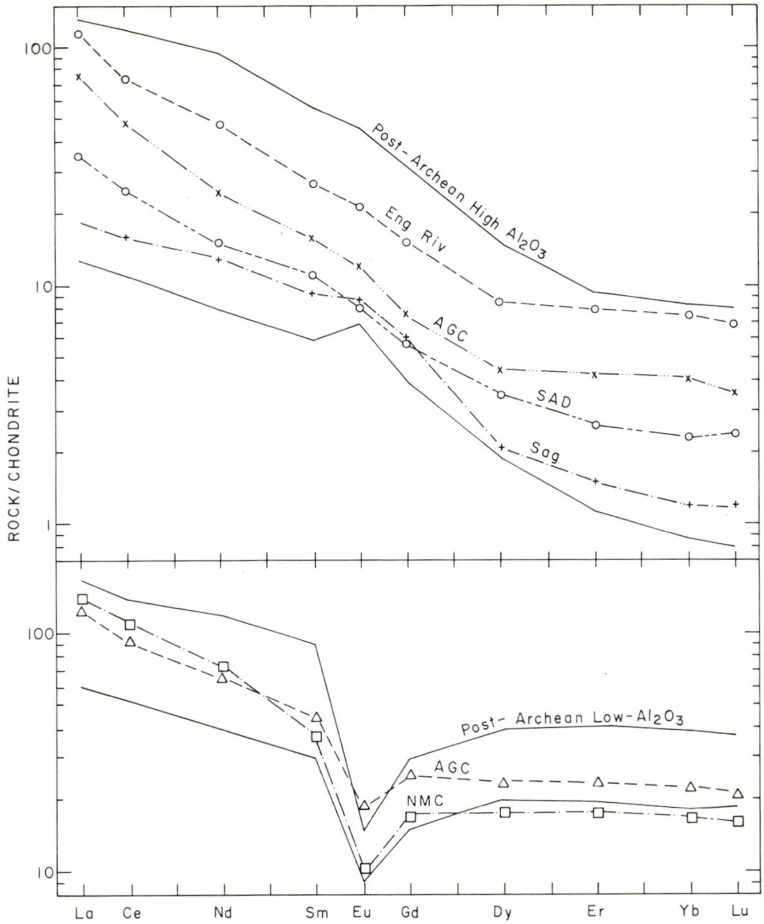

Fig. 5-14. Envelopes of variation of chondrite-normalized REE abundances in post-Archean tonalite-trondhjemite (Table 5-1) compared to average examples of Archean tonalite-trondhjemite. References given in Table 5-1. Abbreviations: *Eng.Riv.* = English River belt; *AGC* = Ancient Gneiss Complex; *SAD* = South African diapirs; *Sag* = Saganaga tonalite; *NMC* = Northern Metamorphic Complex, Wyoming.

Metamorphic Complex in eastern Wyoming (Condie, 1969b, and unpublished data). The Webb Canyon Gneiss is unusual compared to other low-Al_2O_3 trondhjemites in that it has REE contents 100—300 × chondrites (Barker et al., 1979). In the Ancient Gneiss Complex, high-Al_2O_3 gneisses greatly dominate.

An analogous grouping of post-Archean sodic granitic rocks has also been recognized (Table 5-1) (Barker et al., 1976a; Arth et al., 1978; Condie, 1978). Although trace element data for post-Archean tonalites and trondhjemites are not numerous, existing data as summarized in the two post-Archean averages

TABLE 5-1

Average compositions (oxides in wt.%, trace elements in ppm) of Archean and post-Archean tonalite and trondhjemite

	Archean			Post-Archean	
	gneiss, high-Al_2O_3	gneiss, low-Al_2O_3	high-Al_2O_3 diapir	high-Al_2O_3	low-Al_2O_3
SiO_2	69.4	74.5	69.1	69.7	76.1
TiO_2	0.35	0.39	0.29	0.35	0.17
Al_2O_3	15.8	14.2	15.9	15.4	13.3
Fe_2O_3	1.18	0.36	0.72	0.80	0.79
FeO	1.79	1.92	1.34	2.02	1.85
MgO	1.14	0.45	1.14	1.04	0.39
CaO	3.37	2.43	3.32	2.52	1.13
Na_2O	4.68	4.08	5.28	4.70	3.83
K_2O	1.58	1.95	1.35	2.12	1.16
MnO	0.04	0.05	0.04	0.07	0.07
P_2O_5	0.11	0.03	0.09	0.14	0.05
H_2O	0.54	0.37	0.75	0.68	0.85
K_2O/Na_2O	0.34	0.48	0.26	0.45	0.30
Sc	5	3	5		
Cr	12	8	14	30	1
Ni	13	7	15	25	
Co	5	2	7		1
Rb	44	75	45	72	30
Sr	460	110	470	530	150
Zr	175	290	90		220
Ba	400	420	350	575	440
La	25	45	15	22	35
Ce	42	91	18	41	84
Nd	15	42	8.5	16	42
Sm	2.9	7.6	1.7	3.0	7.2
Eu	0.82	1.0	0.50	0.93	0.76
Gd	1.9	5.2	1.5	2.6	5.1
Dy	1.4	6.7	0.90	1.8	7.0
Er	0.86	4.0	0.40	0.98	4.0
Yb	0.82	4.0	0.46	0.99	4.4
Lu	0.12	0.54	0.07	0.15	0.75
Th	7	12	4	5	
K/Rb	290	216	250	244	320
Rb/Sr	0.12	0.68	0.10	0.14	0.20
Ba/Sr	0.87	3.8	0.75	1.1	2.9
$(La/Sm)_N$	4.7	3.2	4.8	4.0	2.7
$(Yb/Gd)_N$	0.54	0.96	0.38	0.47	1.1
Eu/Eu*	1.1	0.50	0.96	1.0	0.39

N = chondrite-normalized ratio.

Chief references: Archean — high-Al_2O_3: average for Ancient Gneiss Complex type A

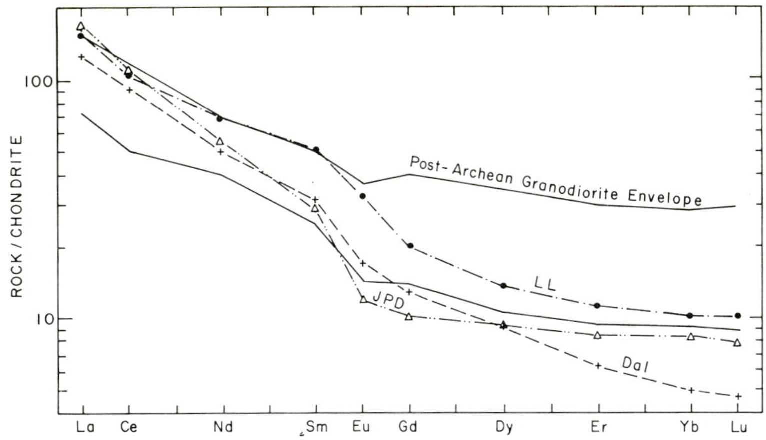

Fig. 5-15. Envelope of variation of chondrite-normalized REE abundances in post-Archean granodiorite compared to three average Archean granodiorites. References given in Table 5-2. Abbreviations: *LL* = Louis Lake batholith; *JPD* = Johannesburg-Pretoria dome; *Dal* = Dalmein-type plutons from South Africa.

(Table 5-1) suggest that post-Archean high-Al_2O_3 rocks may be higher and low-Al_2O_3 rocks lower in some transition metals and perhaps in K_2O and Rb than in corresponding Archean categories.

Granodiorite

Granodiorite is a minor rock type in Archean low-grade granitic terranes. It occurs as a minor component in gneissic complexes and comprises most of some plutons and batholiths like the Louis Lake batholith in Wyoming (Lo, 1970; Condie and Lo, 1971), the Dalmein pluton in the Barberton area (Hunter, 1973; Condie and Hunter, 1976), and the Preissac-Lacorne batholith in Quebec (Dawson, 1966). It commonly grades into tonalite on the low-alkali side and quartz monzonite on the high-alkali side. On the CaO-Na_2O-K_2O plot (Fig 5-3), grandiorites tend to bridge a gap between the more abundant tonalite-trondhjemite and quartz monzonite-granite groups. On the Ba-Rb-Sr plot, granodiorites exhibit low Rb and Ba/Sr ratios higher than

---

TABLE 5-1 (continued)

(Hunter et al., 1978); Northern Light Gneiss (Arth and Hanson, 1975); average gneiss from the English River belt (Chou et al., 1977; Breaks et al., 1978); southern Bighorn Mountains (Heimlich, 1971; K.C. Condie, unpublished data). Low-Al_2O_3: Ancient Gneiss Complex type B (Hunter et al., 1978); Northern Metamorphic Complex, Wyoming (Condie, 1969b, and unpublished data). Diapirs: average tonalite diapir, Barberton region, South Africa (Condie and Hunter, 1976); Saganaga tonalite, Minnesota (Arth and Hanson, 1975); Sesombi tonalite, Rhodesia (Harrison, 1970; K.C. Condie, unpublished data). Post-Archean — Barker et al. (1976a); Condie (1978); Arth et al. (1978).

TABLE 5-2

Average compositions (oxides in wt.%, trace elements in ppm) of Archean and post-Archean granodiorite

	Dalmein type	Louis Lake pluton	Johannesburg-Pretoria dome	Post-Archean average
SiO_2	70.8	65.0	72.8	66.9
TiO_2	0.30	0.69	0.24	0.57
Al_2O_3	14.5	15.4	14.2	15.7
Fe_2O_3	0.88	1.63	0.54	1.33
FeO	1.23	2.94	1.11	2.59
MgO	0.47	1.92	0.37	1.57
CaO	2.03	4.21	1.48	3.56
Na_2O	4.83	4.37	4.18	3.84
K_2O	3.35	2.17	4.14	3.07
MnO	0.04		0.05	0.07
P_2O_5	0.20		0.07	0.21
H_2O	1.2		0.77	0.65
K_2O/Na_2O	0.69	0.50	0.99	0.80
Sc	5			12
Cr	7		7	20
Ni	7	11		15
Co	5		19	10
Rb	88	70	268	110
Sr	540	957	221	450
Zr	120	329		130
Ba	750	1470	590	600
La	41	52	57	36
Ce	82	94	95	47
Nd	30	42	34	26
Sm	5.8	9.2	5.3	6.8
Eu	1.2	2.3	0.81	1.7
Gd	3.2	4.5	2.6	7.4
Dy	2.9	4.5	3.0	3.2
Er	1.3	2.3	1.7	4.8
Yb	1.0	2.0	1.7	3.6
Lu	0.16	0.34	0.27	0.55
K/Rb	330	262	128	231
Rb/Sr	0.18	0.07	1.2	0.24
Ba/Sr	1.5	1.5	2.7	1.3
$(La/Sm)_N$	3.9	3.1	5.9	2.9
$(Yb/Gd)_N$	0.39	0.55	0.81	0.61
Eu/Eu*	0.85	1.0	0.71	0.75

N = chondrite-normalized ratio.

Chief references: Nockolds (1954), Lo (1970), Condie and Lo (1971), Anhaeusser (1973b), Hunter (1973), Condie and Hunter (1976), Glikson (1978), and K.C. Condie (unpublished data).

high-Al_2O_3 trondhjemitic rocks (Fig. 5-13). REE patterns are enriched in light REE, depleted in heavy REE, and may show minor negative Eu anomalies (Drury, 1979) (Fig. 5-15). Compared to an average composition of post-Archean granodiorite (Table 5-2; Fig. 5-15), Archean granodiorites are lower in some transition metals (Sc, Cr, Mn) and exhibit steeper light- and heavy-REE patterns.

Quartz monzonite-granite

Quartz monzonite and granite are the most widespread rock types in most granitic plutons and in some batholiths in Archean granite-greenstone terranes, as previously discussed. Average compositions of three batholiths composed principally of quartz monzonite and of two granite types from plutons in South Africa are given in Table 5-3. Although granite and quartz monzonite may grade into rocks of syenite or granodiorite composition, many plutons are composed exclusively of granite or quartz monzonite of rather uniform composition. The average compositions given in Table 5-3 are quite similar except for Sr, Ba, and REE contents. They exhibit a wide distribution on the Ba-Rb-Sr diagram (Fig. 5-13) in which they define a trend of Ba enrichment followed by Rb enrichment. Most rocks have low Sr contents. REE patterns show light-REE enrichment, variable negative Eu anomalies ($Eu/Eu^* = 0.2$—0.7), and variable heavy-REE depletion (Fig. 5-16). In general, Archean granite and quartz monzonite are similar in composition to post-Archean counterparts (Table 5-3). Existing data, however, suggest that they may have slightly higher Co and Cr contents and lower Zr contents than post-Archean varieties. In terms of REE distributions, the Archean varieties tend to have steeper light-REE patterns ($La_N/Sm_N > 3.5$), and in some cases steeper heavy-REE patterns, than post-Archean varieties (Fig. 5-16).

Alkaline plutonic rocks

Alkaline plutonic rocks in granite-greenstone terranes occur as small parts of batholiths and as small, generally post-tectonic plutons. They are an extremely minor component in such terranes. Average compositions of two Archean syenodiorites and a syenite are given in Table 5-4. They exhibit variable Al_2O_3, MgO, K_2O, Rb, K/Rb, and Rb/Sr and have very low Rb contents compared to most granites and quartz monzonite (Fig. 5-13). Available REE data suggest very similar, strikingly fractionated REE patterns for these rocks (Fig. 5-17). Compared to the post-Archean syenite average given in Table 5-4, Archean alkaline granitic rocks are higher in MgO, P_2O_5, Sr, and transition metals and lower in K_2O. They also exhibit lower Rb/Sr and Ba/Sr ratios. REE patterns in the Archean alkaline rocks differ significantly from those in post-Archean alkaline rocks (Fig. 5-17).

TABLE 5-3

Average compositions (oxides in wt.%, trace elements in ppm) of Archean and post-Archean granites and quartz monzonites

	Archean					Post-Archean	
	Giants Range (qm)	Laramie (qm)	Lochiel (qm)	Mpageni type (gr)	Sicunusa type (gr)	quartz monzonite	granite
SiO_2	73.0	72.9	71.3	70.3	73.8	69.2	72.1
TiO_2	0.19	0.30	0.32	0.46	0.26	0.56	0.37
Al_2O_3	14.9	14.2	14.4	14.0	13.2	14.6	13.9
Fe_2O_3	0.54	0.73	0.59	1.03	1.62	1.22	0.86
FeO	1.01	1.47	1.73	1.83	0.53	2.27	1.67
MgO	0.49	0.49	0.57	0.63	0.35	0.99	0.52
CaO	1.10	1.19	1.33	2.04	1.11	2.45	1.33
Na_2O	3.90	3.45	3.92	3.57	3.20	3.35	3.08
K_2O	4.47	4.60	4.59	5.18	5.15	4.58	5.46
MnO	0.03		0.08	0.05	0.05	0.06	0.06
P_2O_5	0.11		0.31	0.12	0.09	0.20	0.18
H_2O			0.66	0.79	0.60	0.54	0.53
K_2O/Na_2O	1.2	1.3	1.2	1.5	1.6	1.4	1.8
Cr		4	8	7	5	3	2
Co		2	6	3	4	2	2
Rb	192	184	196	230	270	200	250
Sr	202	112	122	306	81	100	100
Zr	225	121	97	161		300	200
Ba	676	632	500	1150	450	700	500
La		54	70	129	83	60	45
Ce	75	120	131	229	159	120	90
Nd	25	43	48	78	96	48	42
Sm	3.6	7.5	9.1	14	12	12	10
Eu	0.61	0.76	1.5	1.9	0.78	1.6	0.70
Gd	2.4	4.0	6.5	9.0	10	7.5	7.5
Dy	1.4	4.8	5.8	8.8	11	10	8.0
Er	0.61	2.6	2.8	4.2	6.4	7.2	4.6
Yb	0.61	2.4	2.5	3.5	5.8	7.0	4.0
Lu	0.11	0.41	0.36	0.54	0.85	1.2	0.7
K/Rb	198	250	168	187	158	190	181
Rb/Sr	1.1	1.6	1.9	0.75	3.3	2.0	2.5
Ba/Sr	3.4	5.6	4.1	3.8	5.6	7.0	5.0
$(La/Sm)_N$		3.9	4.2	5.1	3.8	2.8	2.5
$(Yb/Gd)_N$	0.32	0.75	0.48	0.48	0.72	1.2	0.66
Eu/Eu*	0.63	0.45	0.61	0.72	0.20	0.52	0.25

gr = granite; qm = quartz monzonite; N = chondrite-normalized ratio.

Chief references: Nockolds (1954); Condie (1969b); Sims and Viswanathan (1972); Hunter (1973; 1974b); Arth and Hanson (1975); Condie and Hunter (1976); Glikson (1978); K.C. Condie (unpublished data).

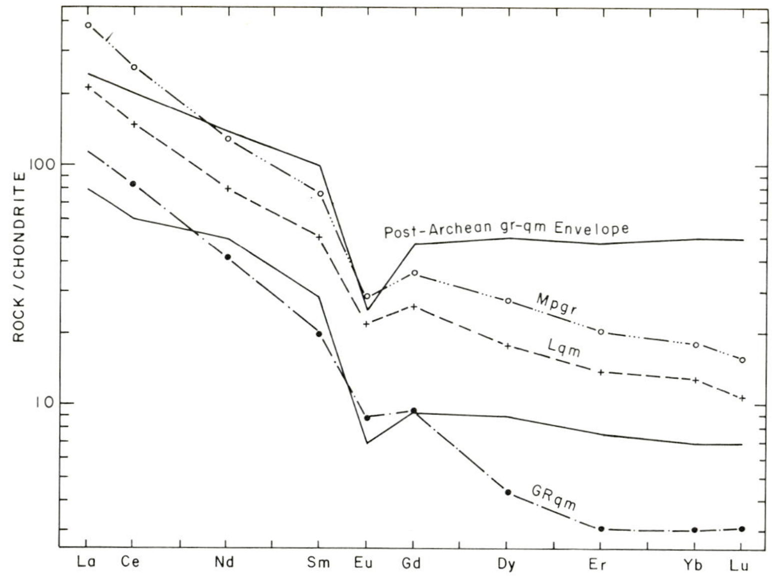

Fig. 5-16. Envelope of variation of chondrite-normalized REE abundances in post-Archean granite and quartz monzonite (gr, qm) compared to three Archean varieties. References given in Table 5-3. Abbreviations: *GRqm* = Giants Range quartz monzonite; *Lqm* = Lochiel quartz monzonite; *Mpgr* = Mpageni-type granite from South Africa.

The Archean rocks exhibit more fractionated light REE ($La_N/Sm_N > 3$) and significant depletion and fractionation of heavy REE ($Yb_N/Gd_N \leqslant 0.2$). Negative Eu anomalies, which characterize post-Archean rocks, are also missing from the Archean rocks.

Intra-pluton compositional variation

Several detailed investigations and a number of general studies are available of compositional variations within Archean plutons. The average composition together with the standard and relative deviations of major and some trace elements in the Laramie batholith ($\sim 2000\ km^2$) from eastern Wyoming are summarized in Table 5-5. SiO_2 and Al_2O_3 have relatively small dispersions ($C < 5\%$), CaO and Sr large dispersions ($C \simeq 75\%$) and the remainder of the elements, intermediate dispersions (Condie, 1969b). Of the element ratio variations, Ca/Sr, Rb/Sr, and Na/K are large. Dispersion of elements that follow each other geochemically generally increases as concentration decreases (viz., Rb > K, Ni > Fe, and Zr > Ti). Although compositional zonation was not detected in the Laramie batholith, many Archean plutons are compositionally zoned (Webber, 1962; Dawson and Whitten, 1962; Paulus and Turnock, 1971; Catherall, 1973; Wolhuter, 1973a,b). Usually plutons show a zonation from a felsic center to more mafic borders. The Lake Dufault granodiorite pluton in Quebec appears to be zoned only in

TABLE 5-4

Average compositions (oxides in wt.%, trace elements in ppm) of Archean and post-Archean syenite and related rocks

	Icarus syenodiorite	Bosmankop syenite	Giants Range syenodiorite	Post-Archean syenite
SiO_2	55.5	65.8	56.5	61.9
TiO_2	0.75	0.69	0.63	0.58
Al_2O_3	13.8	14.9	18.9	16.9
Fe_2O_3	3.46	2.33	2.77	2.32
FeO	2.97	1.61	3.22	2.62
MgO	6.02	1.14	3.18	0.96
CaO	7.55	2.29	4.92	2.54
Na_2O	4.53	4.75	5.90	5.46
K_2O	4.04	4.67	2.60	5.91
MnO	0.10	0.08	0.09	0.11
P_2O_5	0.59	0.37	0.40	0.19
H_2O	0.39	0.85		0.53
K_2O/Na_2O	0.90	0.98	0.44	1.1
Sc		8		2
Cr		8		2
Co		10		1
Ni		6		4
Rb	102	196	40	110
Sr	1870	1260	1050	200
Zr		350		500
Ba	1730	1500	1930	1600
La		128		85
Ce	192	270	161	200
Nd	95	87	74	90
Sm	16	14	11	18
Eu	4.1	3.9	2.7	1.3
Gd		9.9	7.5	11
Dy	4.8	5.7	3.5	15
Er	1.6	2.0	1.4	7.0
Yb	1.1	0.95	1.2	6.4
Lu	0.17		0.19	1.1
K/Rb	328	198	670	446
Rb/Sr	0.06	0.16	0.04	0.55
Ba/Sr	0.93	1.2	1.3	8.0
$(La/Sm)_N$	3.1	5.0	3.6	2.6
$(Yb/Gd)_N$	0.16	0.12	0.20	0.72
Eu/Eu*	1.1	1.0	0.92	0.28

N = chondrite-normalized ratio.

Chief references: Nockolds (1954); Turekian and Wedepohl (1961); Goldich et al. (1972); Arth and Hanson (1975); Glikson (1978); and miscellaneous sources.

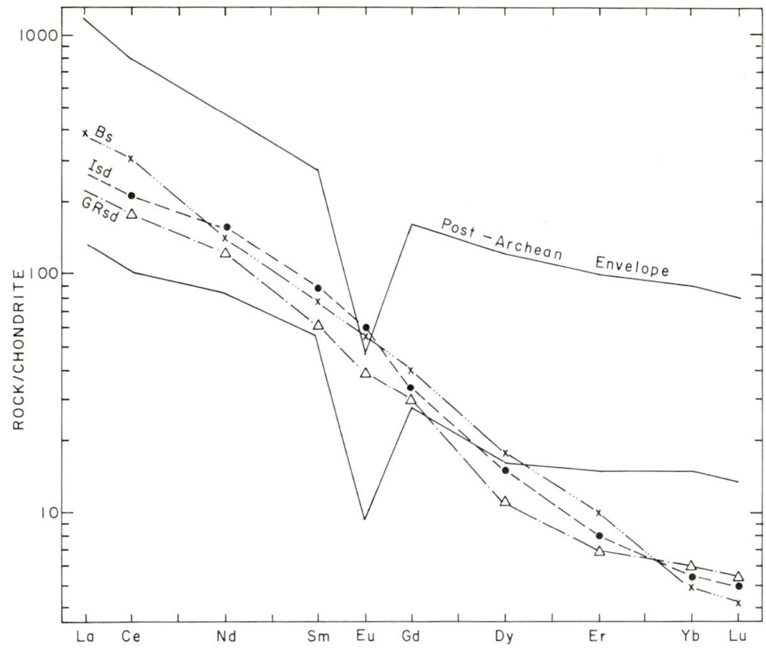

Fig. 5-17. Envelope of variation of chondrite-normalized REE abundances in post-Archean syenites and related rocks compared to three Archean alkaline plutonic rocks. References given in Table 5-4. Abbreviations: *GRsd* = Giants Range syenodiorite; *Isd* = Icarus syenodiorite; *Bs* = Bosmankop syenite.

the western part (Webber, 1962). The Opemisca Lake pluton in Quebec is zoned from a granodiorite core to a syenitic margin (Wolhuter, 1973b): hornblende increases and quartz decreases from core to margin. The Ross River pluton in Manitoba becomes more enriched in CaO and FeO from center to margin (Paulus and Turnock, 1971). Such zonation is commonly interpreted to reflect contamination of the outer parts of the pluton with mafic country rocks (Goldich et al., 1972; Hanson, 1972). An increase in mafic inclusions in pluton border zones supports such an interpretation. Some plutons, however, exhibit an increase in SiO_2 (and quartz) and decrease in K_2O (and K-feldspar) towards the margins as exemplified by the Lacorne-LaMatte-Preissac granitic complex in Quebec (Dawson and Whitten, 1962; Dawson, 1966). Trend surfaces of various elements in this complex indicate that it is composed of several distinct plutons.

Although as previously discussed, some batholiths and plutons exhibit only a limited compositional variation, others vary over a broad range and are often characterized by a calc-alkaline trend (Fig. 5-3). Examples of trends in five batholithic complexes are plotted on an AFM diagram in Fig. 5-18. The Louis Lake and Rainy Lake batholiths exhibit a wide range in composition showing typical calc-alkaline differentiation trends. The Lochiel,

TABLE 5-5

Average composition and variation (oxides in wt.%, trace elements in ppm) in the Laramie batholith, Wyoming (from Condie, 1969b).

	Mean of 75 samples	Standard deviation	Relative deviation, C(%)
SiO_2	72.9	3.1	4.3
TiO_2	0.30	0.13	44
Al_2O_3	14.2	0.5	3.7
$Fe_2O_{3(T)}$	2.36	0.81	34
MgO	0.49	0.20	42
CaO	1.19	0.87	73
Na_2O	3.45	1.10	32
K_2O	4.60	1.17	25
Mn	138	63	46
Ni	6.9	2.9	42
Rb	175	61	35
Sr	183	138	76
Zr	142	70	50

Laramie, and Closepet batholiths, on the other hand, exhibit only limited compositional variations within the granite-quartz monzonite range. Results from some batholithic complexes, such as the Giants Range batholith in Minnesota (Sims and Viswanathan, 1972), indicate that all of the compositional variants within a batholith may not belong to the same magma series.

ORIGIN

The origin of gneissic complexes in granite-greenstone terranes is a subject of considerable discussion and controversy (Glikson, 1979a). It can be divided into three basic problems: (1) the nature of the parent rock for the gneisses (both igneous and sedimentary precursors have been proposed); (2) the role of the K-metasomatism in gneiss production; and (3) the source and mode of production of the parent rock. The last question will be taken up in Chapter 9. Various evidences, field, petrographic, geochemical, and isotopic have been cited to favor igneous or sedimentary precursors for Archean gneisses. Among the criteria used to support a sedimentary precursor are apparent detrital zircon shapes, uniform layering and relict bedding in gneisses, the presence of sedimentary rock inclusions, and the similarity in composition of some gneisses to various clastic sedimentary rocks. Poldervaart (1955a, 1956) suggested that zircon shapes could be used to distinguish

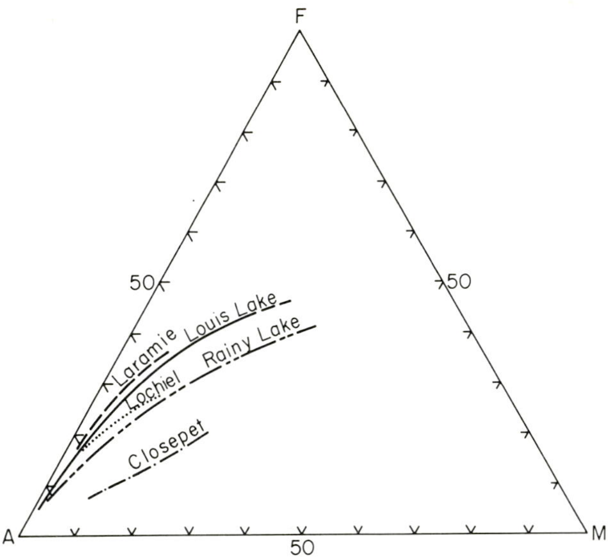

Fig. 5-18. AFM diagram showing trends defined in five Archean batholiths. Principal sources of data: Condie (1969b); Rao et al. (1969); Lo (1970); Hunter (1973, 1974b); Sutcliffe (1978).

para- from orthogneisses. Eckelmann and Poldervaart (1957), Malcuit and Heimlich (1972), and Harris and Goodwin (1976) have employed this method in Archean gneissic terranes. Rounded, often dark-colored zircons are generally interpreted as detrital, while euhedral, often light-colored zircons are interpreted as relict igneous zircons or metamorphic zircons. However, more recent studies of zircons (Reid et al., 1975), indicate that rounded zircons can occur in clearly intrusive granitic plutons. Field relationships, for example, in the Beartooth Mountains indicate that Archean gneissic rocks originally interpreted as metasediments based on zircon shapes (Eckelmann and Poldervaart, 1957), are part of an intrusive pluton. The very uniform layering and banding in some Archean gneisses (Reed, 1963; Breaks et al., 1978), clearly suggest a sedimentary parentage for such gneisses. In some parts of the English River belt it is possible to observe all gradations between graded graywackes and layered gneiss derived from the graywackes (Harris and Goodwin, 1976; Breaks et al., 1978). The presence of inclusions of sedimentary rock in gneissic terranes should not be accepted as evidence for a sedimentary parent for the gneisses in that, as previously mentioned, mafic volcanic inclusions far outnumber sedimentary inclusions. The heterogeneity of inclusion distributions and the fact that trains of inclusions exist between greenstone belts could better be cited as evidence for an intrusive origin for gneisses. A sedimentary origin for the Peninsular Gneiss Complex in India was suggested in Rao et al. (1974) based on a similarity in major element composition to shale and arkose. However, because many known igneous rocks exhibit similar compositional features, such a conclusion does not seem justified. Recent studies of the oxygen isotopic composition of Archean gneisses indicate that it may be possible to distinguish para- from ortho-gneisses by this method in terranes up to middle amphibolite-facies grade (Longstaffe and Schwarcz, 1977; Longstaffe et al., 1978; Longstaffe, 1979). Available data, however, do not indicate a relationship between gneissic precursor and the high- and low-Al_2O_3 compositional groups of tonalite-trondhjemite discussed above.

An igneous precursor for Archean gneisses may be either a volcanic succession dominated by dacitic rocks or intrusive plutons. An igneous origin for most Archean gneisses is supported by their Al/Na + K + 2Ca atomic ratios ($\lesssim$ 1.1) which fall dominantly in the igneous source category of Chappell and White (1974). Evidences which have been cited to support an intrusive origin for many if not most Archean gneissic complexes are as follows (Phaup, 1973; Pichamuthu, 1976): discordant, clearly intrusive contacts; inclusions of greenstone and in particular trains of inclusions leading away from and connecting greenstone belts; and the presence of contact metamorphic aureoles in greenstone belts adjacent to gneisses. Geochemical model studies (Chapter 9) and low $\delta^{18}O$ values and initial $^{87}Sr/^{86}Sr$ ratios (Arth and Hanson, 1975; Hunter et al., 1978) also support an igneous, although not necessarily plutonic, origin for most Archean gneissic complexes. The intimate interlayering of tonalitic (or trondhjemitic) gneiss and amphibolite in many gneissic terranes has been interpreted to favor a volcanic parent (dominated by dacite) for these terranes (Hunter, 1970; Goldich et al., 1972; Hunter et al., 1978).

Existing data seem to indicate that most Archean gneissic complexes had an igneous origin and perhaps the majority of these represent intrusive plutonic complexes. In some regions, such as in parts of the sedimentary-plutonic superbelts in Canada, it appears that paragneisses, derived chiefly from recrystallized graywackes, dominate.

The role of granitization of K-metasomatism in the formation of tonalitic gneiss complexes appears to have been minor in most granite-greenstone terranes. The chief evidence for such metasomatism is generally ascribed to late, microcline megacrysts which can be locally abundant in portions of some granite-greenstone terranes (Hunter, 1970, 1973). This is unlike gneissic complexes in some Archean high-grade terranes in which K-feldspar is widely distributed. Examples are Southwest Greenland and Labrador where the average composition of most of the gneiss terrane ($\gtrsim$ 80%) is granodiorite rather than tonalite (McGregor, 1973; Bridgwater and Collerson, 1976, 1977). In these areas, and in the more localized occurrences in granite-greenstone terranes, K-feldspar occurs as late, randomly dispersed megacrysts, as bands concentrated along foliation planes, and as pegmatitic components. Although textural and field relationships clearly indicate the K-feldspar is late syn-tectonic to post-tectonic in age, the relative roles of late magmatic injection or anatexis and K-metasomatism (perhaps significantly younger than the igneous event) is a subject of current discussion and disagreement (Bridgwater and Collerson, 1976, 1977; Glikson, 1977b).

Archean plutons and batholiths appear to have been emplaced chiefly by vertical forces as evidenced by the distribution of foliation and lineation. Single and multiple diapirs are emplaced syn-tectonically whereas homogeneous, largely discordant plutons reflect post-tectonic emplacement (Viljoen and Viljoen, 1969f; Hunter, 1973; Schwerdtner, 1976). Some batholiths, like the Lochiel in South Africa, appear to have been emplaced as subhorizontal sheets fed by vertical dike systems (Hunter, 1973). Small quartz monzonite plutons in Manitoba were emplaced as structural discontinuities (Ermanovics, 1971). In zoned or composite bodies, the order of emplacement is from more mafic to more felsic. The origin of zoned Archean plutons has been discussed by several investigators and several mechanisms for the production of zonation have been proposed. Some investigators have suggested successive intrusions along the same axis of more to less mafic magma with time (Brownell, 1941; Heimlich, 1965, 1966). Later, more felsic intrusions may have been produced by fractional crystallization of earlier magmas (Dawson, 1966; Condie and Lo, 1971). Gradational contacts between successive zones would appear to necessitate earlier zones not being completely solid as later, central zones are intruded. As previously indicated, the mafic nature of marginal zones in some plutons has been related to varying degrees of digestion of mafic rocks in adjacent greenstone belts (Dawson, 1966; Wolhuter, 1973a). Dawson (1966) suggests that the outer syenodiorite and monzonite phases of the Preissac-Lacorne batholith in Quebec resulted from contamination of a parent granodiorite magma with mafic country rocks. Brownell (1941) suggests that the outer syenodioritic portions of the Falcon Lake Stock in Manitoba were produced by metasomatism on granodiorite by K-rich fluids derived from later felsic intrusion in the center.

The emplacement history of Archean batholiths is poorly known and available studies indicate very complex histories. Recent studies of the Rainy Lake batholithic complex in Ontario suggest three evolutionary stages (Sutcliffe, 1978):

(1) Initial intrusion of tonalite-granodiorite gneissic diapirs which may represent remobilized gneissic complexes.

(2) Emplacement of the Jackfish Lake—Weller Lake pluton along the interface between the diapirs and surrounding greenstone belt rocks (the pluton appears to have ascended as a conformable, near vertical sheet).

(3) Late, discordant biotite granite (with pegmatites) plutons invade and fragment earlier granitic rocks.

Detailed studies of the Vermilion batholith in Minnesota reveal the following stages of development (Southwick, 1972, 1978):

(1) Early emplacement intrusion and extrusion of mantle-derived trondhjemite-granodiorite magmas.

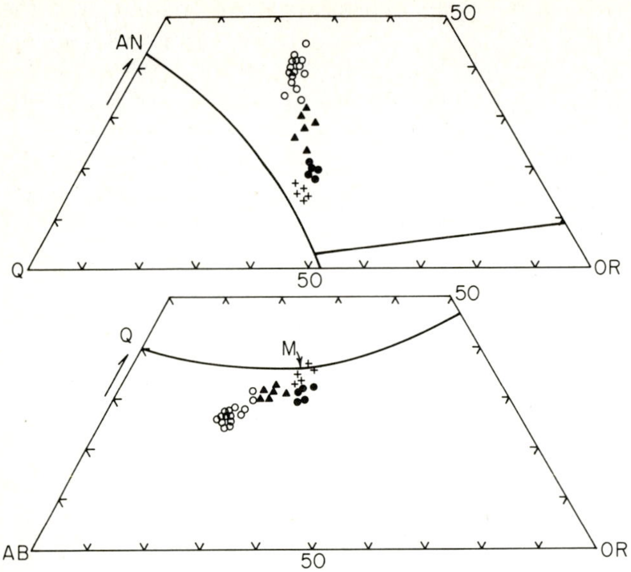

Fig. 5-19. Projection of individual samples from the Louis Lake batholith (Wyoming) on the Q-Ab-Or and Q-An-Or faces of the Ab-Or-An-Q system at 1 kbar P_{H_2O} (after Lo, 1970). Experimental references: Tuttle and Bowen (1958), Luth et al. (1964) and James and Hamilton (1969). Symbols: ○ = granodiorite; ▲ = quartz monzonite; ● = granite; + = aplite; M = minimum at P_{H_2O} = 1 kbar.

(2) Diapiric injection of the Lac LaCroix granite and associated migmatization with emplacement resulting in flattening of early folds and some refolding.

(3) Volatile-rich fluids accumulate in the roof zone of the granite resulting in pegmatite formation.

Consideration of norms of Archean granitic rocks in light of experimental data in the system Ab-Or-Q-An can be informative in terms of pluton origin. Data from small homogeneous granite-quartz monzonite plutons and from large parts of some quartz monzonite batholiths cluster near the minimum in the system Ab-Or-Q at low water pressures ($\leqslant$ 5 kbar) (Condie, 1969b; Gewald and Pirajno, 1973; Hunter, 1974a). This not only favors an origin by magmatic processes, but indicates these bodies represent either first liquids produced by partial melting of crustal rocks or residual liquids remaining after fractional crystallization of more mafic parent magmas. The near absence of earlier more mafic cumulates associated with these bodies and geochemical and isotopic results discussed in Chapter 9 favor a crustal partial melting origin. The absence or sparsity of pegmatites and aplites in many of these bodies also supports low water contents for the magmas. Norms from tonalite and trondhjemite plutons commonly cluster in the plagioclase volume near the Plg-Q cotectic or the Plg-Q boundary surface at low water pressures. The close clustering of norms from individual plutons (Wolhuter, 1973b; Hunter, 1974a) indicates a lack of fractional crystallization in these bodies. Data from some batholiths, however, support a fractional crystallization origin as exemplified by the Louis Lake batholith (Lo, 1970; Condie and Lo, 1971). The norms of the major compositional variants in this body are plotted on Or-Ab-Q and An-Q-Or projections of the Ab-Or-An-Q system in Fig. 5-19. There is a clear progression consistent with fractional crystallization from granodiorite, through quartz monzonite, to granite and aplite with the granite and aplite points clustering above the minimum at 1—5 kbar P_{H_2O}. Geochemical model studies also support an origin for the observed rock types in this body by fractional crystallization (Condie and Lo, 1971).

Chapter 6

STRUCTURE AND METAMORPHISM

INTRODUCTION

The geologic history of Archean granite-greenstone terranes is complex involving periods of deformation, metamorphism, and plutonism (Anhaeusser et al., 1969; Goodwin et al., 1972). Despite this complexity certain patterns are repeated, as pointed out in Chapter 1. In terms of structure, the earliest folds in greenstone belts are usually the largest with wave lengths of kilometers to tens of kilometers. These folds are isoclinal and generally have steeply dipping axial planes and moderate to steep plunges. In some belts, the early folds are cut by cross-folds oriented at steep angles. Sedimentary rocks are generally more tightly folded than volcanic rocks and sills (McGlynn and Henderson, 1972). Late folds are small (centimeters to meters) and often involve conjugate sets. Steeply plunging stretch lineations are common in foliation planes both in greenstones and granitic rocks near greenstone-granite contacts in which granites are syntectonic. Folded arms of greenstone belts may extend into surrounding gneissic terranes where they become broken and sometimes partially granitized. Such arms can be traced for many kilometers into gneissic terranes by trains of inclusions. Goodwin (1965) suggests that folds become more frequent and have shorter wavelengths as the margins of major volcanic complexes are approached. Various penetrative fabrics may develop during folding, the major fabrics usually associated with the early periods of deformation. Non-penetrative cleavages commonly develop during the late stages of deformation.

Faults in greenstone belts are of diverse types and ages. Most faults can be traced with confidence for only a few kilometers. A few major faults, such as the Larder Lake Break in the Superior Province (Wilson, 1962), can be traced for nearly 100 km. Most faults in greenstone belts are parallel or subparallel to major folds and record dip-slip or transcurrent motions. Major faults have steep dips and some have associated shear zones up to several hundred meters wide (Henderson and Brown, 1966; Stone, 1976). As indicated by some structural studies (Coward, 1976; Gorman et al., 1978), however, thrust faults may be more common in greenstone belts than suggested by the literature. Faulting appears to have gone on throughout the polyphase deformational history of greenstone belts, with early faults being reactivated during later periods of deformation.

Archean granite-greenstone terranes exhibit varying degrees of metamorphism (Fraser and Heywood, 1978). Evidences for regional, contact and retrograde metamorphism are found in most greenstone belts and in addition, many rocks have undergone pre- or post-metamorphic alteration. Regional metamorphic grade is generally of the greenschist or amphibolite facies although terranes metamorphosed to the prehnite-pumpellyite and granulite facies are found in some areas. Primary textures and structures range from well-preserved in low-grade terranes to absent in highly metamorphosed and sheared terranes. Contact metamorphism is common around the margins of many granitic plutons and may be syn- or post-regional metamorphism. Contact aureoles are up to several kilometers wide, discontinuous, and exhibit grades of metamorphism from the hornblende to the pyroxene hornfels facies (Ayres, 1978). Retrograde metamorphism, although widespread in most greenstone terranes, is of minor importance, generally characterized by incipient chloritization of biotite and hornblende and saussuritization of plagioclase.

It is noteworthy that there does not seem to be any relationship between apparent stratigraphic thickness in greenstone belts and metamorphic grade (Goodwin et al., 1972; Binns et al., 1976). As many as three periods of metamorphism (including both regional and contact) are recorded in some greenstone belts and both syn- and post-tectonic types are present. Prismatic metamorphic minerals and other linear features in greenstone rocks are often aligned parallel to intrusive contacts and steeply dipping lineations are common near such contacts (Anhaeusser et al., 1969). Metamorphic grade changes from low (greenschist facies) to high (amphibolite facies) in going from the center to the edges of many individual belts (Engel, 1968; Ayres, 1978). However, as illustrated in later discussions, there is not a simple relationship between metamorphic grade and distance from the center of a belt. In terms of facies series (Miyashiro, 1973), low-pressure (andalusite-sillimanite) to less commonly medium-pressure (kyanite-sillimanite) types characterize Archean granite-greenstone terranes (Shackleton, 1976).

Although few studies of the effects of progressive metamorphism on the composition of rocks in Archean greenstone belts are available, other investigations of progressive metamorphism are pertinent in this regard. Extensive data from regional sampling of the Canadian Shield (Eade and Fahrig, 1971, 1973) suggest that Si, Na, K, H_2O, Rb, Cs, Th, and U are lost and Mg, Fe, and Ca may be enriched in going from the amphibolite to the high-pressure granulite facies. Other elements exhibit irregular behavior (viz. Pb, Sr, Zr, Cr) or show şmall, non-systematic changes with increasing grade of metamorphism (viz., Ni, Co, Cu, Zn, V, Sc, Ti). The most rewarding studies of chemical changes as a function of increasing metamorphic grade in the same rock type are those of Engel and Engel (1958, 1962) and Schwarcz (1966). The Engel's studies of Precambrian paragneisses and interlayered amphibolites in the Adirondack Mountains in New York indicate decreases in Si, K, H_2O, F, Cl, and Fe^{3+} and increases in Ca and Mg in both rock types in going from the upper amphibolite to the lower granulite facies. Schwarcz (1966) studied meta-arkoses in Southern California ranging from the lower to the upper hornblende hornfels facies. His results indicate a decrease in Fe^{3+} and possibly slight increases in Mn, Ti, Sr, Co, La and Y with increasing grade.

The effects of retrograde metamorphism in Precambrian gabbros from southern Norway have been defined by sampling on a scale of centimeters to meters (Elliot, 1973; Field and Elliot, 1974). The results of this study indicate that retrograde metamorphism was not isochemical, but involved enrichment of K, H_2O, P, Fe^{3+}, Cl, V, and Rb, and losses of Ca, Fe^{2+}, S, and Zn. Recent studies of retrograded granulite-facies gneisses from Scotland also indicate remobilization of major elements during retrograde metamorphism (Beach and Tarney, 1978).

Although existing results are not adequate to define unambiguously the behavior of most trace elements during progressive metamorphism, they clearly show that metamorphism can have an effect on rock composition.

AREAL STUDIES

The Abitibi belt, Canada

Studies of metamorphic mineral assemblages in the Larder Lake area of the Abitibi greenstone belt in Ontario have been informative in understanding the complex relationships between regional and contact metamorphism (Jolly, 1974; Pearce and Birkett, 1974). Jolly (1974) has divided the area into six distinctive zones based on metamorphic mineral occurrences (Fig. 6-1). The chlorite zone occurs in sediments remote from plutons and is characterized by quartz-albite-calcite-white mica-chlorite-sphene assemblages. In the prehnite zone, prehnite makes its appearance chiefly in fractures but also in volcanic rocks. Volcanics north of the Larder Lake fault contain the mineral pair prehnite-pumpellyite composing up to 95% of the mafic volcanics. It is noteworthy that zeolites are not observed in any of the low-grade rocks. Around some of the granitic plutons, actinolite replaces some of the prehnite and pumpellyite and in the actinolite zone, actinolite, epidote and stilpnomelane are the characteristic minerals and prehnite and pumpellyite are absent. Rocks containing hornblende have a very limited distribution around several small plutons. This general pattern of metamorphic zonation is repeated in the Abitibi belt as a whole (Jolly, 1978). Most of the belt if metamorphosed to the greenschist facies with lower-grade rocks limited to an area in the center of the belt between Noranda and Kirkland Lake and higher-grade (amphibolite-facies) rocks found around the margins of granitic plutons. Compared with thicknesses of young sediment-volcanic successions that have undergone low-grade metamorphism, it would appear that the thickness of the Abitibi Group was $\lesssim 12$ km at the time of prehnite-pumpellyite metamorphism (Jolly, 1978). This is clearly in conflict with estimates of stratigraphic thicknesses which range up to 18 km

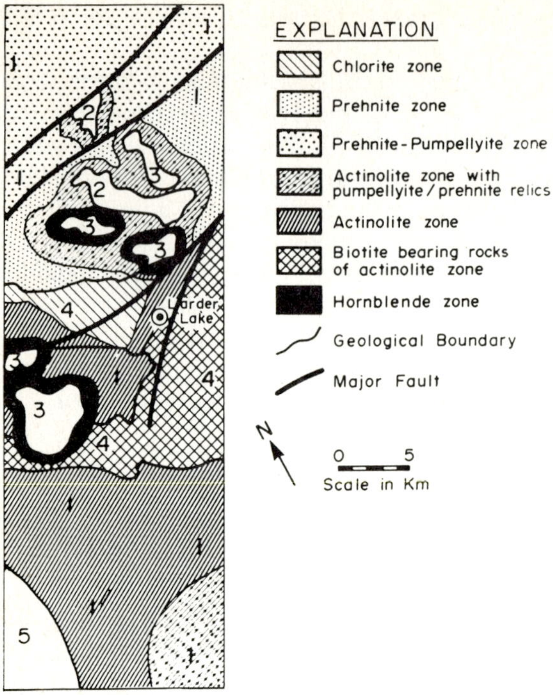

Fig. 6-1. Metamorphic zonation in the Abitibi belt in the vicinity of Larder Lake, Ontario (from Jolly, 1974).

(Table 2-1). It is probable that the stratigraphic thicknesses reflect structural thickening or topographic irregularities produced during or after volcanism.

A summary of the metamorphic history of the Abitibi belt is given in Table 6-1. The two earliest periods of regional metamorphism (M_1 and M_2) are not well known. The metamorphism of the Pontiac Group is the medium-pressure type and differs from the other metamorphism which is the low-pressure type. The major period of low-grade regional metamorphism (M_3) accompanied the eruption and burial of the Abitibi Group. Local contact metamorphism (M_4) accompanied emplacement of syenitic plutons at 2.5—2.6 b.y.

The Slave Province

The Yellowknife greenstone belt in the Slave Province (Fig. 1-9) has been the subject of several geological studies (examples are Henderson, 1943; Henderson and Brown, 1966; Fyson, 1975, 1978; Drury, 1977). Three periods of folding, two periods of metamorphism, and one or more periods of faulting have been recognized in the Yellowknife Supergroup (Figs. 6-2 and 6-3). Development of metamorphic minerals is related chiefly to

TABLE 6-1

Summary of metamorphic history of the Abitibi belt (modified after Jolly, 1978)

Unit/event	Rocks formed	Age (b.y.)	Metamorphism
Kenoran orogeny and post-tectonic plutonism	granitic plutons	2.5—2.6	M_5 contact metamorphism to greenschist and amphibolite facies
Tamiskaming Group	graywacke, conglomerate		
	Unconformity		
Small plutons	syenite and related rocks		M_4 local contact metamorphism
Abitibi Group and early plutonism	volcanics and minor sediments; granitic plutons	~ 2.7	M_3 regional metamorphism to prehnite-pumpellyite and greenschist facies
	Unconformity		
Pontiac Group	clastic sediments		M_2 regional metamorphism to amphibolite facies; intermediate-pressure type
Orthogneiss	tonalite-trondhjemite, amphibolite	2.8—3.0	M_1 regional metamorphism to amphibolite facies
Earlier crust in part sialic		> 3.0	

progressive regional metamorphism to the amphibolite facies. Emplacement of late granitic plutons are thought to have imposed only local contact metamorphic conditions (Kamineni, 1973). The earliest folds (F_1) are broadly open folds which extend for 5—20 km in length and are 1—15 km apart (Fyson, 1975). Except for possible quartz veins along bedding, no metamorphic fabrics are associated with F_1 folds. These folds are markedly discordant to the north-south margins of intrusive granitic plutions. The fact that they appear to diverge around the granodiorite-gneiss complex in the northeastern part of the area (Fig. 6-2) suggests that this complex influenced the folding and hence was emplaced before or during folding. Drury (1977) recognizes an earlier suite of isoclinal folds in the graywackes east of Yellow-

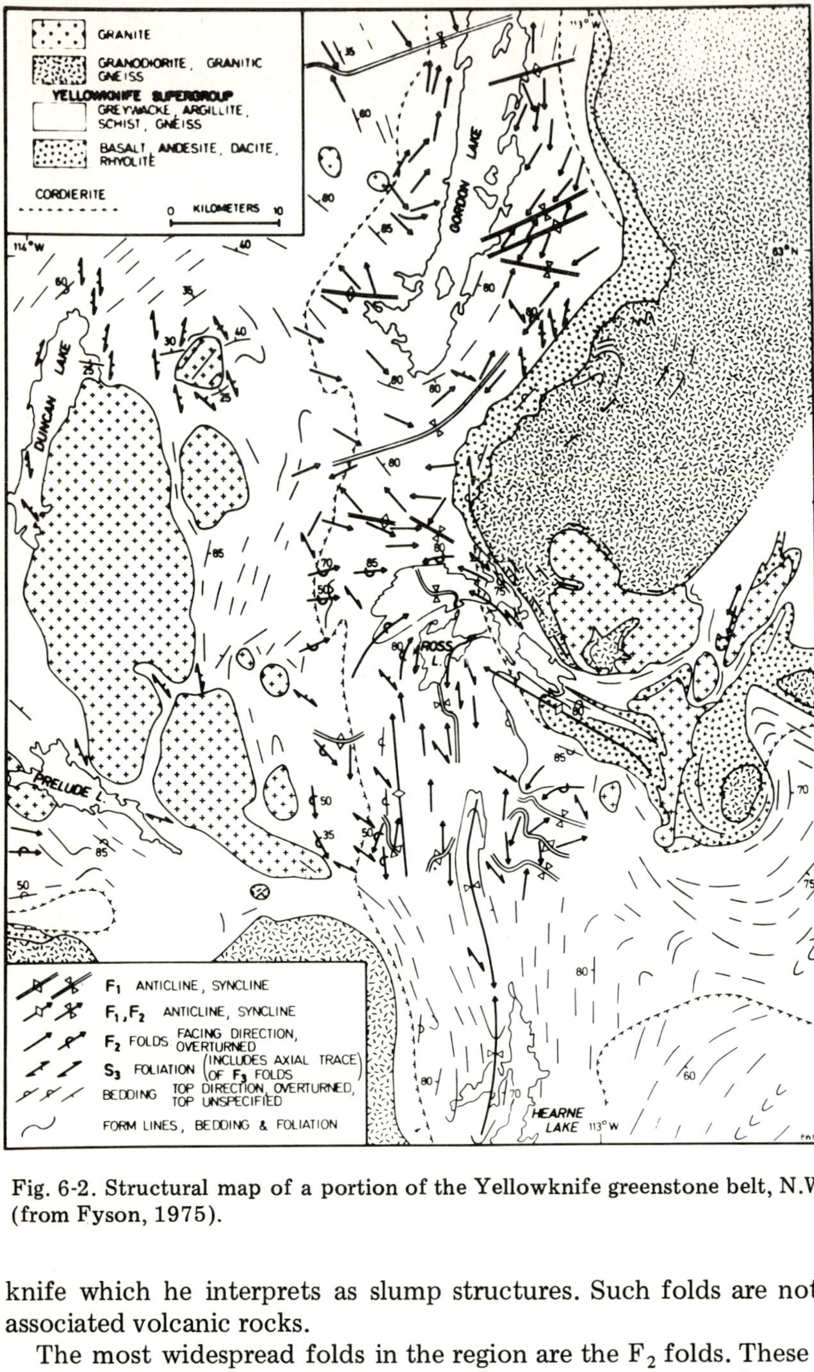

Fig. 6-2. Structural map of a portion of the Yellowknife greenstone belt, N.W.T., Canada (from Fyson, 1975).

knife which he interprets as slump structures. Such folds are not found in associated volcanic rocks.

The most widespread folds in the region are the F_2 folds. These folds vary considerably in size, shape, and orientation and generally have steep axial

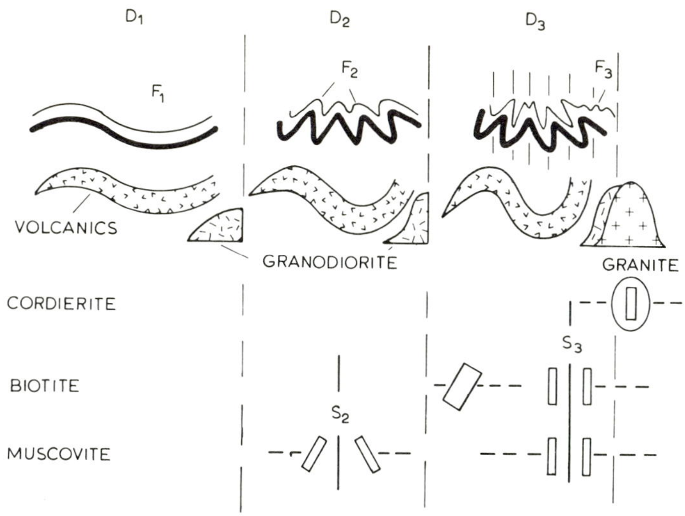

Fig. 6-3. Relative age and geometric relations of folds, metamorphic minerals, and plutons in the Ross Lake area, Yellowknife region, Canada (from Fyson, 1975).

planes. Many are overturned, often towards late granitic plutons, which is inconsistent with the F_2 folds resulting from diapiric emplacement of these plutons. The fold patterns, however, are consistent with uplift of older granodiorite gneiss complexes and gravitational sliding away from such uplifts into depressions (Fig. 6-3). Isoclinal F_2 folds in pelitic units are accompanied by axial surface foliation (S_2) defined by aligned quartz and muscovite (± chlorite). Lenses and layers of quartz alternate with mica-rich S_2 layers. The age of widespread shear zones in the Yellowknife area is unknown, but they also may represent F_2 features (Drury, 1977).

F_3 folds and associated axial-planar S_3 foliation occur scattered throughout the area in thin-bedded sediments. These folds range from open to tight, have subvertical axes, and limbs rarely exceed a few tens of meters in length. The F_3 folds bend S_2 foliation that is axial planar to F_2 folds. The S_3 foliation varies in style from crenulated muscovite layers to a coarsely crystalline schistosity defined by aligned muscovite and biotite. This coarse schistosity increases proportionally towards the cordierite isograd (Fig. 6-2). At least two generations of biotite are recognized, one syn-D_3 and one, possibly just pre-D_3. Cordierite and andalusite crystals are often flattened in S_3 planes indicating a syn-D_3 age. Near some of the granitic plutons, however, such as within a few hundred meters of the pluton east of Prelude Lake (Fig. 6-2), cordierite, andalusite, and sillimanite cross S_3 foliation. This indicates these minerals were formed in contact aureoles around plutons that were emplaced after D_3 (Henderson, 1943; Kamineni, 1973). Cataclastic foliation within

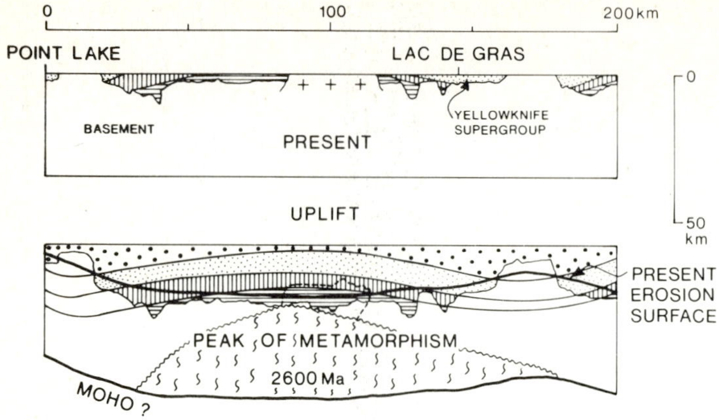

Fig. 6-4. Reconstruction of crustal cross-section today and at 2.6 b.y. between Point Lake and Lac de Gras in the Slave Province (from Thompson, 1978). Key: widely-spaced dots = unmetamorphosed; closely-spaced dots = low grade; vertical lines = medium grade; horizontal lines = high grade; squiggles = partly melted sialic basement.

the marginal areas of some plutons is parallel to S_3 in adjacent sediments, possibly reflecting a continuation of D_3 after pluton emplacement.

Many features of the F_1 and F_2 folds in the Yellowknife belt appear to be related to diapiric uplift of the granodiorite-gneiss complexes. The folds have developed in response to predominantly vertical forces and gravity sliding from high to low areas. The F_3 deformation, on the other hand, developed in response to horizontal forces. Fyson (1975) suggests that the F_3 deformation reflects a transition from a tectonic regime controlled by localized density contrasts in the crust to a widespread, largely compressional regime.

Regional metamorphism in the Slave Province is characterized by the assemblage cordierite-biotite-andalusite-sillimanite (± staurolite) which is indicative of the low-pressure series (Miyashiro, 1973). Kyanite occurs with cordierite, staurolite, and/or sillimanite at a few localities indicating the presence, locally, of the medium-pressure series. Thompson (1978) interprets the patchy distribution of metamorphic grade in the Slave Province to result from differential erosion of crust that was subjected to an irregular distribution of thermal domes and depressions as reflected by the isograd patterns. Estimated *P-T* curves derived from traces across the erosion surface imply post-metamorphic uplift ranging from < 5 km in low-grade areas to 12—15 km in high-grade areas. These results are used to construct a cross-section of part of the Slave Province as it is now and as it was during metamorphism at ~ 2.6 b.y. (Fig. 6-4). Present crustal thickness is estimated from available seismic data. The results suggest the presence of a major thermal dome in the center of the section and an Archean crustal thickness of 40—50 km.

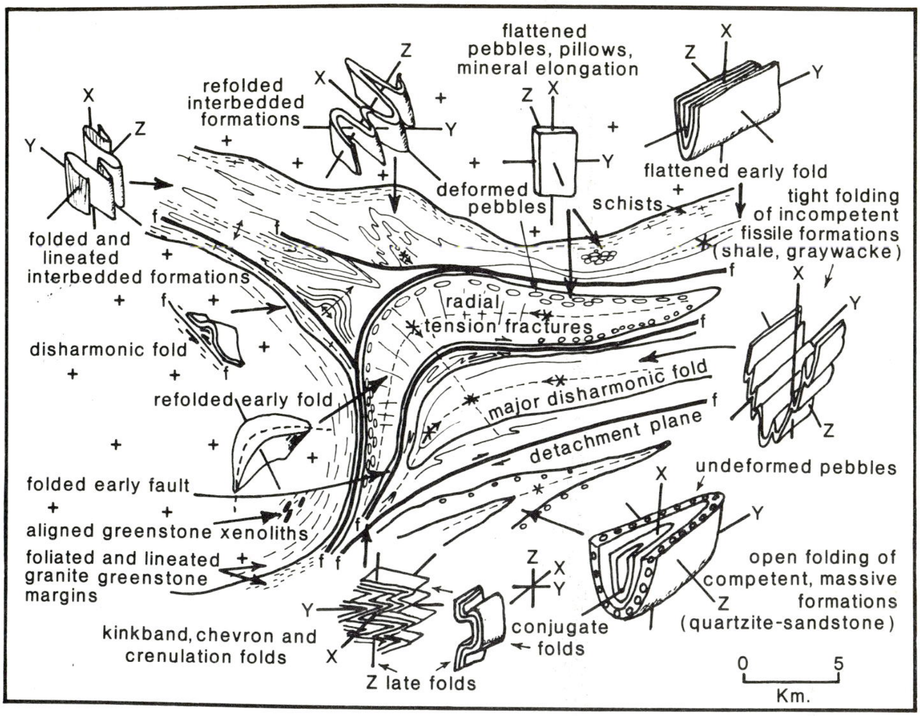

Fig. 6-5. Schematic diagram of the main tectonic elements and strain indicators along the northwest flank of the Barberton greenstone belt (from Anhaeusser, 1975; reproduced with permission of Annual Reviews Inc.).

The Barberton belt, South Africa

The structural investigation by Ramsay (1963) of the Barberton greenstone belt in South Africa is one of the first detailed structural studies of a greenstone belt. Ramsay established the presence of three major periods of deformation in this region. Subsequent studies of Anhaeusser (1974) have provided further detail in the northern part of the greenstone belt. The major structural elements and strain indicators in the Barberton belt are shown schematically in Fig. 6-5. The first deformation in the Barberton area produced a series of large NE-SW-trending folds of which the Eureka Syncline has been most extensively studied. This fold has a curved axial plane which dips steeply to the south, southeast, or east and the plunge of the syncline is to the west, southwest, or south at high angles. The maximum outcrop width of 3 km is where the change in axial-plane strike occurs.

Slaty cleavage and various lineations were produced in the greenstone belt during the second period of deformation. The cleavage cuts across the

Eureka Syncline and related folds with cleavage and bedding planes generally intersecting at low angles. The cleavage is best developed in shales and is not apparent in most of the quartzites. Anhaeusser (1974) has shown that the cleavage poles define a great circle on stereographic projections thus indicating that the cleavage formed during the refolding of the early folds (Fig. 6-5). The only large folds produced during the second deformation occur in and south of the Ulundi Syncline. Steeply plunging lineations are common in the form of small-scale fold axes and alignment of platy inclusions, mineral grains, and elongated pebbles. In general, both cleavage and lineations occur in the outer portions of the Kaap Valley and Nelspruit diapiric plutons which are intrusive into the greenstones. This relationship strongly suggests that the second-stage deformation was related to emplacement of the diapiric plutions. The following sequence of events is suggested by Anhaeusser (1974) for the second period of deformation:

(1) Compression from the NW-SE producing the slaty cleavage which is superimposed obliquely across the earlier folds.

(2) Deformation along the northern flank of the Eureka Syncline (closest to the deforming forces) resulting in well-developed planar and linear fabrics in the rocks.

(3) Final emplacement of adjacent granitic plutons causes refolding of first-generation folds and of the slaty cleavage and produces steeply dipping foliation and lineations in rocks adjacent to the plutons and in the marginal portions of the plutons.

A third period of deformation is recorded by the presence of small crenulation and chevron folds which are related to an almost vertical stress field. Also present are conjugate folds. These third-period folds, which occur on scales of centimeters to meters, deform all earlier structures.

Faults in the Barberton belt are of several types and developed at different times; some were reactivated during later deformation. The major strike faults, as illustrated by the Sheba fault, occur along the overturned limbs of the first-stage folds (Fig. 6-5). Most evidence suggests they are high-angle thrusts. A number of minor faults and shear zones were produced during the second deformation and represent tension fractures which strike dominantly in a northwesterly direction.

Deformed pebbles in conglomerates of the Moodies Group have been used to estimate strain at a number of localities (Ramsay, 1963; Gay, 1969; Anhaeusser, 1974). The results of these studies in the Eureka Syncline area are as follows (Anhaeusser, 1974): (1) pebbles along the northern and western limbs of the Eureka Syncline are more deformed than elsewhere within the structure; (2) the greatest amount of pebble elongation occurs in conglomerates adjacent to intrusive diapiric plutons; and (3) pebble deformation along the southern limb of the syncline is minor or non-existent.

Ramsay (1963) suggests the following deformational history of the Barberton belt:

Stage 1. NW-SE-directed compressive forces produce a series of major folds striking NE-SW or NNE-SSW and associated high-angle faults.

Stage 2. (1) Continued compression in the same general direction produces slaty cleavage which is superposed on the stage-1 folds.

(2) New folds develop locally.

(3) Regional low-grade metamorphism is associated with this stage of deformation.

(4) Emplacement of diapiric plutons causes refolding of stage-1 folds, tensional faulting, and produces steeply dipping foliation and lineations adjacent to and within the marginal regions of the plutons.

Stage 3. Folding of the slaty cleavage by large and small-scale folds with maximum compression in a vertical direction.

The similar orientation of the maximum compressive stress axes for the first- and second-stage deformations suggests they were part of the same orogenic movement. The fact that the maximum stress axis of the third-stage deformation is approximately normal to earlier deformations suggests that some time lapsed between the first two deformations and the third one.

Greenstone belts of the Norseman area, Western Australia

Detailed mapping of a greenstone terrane in the vicinity of Norseman in the Yilgarn Province in Western Australia has revealed the presence of four major periods of deformation (Fig. 6-6) (Archibald et al., 1978). The area can be divided into two structural-metamorphic domains (Fig. 6-6B). The static domain is characterized by deformations which predate metamorphism and the dynamic domain by synchronous deformation and metamorphism. Within the static terrane, metamorphic grade increases from north to south from the lower-greenschist to the lower-amphibolite facies. Within the dynamic domain, the grade ranges from mid-amphibolite facies around the Widgiemooltha dome to high-amphibolite facies around the Pioneer dome. Pelites in the static domain contain andalusite and sillimanite and those in the dynamic domain contain staurolite and almandite suggesting the low- to medium-pressure facies series. The deformational events are schematically illustrated in Fig. 6-7. The first deformation (D_1) is manifest by isoclinal folds in the static terrane and by younging reversals within sequences folded by the second-phase deformation (D_2) in the dynamic terrane. Geometric analysis of the F_1 folds is uncertain due to the lack of a tectonic fabric and rarity of small folds. They are tentatively interpreted in terms of recumbent folds, possibly nappes. Metamorphic fabrics are not associated with D_1 and only minor low-grade metamorphism, as evidenced by the development of slaty cleavage in pelites, appears to have accompanied D_2 (Fig. 6-7).

The second deformation (D_2) is widespread and is characterized by regional N- to NNW-trending slaty cleavage that is axial planar to large folds. In the dynamic domain, F_3 folds also have a metamorphic foliation parallel

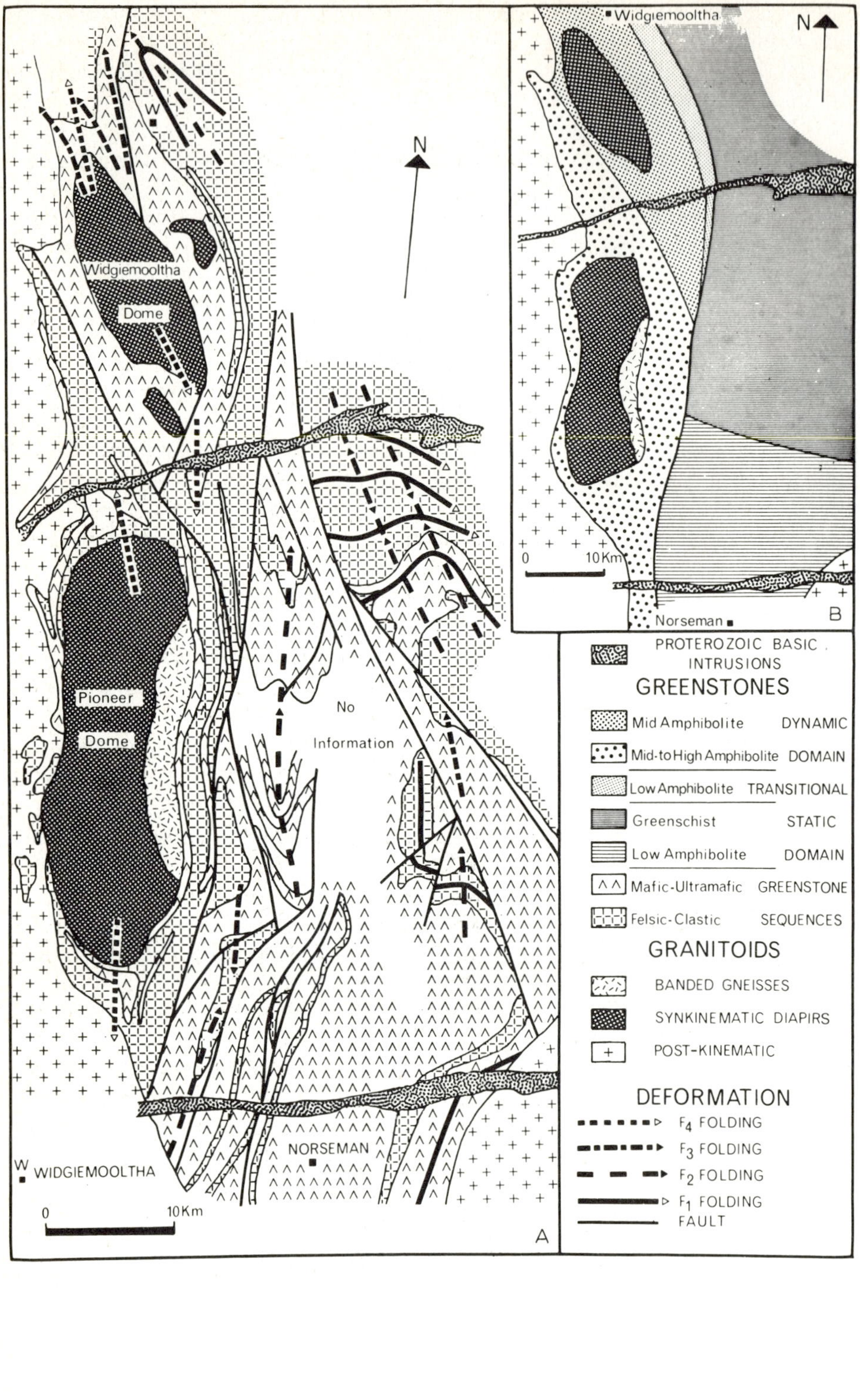

Widgiemooltha
N
W
Widgiemooltha
Dome
Pioneer
Dome
No
Information
NORSEMAN
W WIDGIEMOOLTHA
0
10Km
A
Widgiemooltha
N
0
10Km
Norseman
B
PROTEROZOIC BASIC INTRUSIONS
GREENSTONES
Mid Amphibolite
Mid-to High Amphibolite
DYNAMIC DOMAIN
Low Amphibolite
TRANSITIONAL
Greenschist
Low Amphibolite
STATIC DOMAIN
Mafic-Ultramafic
Felsic-Clastic
GREENSTONE SEQUENCES
GRANITOIDS
BANDED GNEISSES
SYNKINEMATIC DIAPIRS
POST-KINEMATIC
DEFORMATION
F4 FOLDING
F3 FOLDING
F2 FOLDING
F1 FOLDING
FAULT

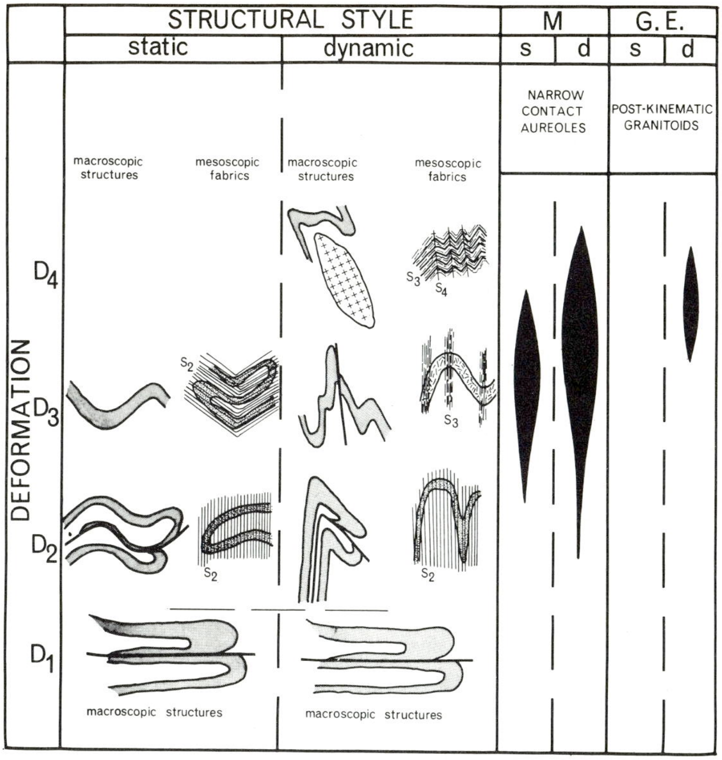

Fig. 6-7. Schematic diagram showing the deformational history of greenstones in the Norseman area, Western Australia (from Archibald et al., 1978). M = metamorphism; $G.E.$ = granite emplacement; s = static domain; d = dynamic domain.

to their axial planes. In the static domain, F_3 folds are open folds and metamorphism appears to be approximately coincident with D_3. Synkinematic plutons and associated contact metamorphism accompanied D_3. Textural studies indicate that metamorphic minerals associated with D_3 grew simultaneously in static and dynamic domains and that this represents the major period of regional metamorphism. D_4, which is confined to the dynamic terrane, is characterized by folding of the F_3 cleavage. Late syn-tectonic plutons were emplaced during the waning stages of D_4. The last metamorphic episode recorded is the development of narrow contact metamorphic aureoles around post-tectonic plutons. The youngest structures in the area

---

Fig. 6-6. Structural-metamorphic maps of the Widgiemooltha-Norseman area, Western Australia (from Archibald et al., 1978). A. Distribution of litho-stratigraphic units and major generations of folds. B. Structural-metamorphic domains.

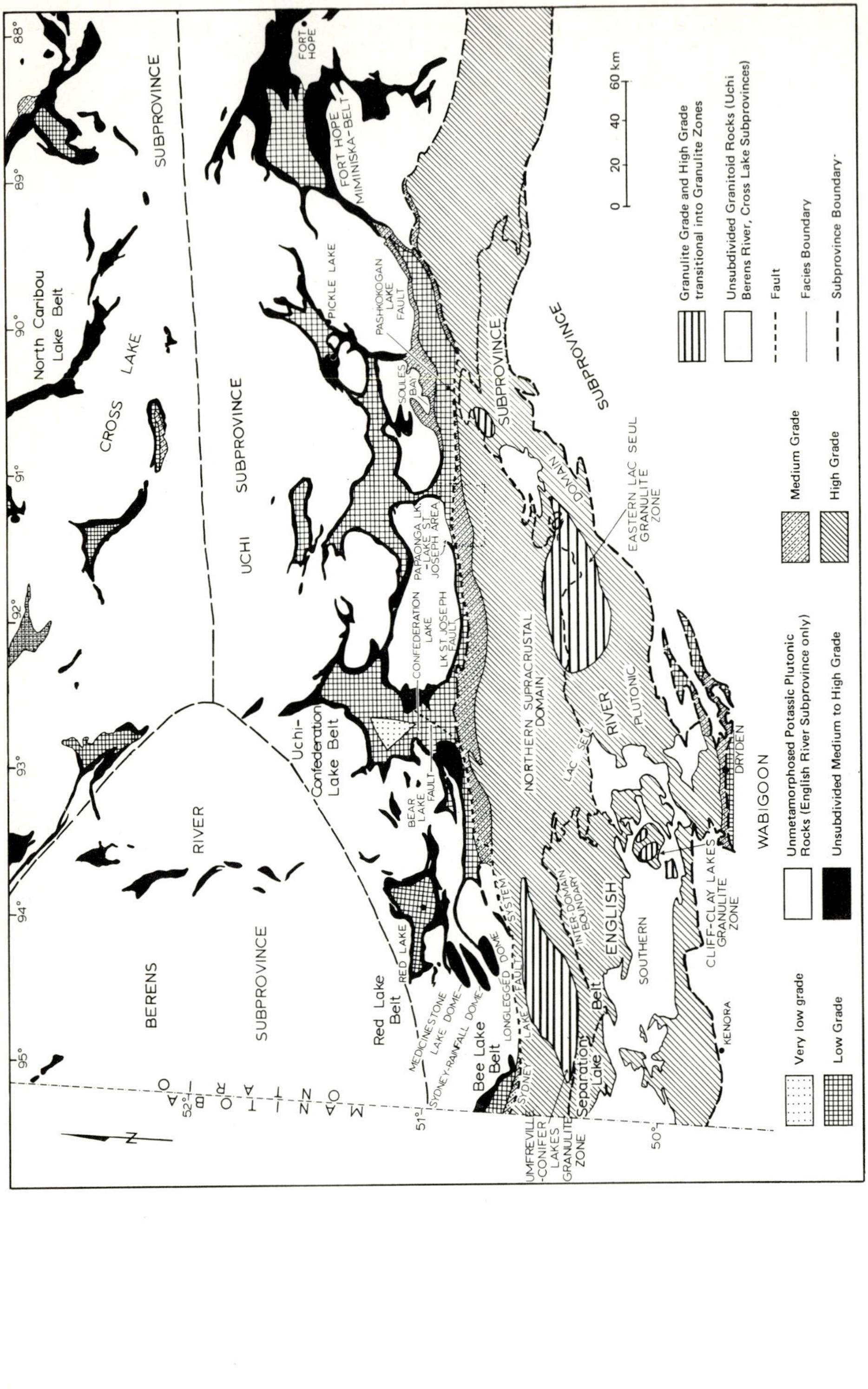
N
BERENS
SUBPROVINCE
RIVER
NORTH CARIBOU LAKE BELT
CROSS
LAKE
SUBPROVINCE
UCHI
SUBPROVINCE
Red Lake Belt
RED LAKE
Uchi-Confederation Lake Belt
BEAR LAKE FAULT
CONFEDERATION LAKE
LK ST. JOSEPH FAULT
PAPAONGA LK.-LAKE ST. JOSEPH AREA
SOULES BAY
PICKLE LAKE
PASHKOKOGAN LAKE FAULT
FORT HOPE MIMINISKA BELT
FORT HOPE
MEDICINESTONE LAKE DOME
SYDNEY-RAINFALL DOME
Bee Lake Belt
LONGLEGGED DOME
SYDNEY LAKE FAULT SYSTEM
UMFREVILLE-CONIFER LAKES GRANULITE ZONE
Separation Lake Belt
NORTHERN SUPRACRUSTAL DOMAIN
INTER-DOMAIN BOUNDARY
LAC SEUL
EASTERN LAC SEUL GRANULITE ZONE
DOMAIN
ENGLISH
RIVER
PLUTONIC
SOUTHERN
SUBPROVINCE
CLIFF-CLAY LAKES GRANULITE ZONE
KENORA
DRYDEN
WABIGOON
MANITOBA
ONTARIO
95°
94°
93°
92°
91°
90°
89°
88°
52°
51°
50°
0 20 40 60 km
Very low grade
Low Grade
Unmetamorphosed Potassic Plutonic Rocks (English River Subprovince only)
Unsubdivided Medium to High Grade
Medium Grade
High Grade
Granulite Grade and High Grade transitional into Granulite Zones
Unsubdivided Granitoid Rocks (Uchi Berens River, Cross Lake Subprovinces)
Fault
Facies Boundary
Subprovince Boundary

are transcurrent and normal faults, some of which cut the youngest granitic plutons.

The deformational history of the Norseman greenstones appears to reflect early dominantly compressional forces (D_1, D_2) and later dominantly vertical forces (D_3, D_4) related to emplacement of granitic diapirs.

The English River Superbelt, Canada

Recent studies in the metasedimentary terrane of the English River Superbelt in northwestern Ontario (Fig. 1-6) have been instructive in terms of understanding Archean metamorphism (Harris, 1976; Harris and Goodwin, 1976; Thurston and Breaks, 1978). The English River Superbelt is gradational, in part, into the Uchi Superbelt on the north with metamorphic grade decreasing in going south from the Uchi belt (Fig. 6-8). The greenstone belts in the Uchi Superbelt are characterized by greenschist-facies mineral assemblages which change to amphibolite-facies assemblages around their margins in response to heat derived from intrusive granites. Metamorphic grade increases to amphibolite facies and locally to granulite facies in the northern part of the English River belt. Mineral assemblages in metagraywackes indicate that metamorphism was low- to medium-pressure types characterized by andalusite at low grades and sillimanite or rarely kyanite at higher grades. The increase in metamorphic grade in the English River Superbelt is consistent with the model of Richardson (1970) which would suggest that a thermal dome or anticline was present beneath this superbelt during metamorphism. In this model, isotherms are displaced upwards and geothermal gradient steepened over the thermal anticline.

The metamorphic and deformational history for part of the English River Superbelt has been described by McRitchie and Weber (1971) and is summarized in Table 6-2. The earliest metamorphism (M_1) produced porphyroblasts of staurolite, andalusite, biotite, and almandite. These porphyroblasts were rotated during D_2. Metamorphism M_2 resulted in coarsening of metasediments and growth of muscovite and biotite parallel to axial planes of D_2 folds. M_3 metamorphism was retrograde and is characterized by the sericitization of plagioclase and andalusite, chloritization of biotite and amphibole, and pinitization of cordierite. M_4 and M_5 are represented by local retrograde metamorphism along shear zones and faults.

Greenstone belts in eastern Manitoba

A summary of the deformational, plutonic, and metamorphic history in the Island Lake greenstone belt in eastern Manitoba is given in Fig. 6-9.

Fig. 6-8. Generalized distribution of metamorphic zones in the English River and Uchi Superbelts (subprovinces) in northwestern Ontario (from Thurston and Breaks, 1978).

TABLE 6-2

Summary of metamorphic and deformational events in the English River Superbelt (after McRitchie and Weber, 1971; Thurston and Breaks, 1978)

Deformation	Metamorphism	Fabric	Comments
D_5	M_5	S_5	late transcurrent faulting and minor recrystallization in fault zones
D_4	M_4	S_4	late-stage myonitization along transcurrent faults and development of retrograde muscovite and chlorite in shear zones
D_3	M_3	S_3	large-scale concentric S-folding and development of incipient strain-slip cleavage S_3; development of muscovite in D_3 axial planes
D_2	M_2	S_2	regional asymmetric Z-folding associated with pluton emplacement; growth of micas in axial planes of D_2 folds; major migmatization of metasediments
D_1	M_1	S_1	isoclinal folding of greenstone belts, development of major foliation S_1, and major regional metamorphism M_1; beginning of migmatization of metasediments
		S_0	original sedimentary and volcanic fabrics

A similar sequence of events has been proposed for the Bigstone Lake and Stevenson Lake greenstone belts west of the Island Lake belt (Park and Ermanovics, 1978). Subsidence of greenstone volcanics and sediments together with uplift of older granitic rocks produced homoclines and structural basins in this region (D_0). The granitic rocks may represent reactivated diapiric plutons (see Chapter 5). F_1 folds were overturned towards the interior of the Island Lake belt, perhaps accompanied by gravity sliding from the margins of the belt. Strain fabrics (S_1) are not present in the center of the belt consistent with D_1 representing soft-sediment deformation. During the late stages of D_1, granitic plutons were intruded across early F_1 folds. The D_2 deformation changed in response to a north-south compression. F_2 folds formed accompanied by an east-west S_2 foliation which penetrated some granitic terrane. In high strain areas of F_2 folding, northwest-trending F_1 folds were reoriented in east-west directions. Cordierite began to develop late in D_2 and continued to form into D_3. Upright F_3 folds and S_3 foliation

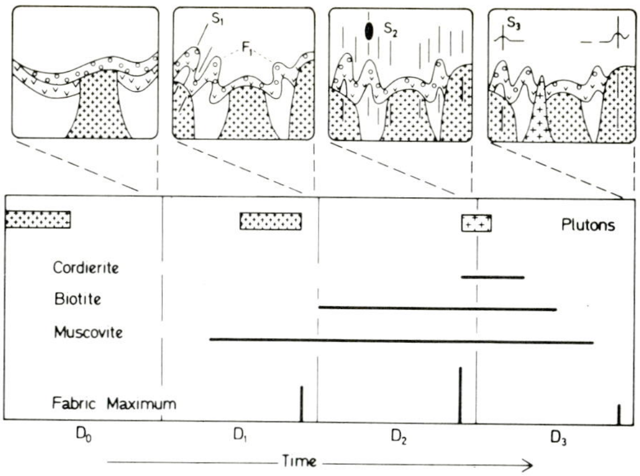

Fig. 6-9. Schematic diagram showing the deformational, metamorphic, and plutonic events in the Island Lake greenstone belt, Manitoba (from Fyson et al., 1978). Time divisions of arbitrary width. Ellipse represents strain in pebbles after D_2.

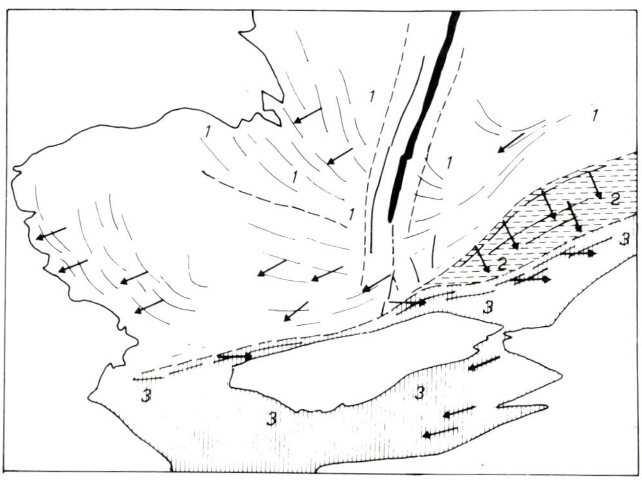

Fig. 6-10. Map showing the three structural domains in Rhodesia, Botswana, and South Africa area (from Coward, 1976).

developed during lower temperatures and trend in a general northerly direction. Post-tectonic plutons were emplaced late during D_3.

It is clear that D_2 and D_3 structures cannot be related to upwelling diapiric plutons, a mechanism commonly proposed to account for greenstone belt structural features (Anhaeusser et al., 1969). They appear to have developed from large-scale regional compression.

The Agnew greenstone belt, Western Australia

The Agnew greenstone belt in Western Australia consists of a typical bimodal greenstone succession unconformably overlain by conglomerate and arkose. Platt et al. (1978) have recently summarized the deformational and metamorphic history of this area. The first event recorded is the intrusion of the Lawlers tonalite into the greenstone succession which predates both of the major periods of deformation. Next, the greenstone-tonalite terrane was uplifted, eroded and then sank forming a basin that received continental sediments. The first period of deformation (D_1) is characterized by isoclinal folding and development of schistosity (S_1) in the supracrustal rocks and in the Lawlers tonalite. Axial planes of folds were gently dipping during this deformation and regional metamorphism (M_1) probably did not exceed greenschist facies grade. D_1 was followed by intrusion of minor leucogranite along the tonalite-greenstone contact.

The final deformation (D_2) in the Agnew belt produced large-scale NNW-trending folds, a northerly trending shear zone, and a steep NNW-trending foliation. D_2 is characterized by ENE-WSW shortening and right-lateral ductile shear. M_2 regional metamorphism ranged from upper greenschist to lower amphibolite facies.

Greenstone belts in the southwestern part of the Rhodesian Province

Although some structural studies of greenstone belts emphasize the importance of vertical forces, recent investigations of greenstone belts in Botswana and southwestern Rhodesia have shown that compressive forces may also be important (Coward and James, 1974; Coward, 1976; Coward et al., 1976a). Detailed strain studies have also been made in this region. Coward (1976) has suggested that this region can be divided into three structural domains (Fig. 6-10). Domain 1 is the granite-greenstone terrane and is characterized by steep foliation and down-dip lineations in the west; in the south and east the foliation curves and lineations plunge northeast or southwest. Major shear zones cut across the regional foliation. Coward et al. (1976b) consider the arcuation of the foliation to indicate movement of the Rhodesian Province to the southwest relative to the Limpopo mobile belt (structural domains 2 and 3) to the south. Fold hinges, lineations, and the maximum extension direction in domain 2 are nearly normal to those in domain 1 and plunge to the southeast. Domain 3 cross-cuts and is younger than domains 1 and 2 and is characterized by mylonite zones formed in response to right lateral shearing.

Deformation in domain 1 can be divided into four stages (Coward, 1976).

(1) Pre-cleavage regional deformation prior to emplacement of diapiric plutons. The Tati, Vumba, and part of the Matsitama greenstone belts in Botswana represent remnants of an extensive sheet which is overturned to

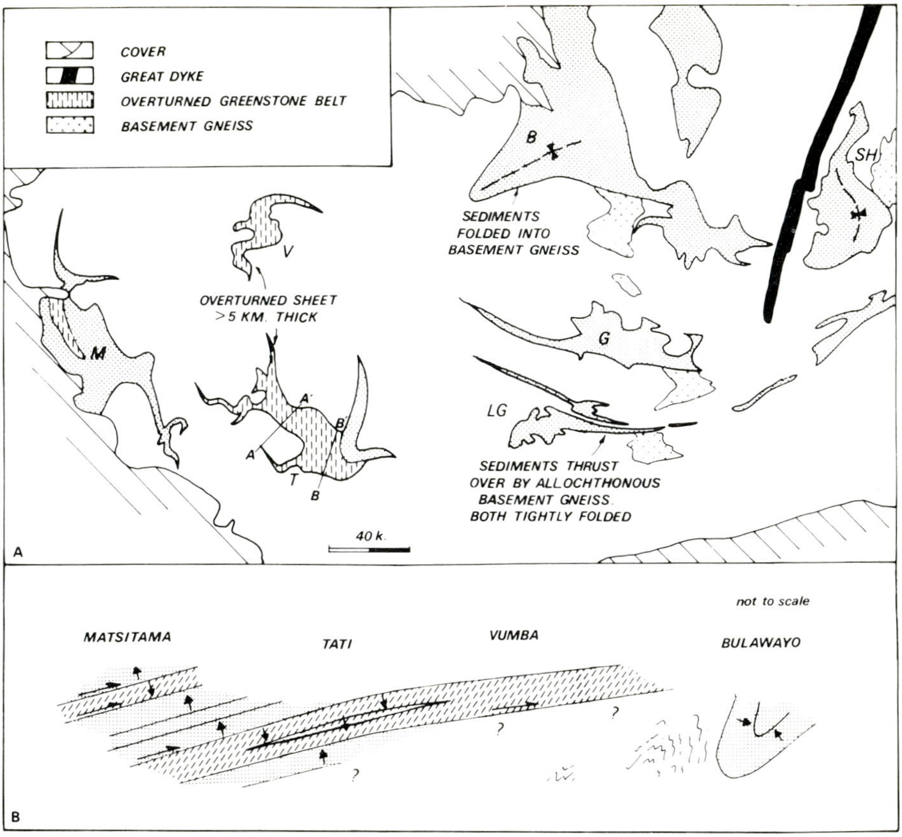

Fig. 6-11. A. Map of the southwestern part of domain 1 (Fig. 6-10) showing autochthonous and allochthonous greenstone belts (from Coward, 1976). Greenstone belts: M = Matsitama; V = Vuma; T = Tati; B = Bulawayo; G = Gwanda; LG = Lower Gwanda; SH = Shabani. B. Schematic section through the greenstone belts from Matsitama to Bulawayo before emplacement of diapiric granites. Arrows indicate younging directions.

the northeast (Fig. 6-11). Results indicate the overturning predates both the main cleavage and the intrusion of granitic diapirs. Perhaps the most carefully documented case of a large nappe structure is in the Selukwe greenstone belt in Rhodesia (Stowe, 1968b, 1974). The stratigraphic succession is inverted in the nappe which is thrust over the older gneissic complex. The nappe is about 10 km wide and can be traced in a northwesterly direction for about 60 km. Several major mylonite zones have been described within this structure. The rocks in the Selukwe belt may have been transported for more than 50 km from the south or southwest and the tonalitic basement gneiss appears to be thrust over the Lower Gwanda greenstone succession (Fig. 6-11). The Bulawayo, Fort Rixon, Shabani, and Fort Victoria greenstone belts, although folded, appear to be autochthonous.

(2) Deformation associated with emplacement of granitic plutons. Granitic plutons are diverse and have been emplaced at various times during the deformational history in the Rhodesian Province. Some are elliptical in shape and parallel the foliation in greenstone belts; others post-date all deformation. Syn-kinematic plutons often produce steep foliation and lineation in contact zones with greenstones as described in the Barberton belt.

(3) Regional deformation producing the main fabric. During this time, few major structures formed but a widespread cleavage was produced in both greenstone and granite terranes. Strain measurements suggest that greenstone belts were compressed by up to 65% during this time (Fig. 6-16).

(4) Late deformation. The late phases of deformation are characterized by crenulations and tight folds which deform the earlier fabrics. Late deformation at Selukwe is along NNE-trending fold axes and along shear zones with the same trend.

The results of the structural studies in the Rhodesian Province clearly show that the early stages of deformation in this region are characterized by imbrication and overturning of nappes directed chiefly in a northeasterly direction. Later emplacement of diapiric granites appears to have produced structures of more local extent. Production of the major cleavage, which extends into both greenstone and granite terranes, involved a considerable amount of shortening in a NE-SW direction. Regional metamorphism in the Rhodesian Province is described in a later section.

Conclusions and discussion

Although each greenstone belt has its own peculiar deformational history, some overall evolutionary patterns emerge from existing studies. Results suggest that the major folds in greenstone successions develop in response to primarily vertical forces associated, perhaps, with the emplacement or reactivation of granitic diapirs between and among gravitationally collapsing greenstone belts. Exceptions occur in some greenstone belts, such as those in southwestern Rhodesia and eastern Manitoba, where horizontal forces appear to have been responsible for part of the deformation. Typical greenstone belts have undergone two periods of deformation that are characterized by large-scale folding and the development of penetrative foliations. Widespread burial and regional contact metamorphism and plutonism accompany one or both of these periods of deformation. Later periods of deformation in most greenstone belts are characterized by one or a combination of small-scale conjugate folding, development of non-penetrative fabrics often localized along faults or post-tectonic intrusive contacts, and retrograde metamorphism.

Little is known about the deformational histories of the gneissic terranes surrounding greenstones. On-going studies of granitic complexes in the northwestern Superior Province suggest the presence of two contrasting suites of granitic and gneissic rocks (Schwerdtner et al., 1978): an early suite of foliated and deformed tonalite-trondhjemite and a later suite of massive (undeformed) to foliated tonalite-granodiorite-granite. Rocks of the first suite are commonly migmatitic, crenulated and/or porphyroblastic whereas rocks of the second suite generally exhibit non-metamorphic (igneous) textures and structures. The relationship between the deformational-metamorphic histories of granitic complexes and greenstone belts is at present not understood and is a topic of current investigation.

Gorman et al., (1978) have recently proposed a model for greenstone deformational history based on the laboratory experiments of Ramberg (1971, 1973). They assume a tonalitic crust underlies greenstone belts and that the deformational history is largely controlled by gravitational inversion of the greenstone volcanics by underlying, less dense tonalitic gneisses. The proposed model is illustrated in Fig. 6-12 and summarized in terms of five stages of development: (A) a large shield volcanic complex is formed some 5—7 km thick and $\gtrsim$ 100 km in diameter; (B) the center and edges of this complex sink as remobilized tonalitic basement begins to rise; (C) continued sinking and uplift form a central basin which collects volcanic sediments and marginal synclines develop; (D) continued subsidence leads to shortening of the volcanic pile and infilling of the central basin; and (E) the subsiding greenstones assume the shape of an inverted mushroom, partial melting of the root zones produces calc-alkaline magmas, and reactivated tonalitic gneiss and granitic plutons further compress the volcanic pile.

Some of the structures which are expected to develop along the edges of the volcanic pile are illustrated in Fig. 6-13. Structures formed by compressional forces should characterize the margins of the sinking belt while the central part should be characterized by isoclinal folds and high-angle faults which tend to shear-out anticlines. All of the predicted structures except for extensive thrusting have been recognized in most Archean greenstone belts. Gorman et al. (1978) suggest that enough evidence exists for marginal thrusting in some greenstone belts, that such thrusting may be more common than previously indicated. It is possible also that the down-folded margins of greenstone belts, where evidence of compressive forces should exist, may often be removed by erosion, preserving only the central basin which reflects vertical forces.

Halls (1978) has recently drawn attention to the fact that late Archean mafic dike swarms in the Superior and Slave Provinces in Canada and in the Yilgarn Province in Western Australia trend approximately at right angles to the trend of earlier greenstone belts which they cut. No more than 200 m.y. separates the ages of greenstone belts and dikes. The simplest explanation for this relationship is that the synclinal habit of the greenstone belts developed in response to horizontal compressive forces and that the direction of maximum horizontal stress remained unchanged up to and including the time of dike intrusion.

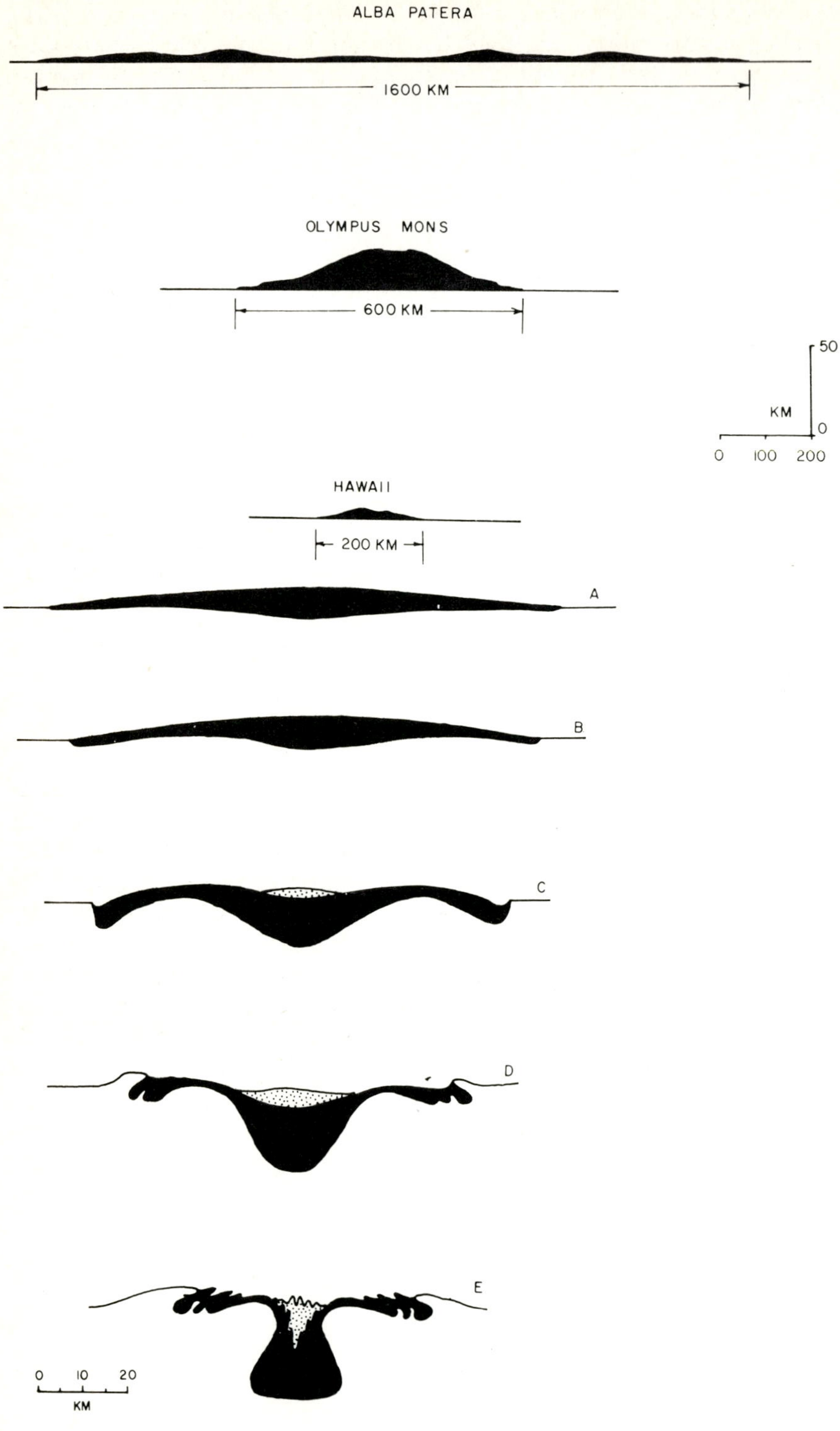
ALBA PATERA
1600 KM
OLYMPUS MONS
600 KM
50
KM
0
0 100 200
HAWAII
200 KM
A
B
C
D
E
0 10 20
KM

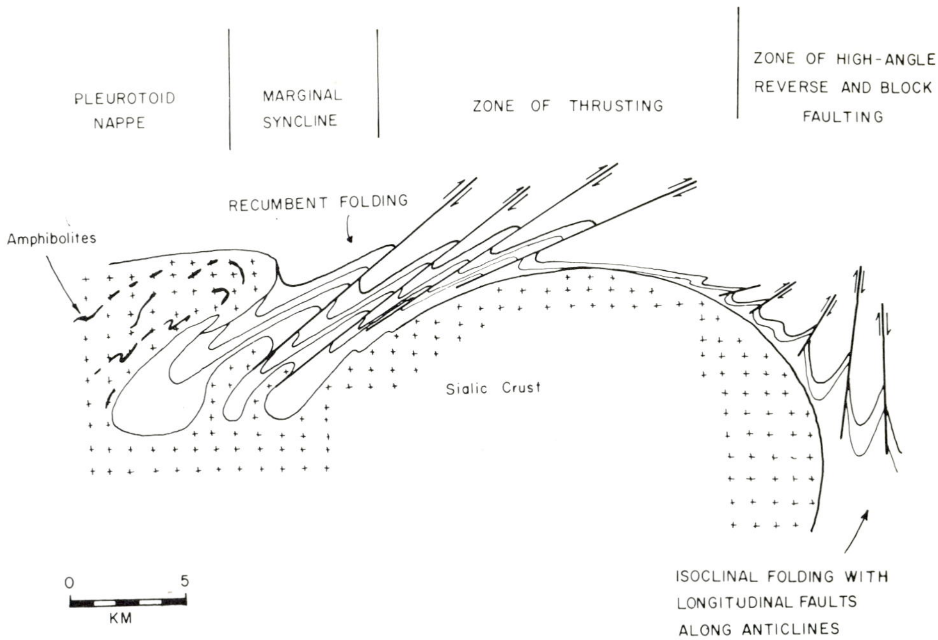

Fig. 6-13. Structures expected along a cross-section from the margin to the center of a subsiding greenstone belt shown in Fig. 6-12 (from Gorman et al., 1978).

Existing data indicate two different patterns of metamorphism in Archean granite-greenstone terranes. One may be considered contact metamorphism, although on a regional scale. It is exemplified by increases in metamorphic grade towards the margins of many, if not most, greenstone belts (such as the pattern in the Uchi Superbelt, Fig. 6-8). This pattern, as will be discussed in a later section in specific reference to the Yilgarn Province (Fig. 6-20), is often quite irregular. Such metamorphism appears to be caused by heat derived from intrusive, chiefly syn-tectonic plutons. The second metamorphic pattern, to be described more fully later in this chapter, is characterized by regional changes in metamorphic grade over hundreds to thousands of kilometers. Examples are the increases in metamorphic grade in going from the Uchi to the central part of the English River Superbelt (Fig. 6-8) and from the central to the southern margin of the Rhodesian Province (Fig. 6-17). The simplest model for this type of change is that proposed by Richardson (1970) in which the increases in metamorphic grade are related to thermal anticlines or domes in the crust and hence reflect progressively steepening geothermal gradients.

Fig. 6-12. Diagrammatic sequence of events in the deformation of a greenstone belt overlying sialic crust (from Gorman et al., 1978).

STRAIN ESTIMATES IN GREENSTONE BELTS

Several estimates of the nature and amount of strain in greenstone belts have appeared in the last few years. Most of these deal with greenstone belts in the southwestern part of the Rhodesian Province (Coward, 1976). Strain measurements are generally made using the methods proposed by Ramsay (1967) and Dunnet (1969). Various primary features in the rocks (such as pebbles, vesicles, pillows, breccia fragments) are measured in deformed rocks. Two measurements of strain are used: e, the finite extension and E_s, the natural strain where:

$$e = (L_1 - L_0)/L_0$$

and:

$$E_s = \ln L_1/L_0$$

and L_1 = length after strain and L_0 = length before strain. E_s is related to the octahedral unit shear γ_0 and to Lode's unit V by the following expressions (Hossack, 1968):

$$\gamma_0 = \frac{2}{3}[(E_1 - E_2)^2 + (E_2 - E_3)^2 + (E_3 - E_1)^2]^{1/2}; \quad E_s = \tfrac{1}{2}(3\gamma_0)^{1/2};$$

$$V = \frac{2E_2 - (E_1 + E_3)}{E_1 - E_3}$$

V can be considered as a measure of the shape of the strain ellipsoid. Values of V range from $+1$ for a uniaxial oblate ellipsoid to -1 for a uniaxial prolate ellipsoid.

One application of strain measurements to Archean greenstone belts is in the Fort Victoria belt in Rhodesia (Coward and James, 1974). The variation in E_s and V along a NW-SE section line (approximately normal to the major fold axes) indicates that the amount of strain in the belt is quite variable. Synclinal zones show intense strain with flattening normal to cleavage ($V = 0.5$—1) ranging from 55 to 80% whereas anticlines exhibit minimal strain. Variation in strain in the Tati belt has also been estimated by Coward and James (1974). Results are summarized on the maps in Fig. 6-14. In terms of V, three zones can be defined as shown in the map. Oblate strain characterizes the eastern arm of the belt ($V > 0.5$) whereas prolate strain characterizes the western part ($V < 0$). Intense strain as measured by E_s is recorded in a narrow zone in the center of the western part of the belt and along the granite contact in the eastern arm.

Similar measurements have been made in the Gwanda belt in Rhodesia (Wright, 1975) and are summarized in terms of E_s and V in Fig. 6-15. The most intense strain in this belt occurs along the southern margin of the belt and is generally paralleled by a decrease in V. The shortening across the belt,

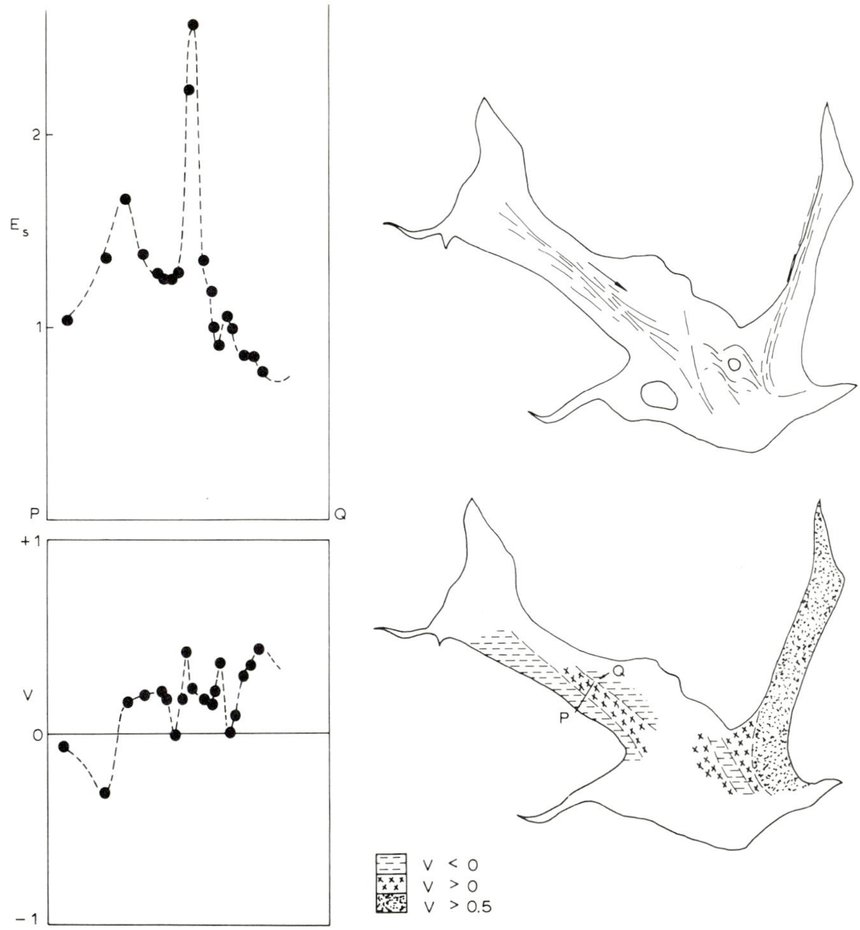

Fig. 6-14. Map of the Tati greenstone belt, Botswana showing main shear zones, variation in Lode's unit *V*, and a strain profile through part of the northwestern arm of the belt (from Coward, 1976).

assuming no rotation, is about 65% on the south and 15% on the north. Some of the apparent increase in shortening in the southern part of the belt may be accounted for by an increase in simple shear along the southern margin.

Average strain measurements from six greenstone belts in the southwestern part of the Rhodesian Province are given in Fig. 6-16. The amount of deformation ranges from over 60 to about 30%. It is noteworthy that the maximum amount of deformation occurs at the northeastern and southwestern extremes of the map area.

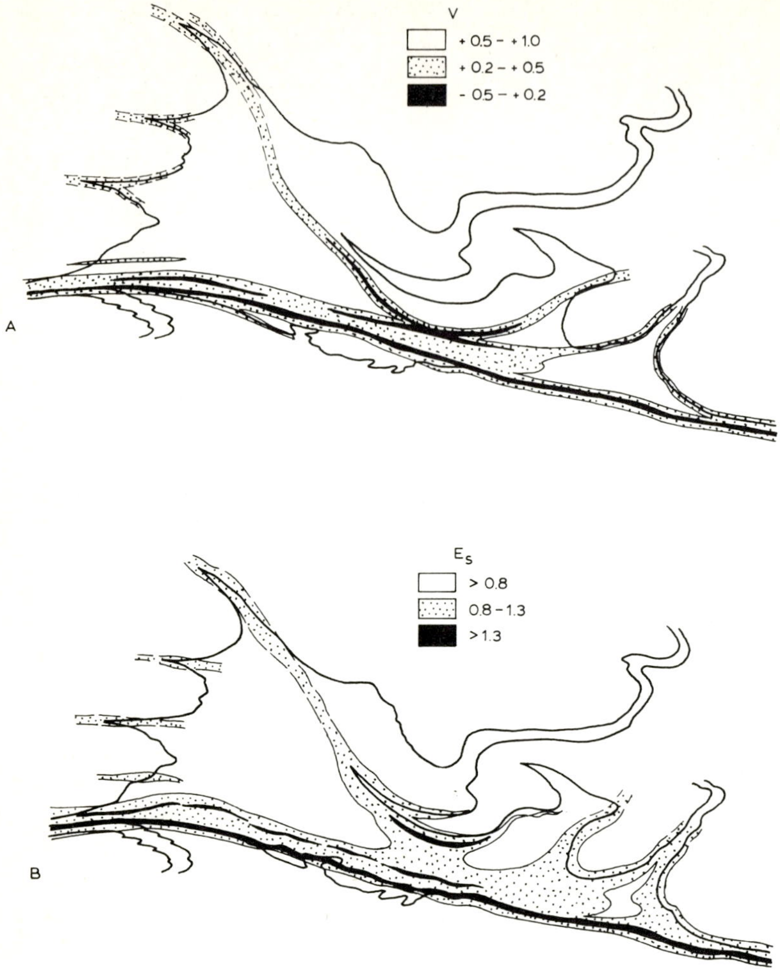

Fig. 6-15. Map of the Gwanda greenstone belt, Rhodesia showing the distribution of strain (E_s) and Lode's unit (V) (from Wright, 1975).

RELATIONSHIP OF LOW-GRADE TO HIGH-GRADE TERRANES

The Rhodesian Province

One of the major problems in understanding the relationship of high-grade to low-grade Archean terranes is that of the distribution of metamorphic facies. Although metamorphic grade is distributed irregularly in any given greenstone belt due to granitic plutonism and its associated contact metamorphism, greenstone belts in the Rhodesian Province seem to share the

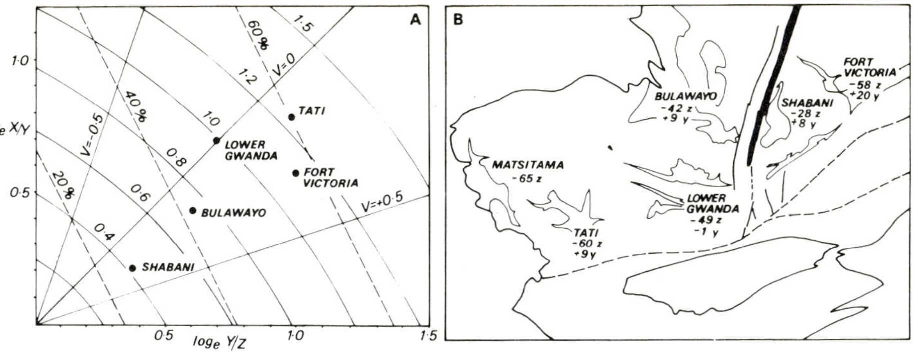

Fig. 6-16. A. Mean strains of five greenstone belts in the Rhodesian Province (from Coward et al., 1976b). Lines of equal value of Lode's unit (V), natural strain (E_s) and percentage shortening in the z direction (dashed lines) are shown. B. Map showing mean strain expressed as percentage shortening in the z direction and percentage elongation in the y direction for greenstone belts in the southwestern Rhodesian Province (from Coward, 1976).

same regional metamorphic imprint (Saggerson and Turner, 1976). The overall grade increases outward from the center of the province in both the upper greenstones (Bulawayan; Chapter 2) (Fig. 6-17) and in the Shamvaian Group. The low-pressure facies series dominates and is characterized by the presence of andalusite and cordierite-anthophyllite in pelitic rocks and lack of garnet in plagioclase amphibolites. Medium-pressure series occurs only in the southwestern part of the province adjacent to the Limpopo mobile belt.

The upper greenstone terranes can be divided into four zones based on metamorphic grade (Fig. 6-17).

Zone 1 — very low grade. Rocks of the Maliyami and Umniati Formations in the Midlands greenstone belt (Harrison, 1970; Bliss, 1970) represent the zeolite, prehnite-pumpellyite, or lower greenschist facies. Prehnite, zoisite, and calcite are all stable phases and zeolite-filled cavities are still preserved.

Zone 2 and 3 — low to medium grade. Greenstone belts are metamorphosed to the greenschist facies with the following representative minerals: chlorite, biotite, muscovite, chloritoid, actinolite, garnet, pyrophyllite, andalusite, and epidote. Kyanite is rare (zones 2b and 3b).

Zone 4 — medium grade. Rocks are metamorphosed to the amphibolite facies with anthophyllite, cordierite, corundum, andalusite, sillimanite, and grunerite as typical minerals.

Zone 5 — high grade. Greenstone belts are metamorphosed to the granulite facies with representative minerals hypersthene, diopside, olivine, brown hornblende, garnet, scapolite, cordierite, and sillimanite. Sapphirine-cordierite-sillimanite assemblages are recorded from at least three localities providing a *P-T* estimate of 750—850°C and 8—10 kbar. Zone 5 grades into the Limpopo and Zambezi mobile belts on the south and north, respectively.

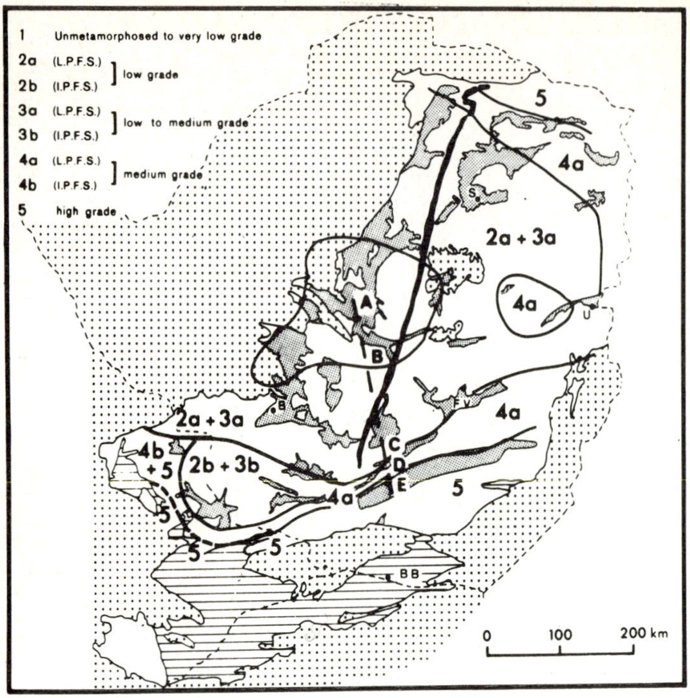

Fig. 6-17. Metamorphic zonation of the Rhodesian Province and Limpopo belt (from Saggerson and Turner, 1976). *L.P.F.S.* = low-pressure facies series; *I.P.F.S.* = medium-pressure facies series: *A-E* = line of section in Fig. 6-19.

Sediments assigned to the Shamvaian Group unconformably overlie the upper greenstone belts and appear to record post-Bulawayan regional metamorphism (Wiles, 1972; Wilson, 1964). A similar distribution of zones is shown by the metamorphic mineral assemblages found in Shamvaian-type rocks and the intensity of contact metamorphism also increases outward from the center of the Rhodesian Province. A similar but less well-defined metamorphic zonation occurs in greenstone belts of the Kaapvaal Province south of the Limpopo belt. Here, the grade increases northward towards the Limpopo belt (Saggerson and Turner, 1976).

The relationships between the Rhodesian Province and the Limpopo mobile belt on the south have been the subject of several investigations (Robertson, 1968; Mason, 1973; Coward et al., 1976b; Key et al., 1976; Saggerson and Turner, 1976). Both the northern and southern boundaries of this belt should be considered rather arbitrary in that the granite-greenstone terrane on both sides appears to grade into the mobile belt. A close relationship exists between the tectonic history of the Rhodesian Province and the Limpopo belt as described in Chapter 10. Mason (1973) has divided the Limpopo belt into three subdivisions (Fig. 6-18). Two

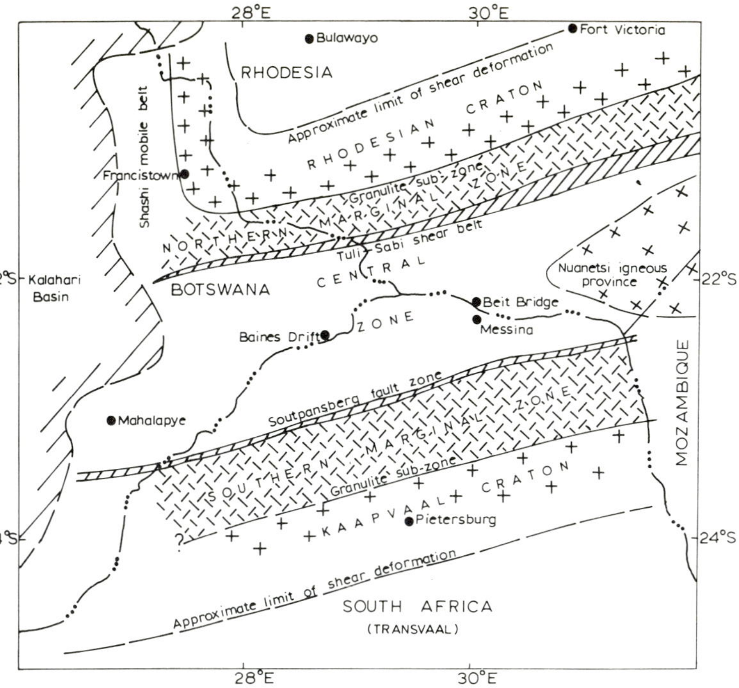

Fig. 6-18. Tectonic subdivisions of the Limpopo mobile belt in southern Africa (from Mason, 1973).

marginal zones are characterized by highly sheared rocks striking parallel to the belt and are composed chiefly of high-grade terranes (zone 5 above). A central zone is comprised of tectonically mixed and structurally complex basement (3.8 b.y.) and supracrustal rocks. The marginal zones are separated from the central zone by shear belts.

Timing of the periods of regional metamorphism in the Rhodesian Province and Limpopo belt are not well known. The overall increase in grade in the Rhodesian Province towards the Limpopo belt, however, suggests a relationship between the two areas. This is true also for the northerly increase in grade observed in the Kaapvaal Province south of the Limpopo belt. Two explanations for the increase in grade outwards from the centers of the Rhodesian and Kaapvaal Provinces towards the Limpopo belt merit consideration: (1) such a zonation reflects differential uplift with the Limpopo belt which represents deeper crustal levels of granite-greenstone terranes; and (2) the zonation reflects an increase in the geothermal gradient towards the Limpopo belt. Although both processes may occur simultaneously, Saggerson and Turner (1976) favor the second explanation. A diagrammatic cross-section from the center of the Rhodesian Province to the

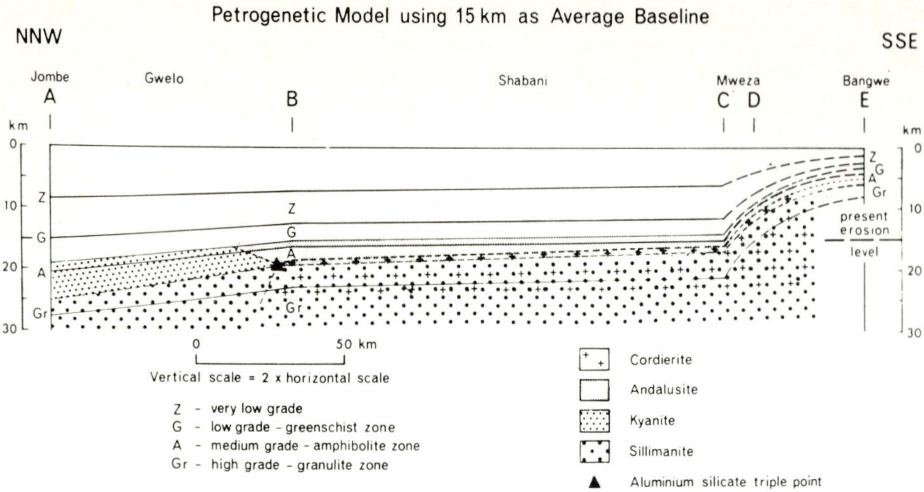

Fig. 6-19. Diagrammatic cross-section across the Rhodesian Province to the Limpopo belt (from Saggerson and Turner, 1976). Line of section noted in Fig. 6-17.

Limpopo belt is given in Fig. 6-19. An average thickness for the upper greenstone belts of 15 km is assumed in the diagram and geothermal gradients corresponding to each facies series were deduced from *P-T* metamorphic phase diagrams (Hietanen, 1967; Richardson, 1970). Intersections of metamorphic isograds with each geothermal gradient are transferred on to the cross-section at each location (*A*, *B*, etc.). The results indicate that a rapid increase in geothermal gradient occurs as the Limpopo belt is approached, defining the northern limb of a thermal anticline. A similar cross-section with a mirror image could be drawn south of the Limpopo belt into the Kaapvaal Province. This model requires a present erosion level of about 20 km throughout.

The distribution of metamorphic facies in granite-greenstone terranes of southern Africa clearly indicates that the high-grade terranes in the Limpopo mobile belt are an important part of the Archean crust and that any evolutionary model for granite-greenstone terranes in this region must also include the Limpopo belt.

The Yilgarn Province

Four types of metamorphic domains have been recognized in the Eastern Goldfields subprovince of the Yilgarn Province (Fig. 6-20). Very-low-grade domains exhibit prehnite-pumpellyite and lower greenschist-facies assemblages while low-grade domains contain greenschist and transitional greenschist-amphibolite-facies assemblages (Binns et al., 1976). Medium-grade domains are low- to mid-amphibolite facies and high-grade domains, mid-amphibolite to upper-amphibolite facies. Facies series are typically low-pressure type with localized occurrences of medium-pressure type. Two styles of metamorphism are recognized: static, where primary textures and structures are well-preserved and dynamic, with well-developed penetrative foliations and lineations. Static metamorphic terranes grade into dynamic types. In general, the distribution of regional metamorphic grade does not correlate well with the distribution of intrusive granites. Superimposed contact metamorphic aureoles, however, do occur around some plutons. Dynamic terranes are characterized by tight folds and more complex polyphase deformation than observed in static terranes. Most of the static metamorphism appears to be post-tectonic while the dynamic is syn-tectonic and neither is related to stratigraphic level exposed. Existing data are not adequate to determine if coeval dynamic and static domains reflect differences in rigidity in a uniform stress field or localization of heat and stress in some areas and only heat in others. It is important to note that within the Eastern Goldfields subprovince, an outward progression in metamorphic grade is not observed as it is in Rhodesia.

Evidence exists in the Eastern Goldfields area that greenstone belts may not have evolved from low to high metamorphic grades (Binns et al., 1976). For instance, low-grade ultramafic and mafic volcanics often contain relics of clinopyroxene while Ca-plagioclase completely recrystallizes to albite-epidote-chlorite-mica and olivine is completely serpentinized. In compositionally equivalent rocks of medium grade, however, Ca-plagioclase often remains as cores and relatively fresh olivine crystals are present. Hence, it would appear that metamorphism is not progressive, but that low- and medium-grade terranes must have formed and stabilized at about the same time. For some reason, medium-grade terranes in this area did not undergo earlier, low-grade recrystallization.

The origin of the irregular distribution of metamorphic facies in the Eastern Goldfields subprovince is not understood. Two possibilities merit consideration: (1) rapid lateral changes in geothermal gradient, and (2) differential uplift in a Basin and Range-type province. If changes in geothermal gradient were responsible, they must occur over distances of 25—50 km which seems remarkably small to sustain major temperature differences. A Basin and Range-type tectonic regime would require tensional forces on a regional scale as well as a crust that behaved as a brittle solid.

The Southwestern subprovince (Wheat belt) in the southern part of the Yilgarn Province is comprised chiefly of high-grade terranes (Fig. 1-15). Three origins for these terranes are possible (Glikson and Lambert, 1973, 1976):

(1) The high-grade rocks represent the lateral equivalents (at the same crustal level) of the greenstone-granite terranes to the northeast in the Eastern Goldfields subprovince.

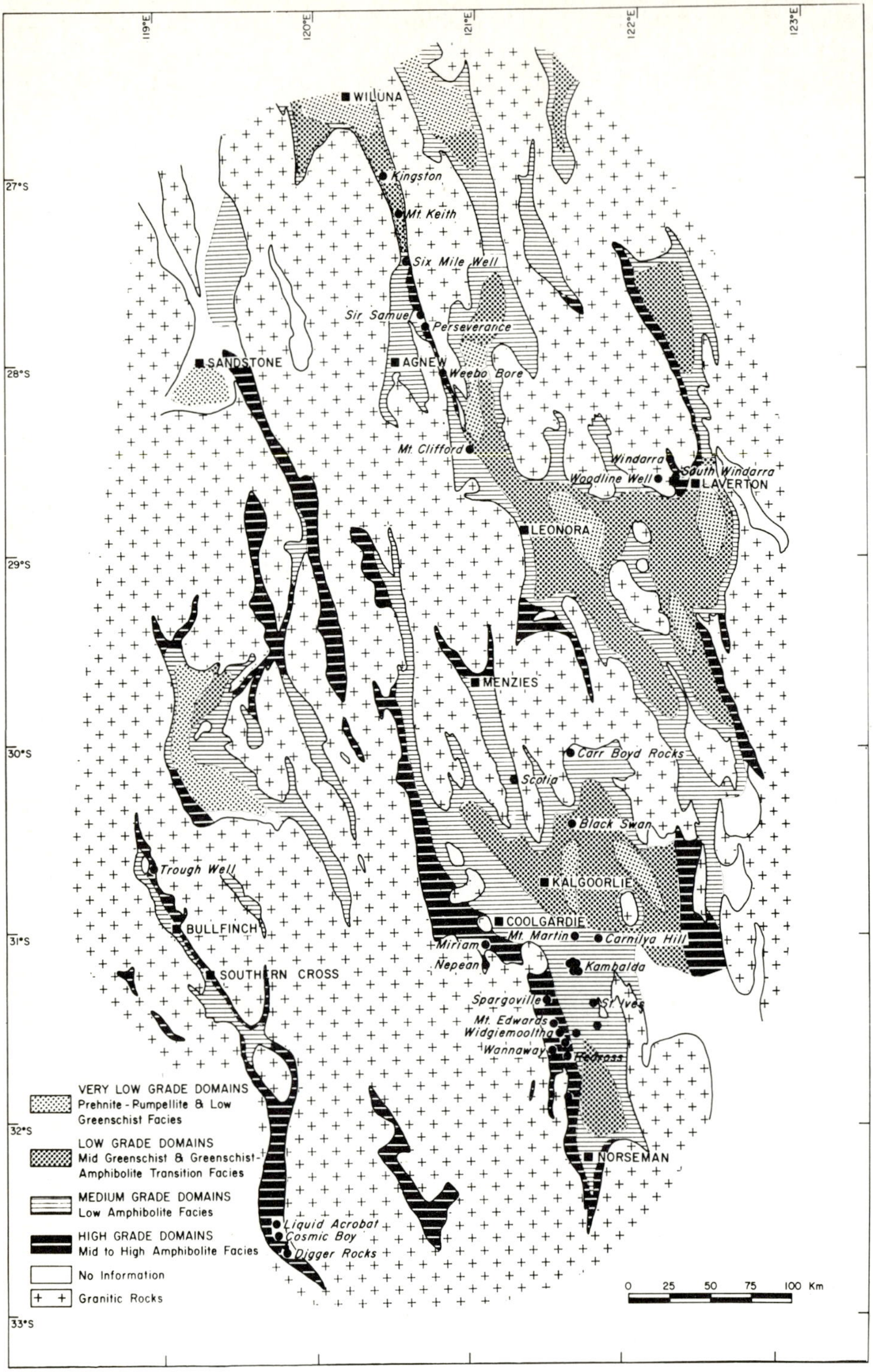
119°E
120°E
121°E
122°E
123°E
27°S
28°S
29°S
30°S
31°S
32°S
33°S
WILUNA
Kingston
Mt. Keith
Six Mile Well
Sir Samuel
Perseverance
SANDSTONE
AGNEW
Weebo Bore
Mt. Clifford
Windarra
South Windarra
Woodline Well
LAVERTON
LEONORA
MENZIES
Carr Boyd Rocks
Scotia
Black Swan
Trough Well
KALGOORLIE
BULLFINCH
COOLGARDIE
Mt. Martin
Carnilya Hill
Miriam
Nepean
Kambalda
SOUTHERN CROSS
Spargoville
St. Ives
Mt. Edwards
Widgiemooltha
Wannaway
Redross
NORSEMAN
Liquid Acrobat
Cosmic Boy
Digger Rocks
VERY LOW GRADE DOMAINS
Prehnite - Pumpellite & Low Greenschist Facies
LOW GRADE DOMAINS
Mid Greenschist & Greenschist-Amphibolite Transition Facies
MEDIUM GRADE DOMAINS
Low Amphibolite Facies
HIGH GRADE DOMAINS
Mid to High Amphibolite Facies
No Information
Granitic Rocks
0 25 50 75 100 Km

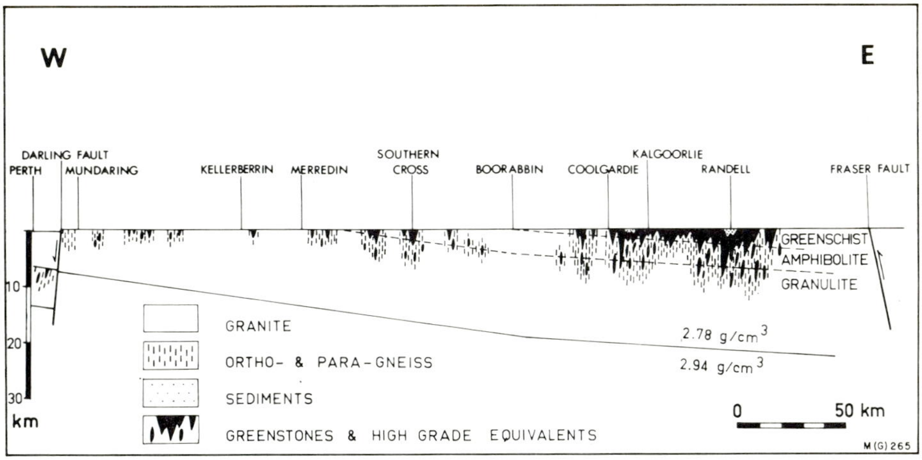

Fig. 6-21. Hypothetical east-west cross-section across the Yilgarn Province (from Glikson and Lambert, 1976).

(2) The high-grade rocks represent basement on which the greenstones formed.

(3) The high-grade rocks are the uplifted root zones of the granite-greenstone terranes.

The first possibility seems unlikely because the metamorphic mineral assemblages in the Southwestern subprovince reflect greater burial depths that those in the Eastern Goldfields subprovince and the greenstone belts become progressively less frequent to the southwest. Because the exposures of contact relations between the greenstone belts and the high-grade terranes are poor, it is difficult to evaluate possibility two. However, as pointed out by Glikson and Lambert (1976), no evidence of a sialic basement exists for the older greenstones in the Eastern Goldfields area.

Glikson and Lambert prefer the third alternative (Fig. 6-21). They interpret the mafic granulites in the Southwestern subprovince as relics of the mafic volcanics in the greenstone belts. Wilson (1969) suggests that granulite-facies supracrustals in the Southwestern subprovince can be traced northward into low-grade greenstone terranes thus supporting explanation three. Examples are the granulites in the Dangin region which can be traced NNW into amphibolites and then into greenschists in the Bolgart and Wongan Hills over a distance of about 150 km. Gravity and seismic data from Western Australia are also consistent with the crust being tilted upward

---

Fig. 6-20. Distribution of metamorphic domains in the Eastern Goldfields subprovince, Western Australia (from Binns et al., 1976).

towards the west (Mathur, 1974). The results of the studies of Binns et al. (1976), however, show that a clear progression in metamorphic grade within the Eastern Goldfields segment of the Yilgarn Province is not present. Also, the presence of highly metamorphosed cratonic sediments in the Southwestern subprovince is not consistent with this terrane representing the root zones of a granite-greenstone terrane (Rutland, 1976).

The Indian Province

A progressive increase in metamorphic grade in going from north to south in the Karnatka subprovince in peninsular India (Fig. 1-16) has long been recognized (Fermor, 1936; Pichamuthu, 1967, 1975). The isograds run at steep angles to the northwesterly strike of the greenstone belts. Greenschist-facies mineral assemblages characterize the greenstone belts from where they emerge from beneath the Deccan Traps for about 300 km to the south (Pichamuthu, 1975). They grade into amphibolite-facies rocks forming a broad east-west band north of Mysore and in the southern part of the province, granulite-facies grade is reached. The southern parts of the Kolar and Sargur greenstone belts approach granulite-facies grade (Pichamuthu, 1962) and available data seem to point to a similar metamorphic zonation as observed in the Rhodesian Province. Whether the charnockite belt which bounds the granite-greenstone terrane on the south is a deeper equivalent of the granite-greenstone terrane, however, is an unresolved question at present. Although many investigators favor this interpretation (Nautiyal, 1966; Naqvi et al., 1978a, b; Ramiengar et al., 1978), the older "Sargur-type" greenstone belts which occur in the charnockite province have lithologic associations (more quartzite and carbonate) quite different from the Dharwar-type belts. Also, this terrane contains many layered igneous complexes not found in the lower-grade terranes (Shackleton, 1976).

The northwestern Superior Province

In the northwestern part of the Superior Province in Canada in the Cross Lake area, a sequence of Archean volcanics and sediments can be traced along strike from greenschist-facies grade, through a migmatitic gneiss zone of probable amphibolite-facies grade, into a granulite-facies terrane (Rousell, 1965). The transition takes place over a distance of about 50 km. The granulite-facies terrane, known as the Pikwitonei subprovince (Chapter 1), contains the following minerals indicative of granulite-facies grade (Ermanovics and Davison, 1976): plagioclase, clinopyroxene, orthopyroxene, garnet, quartz, and hornblende. The rocks are mostly gneisses, generally pale brown and medium grained. A diagrammatic cross-section from the Wawa volcanic belt on the south to the Superior-Churchill provincial boundary on the north is given in Fig. 6-22. The Hudsonian orogeny (1.7—1.8 b.y.) reset

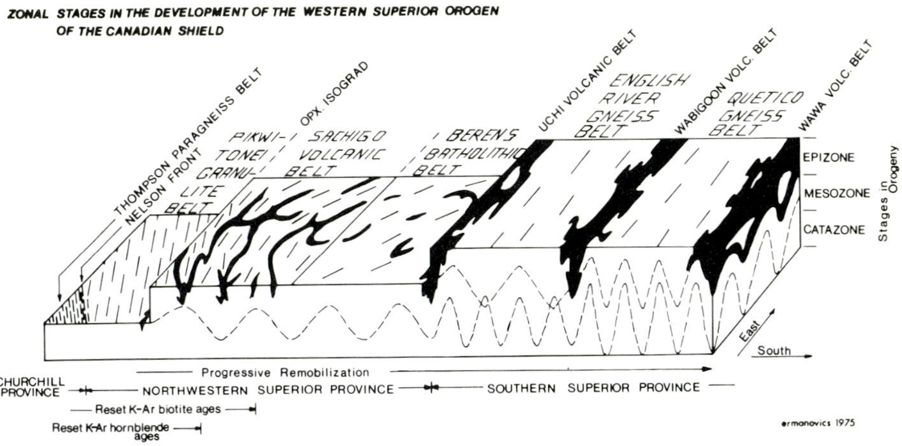

Fig. 6-22. Diagrammatic block diagram showing the major tectonic elements in the western Superior Province (from Ermanovics and Davison, 1976).

K-Ar ages in amphiboles in the marginal zone of the Superior Province on the north. The successive belts from south to north are interpreted by Bell (1971), Ermanovics and Davison (1976), and Weber and Scoates (1978) as progressively deeper levels of exposure of the Superior Province. This relationship suggests that the Superior Province is tipped upwards towards the northwest or that successive fault blocks have been uplifted more and eroded to deeper levels in this direction. Metamorphic grade in the Superior Province also increases along the southwestern margin into the granulite-facies terrane in the Minnesota River Valley and in the northeast where granulite-facies terranes of the Ungava subprovince appear (Fig. 1-6).

ARCHEAN GEOTHERMS

The decrease in abundance of U, Th, and K with time due to radioactive decay suggests that heat coming from within the earth and geothermal gradients have decreased with time (McKenzie and Weiss, 1975; Lambert, 1976). It is possible to monitor this decrease in the continents from the distribution of metamorphic mineral assemblages in space and time. A summary of recent estimates of pressures and temperatures of high-grade Archean metamorphic mineral assemblages are given in Table 6-3. These results are plotted in Fig. 6-23 and lie chiefly in the range of 8—12 kbar and 800—900°C. Recent experimental and thermodynamic evidence also supports such high *P-T* regimes (Newton, 1978). Also shown on the figure are inferred geotherms based on metamorphic mineral assemblages in Archean granite-greenstone terranes. It is clear that most Archean geotherms

TABLE 6-3

Estimates of pressure-temperature conditions in high-grade Archean terranes (from Tarney and Windley, 1977)

Region	P (kbar)	T (°C)	Reference
Scourie, Scotland	15	1250	O'Hara (1977)
South Harris, Scotland	10—13	800—860	Wood (1975)
South Harris, Scotland	9—11	700—800	Dickinson and Watson (1976)
Buksefjord, West Greenland	{ 7 10.5	630 810	Wells (1976)
Fiskenaesset, West Greenland	>7	800—900	Windley et al. (1973)
Sittampundi, India	7—8	850	Chappell and White (1970)
Sittampundi, India	8—10	825—850	Yardley and Blacic (1976)
Limpopo belt, South Africa	>10	800	Chinner and Sweatman (1968)
Enderby Land, Antarctica	10	970	Hensen and Green (1973)

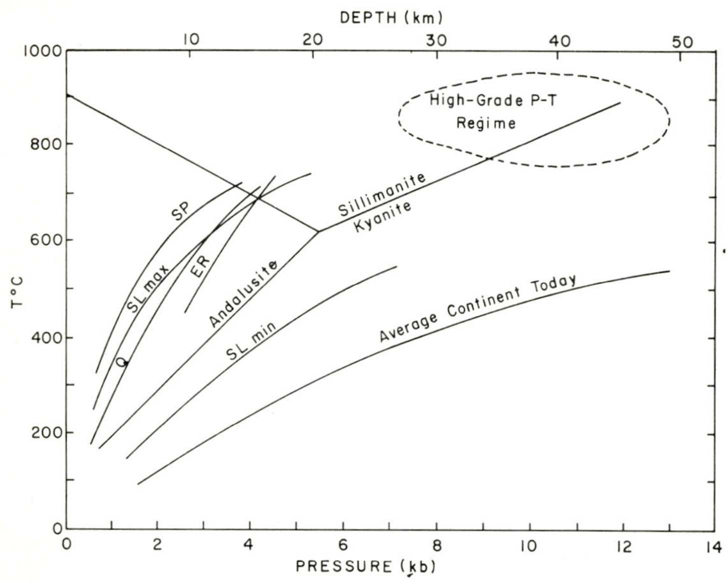

Fig. 6-23. Archean geotherms inferred in granite-greenstone terranes compared to an average continental geotherm today and to the P-T regimes reflected by Archean high-grade terranes (data from Table 6-3). Symbols and references: SP = South Pass greenstone belt (Bayley et al., 1973); ER = English River Superbelt (Thurston and Breaks, 1978); Q = Quetico Superbelt (Pirie and Mackasey, 1978); SL = Slave Province maximum and minimum (Thompson, 1978).

are of the low-pressure type reflecting gradients of the order of 20—30°C/km which is higher than most present continental gradients which average 10—15°C/km (Fig. 6-23). Only in areas of high heat flow today, such as the

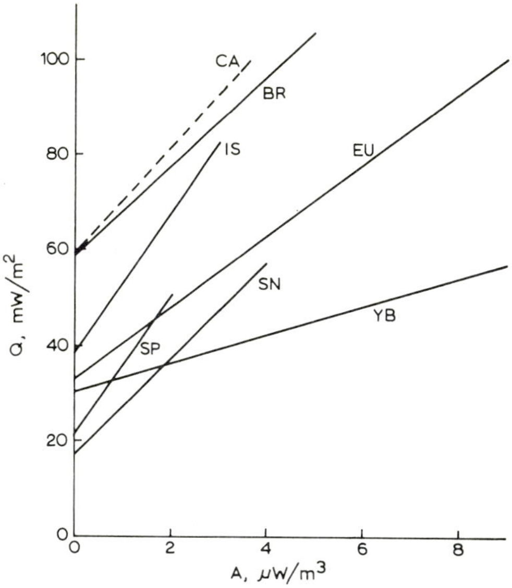

Fig. 6-24. Linear relationship between heat generation (A) and heat flow (Q) for various crustal provinces (from Jessop and Lewis, 1978). Key: BR = Basin and Range, EU = eastern U.S.A., IS = Indian Province, SP = Superior Province, SN = Sierra Nevada Range, YB = Yilgarn Province, and CA = Postulated Superior line during the Archean.

Basin and Range Province, are geothermal gradients comparable to the Archean gradients found. In some granite-greenstone terranes, such as parts of the Rhodesian and Slave Provinces, the presence of kyanite reflects a medium-pressure type of metamorphism with gradients of the order of 15—20°C/km. Archean gradients may change rapidly over small lateral distances as exemplified by the change from low- to medium-pressure metamorphism across the Hackett River gneiss dome in the Slave Province (Percival, 1979). It is noteworthy that most of the granite-greenstone gradients project into the high-grade *P-T* field which is consistent with, although does not necessitate, the model of Glikson and Lambert (1976) suggesting that the two terranes are the depth equivalents of each other. Another notable feature of all Archean terranes is the absence (with one exception in India; Shackleton, 1973a) of the blueschist facies metamorphism. This is most readily explained by the high Archean geotherms which do not pass into the blueschist stability field.

Recent studies of the relationship between surface heat flow and crustal heat generation in the Superior Province support the metamorphic results suggesting steeper geothermal gradients in the Archean (Jessop and Lewis, 1978). On a heat flow versus heat generation plot reconstructed for the Archean, the Superior Province line falls very near the present Basin and Range line with a reduced heat flow value of about 60 mW/m² (Fig. 6-24).

The present-day reduced heat flow for the Superior Province is about 25 mW/m^2 indicating a substantial drop in mantle heat since the Archean. Although both the Superior and Yilgarn Provinces have low reduced heat flow values today (Fig. 6-24), the characteristic depth (b) from the linear Q-A relationship (Roy et al., 1968) differs significantly (14 km for the Superior Province and 3 km for the Yilgarn Province). Jessop and Lewis (1978) suggest that such a difference may be due to the variable preservation of a thin surface layer (2—4 km) with high heat production. The Yilgarn Province would represent an area where this layer is still present (~ 3 km thick), whereas the Superior Province would represent an area where this layer is largely removed by subsequent erosion and the present heat is coming from the underlying, much thicker layer (~ 14 km).

The fact that granite-greenstone geotherms lead into high-grade P-T regimes, appears to be inconsistent with the thermal anticline model of Richardson (1970) as applied, for instance, to the Rhodesian Province and to the English River subprovince discussed previously. This model predicts that geotherms steepen in going from low-grade (granite-greenstone) to high-grade terranes which are characterized by upwardly compressed isotherms (Fig. 6-19). The fact that this is not observed in the geotherms in Fig. 6-23 may be due to a mechanism suggested by Watson (1978). She suggests that rising granites are the primary heat source for metamorphism in granite-greenstone terranes and that these granites transferred heat to shallow depths ($<$ 15 km) in the crust thus steepening the geotherms in the low-grade provinces. The fact that progressive metamorphism of greenstone successions is often spacially related to intrusive granites (Fig. 6-8) supports this idea. Unperturbed granite-greenstone gradients may lie between the aluminium silicate triple point and the minimum gradient for the Slave Province (Fig. 6-23). If correct, this model predicts that during metamorphism kyanite may have been more abundant than sillimanite at depths $>$ 15 km in Archean granite-greenstone terranes.

Chapter 7

MINERAL DEPOSITS

INTRODUCTION

Most Archean mineral deposits occur in or closely associated with greenstone belts and they appear to have been derived directly, or by not more than one intervening stage, from the mantle (Watson, 1973, 1976a). Some of the world's major deposits of Ni, Au, Ag, Cu, and Cr were produced in association with Archean greenstone volcanism. The most striking feature of ore deposits associated with Archean volcanics is their overall similarity in space and time (from ~ 3.8 to 2.6 b.y. ago). Goodwin (1966, 1971) has suggested a model of Archean continental growth in which simple Au-Ag, Ni-Cu, and Fe deposits predominate in lithologically simple and possibly older greenstone belts, whereas complex mineral associations such as Cu-Zn, Ni-Cu, Au-Ag, Fe, and others are characteristic of lithologically more diverse and possibly younger belts. Archean-style mineralization appears to have declined rapidly after the 2.6 b.y. world-wide magmatic event. On some continents, eroded remnants of Archean granite-greenstone terranes are unconformably overlain by late Archean and Proterozoic miogeoclinal sedimentary successions. Gold and uranium are concentrated towards the base of some of these successions (such as in the Witwatersrand and Huronian Supergroups) and appear to represent recycled material from the underlying granite-greenstone terranes.

The following sections summarize the major features of mineral deposits found in Archean granite-greenstone terranes and discuss the origin of these deposits.

MASSIVE SULFIDE DEPOSITS

Zinc-copper

Zn-Cu massive sulfide deposits occur associated with andesitic to felsic volcanics in Archean greenstone successions (Sangster, 1972; Franklin et al., 1975; Sangster and Scott, 1976; Boyle, 1976). They occur in both calc-alkaline and bimodal type belts (Chapter 2). It is noteworthy that such deposits are not important in greenstone belts older than about 2.7 b.y. The host rocks adjacent to massive sulfide deposits are typically felsic pyroclastic rocks, most commonly agglomerates or breccias. In multicycle volcanic

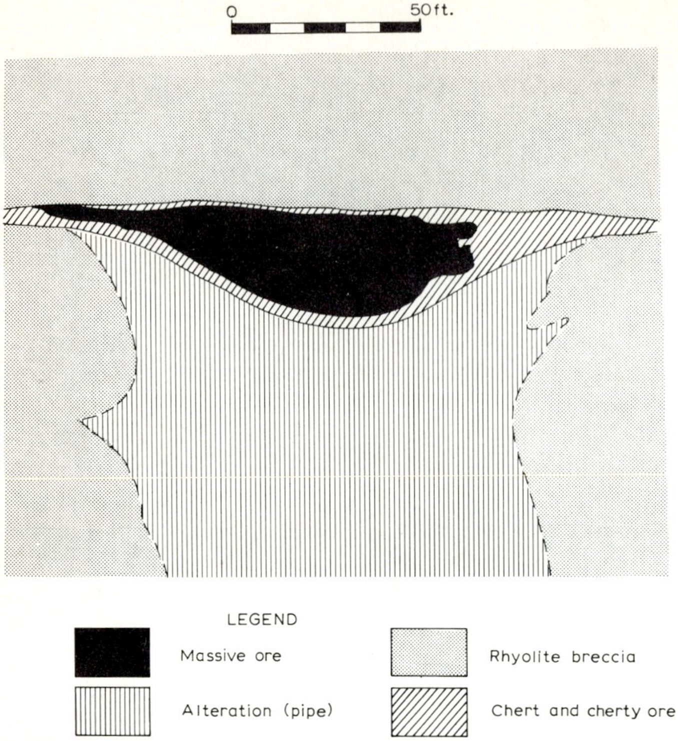

Fig. 7-1. Plan view of the 850 level, Lens B of the Delbridge massive sulfide deposit, Noranda area, Quebec (after Boldy, 1968). Stratigraphic top upwards.

successions, economic mineralization usually occurs associated with only one cycle (Spence, 1967). Some massive sulfides occur associated with felsic porphyry intrusives (Findlay, 1975). In the Abitibi belt, the major deposits occur around the oval-shaped perimeters of the volcanic complexes (Fig. 3-1) where the felsic volcanic centers are located. Shapes of the deposits are variable and they range from discordant to concordant with surrounding rocks. Some are broadly lensoid and have a stratiform appearance (viz., Kidd Creek, Ontario) while others are irregular-shaped. An example of a typical stratiform deposit is the Delbridge deposit in the Noranda area in Canada. This deposit is underlain by an alteration pipe which is discordant to the layering in the host-rock rhyolite breccia (Fig. 7-1). Stratigraphic thicknesses of ore bodies range upwards to about 25 m.

Most of the Zn-Cu massive sulfide ores can be classified into one of two types: massive or stringer ore (Sangster, 1972). Massive ores consist of $\gtrsim$ 50% of sulfides by volume and stringer ores of $\lesssim$ 25%. In massive types, the two longest dimensions of the deposits are chiefly concordant with surrounding host rocks. The contacts of these ore bodies with the hanging wall are generally sharp while footwall contacts are usually gradational.

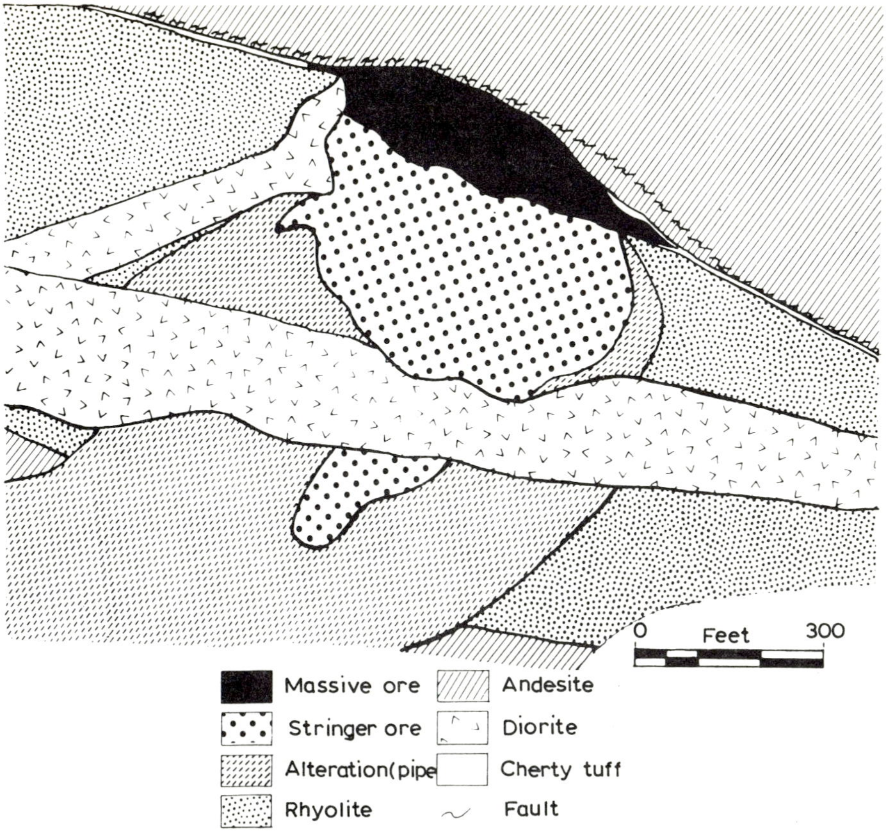

Fig. 7-2. Generalized east-west cross-section of the Lake Dufault massive sulfide deposit, Noranda area, Quebec (after Purdie, 1967; Sangster, 1972). Stratigraphic top upwards.

Stringer ore occurs in the footwall of massive ore and consists of anastomosing sulfide veinlets and irregular replacements (Fig. 7-2) which are generally discordant with host rocks. Stringer ore zones are generally funnel-shaped and grade upwards into massive ore.

Textures in massive sulfide deposits are difficult to interpret and it is often not possible to distinguish an inherited pre-metamorphic texture from a metamorphic texture. Textures within ore deposits tend to reflect those present in the host rocks. Banding in massive sulfide deposits is interpreted as sedimentary layering by some, but some evidence suggests that it is a replacement feature (Boyle, 1976). Volcanic clasts are partly to completely replaced with sulfides in some deposits. Evidence of soft-sediment deformation, although generally not unequivocal, has been described in some deposits (Sangster, 1972; Roberts, 1966). Some ore bodies show a penetrative lineation of fold axes and/or acicular crystals (Martin, 1966; Coats et al., 1970). Ores also increase in grain size during metamorphic recrystallization

TABLE 7-1

Mineralogical zoning of massive sulfide deposits (after Large, 1977)

	Zn-Cu type	Pb-Zn-Cu type	Cu-pyrite type
Hanging wall	pyritic tuff	hematite±pyritic sediment	hematite-py mudstone
	py-sp	barite	py-cpy±sp
	py-sp-cpy	ga-sp-py-barite	py±cpy
	po-py-cpy-sp	py-cpy	
	po-cpy±mag±py		
Footwall	py±mag	py-cpy	py mag?

Abbreviations: py = pyrite; po = pyrrhotite; sp = sphalerite; cpy = chalcopyrite; mag = magnetite; ga = galena.

and deposits which have undergone intense deformation appear to have flowed plastically.

In Archean Zn-Cu deposits, pyrite and pyrrhotite typically comprise at least half of the total sulfides. Sphalerite and chalcopyrite comprise most of the remainder. Minor amounts of tetrahedrite-tennanite, silver, galena, gold, and various tellurides are also often present. The general paragenesis is pyrite and pyrrhotite early and precious metals late. Major elements enriched in the ores are Cu, Ag, Zn, Cd, As, S, and Fe. Archean massive sulfides are generally zoned from the footwall to the hanging wall (column 1, Table 7-1) (Large, 1977). Chalcopyrite is concentrated near the base of the deposit with pyrite-pyrrhotite ± magnetite. Passing upwards, the pyrite/pyrrhotite ratio increases, magnetite decreases, and the top of the deposit is characterized by banded pyrite-sphalerite. The Cu-Fe-Zn ore zone often grades upwards into a laminated ferruginous chert (iron formation or silicified tuff?) containing pyrite. Compared to similar Phanerozoic deposits, Archean ores are notably deficient in Pb and generally contain more Zn than Cu.

Extensive alteration generally occurs on the footwall side of massive ore deposits (Figs. 7-1 and 7-2) (Gilmour, 1965). In some deposits, the alteration zone has been traced for as much as 1000 m (Sangster, 1972). In undeformed deposits, the alteration zone is vertical and pipe-like in shape whereas in deformed deposits it may be strung-out into a subhorizontal position (Fig.7-3). The altered zone generally contains an abundance of chlorite and sericite. At the Mattabi deposit in Ontario, an extensive zone of siderite alteration extends at least 300 m below the ore zone (Franklin et al., 1975). Chemical changes accompanying alteration include increases in Fe, Mg, and S and decreases in Na, K, and Si (Sangster, 1972). Magnesium metasomatism appears to have been important in the formation of most alteration zones and silicification is also characteristic of some.

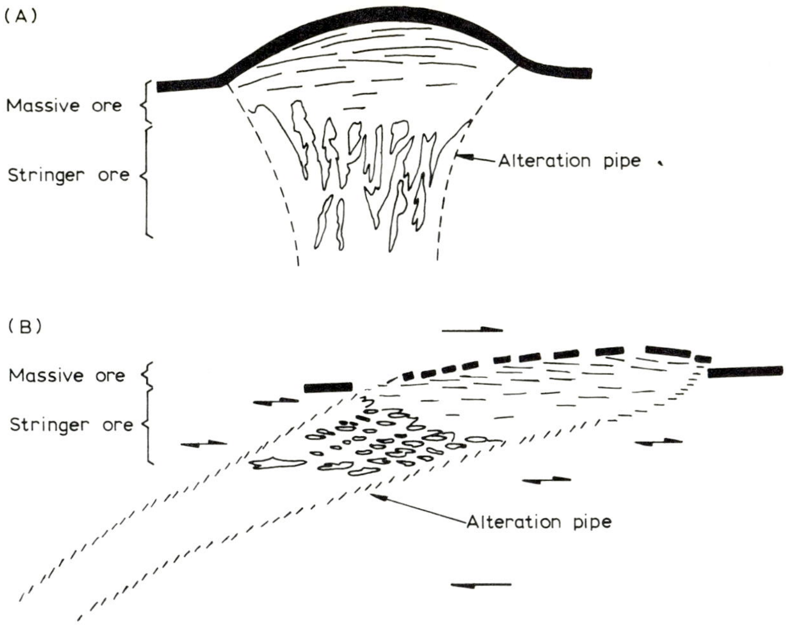

Fig. 7-3. Schematic diagrams of an undeformed massive sulfide deposit (A) and a deposit deformed by right lateral shear (B) (after Sangster, 1972).

In terms of origin and tectonic setting of Archean Zn-Cu deposits, it is of interest to compare these deposits with younger massive sulfide occurrences. Hutchinson (1973) has suggested massive sulfides associated with volcanic rocks can be classified into three groups based on base and precious metal occurrences, associated volcanic rocks, type of sediments, inferred tectonic environment, and distribution in geologic time. The general features of each type of deposit are summarized in Table 7-2. Pb-Zn-Cu type deposits (Tatsumi et al., 1970) have many features in common with Archean Zn-Cu deposits among the more important of which are the similar association with felsic volcanics, occurrence of massive and stringer ores in both types, and the occurrence of a footwall alteration zone. The most notable differences between the Pb-Zn-Cu and Archean Zn-Cu deposits are as follows (Sangster and Scott, 1976).

(1) Hanging-wall alteration is minor to absent in Archean Zn-Cu deposits.

(2) Footwall alteration in Pb-Zn-Cu type deposits is chiefly silicification rather than chloritization and sericitization which characterize Archean Zn-Cu deposits.

(3) Bedded sulfates found in some Pb-Zn-Cu deposits are absent in Archean Zn-Cu deposits.

(4) Bornite, tetrahedrite-tennanite, and galena are common major sulfides

TABLE 7-2

Comparison of volcanogenic massive Zn-Cu sulfide deposits (modified after Hutchinson, 1973)

Type	Precious metal association	Associated volcanics	Type of volcanism	Sediment type	Tectonism	Age	Examples
Zn-Cu	Au, Ag	andesitic to felsic volcanics	dominantly pyroclastic	volcanogenic graywackes chert iron formation	?	chiefly Archean	Noranda, Que.; Sturgeon Lake-Mattabi, Ont.
Pb-Zn-Cu	Ag	andesitic to felsic volcanics	dominantly pyroclastic	volcanogenic graywackes black shales sulfates	convergent plate boundary	post-Archean	Mt. Isa, Australia; Kuroko, Japan; East Shasta, Calif.
Cu-pyrite	Au	ultramafic-mafic volcanics (ophiolites)	subaqueous flows and sills	cherts carbonates	divergent plate boundary	chiefly Phanerozoic	Cyprus; Turkey; California; Philippines

in Pb-Zn-Cu type deposits where they are only accessory minerals in Archean deposits. The small amount of Pb in Archean Zn-Cu deposits is a distinctive feature of these deposits.

(5) Zoning is different in the two types of deposits as illustrated in Table 7-1 (Large, 1977).

The Cu-pyrite type deposits occur in Phanerozòic ophiolite complexes and are associated with mafic to ultramafic rocks rather than felsic or andesitic volcanics. The ores are generally poorly zoned with a simple pyrite-chalcopyrite mineralogy dominating (Table 7-1). The differences between Archean Zn-Cu massive sulfides and Pb-Zn-Cu and Cu-pyrite types are at least as pronounced as the similarities and hence one cannot deduce a Phanerozoic-type tectonic setting for the Archean deposits by comparison to younger deposits.

Generally one of two origins are suggested for massive Zn-Cu sulfide deposits (Boyle, 1976): an epigenetic hydrothermal origin or a syngenetic volcanic exhalation origin. The difference between the two models is primarily one of timing of ore deposition. In the epigenetic hydrothermal model, mineralization occurs by replacement of previously existing rocks whereas in the exhalative model mineralization occurs as a distinct late stage of volcanism. Often, textural and field relations are not adequate to distinguish the two mechanisms. Most recent investigators tend to favor the exhalative origin (Stanton, 1960; Goodwin, 1965; Sangster, 1972; Hutchinson, 1973). Consistent with such an origin is the common occurrence of multiple ore bodies at or near the same stratigraphic level over large areas, the overall concordant relationship between ore zonation and volcanic stratigraphy, the metal zonation of the ore bodies, an abrupt upper boundary of the massive ore, and the lack of hanging-wall alteration. MacGeehan (1978) proposes an origin for the massive sulfides at Matagami, Quebec in which Fe, Mg, Ti, Cu, and Zn are leached from wall rocks, carried upwards forming a chlorite alteration zone, and then deposited at the sediment-seawater interface forming an exhalite ore body. Recent phase-equilibria studies provide additional information bearing on the formation of exhalites. Consideration of equilibria in the system Fe-S-O suggests that Zn-Cu ore bodies are deposited at high temperatures ($> 275°$ C) from mildly acid, reduced chloride solutions which mix with sea water at felsic volcanic centers (Large 1977).

Nickel-copper

Archean Ni-Cu sulfide deposits occur closely associated with ultramafic-mafic sills or volcanic rocks in Archean greenstone belts (Naldrett, 1973; Anhaeusser, 1976a). Major deposits have been described in the Abitibi belt in Canada (Naldrett and Mason, 1968; MacRae, 1969); in the Eastern Goldfields subprovince in Western Australia (Williams and Hallberg, 1972; Nesbitt, 1971; McCall, 1971; Naldrett and Turner, 1977); and in several greenstone

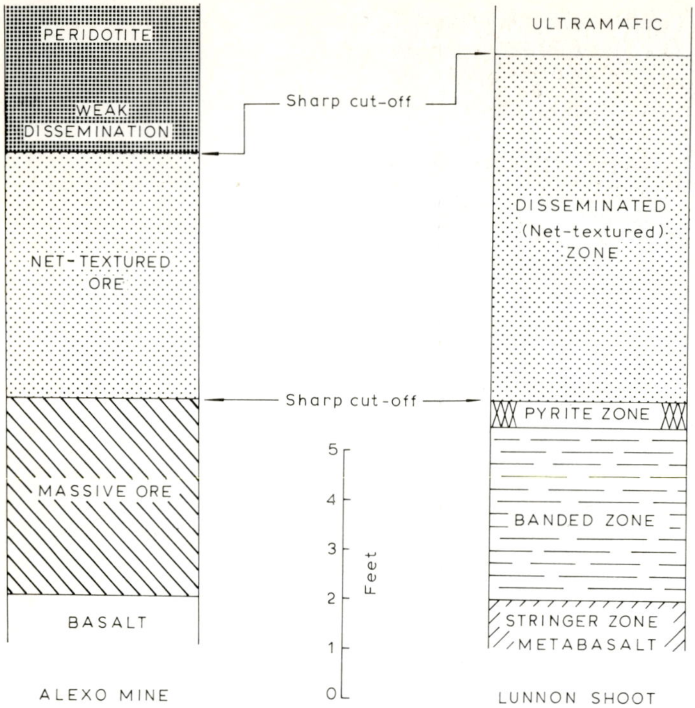

Fig. 7-4. Diagrammatic sections through two typical Ni-Cu massive sulfide deposits (from Naldrett, 1973). Alexo Mine is northeast of Timmons, Ontario, and Lunnon Shoot is near Kambalda, Western Australia.

belts in Rhodesia (Le Roex, 1964; Sharpe, 1964; Viljoen et al., 1976). Ore bodies exhibit a variety of shapes and sizes, most being broadly lensoid or irregular and ranging from a few meters to tens of meters thick. Sulfides are usually concentrated at the base of the ultramafic-mafic host rock suggesting a gravitational settling mode of origin (Naldrett and Gasparrini, 1971). Some, however, occur in shear zones or breccia pipes (Boyle, 1976).

A diagrammatic cross-section through two typical ore horizons is shown in Fig. 7-4. Characteristics common to both sections are as follows (Naldrett, 1973): (1) massive sulfides at the base of the ore zone; (2) a sharp contact separates the massive ore from overlying disseminated ore which consists of a network of interconnected sulfides surrounding euhedral pyroxene and olivine crystals. This texture, known as a *net-texture*, is reminiscent of igneous cumulus textures where the sulfides represent the intercumulus material (Fig. 7-5); (3) the net-textured zones are in sharp contact with overlying peridotite which contains minor sulfides near the contact and grades upward into unmineralized ultramafic rock. The two sharp contacts in Fig. 7-4 are parallel to other stratigraphic contacts in the country rock.

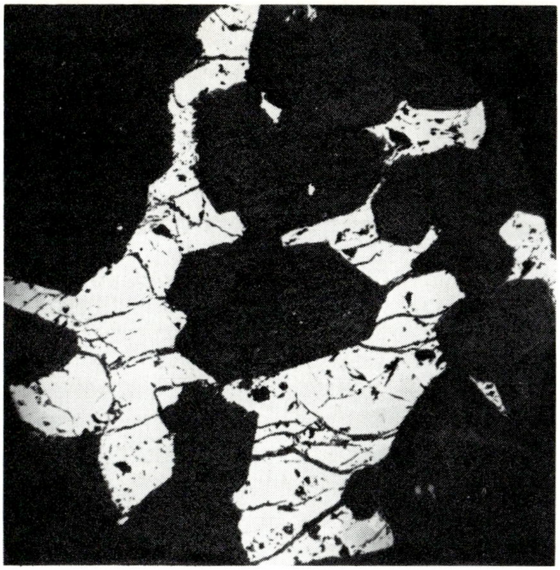

Fig. 7-5. Ni-Cu massive sulfide ore showing net texture (from Naldrett, 1973). Euhedral olivine (now serpentine) surrounded by intercumulus pyrrhotite (gray) and pentlandite (white). × 14, plane polarized light.

The major sulfide minerals in the Ni-Cu ores are pyrite, pyrrhotite, pentlandite, and chalcopyrite. Variable but generally minor amounts of magnetite, arsenopyrite, sperrylite, and other sulfides and arsenides may also be present. The principal elements concentrated in the ores are Cu, Ni, Fe, Co, S, and As. Studies of mineral paragenesis, as illustrated by the Shangani deposit in Rhodesia (Table 7-3), indicate that the sulfides are late magmatic with pyrite and pyrrhotite crystallizing early and chalcopyrite late. Many ores are partly to completely recrystallized by later carbonization or serpentinization processes.

Although the origin of massive Ni-Cu sulfides has been attributed to hydrothermal replacement, serpentinization (or some other secondary process), and to primary magmatic processes (Boyle, 1976), the last process appears to be most important as indicated by the primary igneous textures often preserved. Many investigators favor a model by which an immiscible sulfide melt separates from a primary ultramafic magma and crystallizes as intercumulus material after gravity settling of pyroxenes and olivine (Naldrett, 1973). The sulfide melt being more dense than the silicates would tend to sink to the base of the fractionating sill or flow, thus accounting for the large concentration of net-textured ore at the bases of the ore zones. Sulfur isotope studies indicate a mantle source for the sulfur in Archean Ni-Cu sulfide deposits (Donnelly et al., 1978).

TABLE 7-3

Schematic mineral paragenesis for the Shangani ore deposit, Rhodesia (after Viljoen et al., 1976)

	Magmatic	Serpentinization-carbonization
Decreasing age ↓	chromite-magnetite 1	
	olivine	
	pyroxenes	recrystallization
	pyrite	of sulfides
	pyrrhotite	
	pentlandite	
	chalcopyrite	
	magnetite 2	
		magnetite 3

IRON FORMATION

Several definitions have been proposed for Precambrian iron formation, some narrow and restrictive and others rather broad (Gross, 1965; James and Sims, 1973; Goodwin, 1973). In a general way, iron formation is a dominantly chemical (or biochemical) precipitate consisting of bands of interlayered chert and one or more iron-rich minerals (oxides, carbonates, silicates, or sulfides) (Brandt et al., 1972). It is commonly thin-layered with individual beds ranging up to several centimeters thick. Such beds, in turn, may be laminated on a scale of millimeters. Primary sedimentary structures (other than bedding) are found in some iron formation (Beukes, 1973). Relict oolitic textures, cross-bedding, scour-and-fill structures, slump structures, and rill marks have been reported. Iron formations in Archean greenstone successions range from a few meters to over 100 m in thickness, are broadly lenticular in shape, and are closely associated with volcanic rocks. This type of iron formation is known as the *Algoma type* which is distinct from the *Superior type* that characterizes most Proterozoic occurrences. The Superior-type iron formation is more widespread than Algoma-type iron formation and is associated with miogeoclinal sediments. Because of the close relationship between mineral associations and environment of deposition, iron-formations have been classified into one of four sedimentary facies (James, 1954) which are defined by Eh-pH relationships.

(1) The oxide facies is characterized by alternating bands of chert and magnetite and/or hematite. Magnetite reflects low Eh values and neutral to alkaline seawater. Hematite generally reflects higher Eh and a broad range of pH values. A low P_{CO_2} is necessary to prevent siderite precipitation.

(2) The carbonate facies is characterized by the presence of siderite and reflects strongly reducing conditions with a $P_{CO_2} \gtrsim 10^{-3.5}$ atm (Garrels, 1960). Associated carbonate minerals are ankerite, dolomite, and calcite.

(3) The silicate facies is characterized by iron-rich silicate minerals. Because similar minerals may develop during the metamorphism of iron formation, the identification of primary iron silicates is difficult in greenstone belts.

(4) The sulfide facies is characterized by pyrite and pyrrhotite mixed with carbonates and quartz, interlayered with chert, and by disseminated pyrite in black, carbonaceous shale. The formation of the sulfide facies is favored by a strongly reducing environment with abundant H_2S or HS^-.

Iron formation may grade laterally into banded ferruginous and nonferruginous cherts (Mason, 1970; Beukes, 1973). Such changes are generally interpreted to reflect true changes in sedimentary environment. The most extensive studies of the distribution of iron formation within Archean greenstone belts are those of Goodwin (1962, 1973). Three stratigraphic sections of iron formation in the Michipicoten area in the southern Superior Province are shown in Fig. 7-6. The oxide-facies section is chiefly enclosed in clastic sediments while the other two facies are chiefly enclosed by calc-alkaline volcanic rocks. The three facies grade laterally from one into another and are interpreted by Goodwin to reflect a progressively deepening basin from oxide to sulfide facies (Fig. 4-21). The oxide facies reflects a shelf environment with moderately high P_{O_2} and the deeper parts of the basin reflect more subsidence, greater volcanic input, and more reducing conditions. Dimroth (1975) and Walker (1978) have recently challenged this interpretation of iron formation distribution and suggest that oxide-facies iron formation may actually form in deep rather than shallow parts of Archean basins.

A great deal has been written on the origin of iron formation and many models have been proposed (see, for instance, Gross, 1965; James and Sims, 1973; Goodwin, 1973; Boyle, 1976; Dimroth, 1977; Kimberley, 1978). The close association of Algoma-type iron formation with volcanic rocks suggests a genetic relationship with silica, iron, CO_2, and sulfur being derived from the volcanic sources. This may be accomplished directly by hot spring or fumarole activity or by submarine alteration and leaching of volcanic rocks.

MANGANESE FORMATION

Precambrian sedimentary manganese formations occur in India in greenstone successions of the Karnataka subprovince (Naganna, 1971, 1976). Many of these deposits are probably Proterozoic in age. Manganese formations occur as lenses and pockets associated with phyllites or carbonates in the Dharwar Supergroup. Banded, colloform, and pisolitic structures are

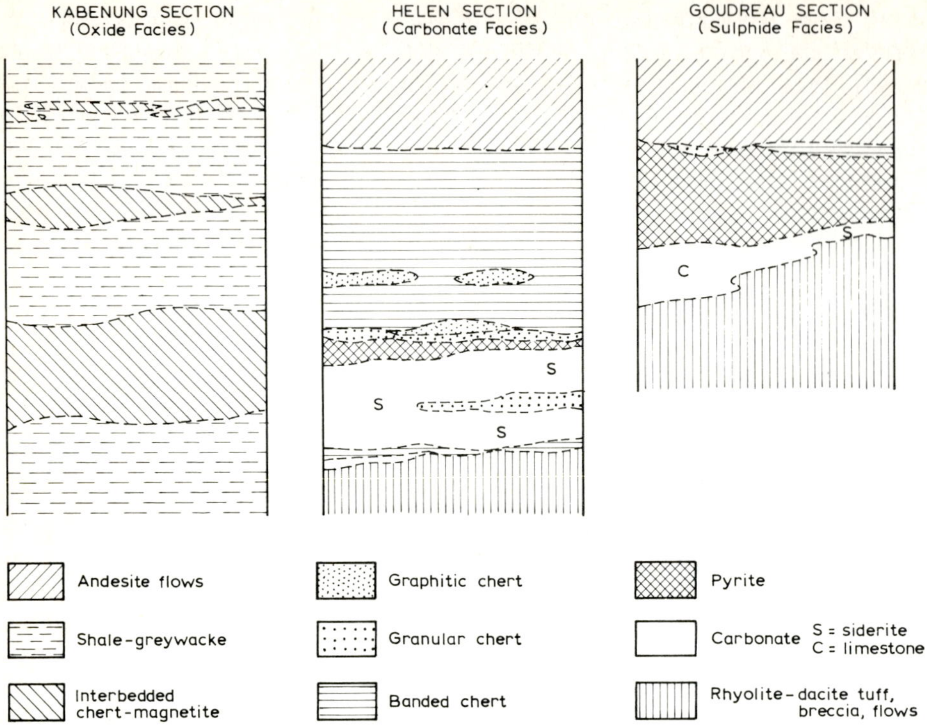

Fig. 7-6. Stratigraphic columns of iron formation in the Michipicoten area, Ontario (from Goodwin, 1973).

common. Major ore minerals are braunite, magnetite, and pyrolusite. Secondary minerals such as cryptomelane and psilomelane often form cavernous and concretionary textures. Naganna (1971) has pointed out two features which favor a primary sedimentary origin for the Indian manganese formations: the manganese formations occur along distinct stratigraphic horizons in the Dharwar Supergroup and the massive primary ores show a crude banding parallel to bedding. Manganese may have been derived from nearby contemporary volcanism or from weathering processes (Roy, 1966; Naganna, 1971).

GOLD DEPOSITS

Gold deposits in Archean granite-greenstone terranes generally fall into one of four categories (Fripp, 1976a): stratiform type, massive sulfides, quartz lode, and disseminations. Placer deposits form a minor fifth category. All occur within greenstone belts although quartz lode deposits may also occur within the margins of surrounding granitic plutons. Stratiform-type deposits are found in banded iron formation where gold occurs in sulfides in carbonate and sulfide-facies iron formations. Gold is included as small grains (< 50 μm in diameter) within pyrite or arsenopyrite (Fripp, 1976b). Individual beds of Au-bearing iron formation are $\lesssim$ 5 m thick and are interlayered with ferruginous carbonate, black argillite, or mafic to felsic volcanics. Gold also occurs in sulfide minerals, some massive Zn-Cu sulfides and such deposits are particularly important in parts of the Superior Province (Hutchinson et al., 1971).

The principal occurrence of Archean gold is in quartz veins, stockworks, and lodes (Boyle, 1961; Stephenson, 1971; Travis et al., 1971; Fripp, 1976a; Anhaeusser, 1976c). Gold-bearing quartz veins range up to about 5 m thick and extend discontinuously along strike for up to 2 km. Most veins are composed chiefly of quartz with small amounts of carbonate and a few percent of sulfides which consist chiefly of pyrite and variable, but small amounts of pyrrhotite, sphalerite, galena, arsenopyrite, stibnite, chalcopyrite, and scheelite. Gold tellurides occur in some deposits. Wall rocks generally exhibit alteration which extends up to several meters away from veins. In mafic rocks the common alteration assemblage is chlorite, carbonate, epidote, tremolite, sericite, albite, and minor sulfide. Regional carbonization is common in some areas such as in the Barberton greenstone belt (Viljoen et al., 1969) and in the Timmins area in Canada (Pyke, 1975). Alteration typically results in losses of Ca, Na, and Mg (± Si and Al), and introduction of variable amounts of H_2O, CO_2, sulfur, and potassium (Boyle, 1961; Bartram and McCall, 1971; Stephenson, 1971). Au also may have been liberated from volcanic rocks and concentrated during such widespread carbonation (Fryer et al., 1979; Kerrich and Fryer, 1979).

Disseminated deposits occur chiefly in clastic sediments (Collender, 1964). The deposits are stratabound, relatively thick, and exhibit gradational contacts with adjacent sediments.

Often there is a close spacial relationship between two or more of the four types of gold deposits suggesting the existence of subprovinces of gold mineralization. Economic gold deposits are chiefly confined to volcanic terranes metamorphosed to the greenschist facies. In the Kaapvaal and Rhodesian Provinces, the quantity of gold decreases outward from the center paralleling increasing metamorphic grades outward from the center (see Chapter 6). These observations suggest that the optimum thermal conditions for gold deposition are roughly those of the greenschist facies (Anhaeusser, 1976a). Most investigators consider all four gold occurrences to be genetically related and the gold to be of volcanogenic origin (Viljoen et al., 1969; Ridler, 1970; Hutchinson et al., 1971; Fripp, 1976a). Existing experimental data on gold solubility suggest that it is carried as complex ions in thermal brines and deposited at temperatures of 300—400° C (Fyfe and Henley, 1973; Fripp, 1976b). A schematic diagram showing the possible

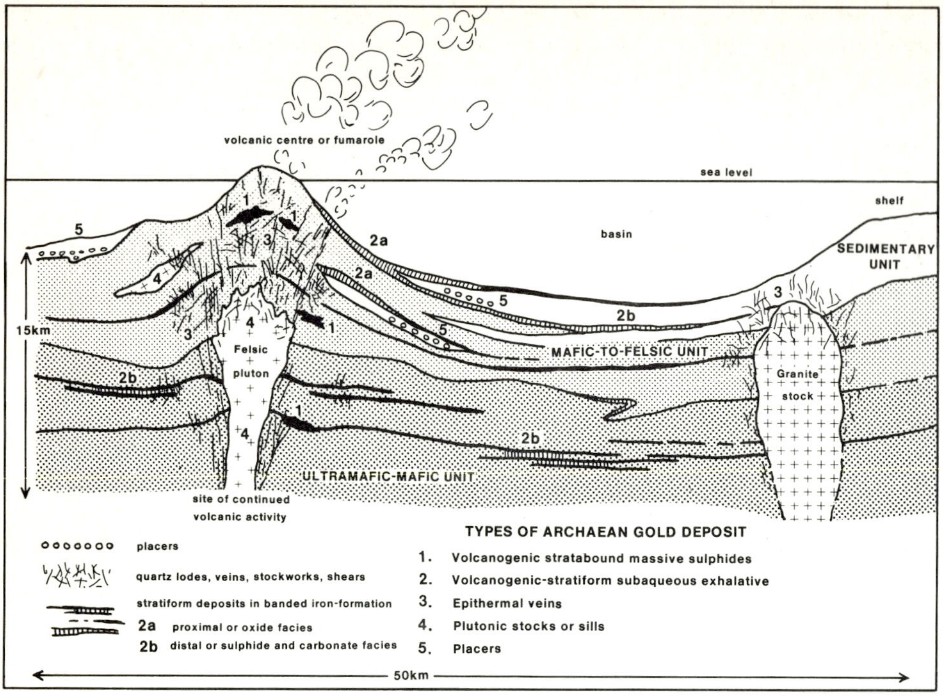

Fig. 7-7. Schematic diagram of an Archean greenstone volcanic complex showing possible relations of gold and sulfide deposits to host rocks (from Anhaeusser, 1976c, modified after Goodwin and Ridler, 1970 and Hutchinson et al., 1971).

relation of gold deposits to Archean volcanic and plutonic rocks is shown in Fig. 7-7.

CHROMITE DEPOSITS

Archean chromite deposits occur in mafic and ultramafic rocks in greenstone belts (Boyle, 1976; Anhaeusser, 1976a). Although chromite occurs in both fresh and altered rocks, relict primary textures suggest an igneous origin for the chromite. One of the largest known deposits is the Selukwe deposit in Rhodesia (Cotterill, 1969). Typical ore bodies at this locality range up to about 15 m thick and can be traced along strike for $\gtrsim$ 300 m. Cumulus textures are well preserved in the ores indicating an origin by crystal settling. Individual bands average about 1 m thick and can be traced for about 100 m. Locally the ore has been sheared and remobilized in fault zones.

The principal minerals associated with Archean chromite deposits are pyroxenes, amphiboles, talc, serpentine, magnetite, carbonate and minor sulfides.

Most deposits, although later faulted and altered, appear to have formed during crystal settling in mafic to ultramafic sills and intrusives intruded into greenstone successions during or soon after volcanic eruptions.

MISCELLANEOUS METALLIC DEPOSITS

As-Sb-Hg deposits, with few exceptions, are of minor importance in Archean granite-greenstone terranes. Their occurrence is similar to that of gold and most deposits that yield As and Sb are also gold producers (Boyle, 1976). The principal ore minerals are arsenopyrite (As), stibnite (Sb) and cinnabar (Hg). One of the largest known Sb deposits in the world occurs in the Murchison greenstone belt in South Africa (Anhaeusser, 1976a; Minnitt, 1975). The stibnite ore bodies at this locality occur as concordant lenses strung out along a strike distance of about 50 km. Host rocks consist of altered volcanics, dolomites, and iron formation. Although many models for the origin of these deposits call upon epigenetic replacement (Anhaeusser, 1976a; Sahli, 1961), recent studies tend to favor a volcanogenic origin (Minnitt, 1975).

Tungsten deposits are widespread but minor in Archean granite-greenstone terranes. They occur in granites and pegmatites, in contact metamorphic aureols, and in quartz veins (Boyle, 1976). Scheelite and wolframite are the principal tungsten minerals. Some Archean quartz-feldspar porphyries contain disseminated Cu, Mo, Au, Ag, and other elements of economic interest (Boyle, 1961, 1976). Disseminated Cu-porphyry deposits similar to those in the southwestern United States have been described at Timmins and at Lang Lake in Ontario (Findlay, 1975; Davies and Luhta, 1978).

NON-METALLIC DEPOSITS

Pegmatites

As discussed in Chapter 5, pegmatites are common in many Archean granitic terranes. Almost all of these are quartz-feldspar-mica pegmatites of no commercial value. Some, however, contain varying amounts of rare minerals that may be of economic value. The chief minerals of potential value reported from Archean pegmatites are spodumene, lepidolite, amblygonite, cassiterite, pollucite, petalite, beryl, tantalite-columbite, wodginite, magnetite, tourmaline, scheelite, bismuthinite, wolframite, monazite, and eucryptite (Boyle, 1976). These minerals concentrate such elements as Li, Rb, Cs, Be, B, Sc, Y, REE, Sn, Ti, Zr, Hf, P, Bi, Nb, Ta, Mo, W, F, Mn, and Fe.

Tin-bearing pegmatites are found in some Archean terranes (Mulligan, 1975; Davies, 1964). The chief tin mineral is cassiterite and it is often associated with Li minerals and magnetite. Li-rich pegmatites are important in parts of the Rhodesian Province (Grubb, 1973). These fall into two categories: the Kamativi type characterized by coarse spodumene and late albite, and the Bikita type characterized by zoning, abundant petalite, and fine quartz-spodume aggregates. Spodumene is mined at the Bikita pegmatite field in southern Rhodesia (Cooper, 1964; Martin, 1964). The main pegmatite, which has a length of 1800 m and a width of 70 m, exhibits a massive lepidolite core; intermediate zones of albite, petalite, spodumene, and pollucite; outer zones of feldspar-quartz-mica with minor beryl; and a border zone of mica and quartz-albite. The Bikita-type pegmatites are interpreted as products of high-temperature ($> 600°$ C) multiple injection, whereas the Kamativi-type pegmatites represent single injections crystallized at temperatures $\lesssim 600°$ C (Grubb, 1973).

Another economically important Archean pegmatite region is the Cat Lake—Winnipeg River and Herb Lake districts in Manitoba (Crouse and Cerny, 1972; Cerny, 1976). The largest pegmatite is the Tanco pegmatite which contains the largest known single source of pollucite. The pegmatite is about 1200 m long and consists of nine zones with 37 minerals. It is interpreted to have crystallized from a melt or supercritical fluid during repeated resurgent boiling of a magma (Crouse and Cerny, 1972).

Corundum and kyanite

Corundum is economically important in Archean greenstone belts of Rhodesia (Morrison, 1972; Anhaeusser, 1976a). Boulder corundum deposits are most important and have the following characteristics: (1) most deposits occur near greenstone-granite or mafic dike-granite contacts; (2) the deposits appear conformable with the greenstone stratigraphy; (3) the corundum occurs as lenses in Al-rich mica schist and contains one or more of the following minerals — andalusite, sillimanite, and kyanite; and (4) associated rocks include ultramafic rocks (talc schists, serpentinites), iron formation, argillaceous sediments, or gneisses. Morrison (1972) concludes that boulder corundum deposits in greenstone belts form by the metamorphism of Al-rich sediments (possibly bauxites). In some greenstone successions in Rhodesia and in many locations in India, corundum occurs as gem quality ruby and sapphire.

Kyanite, besides being associated with corundum, is an uncommon constituent in quartz-sericite-pyrophyllite schists in greenstones (Anhaeusser, 1976a).

Asbestos

Asbestos (chrysotile) occurs in many Archean ultramafic rocks that have been serpentinized (Laubscher, 1968; Viljoen and Viljoen, 1969b; Boyle, 1976; Anhaeusser, 1976b). In only a few areas, however, are fibres developed of sufficient quality to mine. One example is the Munro Mine in the Abitibi belt (Hendry, 1951; Satterly, 1952). Major mines also occur in South Africa and Rhodesia (Anhaeusser, 1976b). Most occurrences are in serpentinized ultramafic sills although some are in volcanic rocks. Anhaeusser (1976b) recognizes three varieties in order of decreasing relative ages: (1) layered complexes associated with the mafic-ultramafic portions of greenstone successions; (2) layered ultramafic bodies associated with the mafic to felsic portions of greenstone successions; and (3) intrusive ultramafic bodies which post-date volcanism but pre-date granite intrusion. Both faulting and folding can control the localization of asbestos development in serpentinized ultramafic rocks.

Magnesite and talc

Magnesite and talc occur as secondary minerals in ultramafic rocks in most greenstone belts (Viljoen and Viljoen, 1969b; Anhaeusser, 1976a). Magnesite deposits are most common in dunites and occur as stockworks or lense-shaped bodies. They appear to have formed by the interaction of CO_2-rich waters with olivine. Talc is formed during regional metamorphism of ultramafic rocks. Major talc deposits occur in contact zones of ultramafic rocks intruded by granite, in fault zones, in folded ultramafic rocks, and in metasomatised ultramafic rocks.

Barite

Small non-economic barite deposits are reported in greenstone belts in Rhodesia, South Africa, and India (Anhaeusser, 1976a; Radhakrishna, 1976). Barite is generally considered to represent a volcanogenic deposit as discussed in Chapter 4.

Chapter 8

ARCHEAN LIFE

INTRODUCTION

Perhaps no other aspect of geology has been the subject of more investigation and general inquiry than that of the origin of life (for summaries see Rutten, 1971; Kvenvolden, 1974). It has been approached from many points of view. Geologists have searched for fossil evidences of the earliest life. Biologists and biochemists have provided a variety of evidence from experiments and models that must be incorporated in any model for the origin of life. One major question that has not been answered completely to everyone's satisfaction is whether life began on the earth or on some other body to be carried later to the earth. The interest in the nature and origin of carbonaceous compounds in Type I carbonaceous chondrites has stemmed in part from this possibility.

Two factors seem to be necessary for the production of living cells (i.e., cells capable of reproducing themselves) (Rutten, 1971):

(1) Free oxygen must not be present in the atmosphere-hydrosphere system. Such oxygen has two deleterious effects on the production of life. First, any organic molecules formed in the presence of free oxygen would be immediately oxidized. Second, oxygen in the atmosphere would produce an ozone layer which prevents ultraviolet (UV) radiation from arriving at the earth's surface and UV radiation appears to be a necessary catalyst to form organic molecules.

(2) The elements (H, C, O, S, etc.) and catalysts necessary for the production of organic molecules must be present.

Most models for the origin of life on earth call upon a primordial "soup" rich in carbon-bearing compounds which form by inorganic processes. Reactions in this "soup" result in increasingly larger and more complex "organic" molecules formed by inorganic reactions (Pirie, 1959). These molecules may grow at the expense of smaller molecules with similar structures. Clays and sulfides may have provided suitable sites for such reactions to begin. During the growth stages, some molecules must have grown into polymers such as peptides which join to form amino acids. The next step in the formation of living cells is the combination of these organic macromolecules into proteins. The proteins in turn must develop in such a way to form membranes which allow living matter to maintain compositional and energy differences from their surroundings (Rutten, 1971). Once membranes have formed, it is possible for metabolism to begin. This involves absorption of food, digestion, and disposal of wastes. In the first large organic molecules this process may involve absorption of globular molecules by a membrane with corresponding transfers and losses to nearby polar molecules. The ultimate stage in biogenesis is the development of the ability to duplicate such that a living cell can perpetuate itself. Mutation would lead to molecules capable of organic photosynthesis. Only after sufficient oxygen had accumulated in the atmosphere would mutation lead to forms capable of respiration.

The earliest experiments dealing with the production of life were carried out by Miller (1953) who sparked a hydrous mixture of H_2, NH_3, and CH_4 to form amino acids. Ponnamperuma (1965) performed similar experiments using UV radiation and reported similar results. Oro et al. (1965) showed that it is possible to synthesize larger "organic" molecules at elevated temperatures (25–150° C) without UV radiation. Fox (1965) successfully synthesized large protein molecules from amino acids in a dry environment at temperatures up to 170° C. These experiments showed that it was possible to produce life in an early reducing atmosphere on the earth. However, if such an atmosphere existed it is likely it was lost very soon after formation of the earth (Walker, 1976) and if life were formed in such an atmosphere, it is likely that it did not survive the catastrophic loss.

Mechanisms for the formation of amino acids have also been sought in a non-reducing atmosphere composed chiefly of H_2O, N_2, CO, CO_2, and H_2. Such mechanisms would be more in line with the probable composition of the first stable atmosphere. Calvin (1965) showed that it is possible to build peptides by dehydrating smaller molecules if HCN is present in the aqueous solution of "soup" in which life forms (beneath a non-reducing atmosphere). Matthews and Moser (1966) showed that such reactions could occur at room temperature in very dilute aqueous solutions. Abelson (1966) has pointed out that such reactions occur only in basic solutions (pH = 8–9).

"Organic" compounds and possible organic remains have been found in carbonaceous chondrites leading to the idea that life exists elsewhere in the solar system and perhaps was created elsewhere and brought to earth. Claus and Nagy (1961) were the first to describe "organic" compounds and "organic" structures from meteorites. Subsequent studies have shown that such materials are common in carbonaceous chondrites and that many of them are contaminants picked up after the meteorite fell on the earth (Anders et al., 1964). Some of the structures, known as "organized elements", appear to be indigenous to the meteorites, however. Whether they represent remains of living organisms or not is a subject of disagreement. Recent experimental studies (Anders et al., 1973) clearly show that it is possible if not likely that the "organic" compounds in meteorites were produced by inorganic reactions at low temperatures, perhaps in the solar nebula. Although recent data seem to cast doubt on the "life in meteorites" hypothesis, the question is definitely not closed.

Iron formation

Several lines of evidence are available for the recognition of living organisms in the earliest preserved Archean rocks (Schopf, 1976): (1) the presence of iron formation; (2) organic geochemical evidence; (3) stable isotope data; (4) microfossil assemblages; and (5) stromatolites. The presence of banded iron formation in Archean greenstone belts is in apparent conflict with the widely held view that free oxygen was absent in the Archean atmosphere. Cloud (1968, 1974) has suggested that this dilemma can be resolved if photosynthetic micro-organisms existed. Free oxygen would be a poison to the first micro-organisms, and hence the early life forms could not survive without an oxygen acceptor. Ferrous iron could represent this acceptor. Cloud suggests that oxygen produced during photosynthesis reacted in solution with available Fe^{2+} to produce Fe^{3+} which was deposited as iron formation. Hence, a delicate balance existed between oxygen production and deposition of iron formation such that oxygen did not collect in the atmosphere in appreciable amounts prior to about 2.0 b.y. This mechanism implies that photosynthetic micro-organisms were present on the earth by about 3.8 b.y., the age of the oldest known iron formation in the Isua greenstone belt in Southwest Greenland (Moorbath et al., 1977b). However, as pointed out by Schopf (1976), the fact that non-biologic sources for minor amounts of oxygen may have been present in the Archean indicates that without other evidence, the presence of Archean iron formation may not be indicative of contemporary life.

Carbonaceous compounds

Carbonaceous compounds in rocks are of two types, extractable (viz., amino acids, hydrocarbons, sugars, etc.) and non-extractable (i.e., those contained in kerogen) (Schopf, 1970; Kvenvolden, 1972). Although the extractable components from Archean sediments are similar or indistinguishable from those in modern organisms, it is not always clear when they were introduced into the rock system. Some may date to the time of sedimentation while others were introduced by secondary processes in the recent geologic past. The insoluble carbonaceous compounds in kerogen, although very likely formed at the time the rock was deposited, may be of biologic or non-biologic origin or both.

Refinement of gas chromatographic and mass spectrometric techniques has made it possible to detect and identify micro-quantities of complex organic compounds in rocks and minerals. It is possible from such results to develop a set of criteria to distinguish between primary and secondary sources for such compounds (McKirdy, 1974). Hydrocarbons (principally alkanes) and very small amounts of amino acids have been extracted from Fig Tree cherts (Schopf et al., 1968). Han and Calvin (1969) report aliphatic hydrocarbons, fatty acids, and n-paraffins from an Onverwacht chert. Straight- and branched-chain alkanes have been found in cherts from the Swartkoppie Formation in the upper Onverwacht Group (Brooks and Shaw, 1971). Because of the mobility of these organic components, however, it is not certain at what time they entered the cherts. Analyses of the insoluble kerogen in Barberton cherts indicate that the Onverwacht samples contain chiefly aromatic degradation products of kerogen while the Fig Tree samples contain an abundance of n-alkanes. These components are generally thought to come from degraded unicellular organisms.

Organic compounds which have survived diagenesis (with little alteration) are sometimes called chemical fossils (Eglinton and Calvin, 1967). Such compounds may be useful as biological markers. Branched alkanes, for instance, are thought to reflect bacterial activity and branched fatty acids and porphyrins reflect evidence of photosynthesis. Nagy et al. (1977) have extracted nitriles and furans from Archean cherts. These are thought to represent the breakdown products of amino acids, peptides, porphyrins, and carbohydrates.

Stable isotope results

Organic photosynthesis fractionates carbon isotopes in that $^{12}CO_2$ is preferred over $^{13}CO_2$. Plants extract CO_2 principally from two sources: atmospheric CO_2 and aqueous carbonate or bicarbonate ions. Fractionation of carbon isotopes are generally expressed as deviations from a carbonate standard (usually the belemnite PDB-1) where:

$$\delta^{13}C‰ = \left[\frac{(^{13}C/^{12}C)_{\text{sample}}}{(^{13}C/^{12}C)_{\text{standard}}} - 1\right] \times 1000$$

$\delta^{13}C$ values are different for atmospheric, seawater, and freshwater CO_2 and carbonate and organisms obtaining their carbon from each of these sources reflects, in part, the $\delta^{13}C$ of the source. Analyses of Archean organic carbon (Hoering; 1962) show low $\delta^{13}C$ values like those observed in modern plants consistent with photosynthesis occurring in the Archean.

Studies of Precambrian carbonates (Schidlowski et al., 1975, 1979) have shown that the average $\delta^{13}C$ value of sedimentary carbonates has remained approximately constant for at least the last 3.7 b.y. This implies that the ratio of organic to carbonate carbon of 1:4 has been maintained for this period of time and that photosynthetic organisms were in existence by 3.7 b.y. Furthermore, if the rather complex model proposed by Schidlowski et al. (1975) is correct, close to 80% of the organic carbon now present in the earth existed by about 3 b.y.

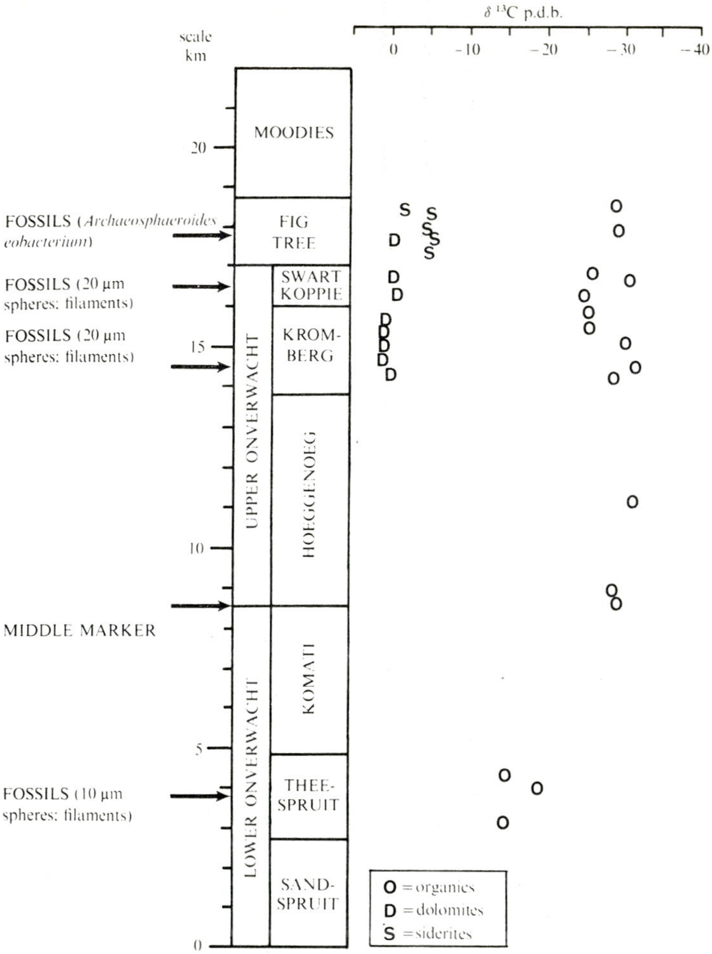

Fig. 8-1. Stratigraphic distribution of microfossils and $\delta^{13}C$(‰) in kerogens from the Swaziland Supergroup, South Africa (from Sylvester-Bradley, 1975).

The distribution of $\delta^{13}C$ values in kerogen and various carbonates in the Swaziland Supergroup in South Africa is shown in Fig. 8-1. Noteworthy is the range of values for the upper Onverwacht and Fig Tree Groups (− 26 to − 33‰) which is similar to the range in modern organic carbon (Oehler et al., 1972). The lower Onverwacht samples (from the Theespruit Formation), however, are distinctly enriched in ^{13}C ($\delta^{13}C = -14$ to -19.5‰).Various explanations have been suggested for this striking difference between lower and upper Onverwacht kerogens as follows (from Kvenvolden, 1974; Sylvester-Bradley, 1975).

(1) A metamorphic effect. Existing experimental data indicate that increasing metamorphic grade tends to deplete kerogen in ^{12}C relative to ^{13}C (Baker and Claypool, 1970; Barker and Friedman, 1969; McKirdy and Powell, 1974). However, this mechanism is not favored for the Barberton samples because there is not a striking difference in metamorphic grade between the upper and lower Onverwacht Group, and $\delta^{13}C$ values in samples collected in a contact metamorphic aureole in the Barberton belt are not affected by an increased metamorphic grade (Oehler et al., 1972).

(2) The hot ultramafic and komatiitic lavas which are important in the lower Onverwacht may be responsible for the anomalous $\delta^{13}C$ values; lower temperature mafic and felsic volcanics are more abundant in the upper Onverwacht (Brooks et al., 1973).

(3) The lower Onverwacht carbon may not be biologic in origin (Oehler et al., 1972); the similarity to $\delta^{13}C$ in carbonaceous chondrites is consistent with this possibility.

Sulfur isotope abundances in 2.7-b.y. iron formation are strikingly similar to those characteristic of modern biological activity and are interpreted to reflect biological reduction of sulfate under anaerobic conditions (Goodwin et al., 1976). The $\delta^{34}S$ values from iron formation in the Isua greenstone belt in Greenland, however, are close to zero and indicate that sulfate-reducing bacteria were not present at 3.8 b.y. (Monster et al., 1979).

Microfossil assemblages

The oldest well-documented assemblage of Archean microfossil-like structures occurs in cherts and other sediments from the Swaziland Supergroup of the Barberton greenstone belt (~ 3.5 b.y.) (Schopf, 1975; Muir and Grant, 1976) and in the Isua Series in Greenland (~ 3.8 b.y.) (Pflug and Jaeschke-Boyer, 1979). Confident recognition of organic structures in such rocks is faced with three major problems (Schopf, 1976): (1) it is easy to contaminate samples with modern micro-organisms during collection or preparation in the laboratory (Cloud and Morrison, 1979); (2) some inorganic structures may be mistaken for organic structures, as for instance the inorganic spheroidal bodies reminiscent of cells described by Engel et al. (1968) from pillow lavas in the Barberton succession; and (3) progressive diagenesis and low-grade metamorphism can produce structures which look organic and such processes can destroy real microfossils. Experimental fossilization studies have documented the production of inorganic fossil-like structures during fossilization (Oehler, 1976). At the present time there is disagreement as to which of the Barberton microstructures are organic and which are not. Recent studies described below, however, seem to leave little doubt that at least some spheroidal structures are of organic origin and represent primitive prokaryotic cells.

Three types of microstructures are reported from the Swaziland Supergroup (Schopf, 1975): rod-shaped, bacterium-like bodies; filamentous, thread-like structures; and spheroidal, unicell-like structures. The rod-shaped forms, first reported by Barghoorn and Schopf (1966), are probably indigenous to the Swaziland sediments. These structures are reminiscent of modern bacteria and range in length from 0.5 to 0.7 μm and in diameter from 0.2 to 0.3 μm (Fig. 8-2A). Rare, filamentous microstructures have been reported in Onverwacht cherts (Brooks et al., 1973; Pflug, 1967) (Fig. 8-2B). These structures are diverse ranging from 4 to 8 μm in diameter and up to 50 μm long. Although interpreted as organic by Brooks et al. (1973), Schopf (1975) indicates that none of the filamentous structures reported thus far provides compelling evidence of an organic origin.

A large number of spheroidal, unicell-like structures reminiscent of modern alga coccoids have been reported in Barberton cherts (Pflug, 1966; Schopf and Barghoorn, 1967; Nagy and Nagy, 1969; Brooks and Muir, 1971; Brooks et al., 1973; Muir and Grant, 1976; Knoll and Barghoorn, 1977) (Fig. 8-3). A similar suite of spheroidal bodies has recently been described from a chert-barite unit from a greenstone succession in the Pilbara Province in Western Australia which is comparable in age to the Barberton belt (Dunlop et al., 1978). The preservation of extremely delicate surface features on Archean spheroids suggests that they are indigenous to the cherts and are not contaminants. Spherical microstructures have been reported from at least four horizons in the Swaziland Supergroup. These bodies generally range from 1 to nearly 200 μm in diameter. Structures in the lower Onverwacht are on the average smaller than those in the upper Onverwacht (Fig. 8-1). Although some investigators still question a biogenic origin for Archean spheroids (Schopf, 1975), recent studies of statistical size distributions and the reporting of spheroids showing cell division indicate a biogenic origin for at least many of these bodies. Muir and Grant (1976) recognize several distinct populations of spheroids in Onverwacht cherts. Studies of spheroidal microstructures in cherts of the Swartkoppie Formation have reported such structures in various stages of binary cell division (Fig. 8-3) (Knoll and Barghoorn, 1977), which clearly points towards an organic origin for these structures.

Several lines of evidence, when considered collectively, indicate that living micro-organisms were present during deposition of the Swaziland Supergroup and hence, that life existed on the earth by at least 3.5 b.y. (Knoll and Barghoorn, 1977): (1) the organic composition of microstructures in cherts and shales in the Barberton belt is similar to that found in younger organisms; (2) the morphology of the microstructures resembles that found in partially degraded microfossils in younger rocks; (3) the size frequency distribution of spheroidal bodies is similar to that of both fossil and modern algal populations; (4) the microstructures occur in a similar lithologic setting to Proterozoic microfossils which are well documented; and (5) the spheroidal microstructures have been preserved in the process of binary cell division.

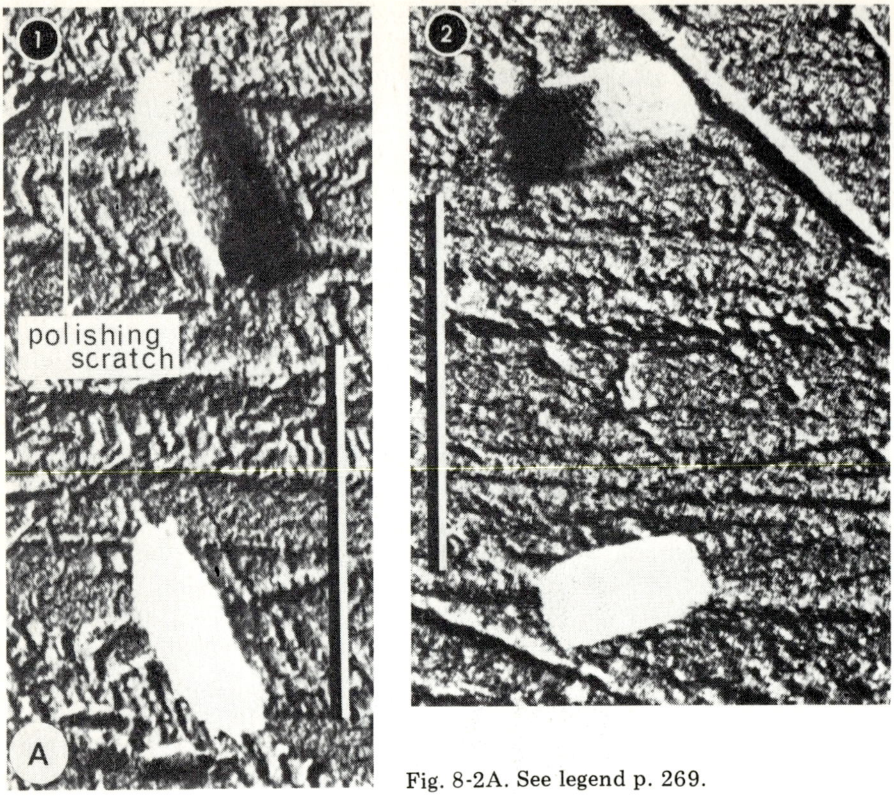

Fig. 8-2A. See legend p. 269.

Archean stromatolites

Stromatolites are finely laminated sediments composed chiefly of carbonates which have formed by the accretion of both detrital and biochemical precipitates on successive layers of micro-organisms. Modern stromatolites are formed chiefly by blue-green algae; a few types are deposited by bacteria. The oldest described stromatolites ($\sim$ 3.0 b.y.) occur in the Pongola Supergroup in South Africa (Mason and Von Brunn, 1977). These stromatolites are found in carbonates interbedded in a section of quartzites, shales and volcanics. The Pongola sediments are thought to represent the earliest phase of the Kaapvaal Basin development which ranged in age from $\gtrsim$ 3.0 b.y. to about 1.8 b.y. and represented chiefly a stable-shelf or miogeoclinal succession (Chapter 1). The first Archean stromatolites described in the literature, however, occur in a dolomite unit within a greenstone volcanic succession at Huntsman quarries north of Bulawayo in Rhodesia (Macgregor, 1940). Several other occurrences in rocks of similar age have recently been reported from nearby greenstone belts in the Rhodesian Province (Bickle et

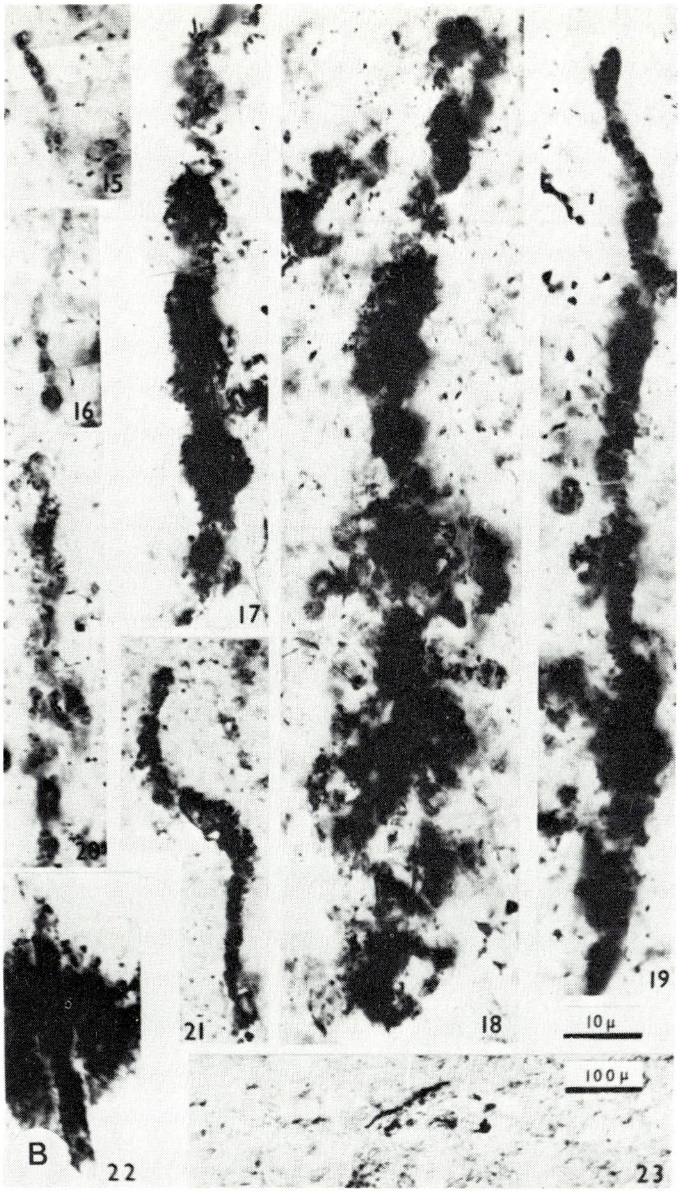

Fig. 8-2. Photomicrographs of microstructures from Barberton cherts. A. Electron micrographs of rod-shaped cells (white areas, below) and their imprints in the chert surface (dark areas, above). Line in each figure represents 1 μm in length. (From Barghoorn and Schopf, 1966; copyright © 1966 by the American Association for the Advancement of Science). B. Non-septate filamentous structures (19—23) and chains of cells (15—18) (from Muir and Grant, 1976; reproduced with permission of John Wiley & Sons, Ltd.).

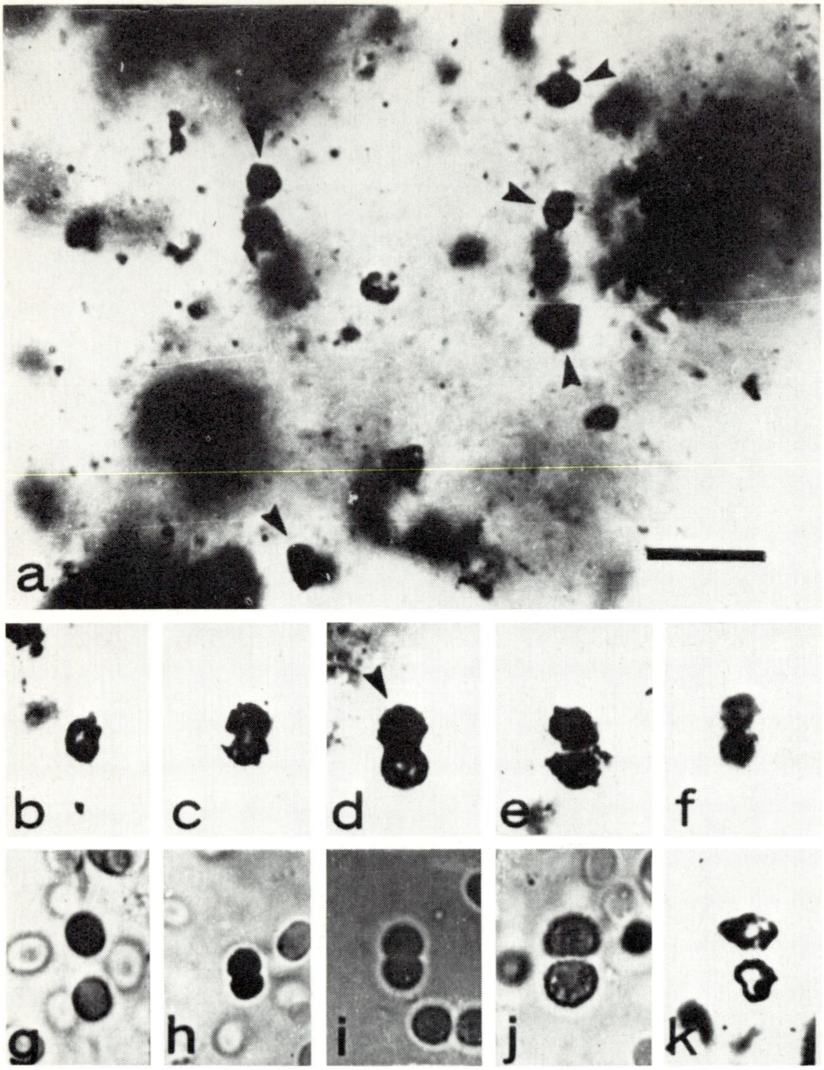

Fig. 8-3. Photomicrographs of spheroidal microstructures from the Swartkoppie Formation in the Barberton area (from Knoll and Barghoorn, 1977; copyright © 1977 by the American Association for the Advancement of Science). Arrows note individual cells. Stages in cell division in the Archean samples in (b) to (e) are compared to modern prokaryotes in (g) to (j). Scale bar represents 10 μm.

al., 1975). Available radiometric ages suggest an age for these stromatolites of about 2.6—2.7 b.y. (Hawkesworth et al., 1975). Stromatolites of similar age are also known from a carbonate unit in a greenstone succession at Steep

Fig. 8-4. Stromatolitic limestone from the Huntsman quarries, Rhodesia (from Schopf et al., 1971).

Fig. 8-5. Stromatolite from the Huntsman quarries, Rhodesia (from Schopf et al., 1971). Thin section shows three zones as described in the text. Vertical scale bar represents 1 cm.

Rock Lake in the Superior Province (Jolliffe, 1955) and from several greenstone localities in the Slave Province (Henderson, 1975b).

The stromatolites from Huntsman quarries near Bulawayo have recently been redescribed in detail (Schopf et al., 1971). Macgregor (1940) originally recognized three types of algal structures: (1) columnar froms 5—7.5 cm in diameter and about 30 cm long with widely spaced laminations; (2) domical

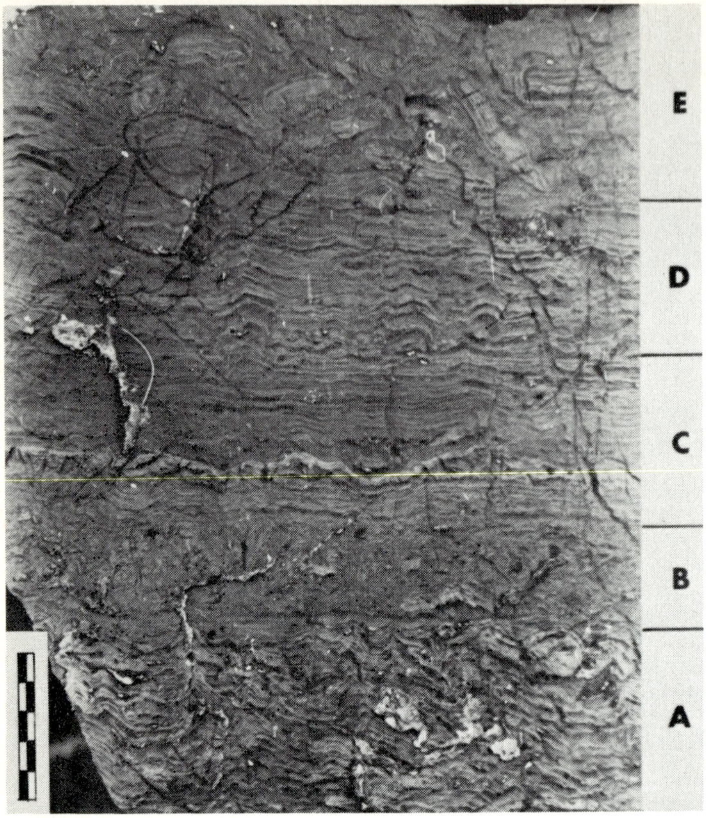

Fig. 8-6. Laminated stromatolitic dolomite from the Snofield greenstone belt, Slave Province, Canada (from Henderson, 1975b). A = irregular wavy lamination; B = non-laminated intraformational breccia; C = flat lamination layer; D = flat to wavy laminations with small hemispheroidal columns; E = intraformational breccia with large, coated laminated clasts (oncolites). Scale bar in centimeters.

structures about 1.2 m in diameter and 80 cm high with widely spaced laminae; and (3) laminated carbonate beds 2.5—8 cm thick having an undulating upper surface and a dentate, second-order organization (Fig. 8-4). Laminations typically contain carbonaceous material. A stromatolite studied in detail by Schopf et al. (1971) exhibits three zones with different microstructure (Fig. 8-5): (1) a lower zone composed of convex-upward laminae and lenses of sparry calcite; (2) a middle zone composed chiefly of sparry calcite lenses; and (3) an upper zone composed almost entirely of undulatory, closely spaced laminations. Spheroidal microfossils have also been reported in samples of this carbonate by Oberlies and Prashnowsky (1968). The $\delta^{13}C$ values of the carbonate are similar to those characteristic of modern limestones of biologic origin (Schopf et al., 1971).

Detailed descriptions of stromatolites from the Snofield Lake area in the northern Slave Province in Canada are also available (Henderson, 1975b). The stromatolitic unit at this locality occurs in a graywacke sequence and is associated with black mudstones. The most common stromatolites are flat laminated forms containing convexities with a relief of about 1 cm (Fig. 8-6). Minor intraformational breccias occur in the unit.

Modern stromatolites occur in a variety of sedimentary environments where the growth rate of micro-organisms exceeds their consumption rate by other organisms. They are found in lakes, marshes, hot and cold springs, tidal flats, hypersaline bays and lagoons, and on the ocean floor (Walter, 1976, 1977). Actual stromatolite shapes are determined by water currents and reaction to sunlight. Although stromatolites range from modern to Archean in age, there are serious limitations in interpreting ancient stromatolites in terms of modern ones (Serebryakov and Semikhatov, 1974; Walter, 1976). First of all, modern stromatolites are still not well understood. Also, stromatolite types are controlled by the availability of specific sedimentary environments which have changed with time. Shallow, stable-shelf marine environments were widespread in the Proterozoic. Such environments are conducive to stromatolite growth and undoubtedly are partly responsible for the abundance of stromatolites in the Proterozoic. The sparsity of Archean stromatolites may be due to a sparsity of stable-shelf environments. The role of burrowing animals which destroy micro-organisms is also important in influencing stromatolite development (Garrett, 1970). Many Archean stromatolites differ from younger stromatolites in being associated with volcanic or volcanoclastic rocks rather than mature sediments.

Chapter 9

MAGMA ORIGIN AND SOURCE

INTRODUCTION

A great deal of progress has been made in recent years in enhancing our understanding of the roles of progressive melting and fractional crystallization in magma production and in learning more about the evolution of upper mantle source areas. Applications of geochemical and isotopic tracer techniques have been particularly fruitful. Both major and trace element modelling have been applied to Archean systems using standard petrologic mixing programs (such as that of Wright and Doherty, 1970), the Rayleigh fractionation law, and equilibrium and fractional melting equations (Neumann et al., 1954; Shaw, 1970; Hanson, 1978; Allègre and Minster, 1978). Geochemical modelling makes use of distribution coefficients (K_d) where K_d is equal to the concentration of a particular element in a solid phase divided by its concentration in a coexisting liquid. They are applicable to both major and trace elements if K_d is approximately constant over the temperature, pressure, and compositional interval being modelled. It is possible to approximate constancy for these parameters in successive small steps between end points. During partial melting in which K_d values are very low ($\leqslant 0.1$), the concentration of the element in the liquid phase varies approximately as $1/F$ where F is the amount of melting (Shaw, 1970). Such an element is referred to as an *incompatible element*. It is possible, using these elements, to place constraints on the composition of magma source rocks. All modelling is dependent upon the concentrations of elements measured in a rock reflecting magma-crystal equilibria. This can be a problem for some elements that are readily remobilized during secondary processes (Condie et al., 1977). Also, in the case of melting models, one is confronted with the problem of identifying rocks which represent liquids that have remained unchanged by fractional crystallization since being extracted from their source.

Sr, Pb, and Nd isotopic systems are particularly sensitive to source area compositions because these systems remain largely unaffected by melting and fractional crystallization processes (Hart and Brooks, 1977). It is possible employing these systems to evaluate the Rb/Sr, U/Pb, and Sm/Nd ratios of the source as a function of time and to identify inhomogeneities in these ratios in the source.

Ultramafic and mafic magmas

Melting and segregation

Although experimental, geophysical, and geochemical evidence indicate a mantle source for ultramafic and mafic magmas, the depths and mechanisms of magma production are subjects of considerable discussion. Many authors, beginning with Viljoen and Viljoen (1969d) have suggested that peridotitic komatiites (PK) represent large degrees (> 50%) of melting of mantle source areas (Brooks and Hart, 1974; McIver and Lenthall, 1974; Green et al., 1975; D. H. Green, 1975). Mysen and Kushiro (1977) have shown experimentally that it is possible to produce liquids that resemble PK in composition by large degrees of melting of lherzolite. Experimental studies of PK indicate that olivine is the only liquidus phase to pressures > 40 kbar (120 km) (Green et al., 1975; Arndt, 1976). Liquidus temperatures may range from 1500–1650°C at 1 kbar to ≳ 1800°C at 35 kbar. Compositions of liquidus olivines in PK from Rhodesia indicate that more Mg-rich magmas cannot be produced by olivine accumulation (Bickle et al., 1977). D.H. Green (1975) has proposed a model for PK genesis involving mantle plumes (diapirs) which rise along adiabats from depths ≳ 200 km with increasing degrees of melting such that PK magmas are produced and extracted from the plumes at shallow depths. This model assumes that liquid and residual solids remain together until shallow depths. Cawthorn (1975) has shown that the temperatures and degree of melting in such plumes are probably less than predicted by a strictly adiabatic cooling curve and to obtain PK by such a mechanism, plumes must originate at depths > 300 km.

D. H. Green (1972) and Arndt (1977b) have suggested that PK are not produced by large degrees of melting because disaggregation of the source rock occurs before 50% melting. Hence, the maximum degree of melting for which magma can remain in contact with residual minerals is an important constraint on the composition of magmas produced by partial melting. Arndt (1977b) has experimentally demonstrated that disaggregation of lherzolite occurs at degrees of melting < 10%. He also shows theoretically and experimentally that settling velocities of olivine in ultramafic magmas change from 10^{-3} to 0.4 cm/s as the degree of melting increases from 15 to 60%. As the degree of melting increases, derivative melts become more mafic, somewhat more dense, and much less viscous. These results do not favor large degrees of melting for production of PK. Arndt (1977b) has proposed a modified plume model in which early, less mafic liquids are tapped off as a plume rises such that PK is produced at shallow levels by only small degrees of melting of residual solid phases in the plume (see Chapter 10). This model is appealing in that it also is consistent with the intimate association of mafic and ultramafic lavas in greenstone belts. It

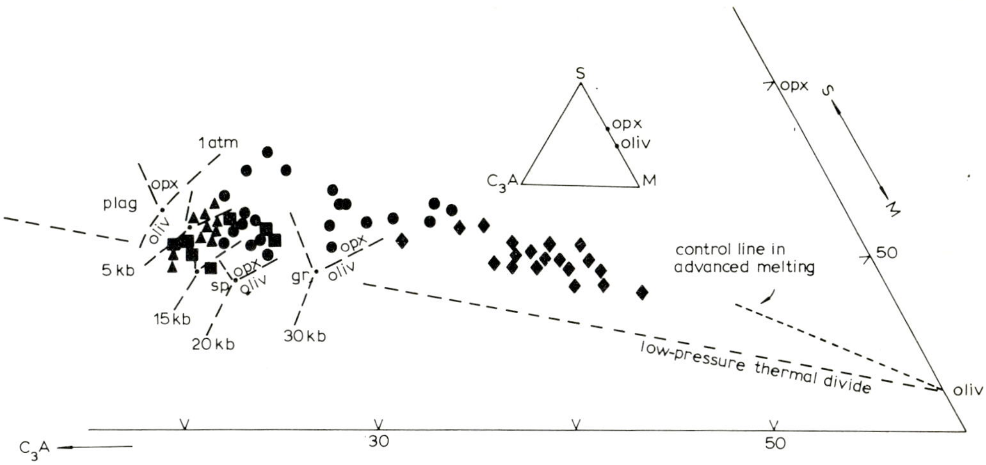

Fig. 9-1. Projection of Onverwacht mafic and ultramafic lava compositions from clinopyroxene into the plane C_3A-M-S (after McIver, 1975). ☆ = garnet lherzolite nodules in kimberlite; ◆ = periodotitic komatiite; ● = basaltic komatiite; ▲ = tholeiite; ■ = metabasalt.

provides a mechanism for obtaining both magma types from the same source by varying degrees of melting. An important consequence of the model is that ultramafic komatiite lavas reflect little about the composition of the mantle source, but only about the composition of residue-rich plumes.

Major element considerations

Chemical variation diagrams show a continuum of compositions within the komatiite and within the tholeiite series (Chapter 3). Although some major element plots show a continuum between the two series (Fig. 3-9), most do not. In some cases, a distinct gap exists between the two series. This indicates that members within each series are genetically related, but that the two series may or may not be related (Arndt et al., 1977). McIver and Lenthall (1974) have suggested that the CMAS system of O'Hara (1968) may be useful in evaluating the origin of Archean ultramafic and mafic lavas. Other variation diagrams have been employed by other investigators. All clearly show that members of the komatiite series ($\geqslant$ 8% MgO) lie on or near an olivine control line which is consistent with experimental data. Komatiites from the Barberton greenstone belt lie along an olivine (± orthopyroxene) control line in the olivine or orthopyroxene volume of the CMAS system (Fig. 9-1). Some basaltic komatiites (BK) lie close to the olivine-orthopyroxene boundary at low pressures ($\leqslant$ 5 kbar) (Nisbet et al., 1977). Archean tholeiites also lie close to this boundary and close to the olivine-plagioclase-orthopyroxene invariant point at low pressures. Garnet is not an important controlling phase at any pressure $\lesssim$ 50 kbar (Arndt, 1976). Consideration of these data on MgO variation diagrams, the MgO-

CaO-Al_2O_3 triangle (Fig. 3-9), or in the system Di-SiO_2-Fo, however, indicates that clinopyroxene becomes an important liquidus phase before orthopyroxene (Nisbet et al., 1977) and experimental results support this observation (Arndt, 1976). These relations are consistent with an origin for komatiites and some tholeiites involving either progressive melting or progressive fractional crystallization of mantle ultramafic rock at varying depths in which olivine and clinopyroxene are the dominant liquidus phases. Experimental studies of Duke and Naldrett (1978) indicate that the sulfide content of komatiitic magmas can also affect composition during fractionation.

The cause of the variation in the CaO/Al_2O_3 ratio in komatiites and, in particular, the origin of the high values in Barberton komatiites is a subject of current discussion. Four explanations have been proposed: (1) garnet fractionation at high pressures; (2) sequential melting of the source; (3) alteration or metamorphism; and (4) a compositionally layered mantle. Garnet fractionation should also result in depletion in heavy REE in derivative melts. Some Barberton komatiites exhibit a small amount of heavy-REE depletion and are consistent with such a mechanism to explain their high CaO/Al_2O_3 ratios (Sun and Nesbitt, 1978). Arndt (1977b) has proposed that sequential melting and magma extraction from rising mantle plumes could produce successive magmas with higher CaO/Al_2O_3 ratios. This would result from the residue after each magma extraction having a higher CaO/Al_2O_3 ratio. Nesbitt and Sun (1976) have suggested that Al loss (or Ca gain) during alteration or metamorphism could explain the high CaO/Al_2O_3 ratios in the Barberton samples. No evidence exists, however, that these rocks are more altered than any other greenstones. Cawthorn and Strong (1975) suggest that melting at varying depths in a mantle in which the CaO/Al_2O_3 ratio increases with depth can explain variations in the CaO/Al_2O_3 ratio in derivative komatiite melts. Nesbitt and Sun (1976), however, have pointed out that such a model requires unreasonably high temperatures ($> 1600°C$) at shallow depths to produce PK magmas. It would appear that some combination of explanations 1 and 2 offers most promise in accounting for variable CaO/Al_2O_3 ratios in komatiites.

Transition trace metals

Ni and Co generally vary systematically with MgO as exemplified by Ni in mafic and ultramafic rocks from Finnish greenstone belts (Fig. 9-2A). Similar relationships are reported by Hawkesworth and O'Nions (1977) and Nesbitt and Sun (1976). In general, it is not possible to distinguish the three igneous rock series by the variation of Ni or Co with MgO. A break in slope occurs between 15 and 20% MgO suggesting a change in minerals controlling liquid composition. It would appear that olivine is the controlling phase at high MgO contents and olivine and clinopyroxene at low MgO contents (Jahn et al., 1979). A significant break at about the

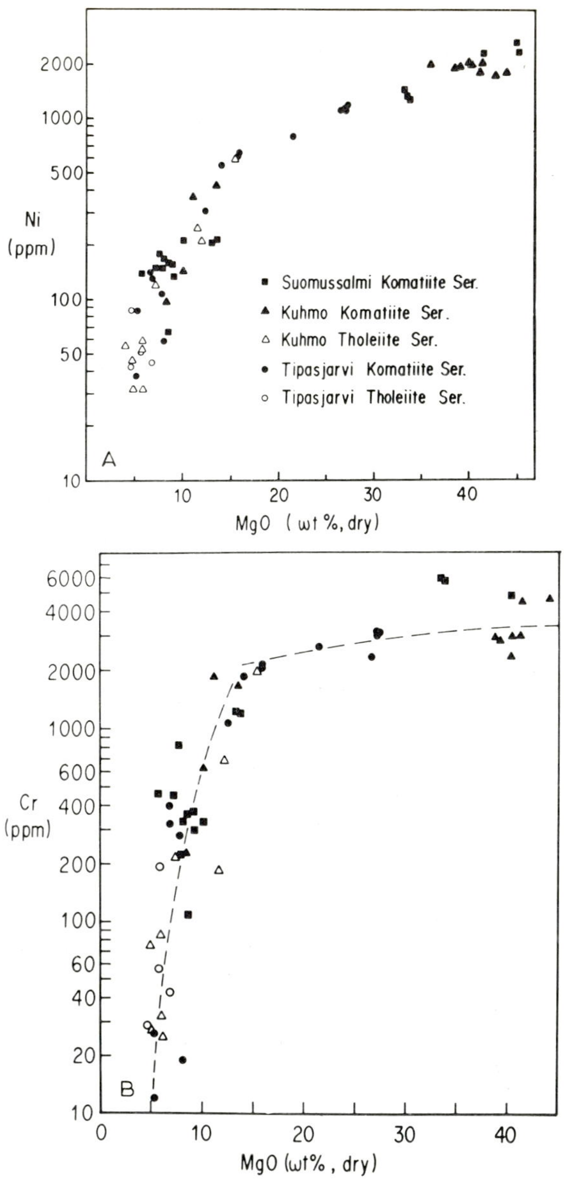

Fig. 9-2. Ni-MgO (A) and Cr-MgO (B) variation diagrams for mafic and ultramafic volcanic rocks from Finnish greenstone belts (from Jahn et al., 1979).

same MgO value occurs on Cr-MgO plots (Nesbitt and Sun, 1976; Jahn et al., 1979) (Fig. 9-2B). Above 20% MgO, Cr is almost constant. Olivine contains very little Cr ($K_d^{Cr} = 0.5$—1.0) and it appears that pyroxene and chromite control the Cr at MgO contents $\gtrsim 20\%$. Below this value, chromite and

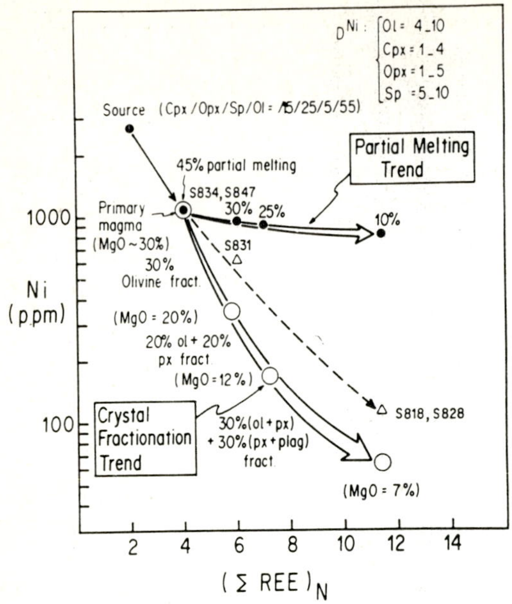

Fig. 9-3. Calculated liquid compositional trends for partial melting and fractional crystallization for the production of BK from the Tipasjarvi greenstone belt (from Jahn et al., 1979). Primary magma is produced by 45% melting of a PK source. The dashed line is the observed trend.

pyroxene are not present and Cr behaves as an incompatible element. In the komatiite and tholeiite series, these results are generally interpreted in terms of progressive melting of an ultramafic source (Nesbitt and Sun, 1976).

Jahn et al. (1979) have compared progressive melting and fractional crystallization models for the production of closely related komatiites and tholeiites in the Tipasjarvi greenstone belt in Finland (Fig. 9-3). Because K_d^{Ni} is sensitive to temperature, a range of values is used for each mineral in the model. The results indicate that the fractional crystallization trend more nearly coincides with the observed trend than the partial melting trend. The partial melting model is not capable of producing magmas with a significant range in Ni content.

Titanium, zirconium, yttrium, and niobium

Nesbitt and Sun (1976) have discussed the relationships of Ti, Zr, Y, Nb, P, and V to magma origin. As exemplified by Ti-Zr and Y-Zr diagrams (Fig. 9-4), these elements exhibit colinear relationships and approximately chondritic ratios (i.e., Ti/Zr = 100—110, Ti/Y = 240—250, Zr/Y $\simeq$ 2.3, Zr/Nb $\simeq$ 18). This suggests that these elements behave as incompatible elements during partial melting and that their ratios represent the ratios of

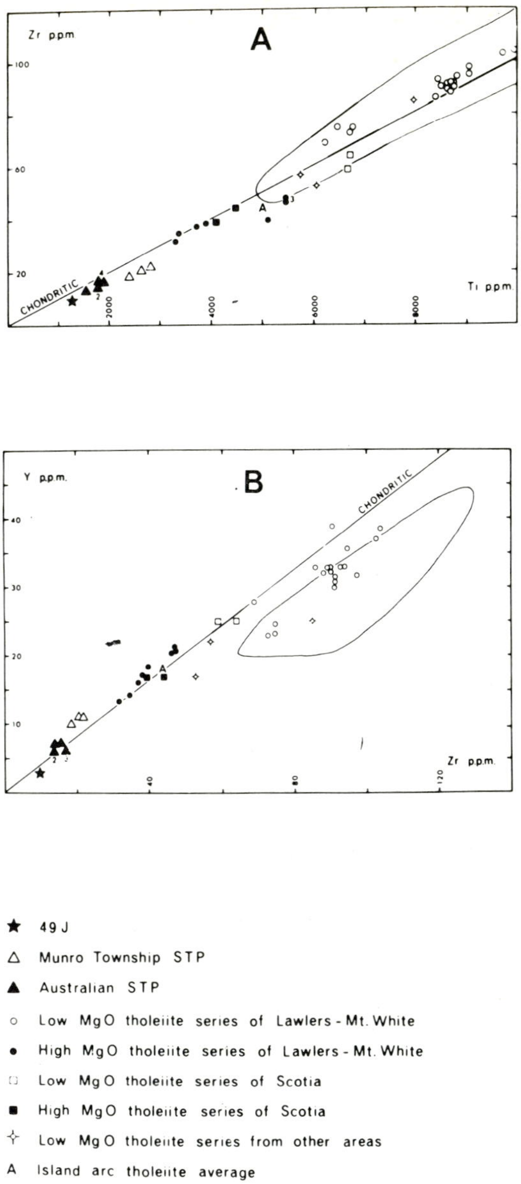

Fig. 9-4. Ti-Zr (A) and Y-Zr (B) plots for Archean PK, BK and mafic volcanic rocks (from Nesbitt and Sun, 1976). The outlined area represents MORB.

the mantle source. On the Y-Zr plot (Fig. 9-4B), members of the tholeiite series and MORB fall on the low-Y side of the line suggesting that Y is being controlled in these magmas by some residual phase(s). Nesbitt and Sun

(1976) suggest that the phase is clinopyroxene that has not been eliminated from the mantle source of these magmas.

Rare earth elements

REE distributions are extremely valuable in placing constraints on the origin of mafic and ultramafic magmas. Existing data suggest that PK and BK can be related by some combination of fractional crystallization and partial melting (Arth et al., 1977; Sun and Nesbitt, 1977; Whitford and Arndt, 1978; Jahn et al., 1979). The fact that MgO content decreases as REE content increases (Fig. 3-19) is consistent with olivine, pyroxene, and plagioclase fractionation since all of these phases have very small distribution coefficients for REE. Differences in light-REE content between the three BK groups (Fig. 3-19) cannot, however, be related by fractionation of these or any other likely source minerals and must reflect varying amounts of depletion in light REE and related LIL elements in the source rocks. It is possible, as pointed out by Sun and Nesbitt (1977) and Arth et al. (1977) that successive tapping of magmas from rising plumes could result in production of komatiites with wide ranges of MgO contents and in light-REE depletion. The earliest liquids would be lowest in MgO and exhibit the least (if any) light-REE depletion. The residual minerals in the plumes would become more Mg-rich and light-REE-poor as successive liquids are tapped off, and hence the liquids would show the same trends with time. Alternately, the varying amounts of light-REE depletion in komatiites may be related to melting of different, unrelated mantle sources that had been subjected to varying degrees of LIL-element depletion during earlier partial melting episodes.

An example of komatiites related by fractional crystallization is shown in Fig. 9-5. These come from a layered komatiite flow (Fred's flow) in the Munro Township (Whitford and Arndt, 1978) (see Chapter 3). Composition of the flow ranges from mafic to ultramafic and spinifex textures are common. All rocks exhibit light-REE depletion, a feature probably inherited from the source. Proportions of fractionating minerals, which are chiefly olivine with smaller amounts of clinopyroxene and plagioclase, are estimated from a major element mixing program and petrographic data. As shown in the figure, the agreement between the calculated and observed REE patterns is good supporting the idea that light-REE-depleted BK can be produced by fractional crystallization of light-REE-depleted PK parental magma. When this result is considered in conjunction with the wide range of komatiitic compositions found in some greenstone belts, it appears that near-surface fractional crystallization may be an important process in producing chemical diversity in komatiitic lavas.

As previously mentioned (Chapter 3), greenstone tholeiites can be classified broadly into two groups (TH1 and TH2) based on REE patterns. Arth and Hanson (1975) have modelled the origin of TH1 from northeastern Minnesota

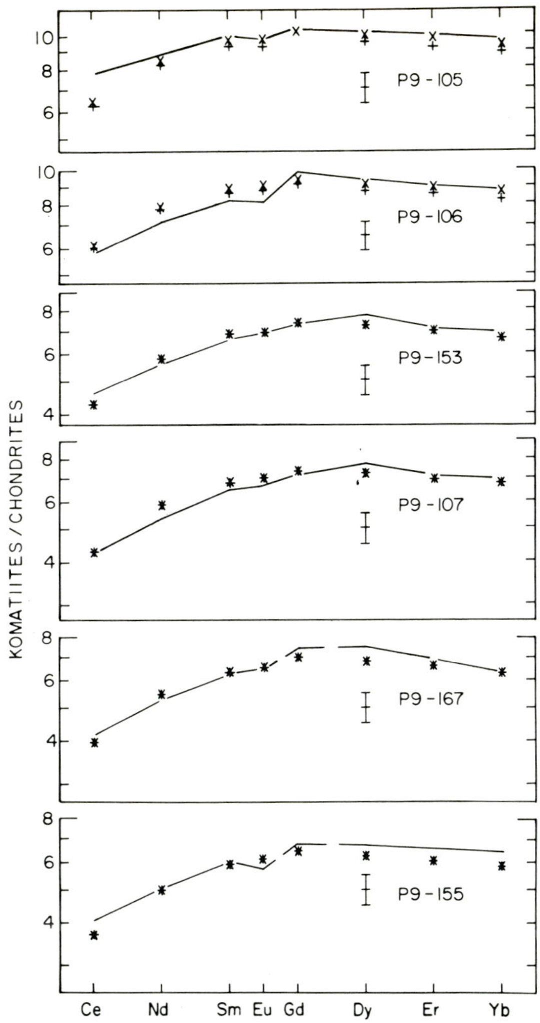

Fig. 9-5. Comparison of observed (solid lines) and calculated chondrite-normalized REE abundances for progressive fractional crystallization of a layered komatiite flow from Munro Township (from Whitford and Arndt, 1978). Liquidus phases are olivine, clinopyroxene, and plagioclase. Proportions and amounts of solid removed at each stage are determined from major element constraints. Symbols: × = surface equilibrium model; + = bulk equilibrium model. 10% error bars are also shown.

by shallow partial melting of plagioclase peridotite and their results are summarized in Fig. 9-6. The close agreement between the calculated (B) and observed (A) REE patterns is consistent with an origin involving 10—25% melting leaving a residue after 20% melting of olivine and orthopyroxene. Alternately, the higher-REE tholeiites (~ 20× chondrites) can be produced

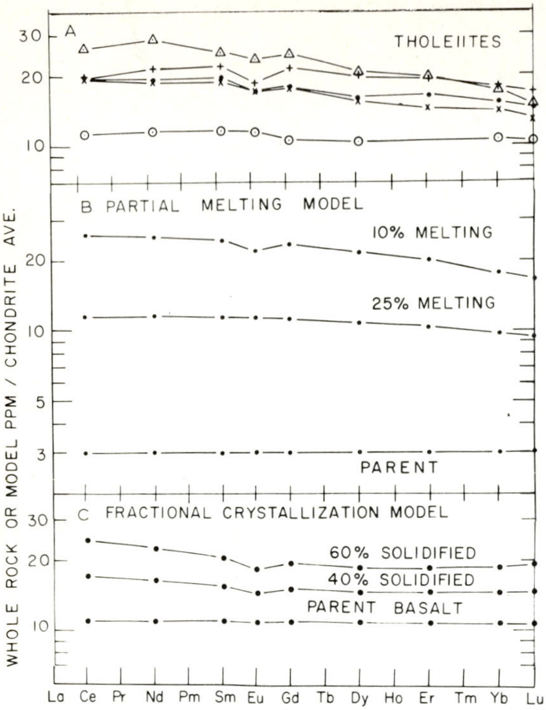

Fig. 9-6. Comparison of chondrite-normalized REE abundances in Archean tholeiites from northeastern Minnesota (A) with partial melting (B) and fractional crystallization (C) models for their production (from Arth and Hanson, 1975).

from the lower-REE tholeiites by 40—60% fractional crystallization of equal amounts of clinopyroxene and plagioclase. The authors prefer a model in which the low-REE tholeiites are produced by 25% melting of peridotite at shallow depth followed by near-surface fractional crystallization to produce the high-REE tholeiites. Condie and Harrison (1976) show that TH1 from the Mafic Formation in the Midlands greenstone belt in Rhodesia can be produced by 30% melting of a lherzolite source in which residual minerals are olivine, clinopyroxene, orthopyroxene, and spinel (in the ratios of 10:4:5:1). Small negative Eu anomalies in some of the rocks can be accounted for by crystallization of plagioclase at shallow depths. TH1a (see p. 96) may be produced either by partial melting of plagioclase lherzolite in which plagioclase is a residual phase or by partial melting of lherzolite already having negative Eu anomalies (Condie and Baragar, 1974). The first possibility, however, fails to explain the lack of a correlation of Eu anomaly size and total REE content with degree of melting (as measured for instance by MgO content). The second has the disadvantage of creating the problem of how Eu depletion was produced in the source. REE data preclude the possibility that

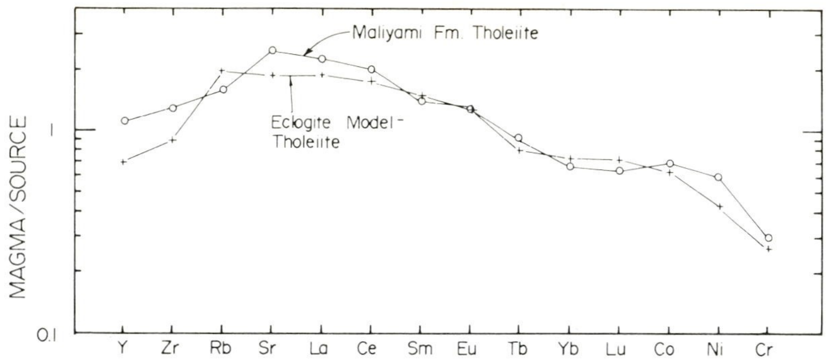

Fig. 9-7. TH1-normalized trace-element patterns in TH2 from the Maliyami Formation in Rhodesia compared to those of a model tholeiite produced by 50% non-modal, equilibrium melting of eclogite of composition TH1 (from Condie and Harrison, 1976).

TH1 or TH2 are related to PK or light-REE-depleted BK by differences in degree of melting or crystallization (Sun and Nesbitt, 1977).

TH2 is characterized by small amounts of depletion in heavy REE (Fig. 3-20). Condie and Harrison (1976) suggest an origin for TH2 in the Maliyami Formation in the Midlands greenstone belt involving about 50% melting of an eclogite source with a composition of TH1 (Fig. 9-7). Garnet amphibolite would also provide a suitable source rock. In light of the studies of Arndt (1977b), it would appear that 50% melting may be high. Subsequent studies have shown that smaller amounts of melting of garnet amphibolite or amphibolite (30—40%) or less likely, of garnet lherzolite (20—30%) are also acceptable mechanisms for the production of Maliyami Formation TH2. Models for the origin of most TH2 seem to require some garnet in the residue to explain the depletion in heavy REE (Arth et al., 1977; Jahn et al., 1979). This implies a melting depth in the mantle of $\gtrsim$ 75 km. It is important that it does not seem likely that TH1 and TH2 can be produced from the same source nor can they be related by fractional crystallization. Although garnet lherzolite may serve as a source rock for both tholeiite types, garnet must be a minor phase that is completely melted in the production of TH1, whereas it must be a residual phase in most TH2 magmas.

Andesitic magmas

As discussed in Chapter 3, Archean andesites broadly fall into three categories (I, II, and III, Fig. 3-30). Types I and II appear to require residual garnet and/or amphibole during their formation (Condie, 1976c; Hawkesworth and O'Nions, 1977; Jahn et al., 1979). These andesite types may be related to tholeiites by either varying degrees of melting or fractional crystallization, or both. An example is illustrated for andesites (and rhyolite)

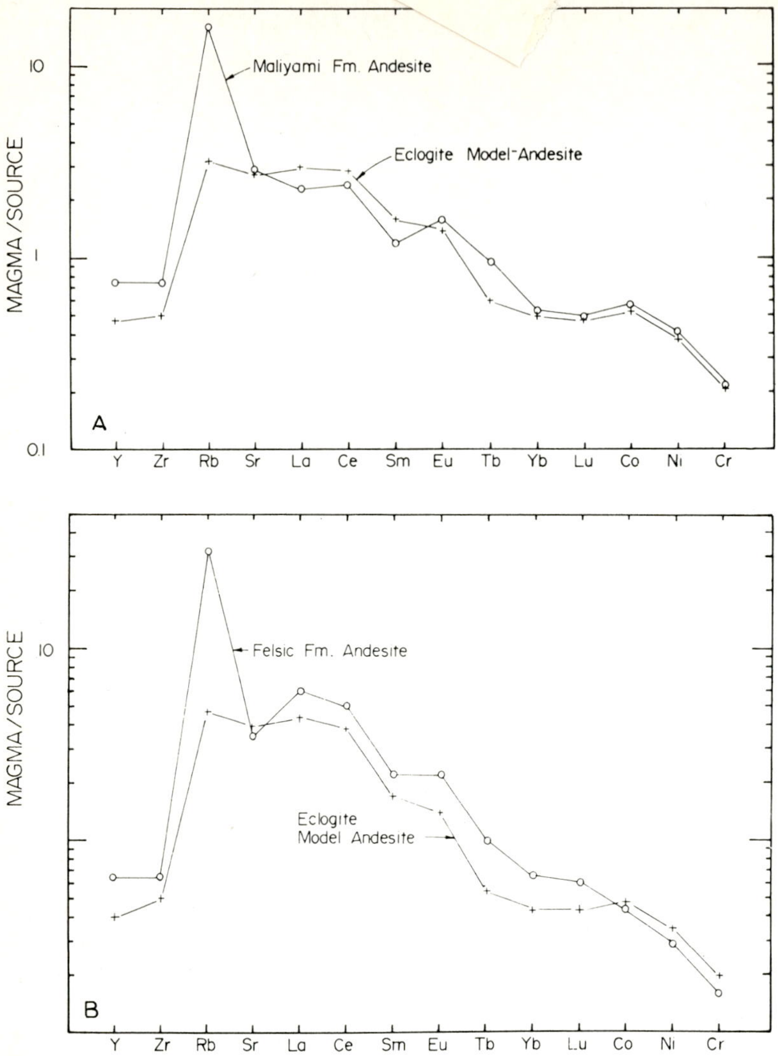

from the Midlands greenstone belt in Rhodesia in Fig. 9-8. Tholeiite TH2 (Fig. 9-7), type I andesite, type II andesite, and FI rhyolite are produced by increasingly smaller degrees of melting of an eclogite or garnet amphibolite source. The mafic source rock has the composition of TH1 from underlying mafic volcanics. Similar results have been reported for types I and II andesite from Kenya (Davis and Condie, 1977). A continuum in compositions between type II andesites and FI dacites and rhyolites in the Marda Complex in Western Australia suggests that these rocks also share a common mafic source (Taylor and Hallberg, 1977). O'Nions and Pankhurst (1978) suggest

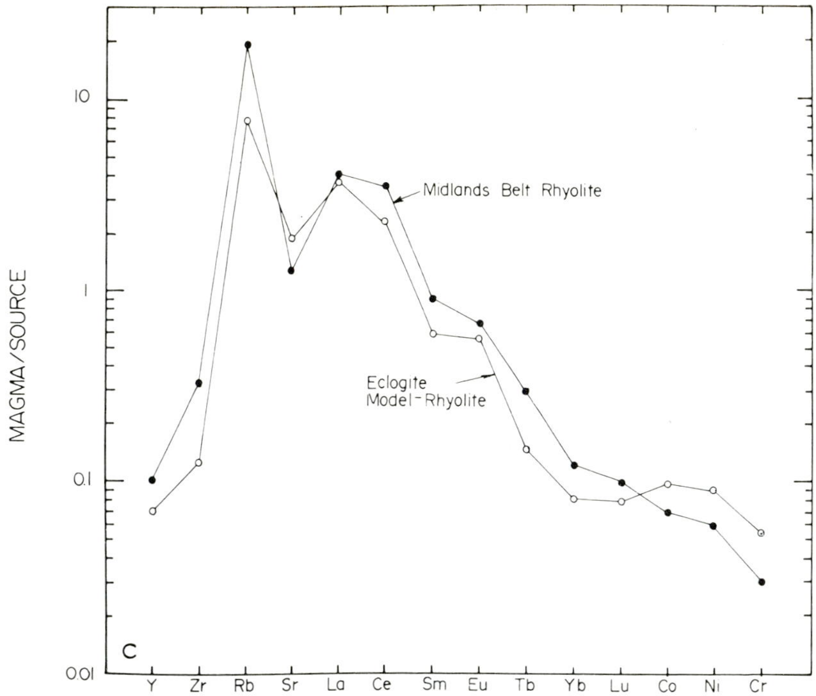

Fig. 9-8. TH1-normalized trace element patterns in andesites type I (A) and type II (B) and in rhyolite type FI (C) from the Midlands greenstone belt, Rhodesia, compared to model magmas produced by partial melting of eclogite of composition TH1 (from Condie and Harrison, 1976). Models are for non-modal, equilibrium melting of 30%, 20% and 10%, respectively.

that the calc-alkaline rocks from the Midlands belt described above may also be related by fractional crystallization of a tholeiite (TH1) magma in which garnet and possibly amphibole are the important liquidus phases. Jahn et al. (1979) have proposed an origin for andesite type II from Finland involving either small amounts of melting of eclogite or garnet peridotite and type II andesites from the Prince Albert Group in northern Canada are explained by partial melting of non-garnet bearing amphibolite (Fryer and Jenner, 1978).

The flat REE patterns of type III andesites preclude garnet fractionation during their production. Fig. 9-9 shows two models for the origin of these andesites from the Abitibi belt (after Condie and Baragar, 1974; Condie, 1976c). Model 1 shows shallow fractional crystallization of average TH1a from the Abitibi belt and model 2 is for partial melting of plagioclase peridotite. With exception of Ni, model 2 agrees best with the observed andesite type III element patterns. Type III andesite appears to be related to tholeiite TH1a by varying degrees of melting of plagioclase peridotite and/or shallow fractional crystallization involving removal of plagioclase and clinopyroxene.

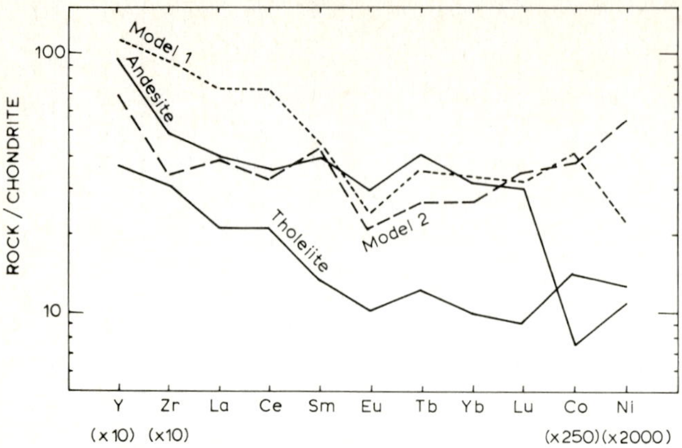

Fig. 9-9. Chondrite-normalized trace-element distribution patterns in TH1a and Andesite type III from the Abitibi belt in Canada compared to two models for andesite production (from Condie, 1976c). Model 1: 75% fractional crystallization of TH1a, plg:cpx = 4/1. Model 2: 10% equilibrium melting of plagioclase peridotite with melting ratios of plg:cpx:opx = 5/3/2.

Felsic magmas

Available geochemical, experimental, and isotopic data indicate a variety of source rocks and mechanisms of magma production for felsic magmas. As previously discussed, the Archean tonalite-trondhjemite gneiss terranes have many different components probably representing different origins (O'Nions and Pankhurst, 1978). The high-Al_2O_3 type (Fig. 5-14) dominates, however. These rocks have similar REE patterns to FI volcanics (Fig. 3-34) and appear to share the same constraints regarding origin. As indicated above, some members of these groups may be related to andesite types I and II by fractional crystallization of a tholeiitic parent magma. Garnet and/or amphibole must be important crystallizing phases to explain the depletion in heavy REE (Arth and Barker, 1976; Frey et al., 1978). Rocks of andesite composition, however, are rare to absent in Archean gneissic complexes. Exceptions are the Granodiorite Suite in Swaziland (Hunter et al., 1978) and a series of closely associated plutonic rocks in the Vermilion district in northeastern Minnesota (Barker and Arth, 1976). The Granodiorite Suite contains rocks ranging from mafic through intermediate to tonalite and trondhjemite in composition. These suites of rocks are similar to trondhjemite suites from Finland which have been modelled geochemically and found to be consistent with an origin by progressive fractional crystallization of a wet tholeiite parent magma involving removal of hornblende, plagioclase, and biotite (Arth et al., 1978).

It has been shown by several investigators that the high-Al_2O_3 trondhjemite-

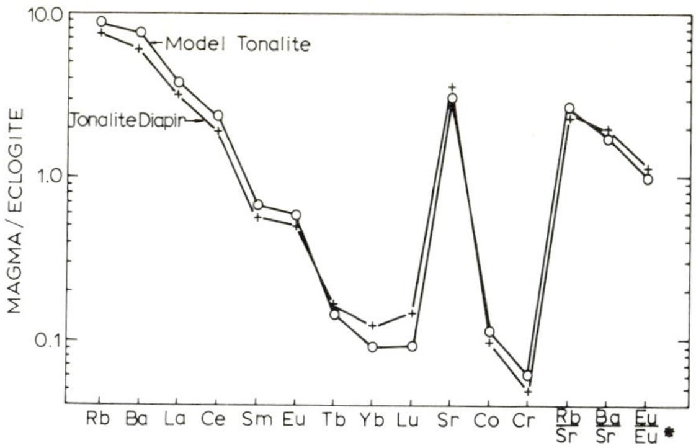

Fig. 9-10. Eclogite-normalized trace element patterns in an average tonalite diapir from the Barberton region compared to those of a model tonalite produced by 10% modal equilibrium melting of eclogite of composition TH1 (from Condie and Hunter, 1976).

tonalite and most FI volcanic rocks can be produced by 10—30% melting of eclogite, garnet amphibolite, or amphibolite (Arth and Hanson, 1972; Hanson and Goldich, 1972; Barker et al., 1976; Condie and Hunter, 1976; Barker and Arth, 1976; Glikson, 1976c; Arth, 1979). The composition of the mafic parent rock is TH1. Experimental studies are also consistent with such an origin for tonalitic liquids at moderate to high water contents (Green and Ringwood, 1968). At lower water contents, andesitic magmas are produced. Fryer and Jenner (1978) have proposed an origin for FI dacite from the Prince Albert Group in Canada involving partial melting of a pyroxene amphibolite in which amphibole, plagioclase, and orthopyroxene are the principal residual phases. An example of a trace element model for an average tonalite diapir from the Barberton area is given in Fig. 9-10. All models share in common the presence of residual garnet and/or hornblende to explain heavy-REE depletion. The presence of basaltic amphibolite in Archean gneissic complexes and the absence or rarity of mafic rocks bearing garnet favors an amphibolite parent for the high-Al_2O_3 trondhjemite-tonalite group (Barker and Arth, 1976). Small positive Eu anomalies occur in some of these rocks (Fig. 5-14) a feature which may reflect plagioclase accumulation (Glikson, 1976c). It is important to point out that an alternate explanation for heavy-REE depletion in the high-Al_2O_3 tonalitic-trondhjemitic rocks is by the loss of a volatile phase in which relatively stable, heavy-REE complexes are concentrated (Collerson and Fryer, 1978). Modelling of the low-Al_2O_3 trondhjemite-tonalite group suggests an origin by partial melting of amphibolite or gabbro in which garnet and hornblende are not residual phases, a feature which is necessary to explain the lack of heavy-REE depletion (Fig. 5-14). Residual phases are primarily pyroxene and plagioclase. Although it is also possible to produce this group by frac-

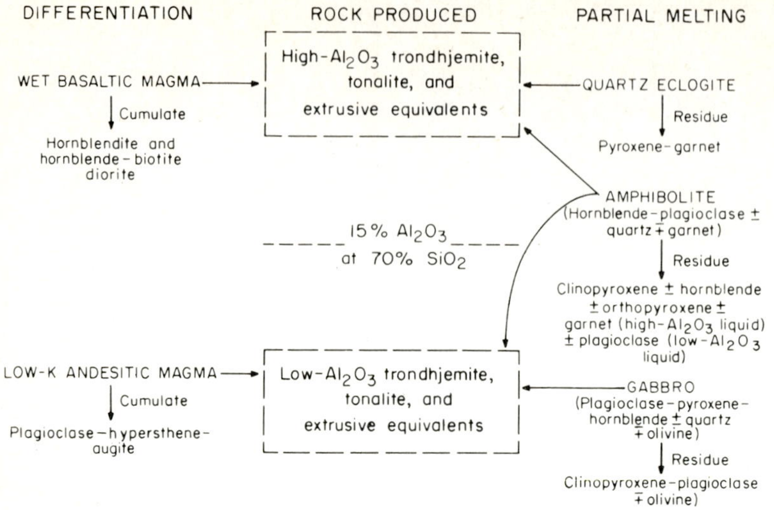

Fig. 9-11. Schematic diagram showing generation of high- and low-Al_2O_3 trondhjemite-tonalite liquids by fractional crystallization and partial melting (after Barker, 1979).

tional crystallization of an andesitic parent magma, the sparsity of rocks of this composition does not support such an origin.

A summary of the modes of production of high- and low-Al_2O_3 trondhjemite-tonalite is given in Fig. 9-11. With the exception of the differentiated trondhjemite suites which are rare in Archean terranes, Archean tonalites and trondhjemites appear to have been produced by partial melting of a mafic (probably amphibolite) parent in the upper mantle (Barker and Peterman, 1974; Barker and Arth, 1976; Hunter et al., 1978; Barker, 1979). Barker and Arth (1976) envision a two-stage model which leads to production of the voluminous bimodal trondhjemite-tonalite and amphibolite association which characterizes much of the granitic terrane in Archean granite-greenstone provinces. The first stage involves production of thick piles of mafic lavas which are metamorphosed to amphibolites near the base. Partial melting of this assemblage, including perhaps garnet-bearing mafic rocks, produces high-Al_2O_3 tonalitic liquids (with residual hornblende, pyroxene ± garnet) or less often at lower water pressures, low-Al_2O_3 tonalitic liquids (with residual plagioclase, pyroxene ± olivine). The model is dependent upon magmas being extracted before 40% melting when the liquids could become andesitic if the water contents were low enough. Basaltic magmatism from the mantle continues as trondhjemite-tonalite plutons are produced and rise through the earlier mafic crust producing the bimodal association.

Both single- and two-stage models have been proposed for the origin of Archean granodiorite magmas. Condie and Hunter (1976) propose a model for Dalmein-type plutons in the Kaapvaal Province involving 50% partial

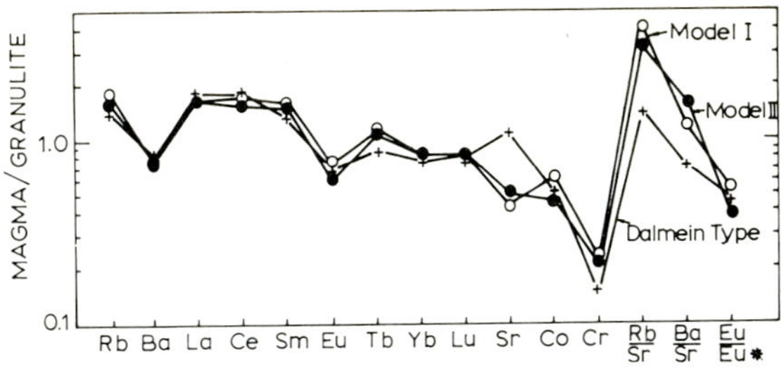

Fig. 9-12. Andesitic granulite-normalized trace element patterns in average Dalmein-type granodiorite compared to those of two model granodiorites produced by 50% modal melting of andesitic granulite composed of plagioclase, orthopyroxene, K-feldspar, quartz, magnetite, biotite, and minor garnet (from Condie and Hunter, 1976).

melting of an andesitic granulite in the lower crust (Fig. 9-12). Minor amounts of residual garnet (or amphibole) are necessary to explain moderate depletion in heavy REE. Glikson (1976c) and Condie and Lo (1971) propose models for the origin of Archean granodiorites by partial melting of quartz eclogite. In addition, Glikson (1976c) proposes a second stage in which plagioclase and amphibole are crystallized from the magma. Unlike most members of high-Al_2O_3 trondhjemite-tonalite group, only minor amounts of residual garnet or amphibole are required in the production of granodiorite, which tends to favor a garnet-bearing andesitic granulite source since garnet and amphibole are major minerals in mafic source rocks.

Archean granites and quartz monzonites and FII volcanic rocks exhibit similar REE patterns mildly depleted in heavy REE with variable negative Eu anomalies (Figs. 3-34 and 5-16). Three major origins have been suggested for these types of magmas: (1) partial melting of graywacke; (2) partial melting of andesitic granulite followed by fractional crystallization; and (3) partial melting of the bimodal trondhjemite-tonalite and amphibolite association. Arth and Hanson (1975) have proposed an origin for Archean quartz monzonites from northeastern Minnesota involving 20—50% melting of short-lived graywacke with quartz, plagioclase, amphibole and garnet as the chief residual minerals. The REE patterns for these models are compared to the quartz monzonites in Fig. 9-13. The residual granulite mineral assemblage does not result in models with REE patterns in as good of agreement as does the amphibolite assemblage. The sparsity of graywackes and derivative paragneisses in Archean granite-greenstone terranes does not, however, seem to favor a graywacke parent as a major source of high-K granitic rocks. Condie and Hunter (1976) have proposed a model for granites and quartz monzonites from the Barberton region involving two stages: first, a granodiorite liquid similar to the Dalmein pluton is produced by partial melting of felsic granulite (Fig. 9-12); this is followed by 70—80% fractional crystallization at shallow crustal levels with variable oxygen fugacities to produce the Sicunusa- and

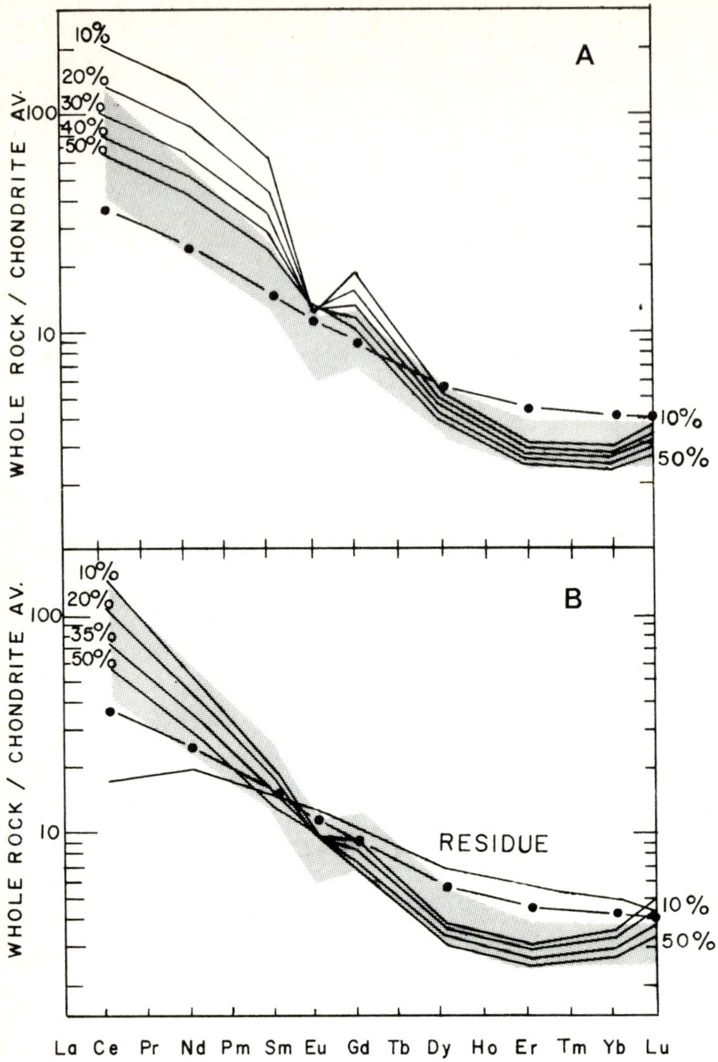

Fig. 9-13. Chondrite-normalized REE patterns for graywacke (●) and quartz monzonites (shaded) from northeastern Minnesota compared to model magmas produced by 10—50% partial melting of the graywacke (from Arth and Hanson, 1975). A. Biotite-bearing source and dry granulite assemblage (both produce identical results). B. Amphibolite-facies, amphibole-bearing residue. Contours represent percent melting.

Mpageni-type granites (Fig. 9-14). A REE model for the production of "average" Archean quartz monzonite involving 30% melting of andesitic granulite in the lower crust is shown in Fig. 9-15. Also shown are REE patterns for average high-Al_2O_3 trondhjemite-tonalite and an average Archean

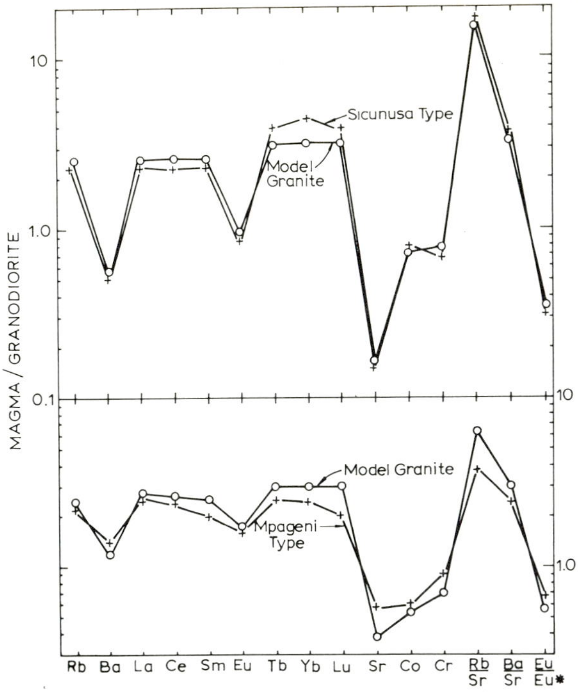

Fig. 9-14. Granodiorite-normalized trace element patterns in averge Mpageni and Sicunusa type granites from the Barberton region compared to model granitic magmas produced by 70% fractional crystallization of average granodiorite (Dalmein-type) parent magma (from Condie and Hunter, 1976).

granulite from Scotland. It is clear that the trondhjemite-tonalite alone cannot serve either as a parent or a residue for the production of quartz monzonite. The andesitic granulite (SG), however, exhibits a REE pattern (with a positive Eu anomaly) not unlike the model residue. Glikson (1976c) has suggested a model for Archean quartz monzonite in the Barberton region in which both the trondhjemite-tonalite and the coexisting mafic rocks are partially melted. The mixture of these two rocks produces a composition not unlike that of the andesitic granulite parent described above; in this regard, the last two models are geochemically indistinguishable. Some granite and quartz monzonites can be produced by partial melting of the tonalite-trondhjemite fraction only. Geochemical results from FII volcanics also favor a crustal source for these rocks (Davis and Condie, 1977; Taylor and Hallberg, 1977; Fryer and Jenner, 1978).

The strongly fractionated heavy-REE patterns in Archean syenites and related rocks (Fig. 5-17) necessitate residual garnet during their formation. Most Phanerozoic counterparts, on the other hand, do not allow residual garnet. Low $^{87}Sr/^{86}Sr$ ratios from Archean syenites in northeastern Minnesota

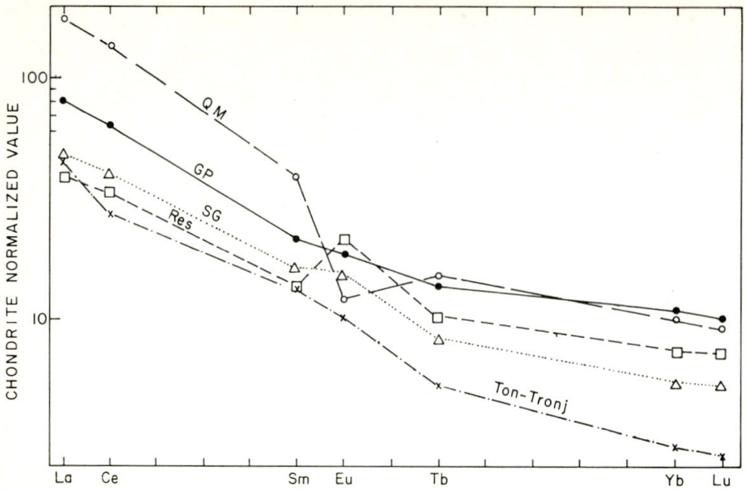

Fig. 9-15. Chondrite-normalized REE distributions in average Archean quartz monzonite (*QM*), high-Al_2O_3 trondhjemite-tonalite (*Ton-Tronj*) and an average granulite from Scourian of Scotland (*SG*) compared to an assumed andesitic granulite parent (*GP*) and calculated granulite residue (*Res*) after 30% non-modal equilibrium melting of an assemblage of plagioclase, quartz, orthopyroxene, and biotite (± garnet, amphibole, and K-feldspar) (K. C. Condie, unpublished results).

seem to require a mantle source (Arth and Hanson, 1975). Most data are consistent with an origin for these rocks involving a very small amount of melting ($\lesssim$ 5%) of eclogite or garnet lherzolite in the mantle (Arth and Hanson, 1975; Glikson, 1976b).

Conclusions and discussion

Fig. 9-16 summarizes the modes of magma production and sources for igneous rocks found in Archean granite-greenstone terranes. The major conclusions are as follows:

(1) PK is produced by partial melting of LIL-element-depleted lherzolite in the upper mantle in which olivine and clinopyroxene are the principal residual phases. It is also possible to produce BK groups 2 and 3 by partial melting of such a source.

(2) BK 2 and 3 may also be related to PK by fractional crystallization in which olivine and pyroxenes are the principal residual phases.

(3) BK 1 cannot readily be related to PK or to BK 2 or 3 by fractional crystallization or partial melting.

(4) TH1 and BK 1 appear to be produced by partial melting of an undepleted lherzolite source in which olivine and pyroxenes are the principal residual phases.

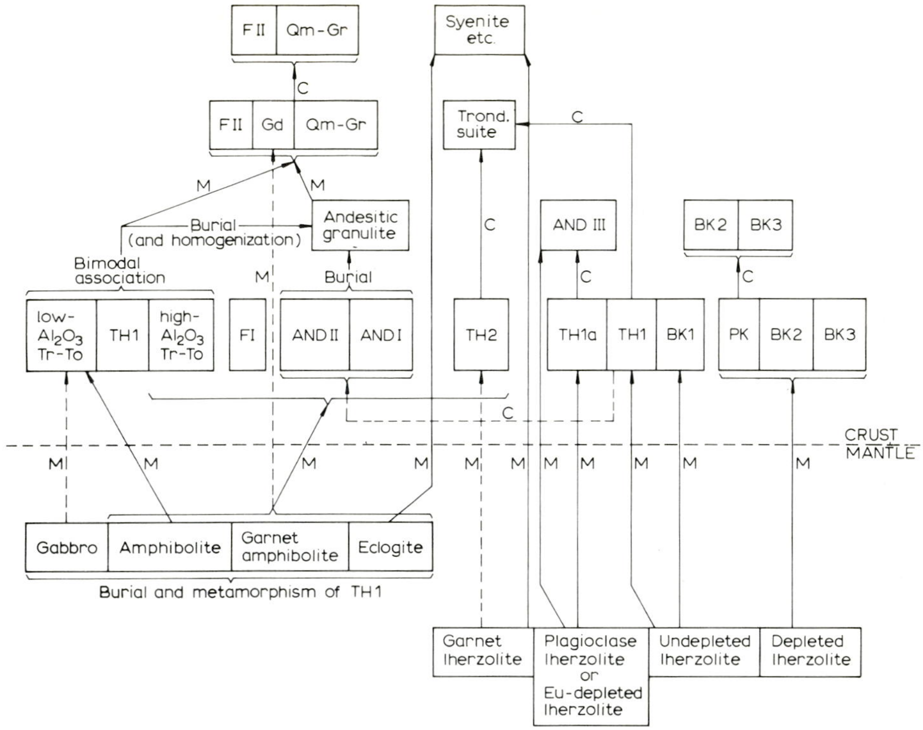

Fig. 9-16. Schematic summary of the possible sources and major modes of production of magmas in Archean granite-greenstone terranes. Key: *M* = partial melting; *C* = fractional crystallization; *Tr-To* = trondhjemite-tonalite; *BK* = basaltic komatiite; *PK* = peridotitic komatiite; *TH* = tholeiite; *AND* = andesite; *F* = felsic volcanics; *Qm-Gr* = quartz monzonite-granite. Solid lines are major trends; dashed lines are possible or minor trends.

(5) Shallow fractional crystallization involving removal of olivine, pyroxenes, and plagioclase is recorded in TH1 by variable REE contents and variable, but small Eu anomalies.

(6) TH1a may be produced by either partial melting of plagioclase lherzolite in which plagioclase is a residual phase or by partial melting of Eu-depleted lherzolite.

(7) TH2 may be produced by partial melting of eclogite, garnet amphibolite, or amphibolite (or less likely, of garnet lherzolite) in which garnet and/or amphibole are residual phases. TH1 and TH2 cannot be related easily by varying degrees of fractional crystallization or partial melting.

(8) Andesite types I and II may be produced in a manner similar to TH2 by smaller degrees of melting. They also may be produced by garnet and/or amphibole crystallization from TH1 magma.

(9) Andesite type III may be produced in a manner similar to TH1a by

smaller amounts of melting and/or by shallow fractional crystallization of TH1a involving removal principally of plagioclase and clinopyroxene.

(10) Most high-Al_2O_3 trondhjemite-tonalite and FI felsic volcanics are probably produced by small amounts of melting of eclogite, garnet amphibolite, or amphibolite (in increasing order of probability) in which amphibole and/or garnet are residual phases. They may be related to TH2 and andesite types I and II by varying degrees of melting.

(11) Trondhjemite suites may be produced by fractional crystallization of a wet TH1 or TH2 magma chiefly by removal of amphibole, plagioclase, and biotite.

(12) Low-Al_2O_3 trondhjemite-tonalite may be produced by partial melting of amphibolite or gabbro in which pyroxenes and plagioclase are the chief residual phases and garnet and amphibole are not residual.

(13) Granodiorite may be produced by partial melting of andesitic granulite, or less likely, of eclogite, garnet amphibolite, or amphibolite.

(14) Most granite, quartz monzonite, and FII felsic volcanics are probably produced by small amounts of partial melting of andesitic granulite and/or the trondhjemite-tonalite and amphibolite bimodal association. Shallow fractional crystallization of granodiorite may also be an important mechanism for production of these felsic magmas.

(15) Syenite and related rocks appear to be produced by very small amounts of melting of undepleted garnet lherzolite and/or eclogite in the mantle.

These results indicate that at least three magma sources are necessary in the evolution of Archean granite-greenstone terranes: an ultramafic, a mafic, and an andesitic or tonalite-trondhjemite source. The relative importances of each of these sources varies with time and location. As pointed out in Chapter 2, ultramafic and mafic rocks decrease at the expense of calc-alkaline rocks as a function of increasing stratigraphic height in most greenstone successions. This implies that ultramafic sources dominate during the early stages of greenstone belt development and mafic and intermediate sources during the later stages. Ultramafic sources dominate throughout the succession in some greenstone terranes (such as western Australia) and mafic sources throughout the succession in others (such as in western Kenya). Existing field data and radiometric dates indicate, however, that in most granite-greenstone terranes, all three sources were available at the same time although not equally important. As previously mentioned and as described in detail in Chapter 10, progressive melting and successive tapping of mantle plumes provides a means of obtaining undepleted to depleted mafic and ultramafic magmas from a similar source. Because mafic source rocks are required for the production of the voluminous trondhjemite-tonalite magmas, two or more magmatic stages must be involved in the production of sialic crust. It is necessary first to produce large volumes of tholeiitic lavas, which sink and are metamorphosed to garnet- and/or amphibole-bearing assemblages that serve as sources for trondhjemite-tonalite melts. The mafic magmas may be produced just prior to trondhjemite-tonalite production, or they may be produced during earlier magmatic episodes.

The general succession from mafic to felsic compositions with time in granite-greenstone terranes appears, on the whole, also to reflect a decreasing thermal gradient and hence a decreasing depth of melting in the mantle. The syenites and related rocks which are very late in the sequence of magmatism appear to have fractionated with garnet lherzolite implying a source depth of $\gtrsim$ 75 km. Earlier magma types reflect either much larger degrees of melting at similar or greater depths or they were produced at depths less than 50 km. Volcanic cyclicity (Chapter 2) necessitates furthermore, that even within the evolution of a specific greenstone belt, magma source rocks must be replenished since the same source rock cannot be used more than once for the same kind of magma (Condie, 1975, 1976c). Any model for the origin and development of Archean granite-greenstone terranes must have a means of replenishing magma sources.

The relative importance of eclogite during the Archean has been discussed by Barker and Arth (1976). With exception of one occurrence in Scotland (Alderman, 1936), eclogite has not been described from Archean terranes. D.H. Green (1975) suggested that Archean geothermal gradients may have been sufficiently steep that they did not pass through the eclogite stability field. Although recent estimates of Archean geotherms (Chapter 6) suggest they were similar to present gradients beneath high heat flow areas, they may have been steep enough that the 10—30% partial melting needed to produce trondhjemite-tonalite liquids was reached before parental amphibolites were converted to eclogites (Barker and Arth, 1976). Amphibolite is an abundant rock type in Archean granite-greenstone terranes and was probably the singly most important mafic source for more felsic magmas.

The origin of the bimodal association in Archean granitic gneiss complexes and the origin of bimodal greenstone belts are important problems related to Archean magma production. Experimental data indicate that increased amounts of water in mafic parent rocks results in the production of tonalitic rather than andesitic melts which are produced in less water-rich systems (Green and Ringwood, 1968; T. H. Green, 1972). It has been suggested that the easiest way to explain the absence or sparsity of rocks of andesitic composition in Archean gneissic complexes is that melting of mafic parent rocks in the mantle occurred under water-rich conditions (Barker and Peterman, 1974). Similar conditions could also apply to the bimodal-type greenstone belts (Chapter 2). This, in turn, would suggest that the calc-alkaline type greenstone belts, which contain andesite, evolved from less water-rich magma source areas. Thus, varying amounts of water liberated by the mantle may have partially controlled the relative abundances of igneous rocks in Archean granite-greenstone terranes.

COMPOSITION AND EVOLUTION OF THE ARCHEAN MANTLE

Introduction

The composition and evolutionary changes in the mantle can be studied from the chemical and isotopic composition of derivative magmas. Because the mantle is probably not an infinite reservoir for LIL elements and because these elements are generally partitioned into the liquid phase during melting, such elements should be depleted from the upper mantle by continuous extraction of magma throughout geologic time (Jahn et al., 1974). On the other hand, most transition metals are partitioned into the residual solid phases during partial melting and hence should become enriched in the mantle sources as a function of decreasing age. The rate of change of composition in the mantle is dependent upon the rate of magma extraction, and hence, the rate of crustal growth, and upon the proportion and distribution of mantle that contributes to magmas (O'Nions and Pankhurst, 1978). In employing mantle-derived igneous rocks to investigate compositional changes in the mantle with time, one is faced with several important problems (Condie, 1976; Hart and Brooks, 1977; Cox, 1978). First is recognition of liquids that were in equilibrium with mantle source material. Subsequent fractional crystallization may modify the composition of mantle melts. Second is estimation of the degree of melting represented by a mantle-derived magma; often it is difficult to distinguish melts that have evolved by progressive melting from those reflecting progressive fractional crystallization. Still another problem is that of selecting elements which are not readily susceptible to remobilization during alteration and metamorphism (Table 3-1).

O'Nions and Pankhurst (1978) have discussed the problem of what volume of the mantle has contributed to the formation of the crust through magma extraction. Employing a composition of the earth as given by Tera et al. (1974) and a crustal composition of Fairbridge (1972), they calculate the composition of the expected residual mantle and compare this to compositional parameters deduced for the present mantle. The calculated Rb/Sr ratio (0.024) is higher than the Rb/Sr ratio in a MORB source ($\sim$ 0.006) which is consistent with the idea that MORB source areas have contributed more to the crust than other mantle sources. Sr and Pb isotope data from igneous rocks of various ages also support this conclusion and suggest that the crust has been extracted non-uniformly from the earth's mantle (as discussed later).

Compositional estimates

In theory, the concentration of major and incompatible trace elements in the upper mantle can be estimated from ultramafic and mafic lavas using

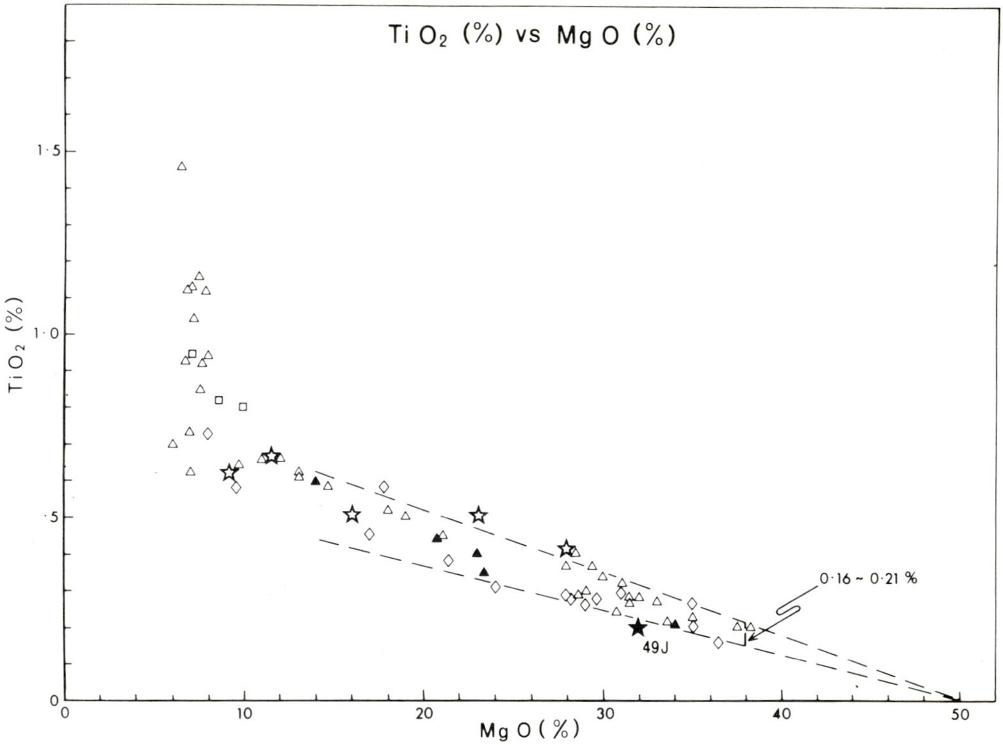

Fig. 9-17. TiO_2-MgO plot for Archean PK, BK, and tholeiites (after Sun and Nesbitt, 1977). Mantle TiO_2 (0.16—0.21%) is estimated by assuming a mantle MgO content of 38%. Symbols defined in Fig. 9-18.

an observed linear relationship between MgO and other elements and from element ratios that are approximately constant over wide degrees of partial melting. A plot of TiO_2 against MgO in Archean komatiites and tholeiites that are thought to represent unfractionated mantle-derived melts produces a linear array of points (Fig. 9-17; Bickle et al., 1976; Sun and Nesbitt, 1977). This intersects the MgO axis at about 50% and suggests that olivine (Fo_{92}) was the major residual phase after partial melting (Sun and Nesbitt, 1977). Bickel et al. (1976) suggest that orthopyroxene also is a residual phase. Experimental evidence, however, as discussed previously tends to support olivine as the sole residual phase for moderate to large amounts ($\geqslant$ 20%) of melting. Making the assumption that olivine is the only residual phase, it is possible to calculate the composition of the mantle source provided the MgO content can be approximated. Values between 38 and 41% have been selected for mantle MgO by analogy with the content in various ultramafic rocks. An estimate of the Archean upper mantle composition based on MgO = 38% is given in Table 9-1 where it is compared

TABLE 9-1

Estimated Archean mantle composition and comparison with other ultramafic compositions (oxides in wt.%, trace elements in ppm) (from Sun and Nesbitt, 1977)

	Archean mantle	Lherzolite, Victoria, Australia	Lherzolite nodule, average	Pyrolite (Ringwood, 1975)
SiO_2	45—52	45.19	45.0	46.1
TiO_2	0.18—0.24	0.06	0.06	0.2
Al_2O_3	3.59—4.94	2.98	2.80	4.3
Fe_2O_3		1.50	1.47	
FeO	8.6—9.7	6.30	6.63	8.2
MnO	0.15	0.13	0.11	—
MgO	38	40.56	40.1	37.6
CaO	2.96—4.07	2.50	2.93	3.1
Na_2O	0.32—0.43	0.19	0.20	0.4
K_2O	0.024—0.033	0.002	0.025	0.03
Rb	0.57—0.78			
Sr	19—26			
Sc	14—20	14.2		
V	83—110	97		
Zr	9.7—13			
Nb	0.53—0.72			
Y	4.2—5.8	2.67		
Yb	0.33—0.51	0.31		
Ni	2000	2100	1960	1600
Cr	3000	3950	3080	2700
Co	100	102		
$(La/Sm)_N$	0.7—1.3	0.45		

N = chondrite-normalized ratio.

to the composition of lherzolite nodules and to Ringwood's theoretical pyrolite. Abundances of Al_2O_3, CaO, Zr, Nb, Y, Sc and V are calculated from various constant ratios of these elements in PK flows (Sun and Nesbitt, 1977). An assumed MgO value of 41% for the upper mantle results in 20—25% difference in the calculated values. The data suggest that for major and many incompatible transition elements, the Archean upper mantle was similar in composition to ultramafic nodules and pyrolite.

Several investigators have presented data which suggest that the Archean mantle was less depleted in LIL elements than the present mantle. Although many, if not most, LIL elements are mobilized during alteration (see Chapter 3), a possible way of overcoming this difficulty is to use averages of large numbers of samples from specific greenstone belts (Hart et al., 1970b; Jahn et al., 1974; Sun and Nesbitt, 1977). This approach assumes that although LIL elements are mobile, individual greenstone belts behave, as a

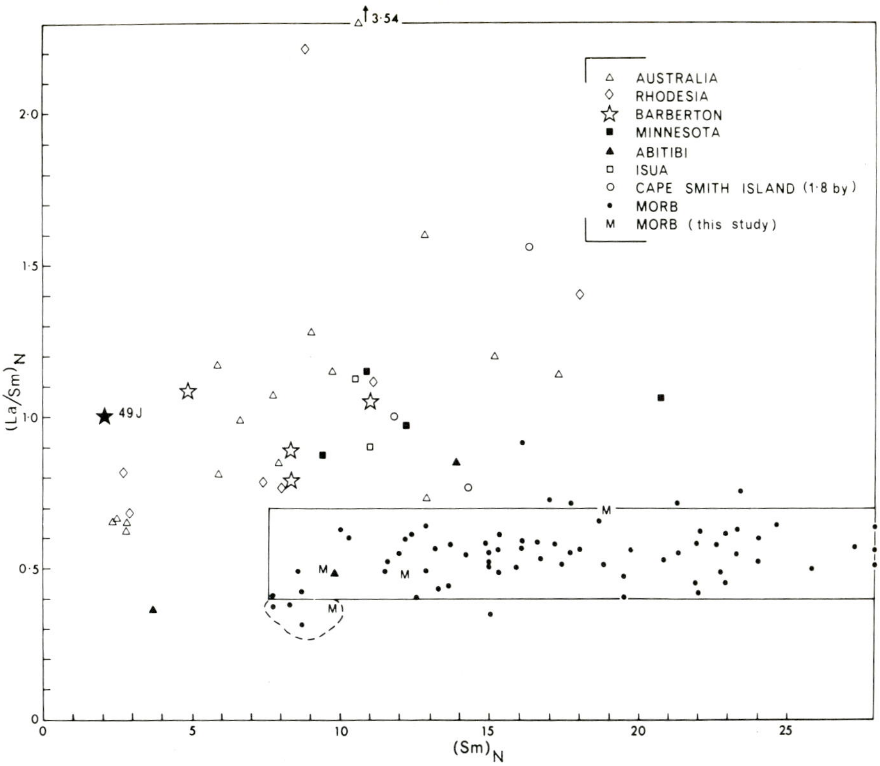

Fig. 9-18. $(La/Sm)_N$ versus Sm_N for Archean PK, BK, and tholeiitic lavas. MORB data are outlined in the rectangle (from Sun and Nesbitt, 1977).

whole, as closed systems. The La/Sm ratio appears to be rather insensitive to alteration (Condie et al., 1977). A plot of $(La/Sm)_N$ versus Sm_N for Archean mafic and ultramafic rocks and MORB is useful in comparing relative depletion in mantle sources since the La/Sm ratio is rather insensitive to progressive melting or crystallization. The following can be concluded from such a diagram (Fig. 9-18): (1) the Archean upper mantle was quite heterogeneous and exhibits varying degrees of depletion in La relative to Sm; and (2) the mantle source area of MORB is less depleted and more homogeneous than Archean mantle sources (Sun and Nesbitt, 1978). Other LIL elements also are consistent with the Archean mantle being less depleted than the present mantle. Ratios of these elements are particularly good monitors of mantle depletion in that they are quite insensitive to variations in the amount of partial melting (Table 9-2). The results clearly indicate that the Archean mantle was higher in $(La/Sm)_N$, and Rb/Sr and lower in Sr/Ba, K/Cs, K/Ba, and K/Rb than the mantle source of MORB.

It is now a clearly established fact that tholeiites and andesites of Archean

TABLE 9-2

Comparison of LIL element ratios in Archean tholeiites and modern MORB (after Sun and Nesbitt, 1977; Hart and Brooks, 1977)

	Archean	MORB
$(La/Sm)_N$	0.7—1.3	0.4—0.7
Sr/Ba	0.3	10
Rb/Sr	0.035	0.008
K/Rb	470	1050
K/Cs	2660	81,000
K/Ba	16	110

age are enriched in most transition trace metals compared to modern counterparts (Glikson, 1971b, 1972b; Condie, 1976c; Gill, 1979). A comparison of chrondrite-normalized values in modern and Archean examples are shown in Fig. 9-19. Cr, Ni, and Co are significantly enriched and Zn, V, and Cu somewhat enriched in the Archean rocks compared to modern rocks (Cu in andesites is an exception). At least three causes might be considered for the Archean enrichment (Condie, 1976c; Nesbitt and Sun, 1976): (1) more accumulation of such phases as olivine and sulfides in Archean magmas; (2) Archean magmas represent a larger degree of melting than their Phanerozoic counterparts; and (3) the mantle has become depleted in transition trace metals with time. There is no petrographic or chemical evidence to support hypothesis number one and it will not be considered further. Higher temperatures in the Archean mantle could result in lowering distribution coefficients for transition metals (which are very temperature sensitive) and also result in larger amounts of melting. The net result would be production of Archean magmas with higher transition metal contents than Phanerozoic counterparts. Incompatible elements, in turn, should be lower in Archean mafic and andesitic volcanics. To some extent this is observed for REE (Fig. 3-20) which are, on the average, lower in Archean tholeiites than in MORB. TiO_2, also, appears to be 40—50% lower in Archean tholeiites than in MORB with similar MgO contents (Nesbitt and Sun, 1976). Melting calculations, however, indicate that a larger amount of melting in the Archean mantle cannot explain the large discrepancy in Ni content observed between Archean and modern tholeiites (Gill, 1979). Naldrett (1973) has suggested a mechanism by which the upper mantle can become depleted in sulfur with time. Experimental data (see Chapter 10) indicate that sulfides in the Archean upper mantle would be completely melted when the silicate fraction is only partially melted. The sulfides may separate as immiscible-liquid droplets and be carried upwards with Archean magmas, depleting the mantle source in sulfur. Because most transition trace metals are, in part, chalcophylic, they

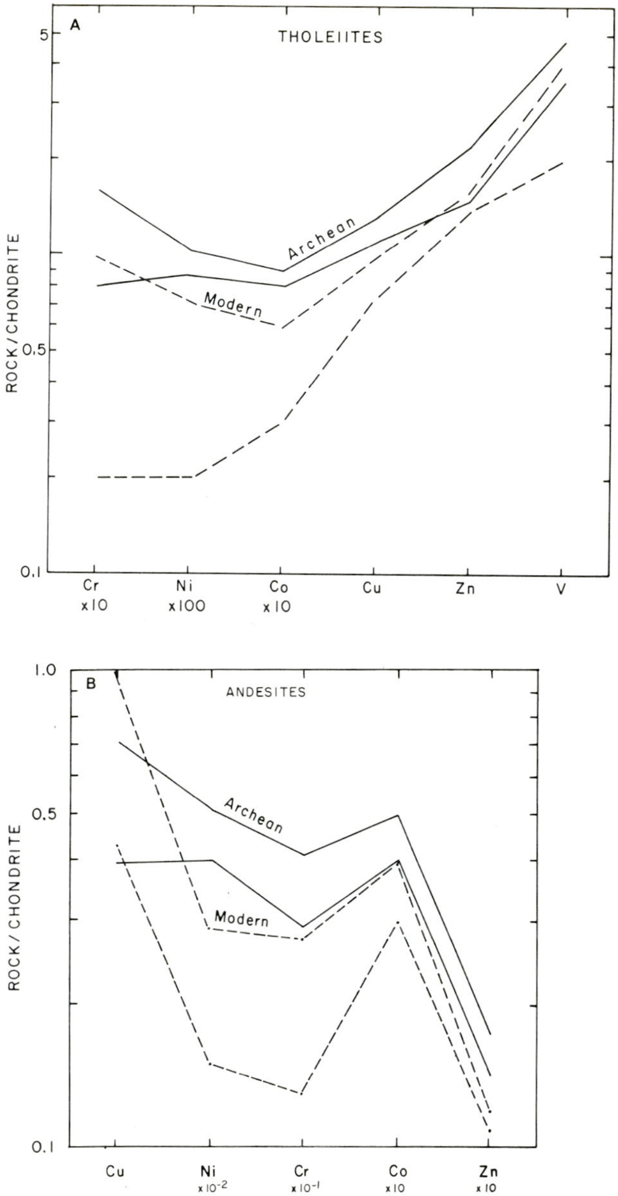

Fig. 9-19. Envelopes of variation of chrondrite-normalized transition metal contents in Archean and modern tholeiites (A) and andesites (B). Data from Tables 3-7 and 3-9.

may also be concentrated in the immiscible-liquid droplets, and carried upwards, thus depleting the source in these elements. The large amount of sulfur in Archean basalts compared to MORB (Naldrett et al., 1978) is con-

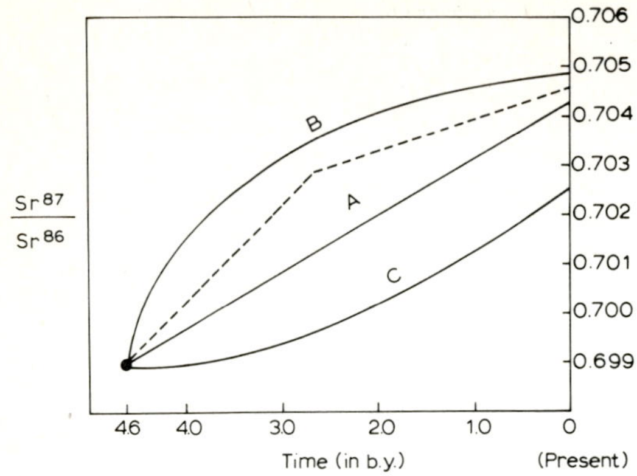

Fig. 9-20. Hypothetical strontium isotope growth curves for the upper mantle (from Jahn and Nyquist, 1976; reproduced with permission, John Wiley & Sons Ltd.).

sistent with this idea as is the relative abundance of Ni sulfide deposits associated with Archean mafic and ultramafic volcanics. To evaluate more fully this theory, however, it will be necessary to have more data on the transition metal contents of post-Archean, pre-modern mafic and andesitic volcanics, which should also reflect transition metal depletion in the mantle.

Strontium isotope constraints

The growth of radiogenic ^{87}Sr in the upper mantle varies as a function of the radioactive decay of ^{87}Rb and of the Rb/Sr ratio. Three basic models can be considered for the continuous evolution of the ^{87}Sr/^{86}Sr ratio in the mantle (Hart, 1969; Jahn and Nyquist, 1976). These are illustrated as mantle growth curves in Fig. 9-20. The slope at any point on the curves defines the Rb/Sr ratio at that point. The implications of each of these curves are as follows (Jahn and Nyquist, 1976).

Curve A. A constant Rb/Sr ratio is maintained in the upper mantle. Four explanations are possible to explain a constant Rb/Sr ratio as a function of time: (1) the upper mantle is an infinite reservoir of Rb and Sr and hence extraction of magmas over geologic time does not change the Rb/Sr ratio of the source; (2) the crust and upper mantle are constantly remixed in such a way as to maintain a constant ratio; (3) preferential Rb loss in the production of new crust is compensated by Rb addition from the lower mantle; and (4) magmas are derived from different volumes of an initially homogeneous mantle with each volume being tapped only once. In terms of our knowledge about the composition of the earth's interior, explanation 1 seems unlikely, and the probable unsubductability of sialic crust renders explanation 2 unlikely (Moorbath, 1977).

Curve B. Rb is extracted from the mantle preferentially to Sr, thus the

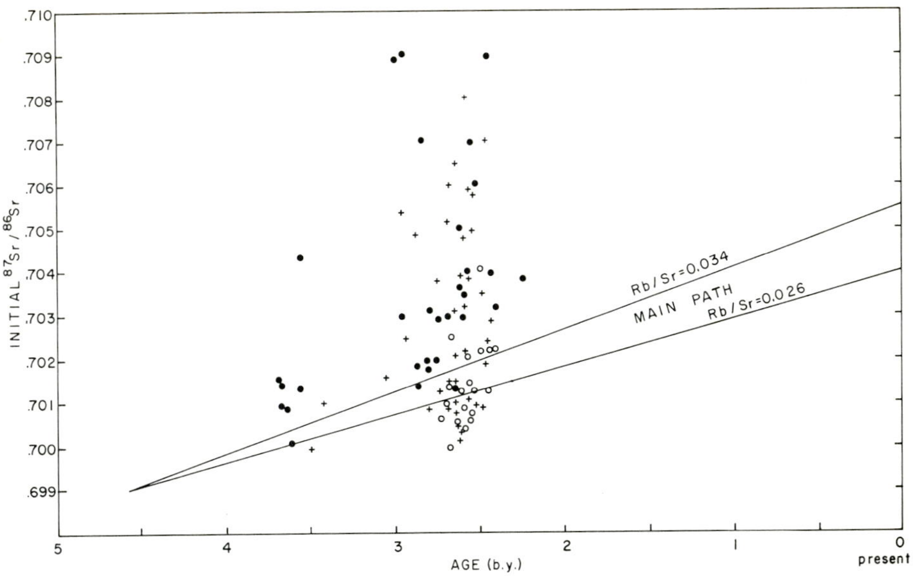

Fig. 9-21. Initial strontium isotope ratios of Archean rocks compiled from many sources. "Main path" after Jahn and Nyquist (1976). Key: ● = high-grade terranes; + = granitic and gneissic rocks from granite-greenstone terranes; ○ = greenstone belts.

Rb/Sr ratio of the upper mantle decreases with time. This model is attractive in terms of available distribution coefficient measurements which indicate Rb is preferentially enriched in magmas relative to Sr.

Curve C. The Rb/Sr ratio of the upper mantle increases with time. This model necessitates that Rb is added preferentially to Sr from the lower mantle or by recycling of crustal material into the mantle.

It is possible that the growth of radiogenic ^{87}Sr in the upper mantle may have been a discontinuous process as illustrated by the dashed line in Fig. 9-20. Although it is possible in theory to distinguish between the alternate models, the widespread scatter of initial ^{87}Sr/^{86}Sr ratios of any given age (as exemplified for instance by the Archean data in Fig. 9-21) and the difficult problem of evaluating the effects of alteration and metamorphism on initial ratios complicates the interpretation of available data. Hart and Brooks (1977) suggest that the early mantle should have been well-mixed by convection and that initial ^{87}Sr/^{86}Sr ratios should be low and of limited variation. Their analyses of clinopyroxene separates (^{87}Sr/^{86}Sr = 0.70114 ± 0.00013) agree well with the average of some 2.7-b.y. Archean greenstone volcanics (0.7011 ± 0.0004). On the other hand, Jahn and Nyquist (1976) suggest a heterogeneous upper mantle that is not well mixed based on the scatter of initial strontium ratios in Archean igneous rocks (Figs. 9-21 and 9-22).

Initial ^{87}Sr/^{86}Sr ratios from Archean rocks are shown in Fig. 9-21 together with the "main path" growth curve of the upper mantle of Jahn and Nyquist (1976). Four observations can be made from the results: (1) although there is a great deal of scatter in the data, many ratios in the 2.5- to 2.8-b.y. age group fall between 0.7005 and 0.7020; (2) most Archean greenstone belts exhibit low ratios ($\lesssim$ 0.702); (3) most Archean high-grade terranes exhibit high ratios ($>$ 0.702); and (4) granitic rocks from granite-greenstone terranes show wide variation. Two explanations of the high initial ratios of the Archean high-grade terranes merit consideration: (1) most high-grade terranes formed earlier than granite-greenstone terranes and their higher initial $^{87}Sr/^{86}Sr$ ratios reflect growth in a crustal environment; or (2) most high-grade terranes were derived from a different mantle source, more enriched in Rb than most granite-greenstone terranes (Clifford, 1974). Calculation of crustal resident times of the high-grade rocks (employing Rb/Sr ratios of 0.2—0.4) indicates that, at most, they resided in the crust for 50—100 m.y. and hence the 2.6- to 2.8-b.y. high-grade rocks cannot represent reworked 3.5- to 3.8-b.y. (or older) sialic crust (Moorbath, 1977). Employing these results, if 2.6- to 2.8-b.y. high-grade terranes are older than corresponding granite-greenstone terranes, they are $\leqslant$ 100 m.y. older. Collerson and Fryer (1978) point out, however, if lower Rb/Sr ratios are used (0.1—0.2), Archean crustal residence times may have approached 400 m.y.

It is of interest to explore the second possibility mentioned above. Initial strontium ratios of Archean granite-greenstone terranes are shown in Fig. 9-22 according to geographic location. The results from West Greenland are also included (Moorbath, 1977). The data from each Archean province show variable amounts of scatter but some tendencies exist for geographic provincialism. In the 2.5- to 2.8-b.y. category, ratios from the Superior Province tend to be low (0.7003—0.7015), those from the Rhodesian Province intermediate (0.7010—0.7015), and those from the Yilgarn Province somewhat high (0.7015—0.7025). The grouping within the Rhodesian Province is particularly tight and lies within the "main path". Other provinces such as the Kaapvaal and Wyoming Provinces, exhibit wide scatter with many ratios $>$ 0.7025. Since there are no province-wide differences in degree of alteration or metamorphism, these results tend to support the conclusion of Jahn and Nyquist (1976) of an inhomogeneous Archean mantle. Hurst (1978a) has proposed that the results from West Greenland and Labrador lie along a growth curve above the "main path" and reflect a mantle source with a lower Rb/Sr ratio (0.014) (Fig. 9-23). Young volcanics (the Svartenhuk tholeiites) lie near the zero-age end of the growth curve. The Greenland growth curve intersects a chondritic growth curve at about 4.45 b.y. At this time, Hurst (1978a) suggests that Rb was lost relative to Sr from the mantle beneath Greenland. It is possible that Rb, together with other alkalies, may have entered the core if it formed at this time. Supporting this idea, Hall and Murthy (1971) have shown that alkali sulfides follow iron under reducing conditions. Moorbath (1978) has criticized the model indi-

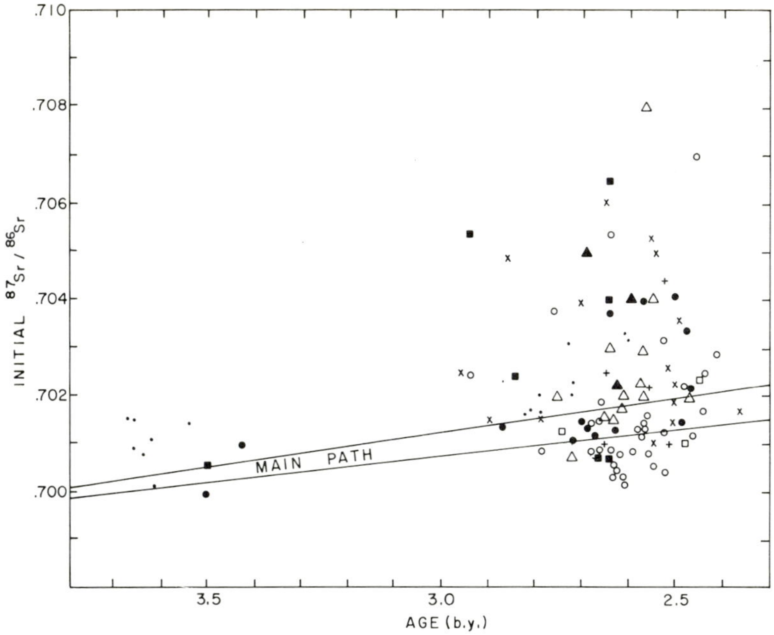

Fig. 9-22. Initial strontium isotope ratios of rocks from Archean granite-greenstone terranes and from West Greenland. "Main path" from Jahn and Nyquist (1976). Key to provinces: ● = Rhodesian; ○ = Superior; △ = Yilgarn; × = Wyoming; + = Slave; ▲ = Liberian; □ = Central African; ■ = Kaapvaal; · = West Greenland.

cating that the Greenland initial strontium ratios may reflect a short-lived crustal residence time (50—100 m.y.) and not the upper mantle source. Hurst (1978b), however, has pointed out several problems if this interpretation is adopted.

Initial $^{87}Sr/^{86}Sr$ ratios from Archean and younger mafic dikes and flows from the Wyoming Province are also shown in Fig. 9-23. As with the Greenland data, the results suggest that mafic magmas have been derived throughout geologic time from a mantle source with a constant Rb/Sr ratio. In this case, however, the growth curve emanates from an initial strontium ratio of 0.699 and reflects at Rb/Sr ratio (0.04) greater than that of the "main path". The fact that K and Ti decrease and Mg increases as a function of rock age in mafic dikes from the Beartooth Mountains has been interpreted by Mueller and Rogers (1973) to reflect a progressively deepening zone of melting resulting from a falling geothermal gradient with time. The strontium isotope results necessitate that such a deepening zone traverse mantle with an approximately constant Rb/Sr ratio caused by one or a combination of explanations 3 and 4 (see above) for curve *A* in Fig. 9-20. It is possible that the strontium isotopic differences between high-grade and greenstone terranes (Fig. 9-21) and the provincial groupings of initial ratios

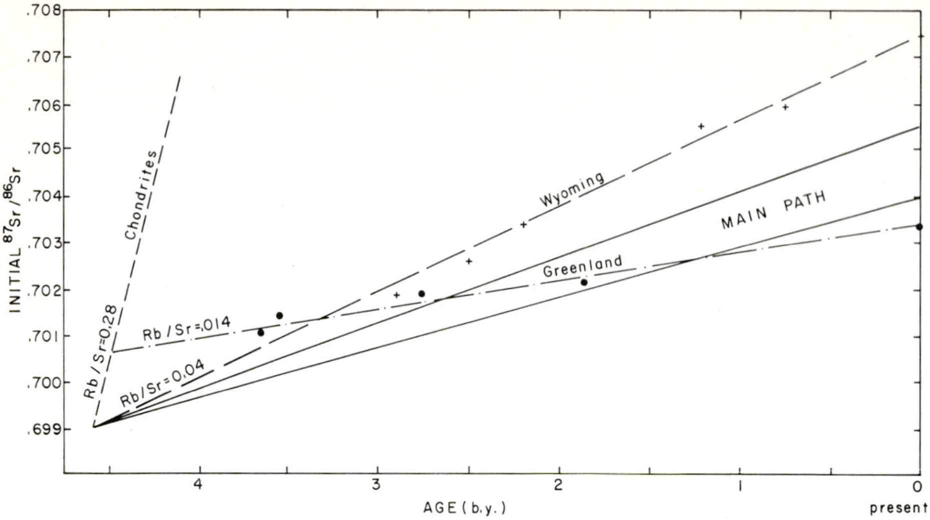

Fig. 9-23. Strontium isotope evolution diagram for West Greenland-Labrador and for mafic rocks from the Wyoming Province (after Condie, 1976b; Hurst, 1978a). "Main path" from Jahn and Nyquist (1976). ● = average values of gneisses from West Greenland and Labrador; + = diabase dikes from Wyoming.

in granite-greenstone terranes also reflect differences in upper mantle growth curves. If this is correct, it is of interest to see how many provinces evolved from a mantle that reflects an early depletion in Rb as exemplified by West Greenland. Although the available data from the Superior and Rhodesian Provinces suggest that they do not reflect such depletion (Fig. 9-22), many more initial ratios from rocks $\gtrsim$ 3.0 b.y. in age are needed from these terranes to evaluate fully this problem. The wide scatter of initial strontium ratios in the Wyoming and Kaapvaal Provinces may, in part, reflect isochrons rotated to high initial ratios in some parts of the provinces. The relatively high initial strontium ratios in most high-grade terranes may reflect, as in West Greenland, early Rb-depletion in their source areas.

In conclusion, available initial strontium ratios from Archean terranes are suggestive of the following:

(1) The Archean mantle was inhomogeneous with respect to the Rb/Sr ratio on a scale of hundreds of kilometers.

(2) A linear growth model for the ^{87}Sr/^{86}Sr ratio in the upper mantle is consistent with the results from West Greenland and from mafic igneous rocks in the Wyoming Province.

(3) Most 2.6- to 2.8-b.y. high-grade terranes are older than associated greenstone-granite terranes and/or they are derived from mantle sources that were depleted in Rb relative to Sr prior to 4.0 b.y.

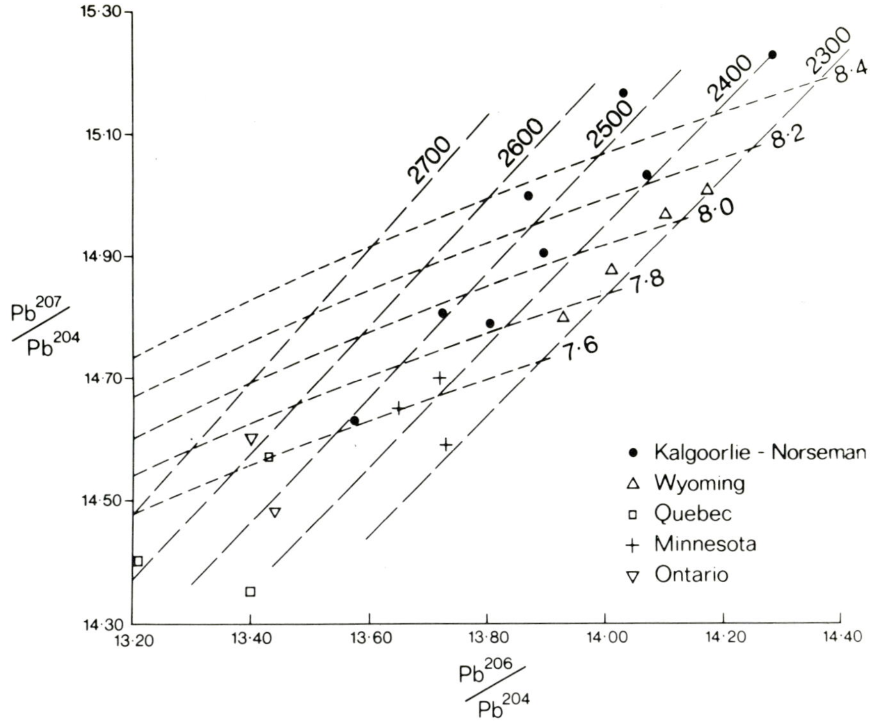

Fig. 9-24. Single-stage growth curves for K-feldspars from Archean granitic rocks for μ values ranging from 7.6 to 8.4 (after Oversby, 1975). Archean isochrons are also shown.

Lead isotope constraints

Lead isotope results from Archean terranes, although much less abundant than strontium isotope data, tend to confirm the heterogeneity of the Archean upper mantle. It is clear from the $^{207}Pb/^{204}Pb$ versus $^{206}Pb/^{204}Pb$ plot in Fig. 9-24 that a wide range of μ (U/Pb) values is necessary in mantle source areas of Archean magmas (Robertson, 1973; Oversby, 1975, 1978). There is also a suggestion of regional provinciality with the Superior Province requiring low upper mantle μ values (7.0—7.7) and the Wyoming and Slave Provinces high values (7.8—8.1). Model lead ages also differ between these areas and are, in general, less than measured radiometric ages (Table 9-3). Oversby (1978) suggests that the Wyoming and Slave Provinces contain significant amounts of reworked material. Alternately they may tap mantle sources with higher U/Pb ratios. Results from the Yilgarn Province in Australia imply high μ values (7.6—8.6) and again model ages are less than measured ages. The very high values ($\geqslant$ 8.2) indicate a crustal origin for some of the rocks. Rocks from the Kaapvaal Province exhibit only a narrow range of model ages ($\sim$ 2.8 b.y.) yet a wide range of μ values (6.9—7.5).

All provinces show in common model lead ages 200—400 m.y. younger

TABLE 9-3

Ranges in model lead ages, measured radiometric ages, and μ values for Archean granite-greenstone provinces (after Oversby, 1978)

Province	Model ages (b.y.)	Measured ages (b.y.)	μ values
Superior	2.4—2.6	2.6—2.75	7.0—7.7
Slave-Wyoming	2.3—2.5	2.6—2.75	7.8—8.1
Kaapvaal	2.77—2.81	3.0—3.3	6.9—7.5
Yilgarn	2.4—2.55	2.6—2.75	7.6—8.6
Pilbara	2.5—2.9	2.88—2.96	7.8—8.1

than measured ages and require mantle source regions with a range of U/Pb ratios. Models for the early evolution of the mantle-crust system which accommodate these requirements are extremely complex and involve two or more magma episodes involving changes in μ in the source (Robertson, 1973; Oversby, 1978). In the early Archean, μ values $\lesssim 7$ must have dominated whereas by late Archean time, high μ values must have been important in most sources. This implies preferential addition of uranium to the source areas (probably from greater mantle depths), since melting and magma extraction lead to a decrease in uranium relative to lead in residual phases.

Neodymium isotope constraints

Because Nd and Sm are not significantly fractionated from one another during most secondary processes, Nd isotope ratios provide a potentially valuable method to study the growth rate of the Archean continental crust (McCulloch and Wasserburg, 1978). Existing data suggest that Archean igneous rocks lie very close to a chondritic growth curve in which the Sm/Nd ratio is 0.308 (DePaolo and Wasserburg, 1976; Hamilton et al., 1978). These results suggest the Archean upper mantle was not very inhomogeneous in regards to the Sm/Nd ratio. Assuming a chondritic mantle source, McCulloch and Wasserburg (1978) have shown that some segments of the continents that record Proterozoic Rb/Sr ages were formed during the Archean. Some of the Churchill Province in Canada, for instance, formed together with the Superior and Slave Provinces at 2.6—2.7 b.y. Recent Sm-Nd studies of Archean gneisses from Scotland suggest that these rocks had a crustal residence time of approximately 200 m.y. (Hamilton et al., 1979).

Origin of heterogeneity in the Archean mantle

It is not difficult to envision processes that can produce heterogeneity in the upper mantle during the Archean; it is difficult, however, to see how such heterogeneities were maintained. Melting and extraction of magmas from the upper mantle would result in widespread inhomogeneities. Recycling of mafic crust by subduction would introduce chemical inhomogeneity at shallow mantle levels. Addition of LIL elements to the upper mantle from degassing of the lower mantle may also have been important in producing compositional heterogeneity.

Although preserved metamorphic mineral assemblages in Archean terranes reflect geothermal gradients in the same range as modern continental gradients (Wells, 1976; Burke and Kidd, 1978), models for radiogenic heat productivity in the earth during the Archean indicate that on the whole, the earth was hotter and gradients steeper than today (McKenzie and Weiss, 1975). This implies very steep gradients under non-continental areas to offset the more normal Archean continental gradients (Burke and Kidd, 1978). Such a large amount of heat in the early earth should have resulted in rapid convection and mixing of the earth. The fact that trace element and isotopic data indicate that the Archean mantle was inhomogeneous is difficult to reconcile with rapid convection and mixing at this time.

Clearly, mantle-wide convection during the Archean seems to be precluded by the geochemical and isotopic data. Some portions of the mantle may have been rapidly convecting, while other segments were rather passive. The fact that compositional provinciality is preserved in Archean provinces suggests that the segments of the mantle beneath these provinces that served as magma sources did not mix with adjacent mantle segments. Perhaps the increased amount of heat generated in the Archean was dissipated by convection cells not overlain by sialic crust.

Chapter 10

ORIGIN AND EVOLUTION OF ARCHEAN GRANITE-GREENSTONE TERRANES

INTRODUCTION

Many factors must be considered in reconstructing the geologic history of Archean granite-greenstone terranes. The geological and geochemical features of these terranes, discussed in previous chapters, provide an important set of boundary conditions for models which attempt to describe their origin and evolution. It is clear also, that models for greenstone belt development cannot be considered in isolation from models for the origin of Archean high-grade terranes. All models of Precambrian crustal evolution are closely allied to and dependent upon the earth's thermal history. The role of plate tectonics in the early history of the earth is a subject of current debate and discussion and recently the expanding earth hypothesis has been receiving more attention. Any workable model for the origin of greenstone belts and associated granites must follow logically from the processes which produced the early Archean crust ($>$ 3.8 b.y.) and hence it is necessary to review models for the origin and growth of the early crust. Each of these subjects will be briefly reviewed before discussing specific models for the origin of Archean granite-greenstone terranes.

THE ARCHEAN THERMAL REGIME

Many models have been proposed for the thermal history of the earth. Most, however, have assumed that conduction is a major means of heat transport (Lubimova, 1958; MacDonald, 1959). It is now clear from our knowledge of sea-floor spreading that convection cannot be overlooked in earth thermal models and indeed most calculations indicate that heat transport by convection greatly exceeds that transported either by conduction or radiation (Elder, 1972; McKenzie and Weiss, 1975). The rate of radioactive heat production (from U, Th, and K isotopes), the distribution of radioactive heat sources with time, and the initial temperature distribution in the earth are three important parameters in any thermal model. Constraints on Precambrian thermal gradients can be estimated from metamorphic mineral assemblages (see Chapter 6).

The initial temperature distribution in the earth is not well known. It is

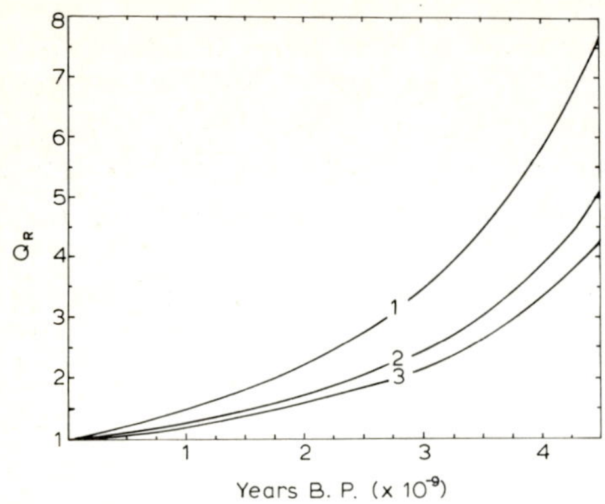

Fig. 10-1. Variation in heat generation in the earth as a function of time (after Dickinson and Luth, 1971; copyright © 1971 by the American Association for the Advancement of Science). Q_R = ratio of heat production at any time in the past to that currently observed. Models: *1* = chondrite, *2* = carbonaceous chondrite, *3* = Wasserburg model with K/U and Th/U = terrestrial values.

dependent upon the timing and the amount of heat contributed by the following (Lubimova, 1958; Runcorn et al., 1977): (1) impacting particles on the accreting earth which is, in turn, dependent upon particle velocity distribution; (2) gravitational energy released by the interior of the earth as it grows; (3) accumulation of radiogenic heat primarily from short-lived radioisotopes such as ^{26}Al and ^{244}Pu; (4) inductive heating resulting from intense solar wind activity; and (5) core formation. Although the relative contributions of each of these heat sources is not well known, it appears that there was sufficient heat available to produce extensive melting of the early earth. It is likely that the early geothermal gradient was adiabatic. Elsasser (1963) and Ringwood (1977) have pointed out that core formation is a highly exothermic process providing enough heat, if completely retained, largely to melt the outer part of the earth. Rapid mixing in the mantle during or soon after core formation was probably adequate to produce an adiabatic gradient in the earth. An upper limit for the surface temperature at this time is about 2000°C, the evaporation temperature of silicates under reducing conditions (Ringwood, 1977). Although estimates of the energy released during core formation range from 250 to 600 cal/g (Birch, 1965; Murthy, 1976), only about 250 cal/g are needed to raise the surface temperature to 2000°C. Such a temperature distribution would result in extensive melting of the earth to depths of about 500 km, with the completely molten zone perhaps, initially extending to the surface.

It is possible to calculate the rate of heat production in the earth as a function of time for various estimates of the U, Th, and K concentrations in the earth. Examples of such rates for three different earth compositions are given in Fig. 10-1 (after Dickinson and Luth, 1971). Q_R is the ratio of radiogenic heat production at any time in the past to that at the present. Recent estimates of the composition of the earth suggest that models 2 or 3 are probably more realistic than the chondritic model 1. It is clear from the figure that heat production increases dramatically in the geologic past and that even in the late Archean (2.5 b.y.) the heat generation was at least twice the present value.

A recent example of the thermal history of the earth was published by McKenzie and Weiss (1975). They develop a model for four initial temperature distributions and assume a constant rate of radiogenic heat generation throughout the earth. Present-day heat generation rate is calculated from oceanic heat flow for a chondritic and a Wasserburg model-earth. Results of two of their models showing surface heat flow E as a function of geologic time are shown in Fig. 10-2. The model in Fig. 10-2A is for an earth with an initial temperature of 1000°C and that in Fig. 10-2B for an earth with an initial temperature sufficiently high to permit world-wide convection. Differences in the models are caused by the time it takes to warm the mantle up to convecting temperatures. It is noteworthy in the 1000°C model that little of the radiogenic heat generated in the first billion years of earth history reaches the surface because of delayed shallow convection (beginning at 4.2 and 3.7 b.y. in the chondritic and Wasserburg models, respectively). If the lower mantle is not convecting in the models, a solid non-convecting core is required, a condition which is unlikely. Three additional sources of heat in the earth are not considered in the model: (1) heat of core formation; (2) heat from solid earth tidal dissipation; and (3) heat associated with crustal formation. Considering all possible sources of heat in the early earth, it is likely that if convection was not occurring as the earth formed, it began soon after.

Estimates of the thermal gradient in the outer 100 km of the earth as a function of time may be approached in two ways. First, is by model studies as described above. A second approach is by the study of regional metamorphic mineral assemblages now exposed at the earth's surface (Wells, 1976; Lambert, 1976). Only a few estimates of P-T conditions in the Archean crust have been made from the results of studies of metamorphic mineral assemblages (Chapter 6). Windley and Bridgwater (1971), Saggerson and Owen (1969), Saggerson and Turner (1976), and Watson (1978) emphasize the low-pressure character of Archean assemblages. Existing data as discussed in Chapter 6 clearly suggest geotherms steeper than those characteristic of average continental crust today.

Theoretical and laboratory model studies of convection have been useful in understanding convection patterns in the earth (Elder, 1972; McKenzie and Weiss, 1975). Convection depends on the combined properties of a fluid (such as viscosity, thermal conductivity, and coefficient of thermal expansion) and can be described by two unitless numbers, the Rayleigh and Reynolds numbers. Employing these numbers it is possible to simulate con-

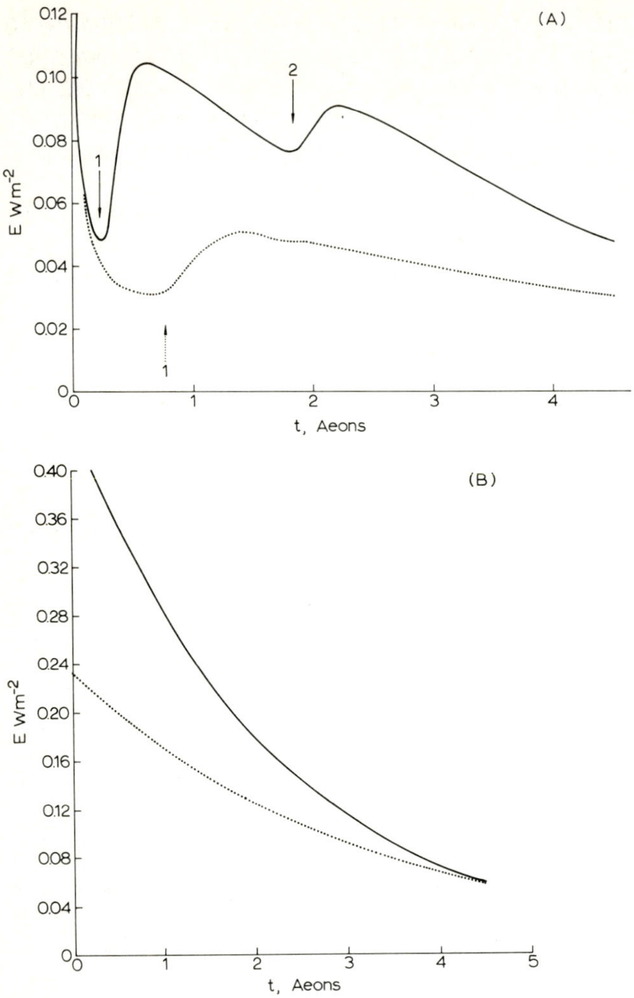

Fig. 10-2. Surface heat flow distribution on the earth as a function of time (from McKenzie and Weiss, 1975). A. Initial temperature of 1000°C. B. Initial temperature sufficiently great to permit convection throughout the earth. Solid line represents a chondritic model and dotted line a Wasserburg model (Wasserburg et al., 1964). Arrows indicate the onset of upper (*1*) and lower (*2*) mantle convection.

ditions in the earth in laboratory models as well as to evaluate theoretical models. Laboratory experiments indicate that the pattern of convection varies with the Rayleigh number. Little is known as yet about possible patterns in the earth, which has a Rayleigh number of 10^6 to 10^7, although it appears that spoke-like patterns (in planview) may be characteristic. The investigations of Elder (1972) and McKenzie and Weiss (1975), however, indicate that two scales of convection may occur in the earth. Small-scale convection cells (with horizontal sizes of a few hundred kilometers) are oriented at right angles to the large-scale convection cells. The existence of the small-scale flow in the Archean, although not yet documented in laboratory experiments for fluids with Rayleigh numbers similar to the earth, is supported by theoretical arguments. During the Archean when the earth was hotter, stresses generated by small-scale flow would be ten times greater than at present and thus could have prevented large plates from forming. The fact that Archean greenstone belts occur on a scale considerably smaller than Proterozoic or Phanerozoic orogenic belts supports a small-scale convective system in the Archean. It is of interest also that some greenstone belts are oriented at approximately right angles to nearby mobile belts consistent with the two scales of convection. An example is the granite-greenstone terrane in the Liberian Province where greenstone belts trend NNE and the bordering Kasila granulite-facies belt trends NW (Fig. 1-14).

Other investigators have related changes in tectonic style with time to changing convective patterns in the earth. Runcorn (1965) suggested that a growing core in the earth resulted in successively smaller and more numerous convection cells with time. Structural trends in Archean provinces when considered on a Pangaeic reconstruction of the continents have also been interpreted to reflect fewer and larger cells in the Archean (Dearnley, 1966; Engel and Kelm, 1972). This approach, however, is faced with two major problems. First, detailed structural trends are not available for many Precambrian provinces, especially in Central Africa. Second, Precambrian terranes are characterized by polyphase deformation and it is not always clear what age should be assigned to a given structural trend. Sutton (1963) proposed that four chelogenic (shield-forming) cycles have occurred in the earth's history, each cycle bounded between the major episodes of orogeny recorded by radiometric dates (i.e., 2.7—3.6, 1.9—2.7, 1.1—1.9, 0—1.1 b.y.). Later the model was modified to suggest that the four cycles are superimposed on an evolutionary earth history changing from mobile to more rigid conditions with time (Sutton, 1967, 1973, 1976). Each cycle records a similar sequence of events and results in the formation of stable cratons. The cycles begin with widespread orogeny in response to many convection cells in the mantle. As time passes, orogeny becomes more and more restricted to the outer margins of continents. In the model, continents disperse and then regroup before the beginning of the next cycle and convective cells increase in size and decrease in number during each cycle.

PLATE TECTONICS IN THE ARCHEAN

The role of plate tectonics in the Archean is a subject of current debate and disagreement. One school of thought proposes that plate tectonics has operated in one form or another from the Archean to the present (Talbot,

1973; Burke and Dewey, 1973; Burke et al., 1976b; Glikson, 1976b; Tarling, 1978). Because more heat was generated in the earth during the Archean, convection would have been more rapid to dissipate the additional heat. This would result in thinner lithosphere ($\leqslant 100$ km) and more rapid spreading and consumption rates (Hart et al., 1970b). Plates also would have been smaller and more numerous than today. The relation between heat loss from the earth, rate of plate creation, and the rate of heat transport to the base of the lithosphere suggests that a significant proportion of heat loss in the Archean must have taken place at divergent and convergent plate boundaries (Burke et al., 1977; Burke and Kidd, 1978). Plates carrying continents would frequently collide and continents would grow by suturing of collided blocks. Likewise, continental rifting may have disrupted and dispersed continents on a rapid time scale.

Another viewpoint adheres to the idea of Green and Ringwood (1968) and Ringwood (1975) that eclogite is the driving force for subduction. In this model, the inversion of mafic rocks in the oceanic lithosphere to eclogite results in an increase in density which drags the lithospheric plates into the mantle initiating subduction zones (D. H. Green, 1975; Baer, 1977). The adherents of this model suggest that because of the high heat flow in "oceanic" areas in the Archean, the eclogite stability field was not intersected and hence subduction did not exist. Instead, thin plates are jostled around on the earth like ice flows on a turbulent ocean. Such plates would be highly deformed especially around their margins. Due to the absence of subduction of mafic crust they could attain considerable thicknesses and be intimately interthrusted with sialic crust.

Wynne-Edwards (1976) has proposed that changing structural and metamorphic styles with time are related to an overall falling geothermal gradient (Fig. 10-3). He suggests that in the Archean sialic crust was relatively thin, discontinuous, and hot. Tonalitic plutons were fluid enough to reach shallow crustal levels, thickening the crust. By Proterozoic time the temperature in the medial to lower part of the crust had cooled to subsolidus temperatures ($\leqslant 700°$C). Although the crust was subsolidus, it still behaved as a ductile solid and underlying convecting systems caused it to thin by flowage with only localized brittle fracturing at shallow levels. Proterozoic mobile belts develop over areas of mantle upwelling and only locally is sialic crust rifted apart. By late Precambrian time, cooling had progressed to a point where the lithosphere behaved as a brittle solid. Fractures propagate through the continents over upwelling mantle opening ocean basins and the temperature in the upper mantle drops sufficiently to stabilize eclogite resulting in the initiation of subduction. Modern plate tectonic regimes come into existence at this time. If subduction did not occur in the Proterozoic, a very thick basaltic oceanic crust must have developed on the roughly 75% of the earth's surface covered by oceans. Such a thick mafic crust would have been recycled through the mantle with the onset of subduction at about 1 b.y.

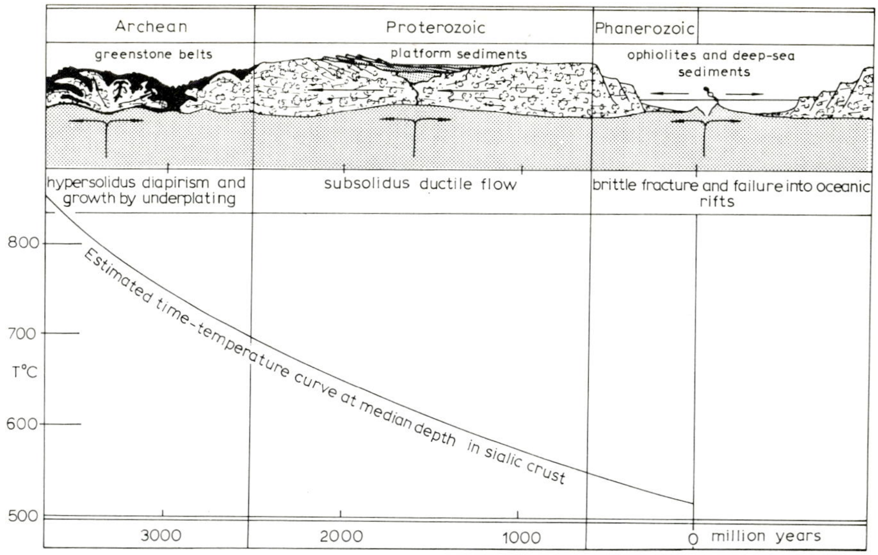

Fig. 10-3. Response of the continental crust to upper mantle convection throughout geologic time (from Wynne-Edwards, 1976).

Whether or not subduction existed in the Archean is dependent upon the manner of heat dissipation. Two driving forces are suggested for subduction — negative buoyancy and viscous drag (Hargraves, 1978). Negative buoyancy resulting from the eclogite inversion may be responsible for modern subduction but was probably not important during the high heat regime in the Archean. Hargraves (1978) indicates that positively buoyant lithospheric slabs 10 km thick with an average density contrast of + 0.25 g/cm^3 could be dragged down into the mantle by a shear stress of only 250 bar which is less than the possible viscous drag force of 500 bar proposed by McKenzie and Weiss (1975) for the Archean. Subduction driven by viscous drag in the Archean may have evolved into buoyancy-powered subduction by late Precambrian time.

Another approach to understanding the role of plate tectonics in the Archean is through paleomagnetism. In principle, it is possible to determine if individual Archean provinces drifted independently of each other by matching apparent polar wander (APW) paths. Paleomagnetic studies of Precambrian rocks are confronted with several problems which thus far have resulted in non-unique interpretations of most Precambrian pole positions (Nairn and Ressetar, 1978). One of the difficult problems with metamorphic rocks is resolving the various magnetizations (hard and soft) and assigning them to the correct radiometric dates. It is also not clear as yet which radiometric dating methods most accurately correspond to magnetization times. Another problem is assigning the correct polarity to a pole location — it may

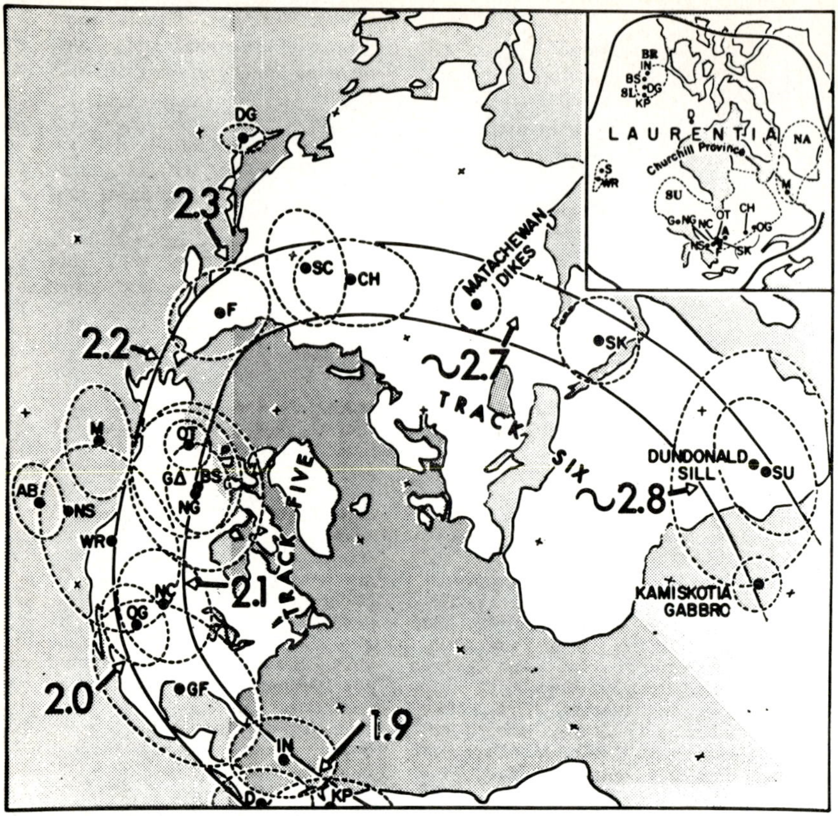

Fig. 10-4. APW path for Laurentia (see inset) for the interval 1.9—2.8 b.y. Dates in b.y. given in bold print. References to data in Irving and Naldrett (1977; copyright © 1977, University of Chicago Press).

be a north or a south pole (Roy et al., 1978). Finally, the shapes of Precambrian APW paths and the problem of loop location and size are also dependent upon the length of the increments of time over which pole positions are averaged and how wide an APW path is drawn through the data (Larson et al., 1973).

Most paleomagnetists interpret the existing data to indicate that adjacent Archean provinces have not drifted independently since about 2.7 b.y. (Piper, 1976a, b; Hargraves, 1976; Irving and McGlynn, 1976; Irving and Naldrett, 1977; McElhinny and McWilliams, 1977; Nairn and Ressetar, 1978). A few investigators, however, using many of the same data, have interpreted results in terms of independent drift of Archean provinces (Burke et al., 1976a; Cavanaugh and Seyfert, 1977). Most investigators favor the previous interpretation as exemplified by results from North America (the Laurentian Shield) in Fig. 10-4. The results, which come from the Superior, Wyoming,

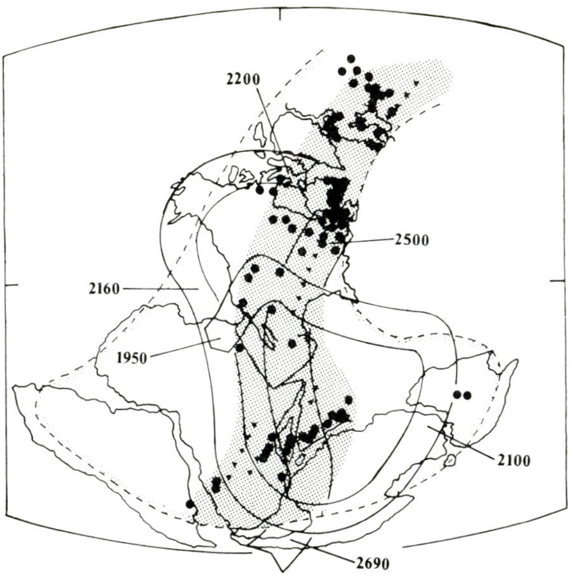

Fig. 10-5. Precambrian supercontinent reconstruction and APW path for the time interval of 1950—2690 m.y. (after Piper, 1976b). Ages given in m.y. Stippled belt shows distribution of Proterozoic anorthosites and high-K granites.

Slave, and North Atlantic Provinces (see inset), suggest that these provinces drifted as part of the same plate between 1.9 and 2.8 b.y. Relict Archean dates in intervening Proterozoic terranes support this interpretation. Similar results suggest that the African Archean provinces define a singular APW path between 1.9 and 2.3 b.y. and that Australian Archean provinces fall on a similar APW path between 1.1 and 2.5 b.y. (McElhinny et al., 1968; McElhinny and Embleton, 1976; McElhinny and McWilliams, 1977). The fact that many geologic features match up across mobile belts which separate adjacent Archean provinces is consistent with the paleomagnetic results indicating an absence of independent provincial drift (Hurley and Rand, 1969; Engel and Kelm, 1972; Shackleton, 1973a, b). Piper (1976a, b) has gone further and suggested that existing Precambrian pole positions can roughly be fit to a singular APW curve from about 1 b.y. back to 2.7 b.y. The results in the southern hemisphere are consistent with a Gondwana reconstruction and those in the northern hemisphere with a Laurasia reconstruction (for North America and Western Europe). The reconstruction indicates that the southwestern coast of North America was connected to Arabia prior to about 1 b.y. This supercontinent reconstruction and APW path for the time interval between about 2 and 2.7 b.y. are shown in Fig. 10-5. Consistent with Piper's interpretation is the distribution of Proterozoic anorthosites and high-K granites (chiefly 1.3—1.7 b.y.) which lie in a broad, northerly trending belt on the supercontinent.

Despite the ambiguities involved in defining precise Precambrian APW paths, several tentative conclusions seem warranted by existing data:

(1) Precambrian APW paths indicate plate motions on a scale at least as great as observed in the Phanerozoic.

(2) Adjacent Archean provinces appear to have drifted as part of the same plate until the beginning of the Phanerozoic and many of these provinces were not broken up and dispersed even during the Phanerozoic.

(3) Uncertainties and inherent errors in paleomagnetic data allow the opening and closing of small ocean basins (500—1000 km across) in Precambrian cratonic areas if such oceans open and close so as to return adjacent age-provinces to their original positions.

(4) By 2.7 b.y., one or at most a few continents existed on the earth's surface.

THE EXPANDING EARTH HYPOTHESIS

The idea of an expanding earth was first suggested by Lindemann (1927). Most evidences for an expanding earth are not very convincing and can be interpreted in more than one way. Egyed (1956) suggested that the continents have become progressively more emergent with time necessitating an increase in the earth's radius of about 0.5 mm/yr. His model is based on data scaled from old paleogeographic maps of the continents and suggests that the continental area of shallow seas has decreased with time. In his calculations, however, he ignores the fact that the continents are only 88% emergent today and that some water is tied up in glacial ice. Recent estimates of shallow continental sea areas, made from more accurate maps than were available to Egyed, indicate a rather constant relationship between sea level and average continental elevation since the Cambrian (Wise, 1973). Also, it appears that the effective volume of oceanic rises may be the single most important factor controlling at least Phanerozoic flooding of the continents (Rona, 1973).

Some of the most convincing lines of evidence for an expanding earth come from geometric arguments (Carey, 1958, 1976). Carey (1958) pointed out that significant gaps left unaccounted for in Phanerozoic Pangaea reconstructions disappear on a globe of smaller radius. He showed this to be true for both oblique stereographic projections and direct global plotting of earlier continental reconstructions. Jordan (1966) suggested that the gravitational constant G varied inversely with the age of the universe and was the prime cause for earth expansion. Dearnley (1966) proposed an expanding earth based on reconstructions of Precambrian orogenic belts. He assumed mountain building to be controlled by a changing mantle convection system as previously mentioned. During the Archean, his model is consistent with bilaterally symmetrical convection (2 cells) with currents rising at the poles and descending at the equator on an earth with a radius of 4400 km.

Another line of reasoning used to support the existence of an earth with smaller radius is the possible existence of a Proterozoic supercontinent (Glikson, 1977c, 1978, 1979b). There is a sparsity of volcanic-sedimentary belts around the edges of the proposed Proterozoic supercontinent and the nature of the crust covering the other three-quarters of the earth's surface during the Proterozoic is unknown. Basically, there seems to be three possibilities for this unknown crust (modified after Glikson, 1979b): (1) the remainder of the earth's crust was sialic and it has been either recycled through the mantle or added to the existing continents by collisions or both; (2) the remainder of the earth's crust was composed dominantly of oceanic crust with or without micro-continents (arc systems) which were added to the supercontinent by collisions; or (3) the radius of the earth was approximately one-half the size of the present earth and a Proterozoic supercontinent covered most of the earth's surface. A major obstacle to the first possibility is the probability, as later discussed, that sialic crust, once formed, is essentially non-destructable. Furthermore, if it were somehow recycled through the mantle, the thickness of the present continents would be two to four times their present thickness. Also, paleomagnetic evidence suggests that most of the supercontinent was intact back to 2.7 b.y. (Piper, 1976b) thus eliminating arc-continent collisions as a major mechanism by which the supercontinent grew in the Proterozoic.

The second possibility presents other problems, the most important of which is the absence of evidence for major Proterozoic arc systems either developed in situ or added to the supercontinent by collisions during the Proterozoic. If three-quarters of the earth's surface were composed of oceanic crust with spreading centers and subduction zones, an appreciable amount of sialic crust would have been produced at the convergent plate boundaries. For instance, in a two-stage melting model with accumulative oceanic ridge length of 20,000 km, a half-spreading rate of 1 cm/yr, an oceanic crust 5 km thick, and an amount of partial melting equal to 30% (Glikson, 1980), the annual rate of production of sialic crust would be 0.3 km^3/yr. If this were added to the supercontinent over a 1.5-b.y. period (assuming a 35-km thickness), it would mean that about 10% of the supercontinent was added by collision during the Proterozoic. Higher spreading rates would result in large amounts of supercontinent growth. It will be of interest to see if future paleomagnetic studies of late Precambrian arc terranes are consistent with such sialic additions. If subduction did not occur on the earth during the Proterozoic as suggested by Baer (1977), the basaltic crust covering two-thirds to three-fourths of the earth's surface would thicken to accomodate new growth at spreading centers. Although this model would account for the sparsity of Proterozoic arcs, it presents another problem. When subduction begins in the late Precambrian, the thick oceanic crust ($\geqslant$ 40 km thick) would descend into the mantle, undergo partial melting and form extensive island arc systems of late Precambrian and early Paleozoic age. If sialic crust is non-subductable, these arcs would eventually be added to the supercontinent or to individual continents as the supercontinent breaks up, beginning in the late Precambrian. Calculations suggest that about 40% of the existing continents would be added at this time to account for subducting the thick Proterozoic oceanic crust. This amount seems unrealistic in terms of available data. The final possibility is attractive in that it avoids problems of what was happening on the other three-fourths of the earth's surface. Glikson (1977c, 1979b) proposes a model whereby active expansion in the earth begins at about 1 b.y.; such a model is consistent with the appearance at about this time of major continental rift systems, ophiolites, and arc assemblages.

At the present time, no clear geological evidence exists to refute an expanding earth. It would appear that plate tectonic processes, mantle convection, and other earth processes can occur with or without an expanding earth. Although in theory it should be possible to detect an expanding earth with paleomagnetic results (Carey, 1976), existing measurements are not sensitive enough to do so. It may be possible in the future, however, to measure expansion from reflected laser beams returned to the earth from reflectors placed on the moon or by measuring the lengths of neutrino beams shot through the earth. The major obstacle at the present to an expanding earth is the inability to define an adequate and plausible mechanism of expansion. Carey (1976) has suggested four possible mechanisms: (1) phase changes at constant mass, (2) a secular decrease in G, (3) a secular increase in mass of the earth, and (4) a secular change in the electron charge/mass ratio. All of these possible causes are speculative at present.

ORIGIN OF THE CRUST

Introduction

The origin of the earth's crust has always been a subject of widespread interest as evidenced by the numerous papers written on the subject. Crustal genesis can be considered in terms of the following (Condie, 1979a): (1) when and by what processes did the crust form; (2) what was the lateral extent of the early crust; (3) at what rate and by what mechanisms did this crust grow; (4) what was the composition of the early crust; and (5) when and how did oceanic and continental crustal types become clearly established. As pointed out in earlier chapters, the oldest preserved fragments of crust are about 3.8 b.y. in age and are comprised chiefly of tonalite-trondhjemite gneissic complexes. Model lead ages (Robertson, 1973), the age of lunar crustal rocks, and consideration of the earth's thermal history as discussed above, however, seem to indicate that earliest terrestrial crust formed prior to 4.0 b.y. The scarcity of rocks older than 3.5 b.y. on the earth may be related to one or a combination of the following factors (modified after Wetherill, 1972):

(1) The moon was captured at about 3.5 b.y. and tidal energies were sufficient to destroy most earlier crustal rocks.

(2) The fragments of pre-3.5-b.y.-old crust are relicts of a once more widespread crust subsequently destroyed (recycled into the mantle) by later geologic processes such as subduction.

(3) The earliest crust (probably basaltic in composition) was recycled through a hot, rapidly convecting upper mantle.

The first possibility is not considered likely in that a capture event is not recognized in the magmatic history of the moon ($\gtrsim$ 3 b.y.) (Wasserburg et al., 1977) and the second possibility is not favored by Sr and Nd isotopic data which suggest that early sialic crust had short crustal residence times (Moorbath, 1977; McCulloch and Wasserburg, 1978). In terms of existing data, the third factor would appear to be most important in accounting for the sparsity of pre-3.5-b.y. crust.

It may be possible to obtain some idea as to how widespread the earth's early crust was from the crusts of other terrestrial planets and in particular, the moon where age relationships are known (Lowman, 1976). The lunar highlands (4.3—4.5 b.y.) appear to represent remnants of the early lunar crust and their widespread distribution suggests that this crust covered the entire moon. Analogous features on Mercury suggest a planet-wide early crust. If the early history of the earth was similar to these bodies, it also may have had a widespread primitive crust. Such a crust, however, may soon have been disrupted and concentrated into nuclei by mantle convection.

Theories for the origin of the crust fall broadly into three categories (Condie, 1976a): inhomogeneous earth accretion, catastrophic models, and non-catastrophic models. Catastrophic models involve an early, rapid melting episode which results in formation of much or most of the crust. Non-catastrophic models, on the other hand, involve more gradual growth of the crust in response to widespread heating of the earth from within.

Inhomogeneous accretion model

The inhomogeneous accretion model for the origin of the planets is described by Turekian and Clark (1969) and Clark et al. (1972). Employing this model, the last compounds to condense from the solar nebula produce a thin veneer on the earth rich in the alkali and other volatile elements which forms or evolves into the first crust. The inhomogeneous accretion model is faced with several difficulties (Ringwood, 1977). One problem with the model is that many non-volatile LIL elements (such as U, Th, and REE), which would be concentrated in the core and lower mantle in an inhomogeneously accreting earth, are today concentrated in the crust. This would seem to necessitate magmatic transfer from within the earth thus producing a crust of magmatic origin. Another line of evidence supporting a magmatic origin for the earth's early crust is that the early crust of the moon ($\geqslant$ 4.3 b.y.) appears to be of magmatic origin and if the earth evolved in a similar manner, it should have developed a crust of similar origin.

Catastrophic models

Many investigators have considered the possibility that the continents were formed in response to impact melting in the upper mantle (Goodwin, 1974, 1977b; Frey, 1977). The record of extensive cratering on the moon early in its history is consistent with such cratering occurring on the earth. Donn et al. (1965) proposed that the first continents were formed by the impact of large meteorites or asteroids of sialic composition. In their model, they propose that the sialic matter was added to the earth's surface to form the first continents. This model, however, seems unrealistic in that there are no known meteorites of sialic composition and, even more compelling, it is likely that such collisions would result in volatilization and/or widespread scattering of the meteorite material rather than concentration into continental nuclei.

Salisbury and Ronca (1966) propose an impact model whereby one or more large meteorite bodies hit the earth's surface transferring some energy but not mass to the earth. They propose the excavation of a crater approximately 10 km deep involving a drop in pressure of > 2500 atm. This sudden release of pressure together with the production of a series of radiating fractures results in partial melting of the mantle beneath the crater. Erosion of the crater rim and intrusion and extrusion of dominantly mafic magmas from below, jointly fill the crater. Fractional crystallization of mafic magmas produces granitic magmas which are intruded at shallow depths and the crater of sediments and igneous rocks isostatically rises forming a protocontinent. The impact may imitate a convection cell beneath the protocontinent, thus thickening the crust and causing it to grow by peripheral magmatic activity. Impact melting may also result in the production of a mantle plume which rises and provides a means for future growth of continental nuclei. One obstacle to the impact model, by analogy again with early lunar cratering, is the fact that lunar cratering apparently did not result in the production of sialic nuclei. The model may be acceptable, however, if the first terrestrial crust was mafic or anorthositic in composition (Frey, 1977). The onset of thermal convection in the earth either before, during, or after impact would disrupt and recycle this early mafic crust so that remnants are not preserved today.

Another catastrophic model relates to early core formation. If the core formed very rapidly by liberation of gravitational energy, this energy would cause extensive melting in the mantle. Such melts would rise towards the earth's surface, many of them reaching the surface forming a crust. The distribution of the crust would be controlled by the specific mechanism of core formation. If the droplet model of Elsasser (1963) is valid, crust would form over major molten iron droplets and the number of crustal nuclei would be equal to the number of droplets. On the other hand, if molten iron moved towards the core as a widely disseminated network of veinlets, the crust may have been rather homogeneously distributed about the globe.

An early and close capture of the moon would result in the transfer of large amounts of angular momentum energy into the earth's interior by tidal forces. This energy would produce widescale melting in the mantle and may result in the formation of a crust. Such a crust may not have been initially stable, however, in that tidal forces would have been strong enough to break it up. Only as the moon retreated from the earth would a crust become stabilized.

Non-catastrophic models

As discussed earlier, it is likely that there was enough heat in the earth at the time the earth formed or soon thereafter to melt the entire earth. Complete melting probably did not occur because the earth has retained volatile elements. If convection was widespread and served to homogenize the heat distribution in the earth, partial melting should occur at the same depth throughout the upper mantle. This, in turn, should result in the formation of an unstable, thin, widespread crust. Such a crust may not have survived for long because surface impact or convective forces or both would serve to disrupt it and recycle it in the partially molten upper mantle.

Partial melting may also occur at inhomogeneities that form during or soon after accretion of the earth. Local concentrations of volatiles (principally H_2O and CO_2), structural discontinuities, or regions with high concentrations of heat-producing elements (U, Th, K) in the upper mantle may result in partial melting, rise of magmas, and production of crustal nuclei. Gradual core formation may also produce inhomogeneous melting in the mantle and a corresponding inhomogeneous distribution of early crust. In any of these cases for inhomogeneous magma production, partial melting may produce a diapir or plume, which because of its lower density, rises towards the surface. Goodwin (1974) has recently suggested the major Archean continental nuclei on the earth were produced in response to rising plumes, which were perhaps triggered by impacting on the earth's surface. It is noteworthy that mantle inhomogeneity models have in common the feature that discontinuous nuclei of crust are produced, with the number and distribution of nuclei dependent upon the number and distribution of melting inhomogeneities in the mantle.

Some models for crustal formation appeal to plate tectonic processes as discussed later. If plate tectonics were operative soon after accretion of the earth, magmas could be produced at plate boundaries as they are today. As previously discussed, the temperature of the earth was probably greater in the early Precambrian than today and hence at least spreading centers must have existed to dissipate the heat. Plate tectonics may have begun in response to a widening zone of partial melting and convection in the mantle. As the upper margin of this zone approached the earth's surface, the overlying rigid shell would break up and plate boundaries would form. If crust formed only at spreading centers, it most likely would have been mafic to ultramafic in composition and it would have been deformed and probably recycled through the upper mantle by convective forces. Later as the temperature falls, sialic crust would form at subduction zones and continents would grow by continent-continent collisions. If this process of crust formation is valid, it has the advantage of providing a mechanism for production of both oceanic and continental crust early in the earth's history. If the model of mesosphere growth of Dickinson and Luth (1971) is correct, the plate tectonics mechanism also provides for early initiation of mesosphere growth by accretion of descending slabs.

COMPOSITION OF THE PRIMITIVE CRUST

Introduction

Perhaps no other subject is more controversial at the present time than that of the composition of the primitive (> 3.8 b.y.) crust. Most models fall into one or a combination of four categories: sialic, anorthositic, andesitic, or mafic (± ultramafic). In part responsible for the diverging opinions are the different approaches to the subject. The most direct approach would be to find and describe a relict of the primitive crust. Although some investigators have not given up on this approach, the chances that a remnant of this crust was preserved seem very small. Another approach is to deduce the composition of the early crust from studies of the preserved Archean crust (Anhaeusser, 1973a; Glikson, 1976b). This approach is hazardous in that the compositions and field relations of crustal rock types in the oldest preserved Archean terranes may not be representative of earlier crust. The oldest known supracrustal rocks in the Isua greenstone belt (~ 3.8 b.y.) contain a mixture of volcanic rocks (mafic and felsic), ultramafics, quartzites, iron formation, carbonates, and pelitic rocks (Allaart, 1976). Another approach has been to assume that the earth and moon have undergone similar early histories and hence to go to the moon where the early record is well preserved to determine the composition of the earth's primitive crust (Windley, 1970). Finally, a third approach is a geochemical model approach based on crystal-melt equilibria and a falling geothermal gradient with time (Shaw, 1972).

Anorthositic models

Geochemical and geochronological studies of lunar samples indicate that the oldest rocks on the lunar surface are gabbroic anorthosites and related high-alumina basalts of the lunar highlands (> 4.0 b.y.) (Taylor, 1975). Some investigators have proposed that a widespread lunar crust formed between 4.4 and 4.0 b.y. This crust appears to have formed in response to one or more periods of catastrophic heating leading to widespread partial melting of the lunar interior and the production of voluminous basaltic magmas which ascend to the surface. Near the surface the magmas cool and undergo fractional crystallization, at which time pyroxenes and olivine sink and plagioclase and some pyroxenes float forming a crust of gabbroic anorthosites. Impact disrupts this crust and produces maria craters. These craters are later filled with basaltic magmas (3.1—3.9 b.y.). The highland areas of the moon represent remnants of the primitive gabbroic anorthosite crust.

Windley (1970) has pointed out that the early Archean anorthosites in Greenland are similar in composition (i.e., high An contents, associated chromite) to lunar anorthosites and not to younger terrestrial anorthositic rocks. It appears clear from the field relationships, however, that these anorthosites are not remnants of an early crust since they often intrude granitic gneisses. If, however, the earth did undergo an early melting history similar to the moon, the earth's first crust may have been composed dominantly of gabbroic anorthosites and the preserved early Archean anorthosites may represent the last stages of anorthosite production which continued after both mafic and sialic magmas were also being produced.

Murthy (1976) has pointed out that the increased pressure gradient in the earth limits the stability range of plagioclase to depths considerably shallower than on the moon. Available experimental data suggest that plagioclase would not be a stable phase at depths > 35 km in the earth. Hence, if such a model is applicable to the earth, the anorthositic fraction, either as floating crystals or as magmas, must find its way to very shallow depths to be stable.

Sialic models

Models for the production of a primitive sialic crust have been proposed by Poldervaart (1955b), Ramberg (1964), Sutton (1971), Fyfe (1974, 1978), Hargraves (1976), and McGregor (1979). Some of these authors emphasize the point that low degrees of partial melting in the mantle will be reached before high degrees and hence granitic magmas should be produced before mafic ones. Others call upon fractional crystallization of basalt to form the sialic crust. Ramberg (1964) develops a model based on analogies to laboratory studies using various substances of different densities and viscosities in which he concludes that granitic magmas once formed, should rise to the surface buoyantly before basaltic magmas because of their lower density.

Shaw (1972, 1976) presents a geochemical model for the formation of a widespread anorthositic and sialic crust. The model is based on measured trace element melt-solid distribution coefficients. He proposes that the earth heated soon after accretion at which time the core formed and the mantle was molten. The mantle cools and crystallizes from the center outwards concentrating LIL elements into a near-surface basaltic magma layer. This layer undergoes fractional crystallization resulting in the accumulation of an anorthositic scum in irregular patches. Impacting meteorites arrive frequently at the surface breaking up the anorthositic veneer. Removal of pyroxenes and olivine from the basaltic magma produces an ultramafic cumulate overlain by a residual granitic magma which crystallizes to form the first stable crust. Such a crust is formed by about 4 b.y.

Two main obstacles face the sialic crustal models. First, the high geothermal gradients in the early Archean probably produced large degrees of melting ($\geqslant 20\%$) of the upper mantle and hence it is unlikely that granitic melts could form. A sialic crust could be produced by fractional crystallization of basaltic magmas as suggested by Shaw (1972). However, if widespread sialic crust were produced in this manner and sialic crust is essentially non-destructable since its lower density results in its resisting mantle recycling, why are there not remnants of the early sialic crust preserved today? As previously discussed (Chapter 9), the earliest preserved sialic crust in Greenland and other areas exhibit initial Sr, Pb, and Nd isotopic ratios that suggest that it was extracted from the mantle with little or no previous crustal history and hence the oldest preserved remnants cannot represent reworked fragments of still older crust. An alternate interpretation of the strontium isotope data involving an appreciable crustal residence time, however, has recently been presented by Collerson and Fryer (1978).

Andesitic models

Taylor and White (1965), Taylor (1967), and Jakeš (1973) were among the first to point out that the average composition of the continents today is similar to calc-alkaline andesite. These authors propose a model for the origin of the continental crust based on a modern island arc setting. The model necessitates, by a strict modern plate tectonic analog, that subduction zones began early in the earth's history. The continents would grow by arc development around the periphery of continents and by arc-continent collisions. A lunar analogy also provides some support for an early crust on the earth of andesitic composition. Because a molten layer in the earth would be considerably enriched in LIL elements and H_2O compared to the moon (Lowman, 1976), fractional crystallization of such a layer would lead to a calc-alkaline fractionation trend rather than a tholeiite-anorthositic trend as observed on the moon.

Mafic models

A mafic primitive crust (or combined ultramafic and mafic crust) was proposed by Gill (1961), Glikson (1972b, 1976a, b), and Anhaeusser (1973a). Such a composition is consistent with the probable large amounts of partial melting that would occur in the earth soon after accretion. Hence, this early crust may be similar compositionally to lower portions of some greenstone belts as pointed out by Glikson (1976b). If subduction did not exist and this crust was formed along an extensive network of ridge systems, it could thicken rapidly.

GROWTH OF THE EARLY CRUST AND LITHOSPHERE

Introduction

Various mechanisms have been suggested for the growth of the crust and lithosphere (Condie, 1976a). Crustal growth can occur both by addition of new material from the mantle, by readdition of crustal material that has been recycled through the mantle, and by redistribution of crustal rocks due to sedimentary and tectonic processes. Net crustal growth, however, includes only additions to the crust of new (unrecycled) material from the mantle. Sr, Pb, and Nd isotopic systems are useful in distinguishing new from recycled crustal rocks (Moorbath, 1977). The major mechanisms of crustal-lithospheric growth which may have operated in the Archean are welding of sedimentary prisms to continental margins, growth by magma additions, interthrusting and stacking of crustal rocks, and thickening in response to a falling geothermal gradient. Continental crust may also grow by continental collisions. Each of these will be briefly considered.

Peripheral welding of sedimentary prisms

As erosion removes material from the continents, it accumulates on the continental shelves and slopes and in oceanic basins near continental margins. Thick sedimentary prisms may accumulate in this manner. Burial leads to metamorphism and perhaps to partial melting, the net result being lateral accretion of the continents. How important this mechanism of growth was in the Archean is difficult to evaluate because the margins of Archean provinces have been intensely reworked during the Proterozoic. However, it may have been quite important adjacent to high-grade Archean provinces where evidence indicates 15—30 km of uplift and erosion has occurred.

Magma additions

Magma from the mantle may be added to the crust and lithosphere by intrusion of sills and plutons, overplating (volcanism), or underplating. Such additions may occur in a variety of tectonic environments. Oceanic crust and lithosphere is added at divergent plate boundaries. Growth of continents and arcs may occur by additions of magma from descending lithospheric slabs and continents and oceanic islands may thicken by addition of magmas from mantle plumes.

Interthrusting and stacking of crustal rocks

From investigations of Archean terranes in Greenland, Bridgwater et al. (1974) have suggested that the early crust may have thickened by the interthrusting and stacking of thrust sheets and nappes of both oceanic and continental rocks. Such intimate intermixing of these rocks implies horizontal compression that is associated with one or a combination of convergent plate boundaries, colliding microcontinents, or closing continental rift systems.

Thickening in response to falling geotherms

The thickness of the lithosphere is critically dependent upon the temperature distribution with depth. Experimental data indicate that the low-velocity zone at the base of the lithosphere is probably caused by incipient melting of ultramafic rocks (Condie, 1976a; Baer, 1977). The present oceanic lithosphere ($\sim$ 75 km thick) reflects a steeper thermal gradient than the continental lithosphere (150—200 km thick). Based on probable Archean temperature distributions, it would appear that the average thickness of the early Archean lithosphere was $\leqslant$ 50 km and that as the temperature gradient fell with time, the lithosphere increased in thickness (Baer, 1977).

Sialic collisions

Continents may grow at convergent plate boundaries by collision and annealing of sialic crustal blocks to the overriding plate (Glikson, 1972b). The simplest mechanism is by an arc continent collision whereby an arc is added to the edge of a continent and the subduction zone relocates at the new continental margin (Dewey and Bird, 1970). The descending plate may become the overriding plate if it carries the continent. Continual sweeping of arcs against a continent can result in substantial growth. Continent-continent collisions would also result in enlarging of continents. As discussed later, it is possible that such a mechanism was largely responsible for the general development of Archean crust.

Thickness of Archean crust and lithosphere

Estimates of the thickness of the Archean crust can be made from three sources of data (Condie, 1973): minimum thickness of stratigraphic sections in greenstone belts, burial depths deduced from metamorphic mineral assemblages, and geochemical indices. Data from all three sources seem to indicate that the Archean continents were close to the thickness of the present continental crust (i.e., 30—40 km). If this is the case, little if any thickening has occurred in stable cratonic areas since their stabilization. Although it has been proposed that thickening of cratons could occur after cratonization was complete by underplating with mafic or granitic magmas (Pakiser and Zietz, 1965), there is no evidence to support such thickening. Evidence at hand suggests that Archean sialic crust thickened to present-day thicknesses as it formed by one or a combination of the growth mechanisms described above. Employing the spacing of Archean volcanic centers in the Abitibi belt in comparison to modern volcanoes, Windley and Davies (1978) suggest that the lithosphere at 2.6—2.7 b.y. was the order of 80—90 km thick which is about half the thickness of present continental lithosphere.

Metamorphic mineral assemblages from some Archean high-grade terranes imply crustal depths at the time of metamorphism of 30—40 km (Table 6-3). Because these areas are underlain today by 30—40 km of continental crust, we are faced with three alternatives to explain the existence of such high *P-T* mineral assemblages at the earth's surface: (1) sialic underplating has kept pace with uplift and erosion (O'Hara, 1977); (2) the Archean crust in these areas was originally the order of 60—80 km thick; or (3) segments of the lower crust have been transported to the surface by low-angle thrust faults. The first mechanism would seem to necessitate a younging of sialic crust with depth, a feature which is not observed in radiometric dates from terranes of increasing metamorphic grade. It is also difficult to explain why the underplating magmatism does not have surface or near-surface manifestations (which should be present in exposed high *P-T* terranes). Regarding the second possibility, Fyfe (1973a, b) has pointed out that it is unlikely that Archean sialic crust could exceed about 30 km in thickness before its base began to melt. Even assuming moderate geothermal gradients for Archean high-grade areas and a dry, intermediate granulite composition for the lower 30—40 km of crust in which minimum granitic melts have been removed, melting would begin by 40—50 km. Davies (1979), however, has recently proposed that the Archean continental lithosphere was approximately 200 km thick and acted as a thermal buffer between the sialic crust and the hot upper mantle and thus prevented the bottom of the crust from being melted. Such thick lithosphere presents problems with alternative 3. Low-angle thrusting of the crust would seem to necessitate decoupling of the crust and lithosphere.

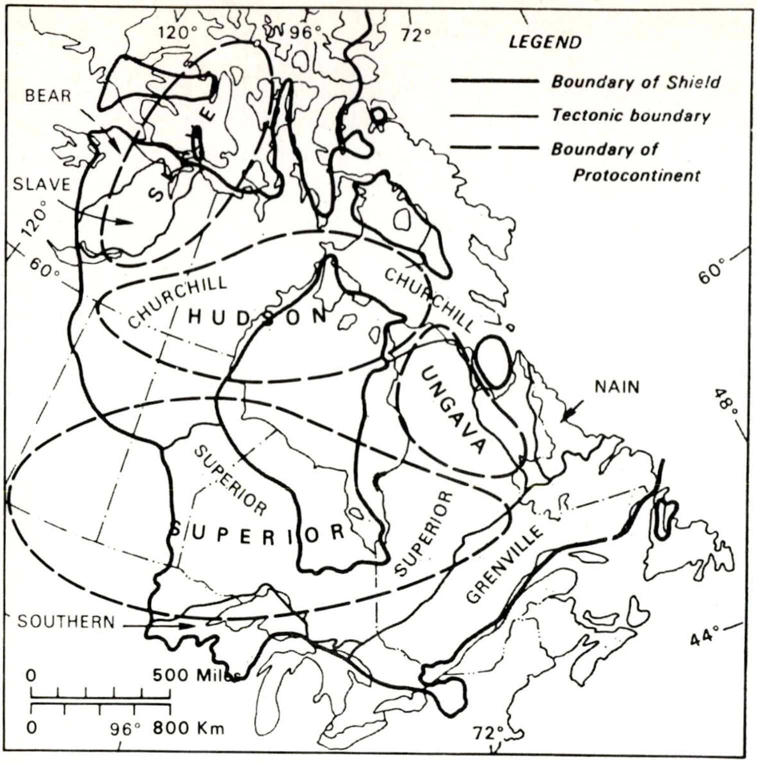

Fig. 10-6. Proposed Archean protocontinents of the Canadian Shield (from Goodwin, 1974).

Protocontinent models

The concept of Archean protocontinents was introduced by Goodwin (1968b). A protocontinent is defined as a positive segment of unstable sialic crust with linear supracrustal belts. Incipient protocontinents are characterized by a random to sub-linear distribution of small cratons and supracrustal belts while a mature protocontinent is characterized by a marginal distribution of long linear belts around larger cratons. Protocontinents grow by marginal accretion of supracrustal belts and by magmatic underplating. With continued growth, protocontinents merge to form large, stable continental areas. In the Canadian Shield, Goodwin (1968b; 1974) recognizes four protocontinents: the Slave, Hudson, Ungava, and Superior (Fig. 10-6). Naqvi et al. (1974) suggest three protocontinents for the Indian Shield.

Three of the protocontinents in Canada cluster around Hudson Bay (Fig. 10-6) and the Hudson Bay area is today underlain by a gravity high suggesting thin crust. Geologic evidence suggests that this area was a topographic

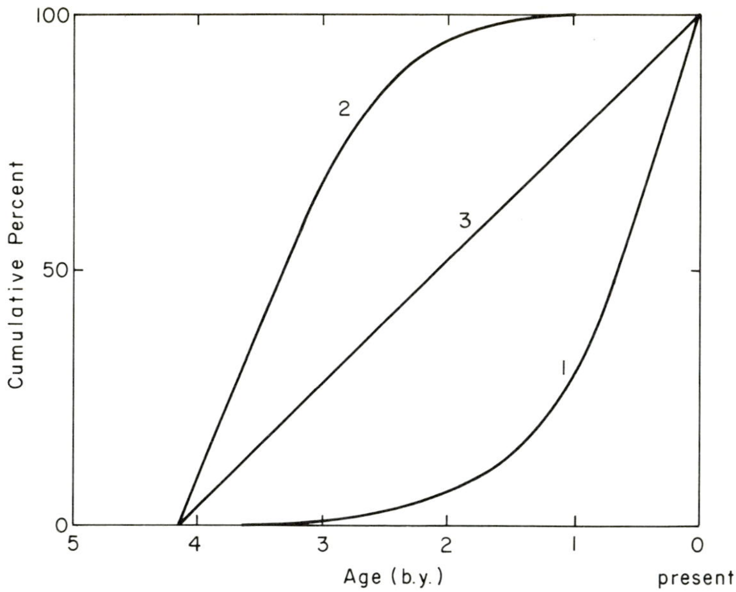

Fig. 10-7. Models for the volumetric growth of the continental crust (modified after Veizer, 1976). Vertical axis is cumulative percent of present-day continental crust.

high during the Late Archean and Proterozoic. Goodwin (1974) interprets the Hudson Bay anomaly in terms of a long-lived family of mantle plumes. During the Archean, protocontinents develop over this plume center and continue to grow from magmas derived from the plumes during the Proterozoic. The protocontinents may grow and merge to form continental cratons. Goodwin suggests the existence of six such cratons which continued to grow and merge by late Precambrian time to form a large supercontinent.

Continental growth rates during the Archean

The distribution of radiometric ages in the continents suggests that the volume of crust increases exponentially with decreasing age, an observation first pointed out by Hurley and Rand (1969). Three models can be considered to explain this observation (Veizer, 1976):

(1) The rate of formation of continental crust has increased exponentially with time.

(2) Rapid growth of the continental crust early in the earth's history with subsequent recycling of crustal material through the mantle.

(3) Linear growth rate of continental crust combined with recycling through the crust.

Each of these models is illustrated diagrammatically in Fig. 10-7. The first model was originally proposed by Hurley et al. (1962) and Hurley and Rand

(1969). They assume that the volume of crust of a given age as deduced from present areal distributions reflects the actual amount of new crust extracted from the mantle over the given period of time. Any crustal material that was recycled through the mantle would also be classified as new crust in this model. Since Hurley and coworkers proposed this model, it has become clear that many relicts of older crust occur in Precambrian and Phanerozoic mobile belts and hence one cannot equate areal extent of "apparent" age provinces on geologic maps with volume of new crust produced during a given increment of time. Perhaps as much as 50% of some Precambrian mobile belts may be composed of crustal rocks older than the dominant radiometric ages from the mobile belt. Veizer (1976) has also pointed out that the calculated Sr isotope growth curves for upper mantle and sialic crust in model 1 are not compatible with reasonable estimates for either of these systems.

Development of the second model was made by Armstrong (1968), Armstrong and Hein (1973), and Fyfe (1978). This model involves steady-state mixing of crust and mantle in subduction zones with the mixing rate decreasing with time as thermal gradients in the earth drop. The results, which involve both Sr and Pb isotopic systems, indicate that most of the continental crust formed between 2.5 and 3.5 b.y. and that any growth after this time has involved chiefly recycling of early crust through a convecting upper mantle. Two lines of evidence, which have become established since Armstrong proposed his recycling model do not support this model. First, isotopic and trace element studies of oceanic sediments and arc volcanics indicate that very little ($<10\%$) continental sediment that is subducted has contributed to the production of calc-alkaline magmas (Church, 1973). Furthermore, the low density of sialic crust should prevent it from being subducted (McKenzie, 1969) or allow only a minor amount of subduction (Molnar and Gray, 1979). Hence, it appears that very little sialic crust is recycled through the mantle today. Isotopic studies of oceanic basalts (both MORB and island basalts) indicate that the mantle source areas are not homogeneous despite a long geologic history of convection (Sun and Hanson, 1975; Hoffman and Hart, 1975) (Chapter 9). Even MORB appear to be derived from a moderately inhomogeneous upper mantle which has undergone major fractionation 1.8—2.0 b.y. ago. The persistence of such heterogeneities in the mantle for long periods of time argues against widespread homogenization by convection in the upper mantle, which is a necessary condition for Armstrong's model.

Veizer (1976) presents a model for linear growth of the continents (model 3) in which sialic crustal material is recycled only in the crust and in which older continental crust is recycled faster than younger. Clearly, the simplest interpretation of the low initial $^{87}Sr/^{86}Sr$ ratios in many sialic rocks is that new sialic crust has been added to the continents throughout geologic time. Some of the high-grade terranes with high initial strontium isotope ratios, however, may have had appreciable crustal residence times (0.5—1.0 b.y.) as pointed out by Collerson and Fryer (1978). The high initial ratios found in some high-K granites and associated volcanics probably represent partially remelted older crustal material. Veizer's model allows for variable amounts of older crustal rock found in mobile belts. Moorbath (1977) suggests that continental crust has grown throughout geologic time, probably episodically, by processes involving essentially irreversible differentiation of the mantle. Existing Sr and Nd isotopic data indicate that the typical time interval of an episode in which new crustal material is extracted from the mantle is of the order of 50—100 m.y. (Moorbath, 1976, 1977; DePaolo and Wasserburg, 1976). The time interval for some high-grade terranes, however, may have been considerably longer ($\gtrsim$ 200 m.y.) (Hamilton et al., 1979). Both the upper and lower sialic crust appear to have grown concurrently with magmas emplaced in the lower crust crystallizing directly to granulite-facies mineral assemblages. The interpretation by most investigators of available Sr, Pb, and Nd isotopic data suggests that at least 50% of the continental crust was produced by 2.5 b.y. (Muehlberger et al., 1967; Moorbath, 1977; Windley, 1977; McCulloch and Wasserburg, 1978).

The freeboard of continents

There is considerable sedimentological evidence to support the idea that the freeboard of continents, which is defined as the relative elevation of continents with respect to sea level, has not varied greatly for 3.7 b.y. The presence of nearshore marine sediments throughout the preserved sedimentary record supports this conclusion. A constant freeboard is important because it provides a means of evaluating other variables. The areas of the earth (A_e), continents (A_c), and ocean basins (A_o); the depth of continents (D_c) and oceans (D_o); the volume of continents (V_c) and oceans (V_o); and the isostatic link (I) relating D_o to D_c are related by four basic equations (after Wise, 1973):

$$(1)\ A_e = A_c + A_o, \quad (2)\ V_o = A_oD_o, \quad (3)\ V_c = A_cD_c, \quad \text{and} \quad (4)\ D_c = ID_o.$$

If A_e is assumed constant, there are seven variables of which three (V_o, V_c, and I) are independent. If a simple relationship for I is assumed ($D_c = 9.1$ km + $4.98D_o$) and A_e is assumed constant, the system is soluble for any time in the past in which two variables are known.

Wise (1974) proposes an internally consistent model of continent-ocean evolution based on the following assumptions: (1) the continents have been at least 90% of their present thickness for 2.5 b.y. (evidence supporting this assumption has been previously discussed); (2) the freeboard of continents and the area of the earth have been constant with time; and (3) there has been a net decrease in the volume of continental rocks or ocean waters recycled back into the mantle over this period of time. This limits changes in area or volume of continents or oceans to $\lesssim$ 20% of their present values. In terms of isotopic constraints, perhaps a more realistic model is the linear growth rate model of Veizer (1976) (model 3, Fig. 10-7) in which the volume of the continents 2.5 b.y. ago was about 50% of the present value. If the continents were 80% of their present thickness at this time, their total area would be 60% of the present area and the corresponding volume of ocean water would be 90% of present.

RELATIONSHIP BETWEEN HIGH- AND LOW-GRADE ARCHEAN TERRANES

Introduction

The problem of the relationship between Archean high-grade and low-grade (granite-greenstone) terranes was briefly discussed in Chapter 6. Models for their relationship fall into three categories (after Shackleton, 1976):

(1) High-grade terranes differ in age and tectonic setting from granite-greenstone terranes.

(2) High-grade terranes are the deep crustal equivalents of granite-greenstone terranes.

(3) High- and low-grade terranes are, broadly speaking, of the same age but represent different tectonic settings.

Age-dependent models

Some investigators have interpreted Archean high-grade terranes as older than granite-greenstone terranes and that each terrane has evolved in different tectonic settings (Hepworth, 1967; Windley and Bridgwater, 1971). Another possibility is that they are younger than granite-greenstone terranes and that they represent uplifted portions of mobile belts between granite-greenstone provinces (Shackleton, 1976). Available radiometric dates, however, do not lend support to either of these models (Fig. 9-21, Table 1-2). Although some high-grade terranes are older than low-grade terranes, many are not. It is possible, as discussed in Chapter 9, however, that many of the 2.6- to 2.7-b.y high-grade terranes with high initial $^{87}Sr/^{86}Sr$ ratios have had crustal residence times of several hundred million years. Another problem with these models is that some high-grade terranes contain remnants of supracrustal successions that resemble greenstone successions (such as the Isua succession in Greenland).

Radiometric dates from closely associated high- and low-grade terranes seem to indicate that such terranes have closely related geologic histories. An example of this is illustrated by the Kaapvaal and Rhodesian Provinces and intervening Limpopo mobile belt in southern Africa (Figs. 1-11; 1-12). A summary of the Archean geologic history of these three provinces is given in

Table 10-1 (after Mason, 1973; Hunter, 1974a, b; Kröner, 1977b; Wilson et al., 1978). Available dates indicate the presence of tonalitic basement rocks in the Limpopo belt by 3.8 b.y. (Barton et al., 1977). Indirect evidence suggests the presence of similar age rocks in the Kaapvaal and Rhodesian Provinces (Kröner, 1977a, b). The oldest greenstone belts formed at $\gtrsim$ 3.5 b.y. in Rhodesia and South Africa, while shallow-water cratonic sediments of the Beitridge Formation were deposited at about the same time in the Limpopo belt. Cratonic sediments of the Messina and Baines Drift Formations were deposited between 3.2 and 3.3 b.y. in the Limpopo belt. While granites were intruded at 2.8—2.9 b.y. in the Rhodesian and Kaapvaal Provinces, regional metamorphism to granulite-facies grade occurred in the Limpopo belt. It is noteworthy that the major period of granulite-facies metamorphism and granitic plutonism in the Limpopo belt at 2.6—2.7 b.y. paralleled the major period of granite-greenstone formation in Rhodesia. It is clear from these results that the Kaapvaal and Rhodesian low-grade provinces developed at the same time as the Limpopo belt and that the Limpopo belt cannot be considered older or younger than the granite-greenstone provinces. The general increase in metamorphic grade from the central portions of the Rhodesian and Kaapvaal Provinces towards the Limpopo belt, furthermore suggests the major heat source for all three provinces lay beneath the Limpopo belt (Chapter 6).

Models dependent upon erosion level

As discussed in Chapter 6, one of the currently popular theories relating high- and low-grade Archean terranes is that they represent, respectively, deep and shallow levels of erosion of the same crust (Windley and Bridgwater, 1971; Glikson and Lambert, 1976; Shackleton, 1976; Goodwin, 1977b; Naqvi et al., 1978a, b). A diagrammatic representation of such a model is given in Fig. 10-8. The diagrams show a greenstone belt at shallow levels passing downwards into a high-grade assemblage consisting of the gneissic equivalents of shallow-level tonalite-trondhjemite containing sheared and stretched inclusions of the greenstone belt rocks. The high-grade terrane is exposed at the surface adjacent to the granite-greenstone terrane by faulting. Despite the appealing attributes of this model, several significant features cannot be explained if the two terranes are simply depth equivalents:

(1) As discussed in Chapter 9, many of the high-grade terranes exhibit high initial $^{87}Sr/^{86}Sr$ ratios suggesting either a different mantle source than that from which granite-greenstone belts are produced or a significant crustal residence time.

(2) Most of the supracrustal rocks preserved in high-grade terranes do not appear to represent the high-grade equivalents of greenstone belts (Sutton, 1976) — quartzites, pelitic schists, and carbonates suggest a cratonic association dominates in high-grade areas.

TABLE 10-1

Summary of major events in the Archean history of southern Africa

Age (b.y.)	Kaapvaal Province	Limpopo Belt	Rhodesian Province
	South		*North*
—		Intrusion of satellites of the Great Dyke (2.46 b.y.)	Intrusion of the Great Dyke (2.46 b.y.)
2.5 —			
—	Intrusion of post-tectonic granites (2.6—2.7 b.y.)	Deposition shallow-water sediments; intrusion of Bulai (~ 2.7 b.y.) and Singelele granites (2.6 b.y.); high-grade metamorphism (2.6—2.7 b.y.)	Intrusion of post-tectonic granites (~ 2.5 b.y.)
—	Cratonic sedimentation (± volcanism) (2.4—2.9 b.y.)		Formation of Bulawayan (upper and lower) and Shamvaian greenstone belts; granitic plutonism; regional metamorphism (2.6—2.7 b.y.)
—	Emplacement of Lochiel- and Dalmein-type granites (2.8—2.9 b.y.)	High-grade metamorphism and plutonism (2.8—2.9 b.y.)	
—			Intrusion of Mashaba, Chingenzi, and related plutons (2.8—2.9 b.y.)
3.0 —		Intrusion of Messina layered complex (3.09 b.y.)	
—		Mafic dike intrusion (3.12 b.y.) Deposition of Messina and related formations (3.1—3.3 b.y.)	
—			Intrusion of Mushandike and Mont D'Or granites (3.3—3.4 b.y.)
—		Intrusion of Zanzibar tonalite (3.33 b.y.)	
—	Formation of tonalite-trondhjemite gneisses (≳ 3.3 b.y.)	Deposition of the Beitridge Formation (3.4—3.5 b.y.)	Thrusting from southwest and regional metamorphism
3.5 —	Formation of Barberton and related greenstone belts (~ 3.5 b.y.)		Formation of Sebakwian-type greenstone belts (≳ 3.5 b.y.)
—			Formation of tonalite gneisses in the Shabani area (≳ 3.5 b.y.)
—		Mafic dike intrusion (3.63 b.y.)	
		Formation of Sand River tonalitic gneisses (~ 3.8 b.y.)	

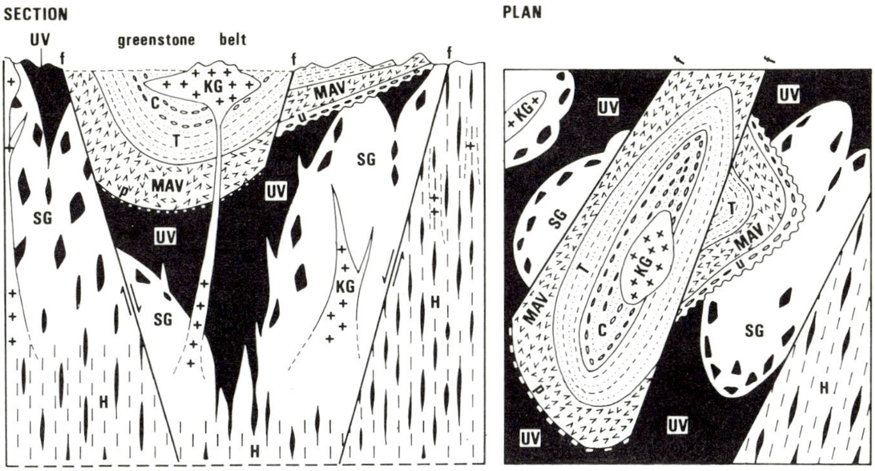

Fig. 10-8. Diagrammatic representation of possible depth relationships between Archean granite-greenstone and high-grade terranes (from Glikson, 1976b). Symbols: UV = ultramafic-mafic rocks; SG = tonalite-trondhjemite; MAV = mafic to felsic volcanics; T = graywacke-argillite; C = conglomerate, arkose, quartzite; KG = high-K granites; H = high-grade gneisses and granulites; u = unconformity; p = paraconformity; f = fault.

(3) The deformation in granite-greenstone terranes reflects dominantly vertical forces (Chapter 6) whereas subhorizontal forces appear to have dominated in high-grade terranes.

(4) While in granite-greenstone terranes there is a sparsity of rocks of andesitic composition (Barker and Peterman, 1974), there is a continuum of calc-alkaline compositions found in some high-grade terranes (Tarney, 1976).

Models dependent upon different tectonic settings

Most data seem to be compatible with high- and low-grade terranes evolving in different tectonic settings (Windley, 1973). The geologic history of the Kaapvaal-Limpopo-Rhodesian Provinces (Table 10-1) is compatible with a tectonically unstable volcanic-plutonic setting for the Kaapvaal and Rhodesian Provinces and a more stable cratonic setting for the Limpopo belt. This idea will be developed further in a later section.

TECTONIC MODELS FOR THE ORIGIN OF ARCHEAN GRANITE-GREENSTONE TERRANES

Density inversion models

Perhaps the earliest model for the development of Archean greenstones was suggested by Macgregor (1951) based on an analogy with mantled gneiss

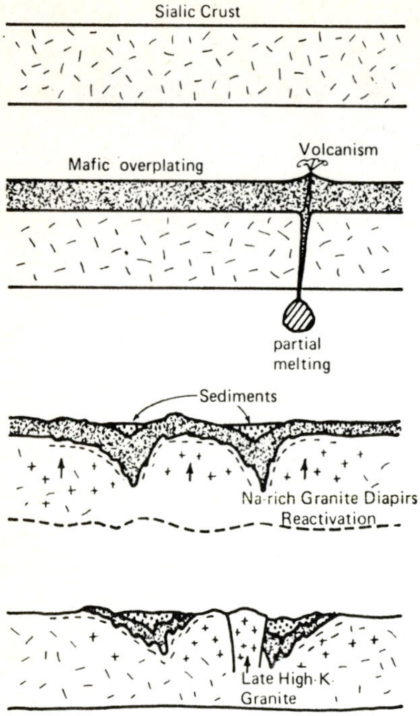

Fig. 10-9. Diagrammatic sequence of events in the density inversion model for the origin of greenstone belts (after Condie, 1976a).

domes in Finland (Eskola, 1948). Eskola proposed that supracrustal rocks are deposited on gneissic basement which later becomes reactivated and moves upwards intruding the supracrustal rocks. Macgregor (1951) suggested that the ovoid "gregarious" batholiths in Rhodesia which now intrude the greenstone belts once served as basement for these belts and were later reactivated and intruded.

Such diapiric intrusion is driven by gravity differences between the more dense greenstones and less dense sialic basement. Both experimental (Ramberg, 1973) and theoretical (Hargraves, (1976) studies confirm the possibility of this mechanism. The model for an individual greenstone belt was discussed in Chapter 6 (Fig. 6-12; after Gorman et al., 1978). A generalized sequence of events in the density inversion model is illustrated in Fig. 10-9. The existence of a sialic crust is assumed in the model. Greenstone volcanics are erupted on top of this crust forming a dominantly mafic veneer. Because the volcanics are heavier than underlying sialic crust, they begin to sink and displace the gneisses which diapirically move upwards. The gneissic basement is conformable with cover rocks at low crustal levels but may intrude them at shallower levels. As the greenstone belts flow off the rising gneissic diapirs, the diapirs are eventually unroofed and serve, together with the greenstones, as source areas for sediments which generally accumulate late in greenstone successions (Chapter 2). Diapirism also results in deformation of the greenstone belts as illustrated in Fig. 6-12. Late, post-tectonic granites may then intrude the deformed gneiss-greenstone complexes.

Talbot (1968) proposed a modification of the model in which the sialic crustal layer is composed of small convecting cells with diameters of the order of 100 km. Greenstone belts would collect at the margins over the down-currents of the cells. He suggests that the higher heat flow in the Archean lowered the viscosity of the sialic crust sufficiently for it to sustain convection. Calculations indicate that the convecting layer would have been $\lesssim$ 40 km thick. Convection also would have occurred well below liquidus temperatures.

Although granitic diapirism appears to have been important in the Archean, density inversion models as a general means of explaining Archean granite-greenstone terranes are confronted with major obstacles. First of all, it is necessary to assume the existence of an earlier sialic crust which may or may not be valid. The most serious problem relates to the age relationships of granitic rocks and greenstones and probable magma sources. Many, if not most, granitic plutons in granite-greenstone terranes are younger than the greenstones and appear to have a mantle source (see Chapter 9). Also, most tonalite-trondhjemite gneisses appear to have come directly from the mantle with little or no crustal residence time. Hence, they could not represent reactivated sialic basement that was older than the greenstone belts by $>$ 100 m.y. Convecting sialic crust models are, in addition, faced with the problem of not explaining the large number of supracrustal inclusions and the complex structural patterns in gneissic complexes. Convection should have led to homogenization of these bodies.

Non-plate tectonic mantle convection models

Fyfe (1973a, b, 1974) and Williams (1977) have proposed models for the evolution of Archean granite-greenstone terranes based on small-scale convection (100—500 km) in the upper mantle. These models are consistent with the theoretical and experimental studies of Elder (1972) and McKenzie and Weiss (1975) which predict small-scale convection in the upper mantle. Fyfe (1974) proposes that a thin (5—10 km thick) continuous granitic layer overlies a zone of partial melting in the upper mantle (Fig. 10-10). Convection cells in the upper mantle have radii of 50—100 km. Basalt penetrates the sialic crust along fractures (*c*) and is terminated from reaching the surface by partial melting of the granitic layer and the formation of granitic domes (*d*). The basaltic magmas then collect and are encapsulated beneath the partially melted sialic layer where they crystallize to granulite mineral assemblages and some fractionate to produce anorthosites (*e*). Above regions of return flow, the crust is thinned and greenstone volcanics are erupted (*f*). Fyfe

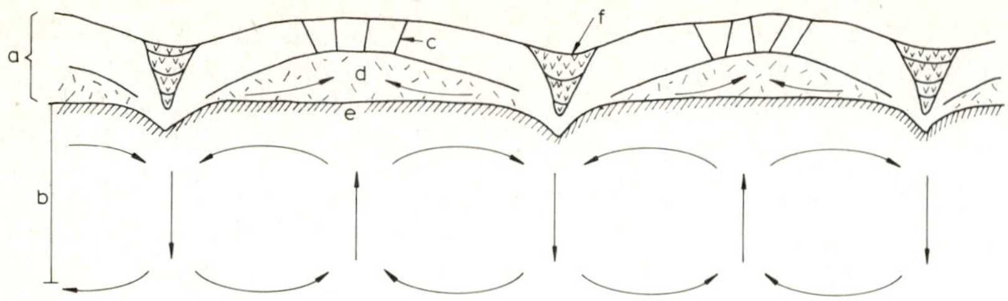

Fig. 10-10. Shallow convection model for the evolution of the Archean crust (after Fyfe, 1974). Symbols: a = sialic crust; b = zone of partial melting; c = mafic dikes, d = granitic domes; e = granulites and anorthosites; f = greenstone belts.

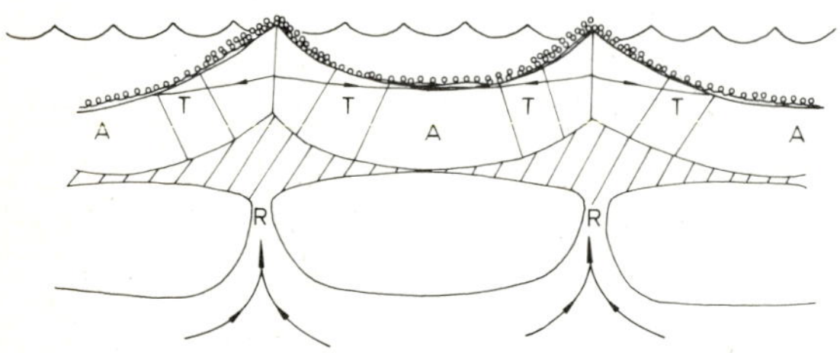

Fig. 10-11. Hot spot model for the Archean crust (from Fyfe, 1978). See text for explanation.

(1978) has recently presented a different version of the model in which a large number of mantle hot spots form at boundaries of convective upcurrents (Fig. 10-11). Most of the ultramafic-mafic magmas produced at these hot spots (R) underplate and uplift the crust leading to possible thrust faulting (T) at shallow levels. Subsidence of the material in the hot spots (which may be considered plumes) terminates the volcanic activity at the surface; thickening of the crust between volcanic centers, due to thrust slices sliding into the basins, results in partial melting of the crust and granite formation (A).

Williams' (1977) model has an additional feature in that it offers an explanation for high-grade mobile belts associated with granite-greenstone terranes. He proposes that mobile belts develop over primary convection upwellings whereas greenstone belts develop over the small-scale secondary upwellings (Fig. 10-12). Experimental studies (Elder, 1972; McKenzie and Weiss, 1975) show that secondary cells circulate at a steep angle to the large primary cells, a feature which is also built into the model. A situation like this is observed in the Liberian and Rhodesian Provinces where the Kasila and Limpopo mobile belts, respectively, cross-cut the adjacent greenstone belts (Figs. 1-12 and 1-14). Relative motions of bordering cratons with respect to mobile belts may be responsible for the development of faults along the margins of the mobile belts.

As with the density inversion models, a sialic crust is envisioned as a precurser to greenstones in the shallow convection models and hence, the same objections regarding ages and sources of granitic magmas are pertinent to these models.

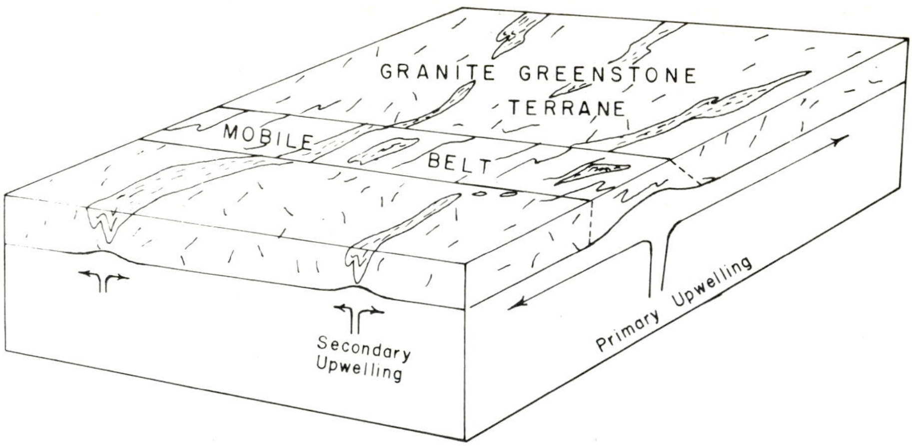

Fig. 10-12. Combined shallow and deep convection model for the evolution of the Archean crust (redrawn after Williams, 1977).

Non-plate tectonic oceanic crust models

Glikson (1971b) has proposed a granite-greenstone model, which he has added to and modified several times (Glikson, 1972b, 1976b, 1977a, 1978, 1979a; Glikson and Lamberg, 1973, 1976), based on an evolving oceanic crust, which at least initially evolves in a non-plate tectonic framework. He presents several lines of evidence which suggest that a primitive mafic-ultramafic oceanic-type crust (represented by lower greenstone belts) existed before granitic rocks (modified after Glikson, 1976b):

(1) Sialic basement is not known to occur beneath lower greenstone successions (Chapter 2).

(2) Sial-derived detritus has not been recognized in lower greenstone belts although it is important in upper greenstones.

(3) Lower greenstones do not contain sialic inclusions.

(4) Experimental and geochemical data (Chapter 9) indicate that tonalite-trondhjemite melts are derived from partial melting of mafic parent rocks, and hence, it follows that a mafic crust must precede a sialic crust.

(5) Both lower and upper greenstone belts contain basalts whose compositions are similar in many respects to MORB (there are, however, also many compositional differences as discussed in Chapter 3).

The model as presented by Glikson (1971b, 1972b) is summarized in diagrammatic form in Fig. 10-13. The first stage is characterized by mega-rippling of the oceanic crust and minor deposition of chert, iron formation, and pelitic sediments derived by erosion of uplifted segments of the oceanic crust. These sediments together with the mafic-ultramafic rocks are the typical assemblages found in lower greenstone belts. The lower portions of the down-ripples invert to eclogite and/or amphibolite assemblages and undergo

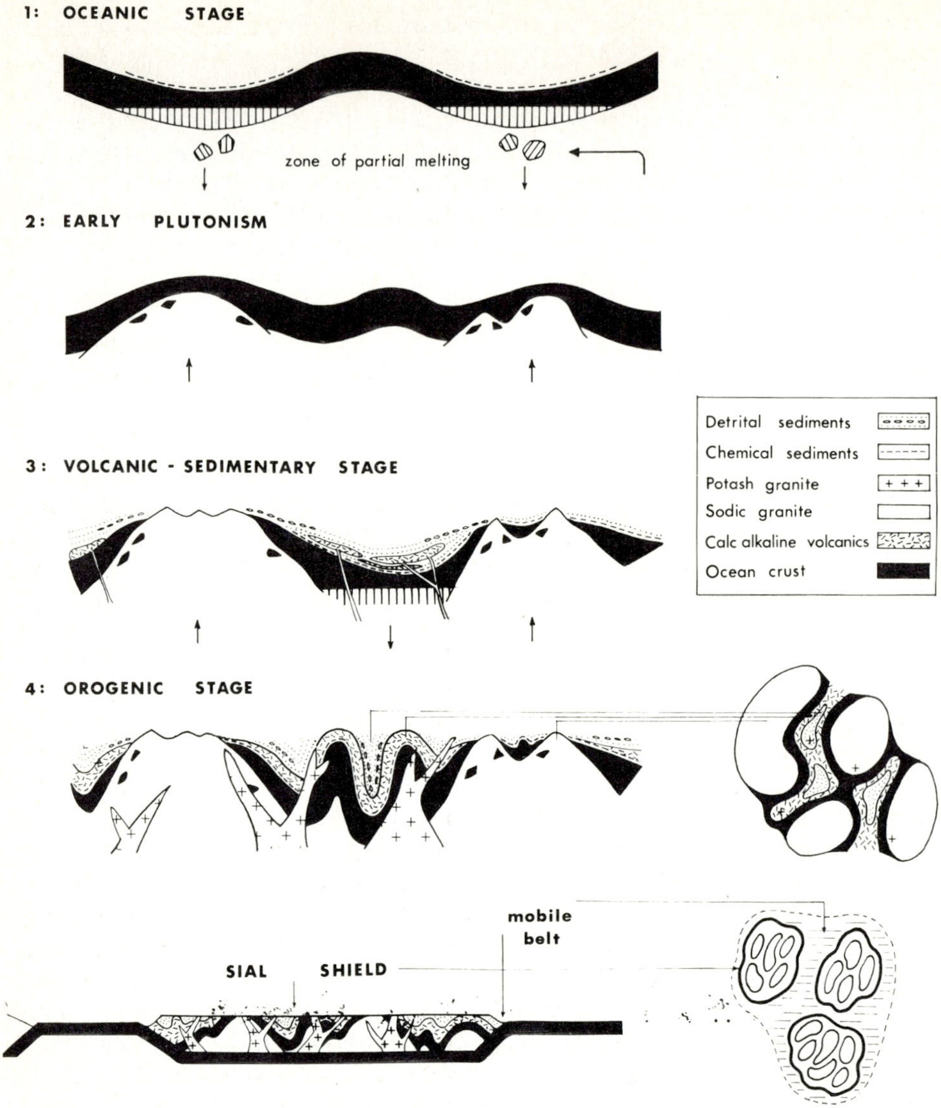

Fig. 10-13. An oceanic crustal model for the evolution of the Archean crust (from Glikson, 1971b, 1972b).

small degrees of melting. These early melts are tonalite-trondhjemite in composition and rise diapirically as plutons through the oceanic crust during stage 2. The isostatic rise of diapirs results in subsidence of oceanic crust between diapirs (stage 3). Continued partial melting of the oceanic crust beneath these troughs gives rise to calc-alkaline magmas which become important at higher stratigraphic levels in some greenstone belts. Sediments accumulate at late stages from erosion of unroofed tonalite-trondhjemite plutons and nearby greenstones. The orogenic stage (stage 4) is characterized by further subsidence of the troughs, folding, and low-grade metamorphism. Small degrees of melting of the base of the sialic crust give rise to granitic magmas which are intruded as post-tectonic plutons. Cooling of the deformed and metamorphosed greenstone successions results in the joining of sialic nuclei into cratonic units (as shown in the inset of Fig. 10-13). The boundaries of the cratons with the surrounding oceanic crust are envisioned as the sites of high-grade mobile belts.

Glikson and Lambert (1976) later presented a revised model based on Archean high-grade terranes being the depth equivalents of granite-greenstone terranes. The example of the Yilgarn Province is described in Chapter 6 and the revised model is summarized in Fig. 10-14. During the first stage a mafic-ultramafic crust develops from magmas derived from rising mantle plumes and a shallow low-velocity zone. As in the first model, downfolding of the mafic-ultramafic crust during the second stage results in partial melting and production of tonalite-trondhjemite diapirs. Continued subsidence results in production of calc-alkaline magmas and sedimentation forming upper (late) greenstone belts. The third stage involves continued production and emplacement of tonalite-trondhjemite and further subsidence of intervening greenstone belts. The final stage, which reflects a rise in geothermal gradient, results in a crust exhibiting a vertical zonation in metamorphic grade. Partial melting in the lower part of this crust gives rise to high-K granites which are intruded at shallow levels.

Glikson's models were some of the first to tie together many fields and geochemical observations from Archean granite-greenstone terranes: the relative abundance of mafic (± ultramafic) rocks in lower greenstone belts; the increase in calc-alkaline volcanics at higher stratigraphic levels in some upper greenstone belts; the relative abundance of clastic sediments in the upper parts of greenstone successions; the production of tonalite-trondhjemite by partial melting of a mafic parent rock; and the production of late high-K granites. He also proposed a mechanism for developing Archean cratons and for relating high- and low-grade terranes. Several problems, however, are associated with the oceanic crustal models. One of the major problems with the revised model is that inclusions of greenstone in tonalite-trondhjemite in granite-greenstone terranes is equated with supracrustal inclusions in high-grade terranes. As previously discussed, the inclusions in high-grade terranes are typically fragments of cratonic sediments and anorthosite, rock types which are scarce in greenstone belts. Also as discussed in the previous section, a depth relationship between high- and low-grade Archean terranes no longer seems tenable. As pointed out by Windley (1977), the Glikson model does not seem applicable to greenstone terranes that have evolved, in part, by compressive forces such as those in the southwestern part of the Rhodesian Province (Chapter 6). Still another problem relates to the volume of tonalitic-

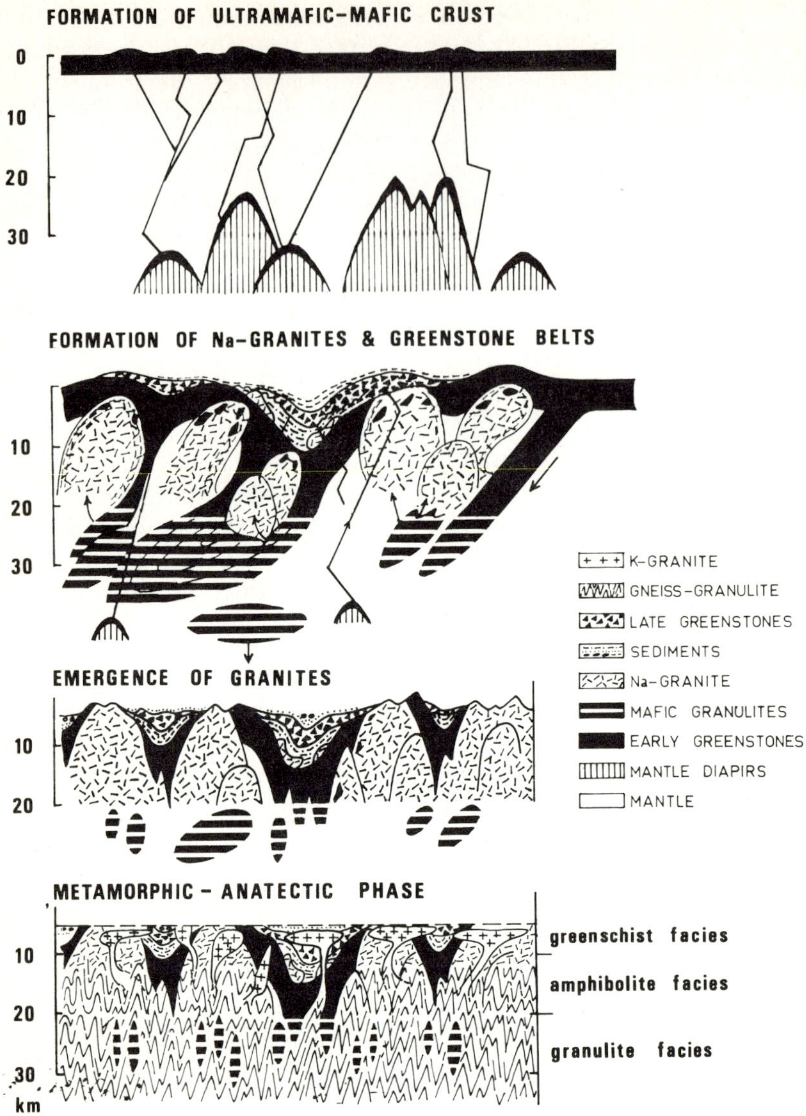

Fig. 10-14. A modified oceanic crustal model for the evolution of the Archean crust (from Glikson and Lambert, 1976).

trondhjemitic magmas produced from the thin oceanic crust (in stages 1—3, Fig. 10-13). Since less than 10% melting of a mafic source is required to produce these sodic granites, a tremendous thickness of oceanic crust or some means of continually creating such crust over the rising tonalitic diapirs is a necessary, but missing feature of the model. Also, in stage 3 (Fig. 10-13), it is not clear why partial melting of the subsiding mafic crust should produce calc-alkaline magmas whereas earlier (and perhaps at the same time) tonalite-trondhjemite magmas are produced from this source. Finally, if lower greenstones are to be equated with oceanic crust, one is faced with the problem that these successions do not resemble Phanerozoic ophiolite successions which are thought to represent fragments of oceanic crust (Coleman, 1977).

Goodwin and Ridler (1970) have presented a model for the development of the Abitibi greenstone belt in Canada in which it is formed on thin oceanic crust between two continental land masses (~ 800 km apart). The model relies chiefly on the shelf-to-basin transitions deduced from lithologic assemblages (see Chapter 4). The major volcanic centers lie along the margins of the basin and shed detritus both inward and outward. As pointed out by Windley (1977), this model could be fitted readily into a continental rift tectonic setting in which the Abitibi basin represents an opening small ocean basin.

Continental rift models

Several investigators have proposed continental rift models for the development of greenstone belts. Such models, of course, presume the existence of older sialic crust. One of the first rift models presented is that of Anhaeusser et al. (1969) and Anhaeusser (1971a). This model begins with a primitive, thin sialic crust overlying a basaltic layer (Fig. 10-15). Downwarps and/or rifts develop on this thin crust and are filled with greenstone volcanics (stage 1). The trough continues to subside as it fills and the sialic crust is elevated around the margins providing a source for sediments which are deposited in the trough (stage 2). During stage 3, continued subsidence and diapiric reactivation of the sialic basement lead to deformation and low-grade metamorphism of the greenstone belt while cratonic sedimentation may occur at the surface. A final stage is represented by granitic plutonism and continued deformation followed by regional uplift and erosion to the present level of exposure (stage 4). This model provides for the overall stratigraphic succession observed in most greenstone belts and the subsequent plutonism and deformation. It is similar in many respects to the density inversion models although it does not necessitate complete overplating of the sialic crust with greenstone volcanics.

Windley (1973) proposed a model whereby greenstone belts develop in proto-oceanic basins similar in size to the Red Sea (Fig. 10-16). The model begins with high-grade sialic crust formed at an earlier time which undergoes rifting (1). As rifting continues, mafic magmas are injected into the rift and extruded in a subaqueous environment and sediments accumulate on the floor of the rift (2). The basin continues to subside and undergoes deformation and low-grade metamorphism (3); tonalitic-trondhjemitic plutons are injected into the sequence. Post-tectonic granites are then intruded and the

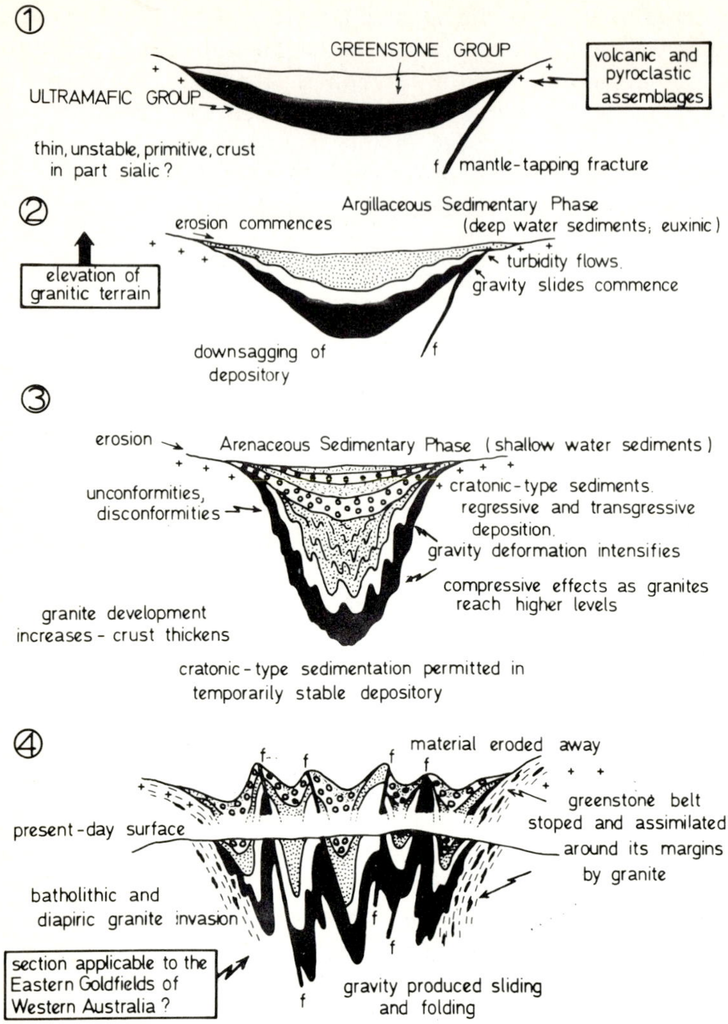

Fig. 10-15. A simplified evolutionary model for the development of Archean greenstone belts (from Anhaeusser, 1971a).

area is uplifted and eroded to the present exposure level (4). This model has the advantage that the size of a greenstone belt can vary according to the amount of opening of the rift.

Hunter (1974a) and Condie and Hunter (1976) propose a continental rift model for the development of the Barberton greenstone belt in South Africa. In this model the rifting occurs in response to an ascending mantle plume. The major stages of development are summarized in Fig. 10-17. The model assumes the existence of sialic crust in this area by 3.5 b.y. The crust is about 20 km thick and the lower part is composed of granulite-facies rocks

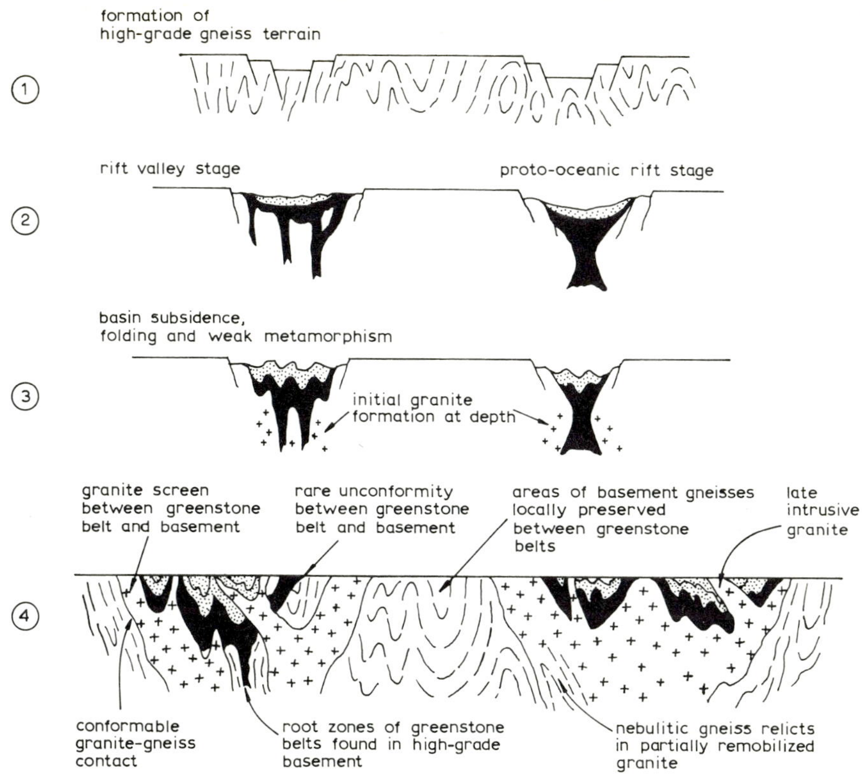

Fig. 10-16. A continental rift model for the development of Archean greenstone belts (after Windley, 1973).

(1). A mantle plume ascends to the base of the lithosphere at ~ 3.5 b.y. and as it spreads laterally a continental rift develops (2). The rift system is partially filled with mafic-ultramafic lavas of the Onverwacht Group which are derived from the plume. A steep geothermal gradient exists beneath the rift decreasing on the flanks. As the plume subsides (3), erosion of the flanks of the rift (which expose initially Onverwacht volcanics and later, granitic rocks) gives rise to the Fig Tree and Moodies sediments. As the rocks in the rift subside, they are deformed and undergo low-grade metamorphism. Sinking mafic rocks in the rift invert to amphibolite (or less likely eclogite) and newly intruded mafic magmas crystallize directly to amphibolite mineral assemblages. Small amounts of melting of this amphibolite under wet conditions produce magmas which give rise to the felsic volcanics found at middle to upper stratigraphic levels in the Swaziland Supergroup. Wet melting conditions are necessary to prevent andesite production (which is rare or absent in the succession). Continued sinking of the Barberton succession and renewed plume activity at 3.1—3.2 b.y. (4) result in renewed partial melting of the amphibolite (again under wet conditions) giving rise to tonalite-

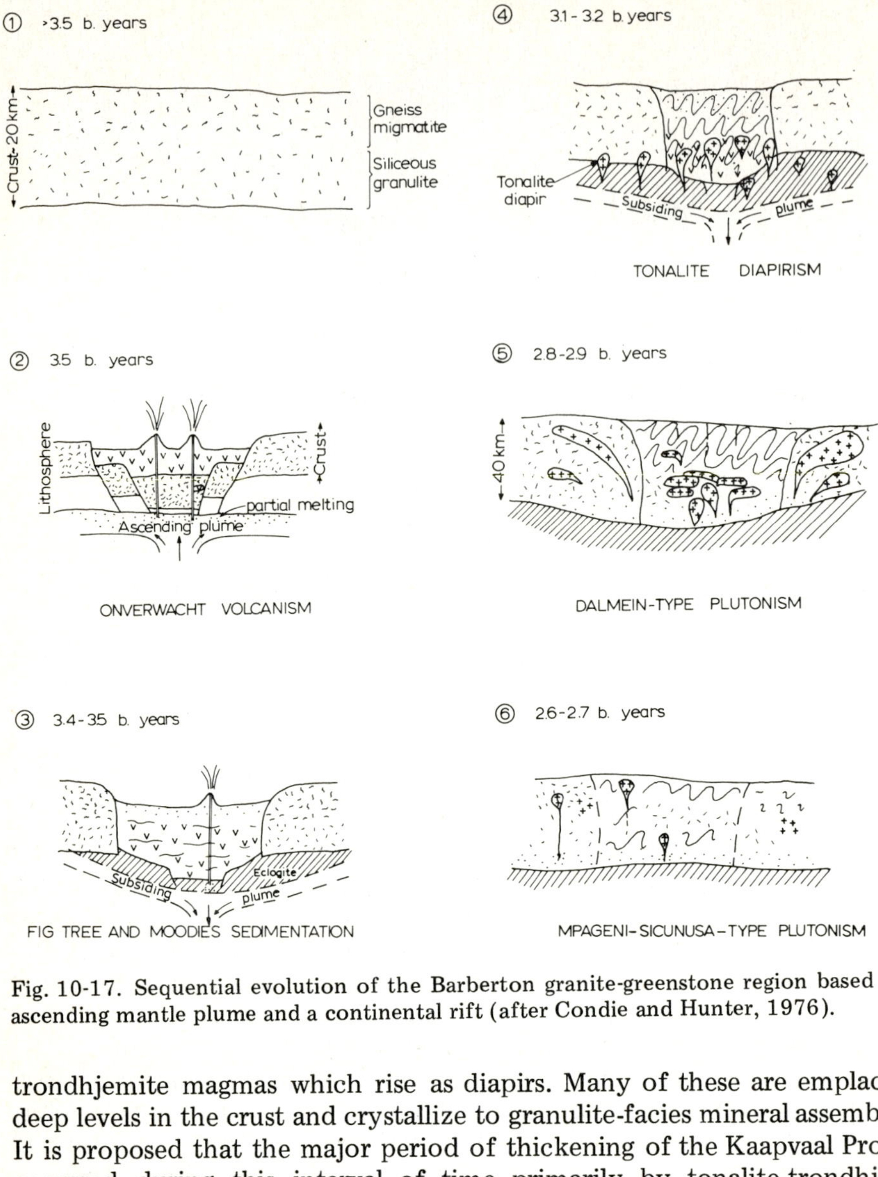

Fig. 10-17. Sequential evolution of the Barberton granite-greenstone region based on an ascending mantle plume and a continental rift (after Condie and Hunter, 1976).

trondhjemite magmas which rise as diapirs. Many of these are emplaced at deep levels in the crust and crystallize to granulite-facies mineral assemblages. It is proposed that the major period of thickening of the Kaapvaal Province occurred during this interval of time primarily by tonalite-trondhjemite underplating and intrusion. Transcurrent faulting in the rift succession is also initiated at this time.

Plume subsidence between 2.8 and 2.9 b.y. (5) results in crustal downwarping and partial melting of the gneisses and granulites in the lower crust begins. Lochiel- and Dalmein-type granitic magmas are produced at this time and rise as plutons or dikes which feed the sheet-like Lochiel batholith. By 2.6—2.7 b.y. (6), cooling progresses to such a point that only small amounts of melting of the crust occurred producing granodiorite or more felsic magmas. The crust had now evolved into a more stable tectonic regime, and the granodiorite magmas, not being disturbed by tectonic activity, underwent fractional crystallization to produce high-K granites which were intruded as post-tectonic plutons. As the Kaapvaal Province became more stable, cratonic sediments of the Pongola and Witwatersrand Supergroups were formed by erosion and deposition of uplifted granite-greenstone terranes (Hunter, 1974b).

Goodwin (1977b), following the ideas of Sun and Hanson (1975), proposes multiple rifting of sialic crust to explain the superbelt pattern in the western Superior Province. Sun and Hanson (1975) suggest that greenstone belts formed in response to the secondary mantle convections previously described. These convection patterns produce an array of "hot lines" over upcurrents which cause rifting of sialic crust, with greenstone belts developing in the rift zones. Note that this is opposite to the model Fyfe (1974) described above in which greenstone belts form over convective down currents. Individual greenstone belts may correspond to "hot spots" on the "hot lines". The latter are responsible for the volcanic-plutonic superbelts. The adjoining sedimentary-plutonic superbelts represent pre-existing fissured sialic crust overlain in part by new clastic sediments derived chiefly by uplift and erosion of the sialic basement. These sediments collect in local basins on the sialic crust which is depressed and undergoes middle- to high-grade metamorphism and partial melting giving rise to migmatites and granites; the granites rise, intruding higher crustal levels. One problem with this model is that the initial $^{87}Sr/^{86}Sr$ ratios indicate that most granites in sedimentary-plutonic superbelts cannot have long crustal residence times and hence cannot be produced by partial melting of sialic crust that is significantly older than the granites. The less-dense sedimentary-plutonic superbelts undergo greater isostatic uplift than adjoining volcanic-plutonic superbelts thus exposing higher metamorphic grades at the surface today.

A mantle-plume mechanism for the production of continental rifts is appealing in that not only can it form the rifts but it provides a mechanism for producing komatiitic magmas without large degrees of melting (which is precluded by experimental data discussed in Chapter 9). Naldrett and Turner (1977) propose the existence of mantle plumes that undergo successive small amounts of partial melting and removal of melts as they ascend. The model is illustrated in Fig. 10-18. A mantle plume originating at A will have undergone 25—30% melting as it rises along an adiabat to point B. At this stage, approximately 20% of the melt separates from the plume and rises along a path perhaps similar to BF, to be extruded at the surface as a mafic lava. The plume, which is now comprised of 5—10% mafic liquid and 90—95% crystals, continues to rise and undergo further melting. By the time the plume arrives at C it is composed of 65—70% residual olivine crystals and 30—35% melt. At this point the melt, which is an ultramafic komatiite, separates either ris-

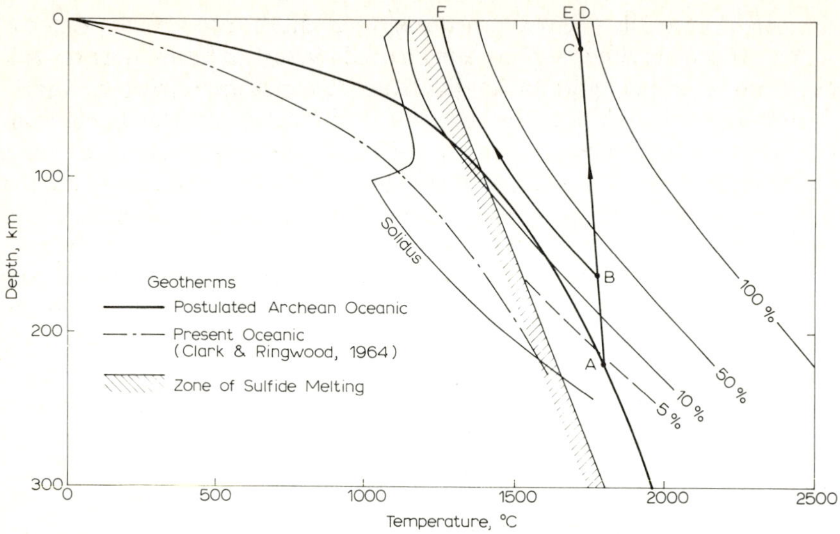

Fig. 10-18. Temperature-depth diagram illustrating a two-stage model for the production of komatiitic magmas from an ascending mantle plume (after D.H. Green, 1975; Naldrett and Turner, 1977).

ing to the surface and being extruded (path *CE*) or collecting in a near-surface chamber and undergoing fractional crystallization to produce the komatiite series. The final residuum may reach the surface at point *D*. Employing this model, no more than 35% melting is necessary to produce ultramafic komatiites, a feature which is consistent with experimental data (Arndt, 1977b; and Chapter 9). Although the model involves only two stages of magma extraction, multiple-stage removal may clearly be involved. The association of nickel sulfide deposits with the komatiitic suite was discussed in Chapter 7. Naldrett (1973) has proposed an explanation for this based on the melting points of sulfides and the ascending plume model. In the model, the proposed Archean geotherm intersects the sulfide melting curve at about 100 km depth (Fig. 10-18). At depths shallower than this, mantle sulfides will be solid, whereas at greater depths they will be liquid. At about 100 km, the mantle consists of about 10% melt. Naldrett and Turner (1977) suggest that the dense liquid sulfides below 100 km percolate downwards displacing the less dense mafic liquid and depleting the mantle in sulfur (and presumably chalcophile elements). The downward movement is thought to be arrested by a decreasing proportion of mantle pore fluid as the degree of mantle melting decreases with depth. As can be seen from the diagram, sulfides tend to be concentrated in the region that was suggested as the source for the mantle plumes (near point *A*). This concentration may account for the close association of nickel sulfides and komatiitic rocks in the Archean.

Condie (1975) shows that a mantle plume mechanism can also explain the following major stratigraphic and compositional features of greenstone belts.

(1) The dominance of ultramafic and komatiitic lavas during the early stages of development of some greenstone belts. This is related to near-surface mantle plume sources employing the melting model described above.

(2) The increasing importance of mafic source rocks as dictated by the increasing importance of calc-alkaline volcanics at higher stratigraphic levels in many greenstone belts. As plumes subside, overlying mafic volcanics sink into the mantle and recrystallize to metamorphic mineral assemblages. In the presence of water, these would be chiefly amphibolite and garnet amphibolite. Partial melting of these mafic rocks above the sinking plume gives rise to TH2 (light-REE-enriched tholeiite) and, with decreasing degrees of partial melting, the calc-alkaline series as described in Chapter 9. Depending upon the amount of water available in the source, andesites may be an important magma type (low-water content) or they may be rare or absent with tonalite-trondhjemite being the dominant magma type (high-water content).

(3) The existence of andesitic or tonalitic granulites during the late stages of evolution of a particular greenstone-granite cycle, which serve as sources for granodiorite and high-K granitic melts. Continued cooling and subsidence of plumes could lead to downwarping of sialic crust (Fig. 10-17, stage 5) and partial melting of granulites in the lower part.

(4) An overall decrease in thermal gradient during a given granite-greenstone cycle (50—100 m.y.) is necessary to explain the increasing importance of calc-alkaline and felsic magmas with time. A continued subsidence in plume activity over this period of time would result in a decreasing geothermal gradient.

(5) The presence of volcanic cycles in greenstone belts necessitates replenishment of fertile mantle source rocks. This may be explained by small episodes of renewed plume activity which results in both direct addition of new magma from the plume (PK, BK, TH1) and production of new mafic sources in the mantle and intermediate sources in the lower crust by the sinking of overlying rocks as each plume episode subsides.

Although continental rift models have provided the first models which accommodate a large number of seemingly unrelated geological and geochemical data, they still are faced with obstacles. They do not, for instance, provide an explanation for the sialic crust which is already present nor do they offer an explanation for Archean high-grade terranes (assuming such terranes are not depth-equivalents of granite-greenstone terranes). Problems are also encountered in providing an adequate deformation mechanism in a rift model (Groves et al., 1978). It is also difficult with such models to explain the relatively large volumes of tonalite-trondhjemite which occur dominantly in areas between, rather than within greenstone belts.

Convergent plate boundary models

The compositional similarities between Archean greenstone belts and modern arc systems have been cited by many investigators as evidence for similar origins (Jahn et al., 1974; Condie, 1976a). This has led to an array of plate tectonic models for greenstone belts in which they are associated in one form or another with convergent plate boundaries. White et al. (1971) were perhaps the first to suggest a convergent plate setting for greenstone belts in the Yilgarn Province. Glikson (1972b) pointed out the striking similarity in sequence of events in many greenstone belts with that observed in the Tertiary succession on the island of Viti Levu in the Fiji Islands, which represents a typical convergent boundary association. Condie and Baragar (1974) pointed out that subduction provided a means of replenishing mantle source rocks during greenstone belt evolution and Condie (1972) presented a plate tectonic model for the South Pass greenstone belt in Wyoming involving an arc-arc collision. Myers (1976) proposed a Himalayan-type collision to account for the thickening of the Archean sialic crust in West Greenland. Talbot (1973) suggested a model in which greenstone belts represent Archean oceanic crust that was not subducted but scraped from leading edges of descending slabs and plastered on overriding plates. As pointed out by Windley (1977), such greenstone belts would occupy the same tectonic position as modern mélanges and it is difficult to see how greenstone belts could be so well-preserved if this mechanism operated.

Burke et al. (1976b) propose a modern plate tectonic model for the Archean operating, however, on a more rapid time scale and involving thinner and more numerous plates than at present. To dissipate the additional heat in the earth during the Archean, they suggest that spreading rates were more rapid or that the total length of oceanic ridge systems was greater than at present. They envision numerous, small continental land masses growing by arc collisions and rarely by continent-continent collisions of the Himalayan type.

One of the first detailed accounts of a convergent plate boundary model was presented by Anhaeusser (1973a) for the Archean crust in southern Africa with particular reference to the Barberton region. This model begins with the initiation of a subduction zone in an oceanic environment (stage 1, Fig. 10-19). The oceanic crust is equated with the Onverwacht Group and similar ultramafic-mafic successions in the lower parts of greenstone sections. Partial melting of the descending slab (stage 2) produces tholeiites and calc-alkaline volcanic and plutonic rocks in an arc system above the slab. Partial melting of mafic rocks in the upper mantle during this stage also produces tonalite-trondhjemite melts which rise as diapirs and underplate the arcs and the oceanic crust during stage 3. Continued uplift exposes both greenstones and plutonic rocks which are eroded and collect in sedimentary basins (analogous to the Fig Tree and Moodies basins). During stage 4, diapirism

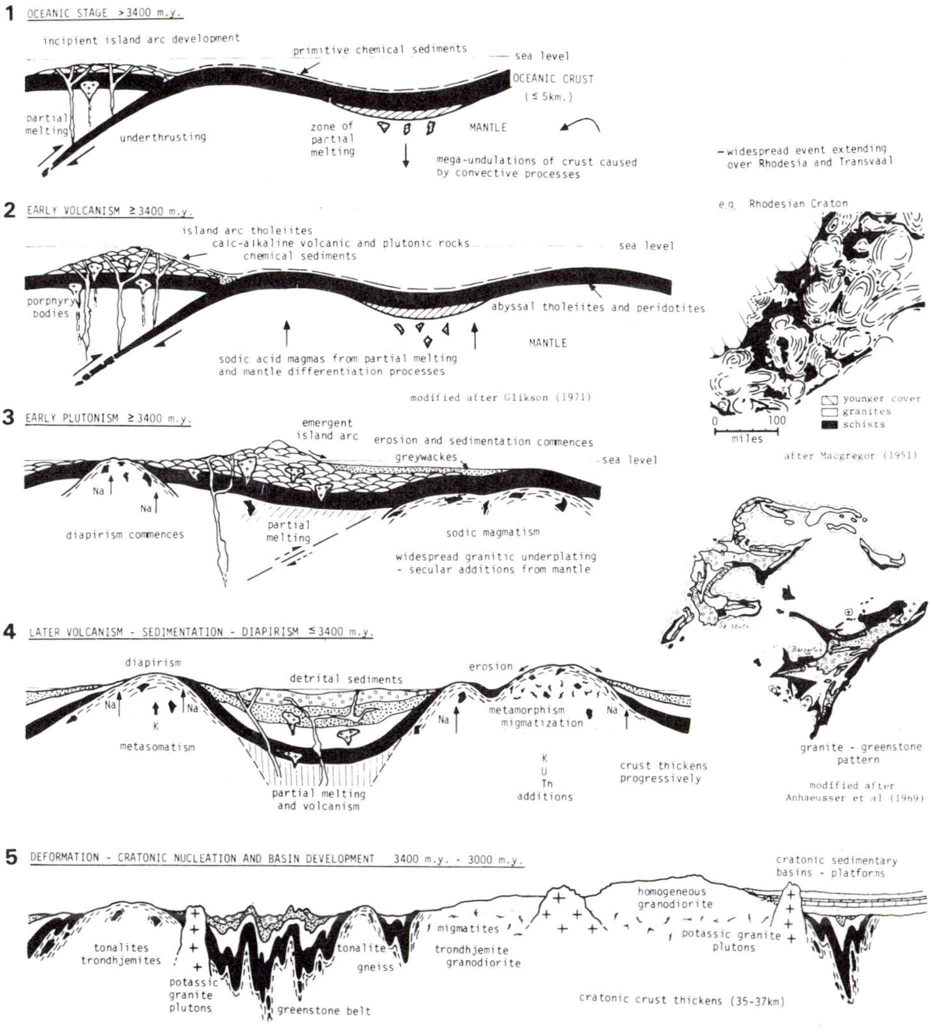

Fig. 10-19. Diagrammatic stages in the Archean crustal development in southern Africa (from Anhaeusser, 1973a).

and sedimentation continue and late volcanism commences. This is followed by the intrusion of post-tectonic granites, regional uplift and erosion, and cratonic sedimentation over much of the Kaapvaal Province (stage 5). One of the major problems with this model is that it does not provide sufficient subducted or downsagged mafic crust to account for the large volumes of tonalite-trondhjemite derived therefrom.

Condie and Harrison (1976) propose a model for the Midlands greenstone succession (Fig. 2-2B) in Rhodesia involving a back-arc basin (Fig. 10-20).

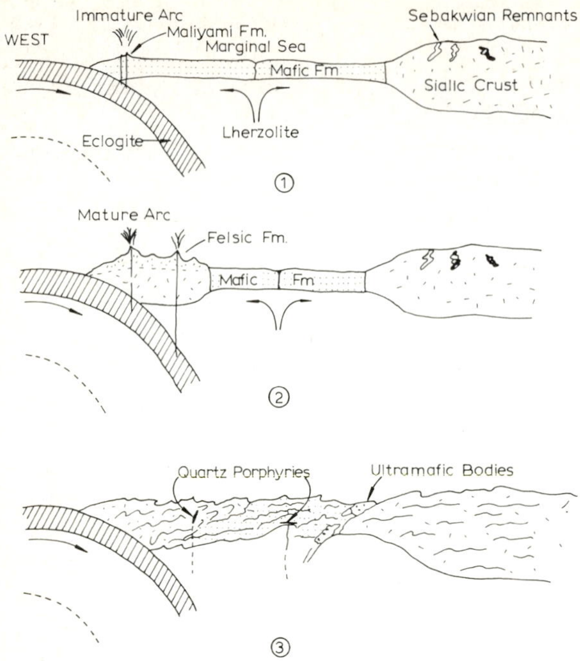

Fig. 10-20. A plate tectonic model for the development of the Midlands greenstone belt, Rhodesia (from Condie and Harrison, 1976).

The authors suggest that the lowest Bulawayan Formation, the Mafic Formation (which is geochemically similar to MORB), is produced at a spreading center in a back-arc basin and that the overlying Maliyami Formation, which includes TH2 and calc-alkaline volcanics, forms in the adjacent immature arc system by partial melting of eclogite in a descending slab (1). Older sialic crust to the east provides only minor detrital input into the back-arc basin. As the arc system matures and thickens primarily by tonalite-trondhjemite underplating, calc-alkaline magmas of the overlying Felsic Formation are formed in the arc (2). Finally, an activation-type orogeny occurs and the arc system and back-arc basin are compressed and welded to the Rhodesian craton (3). During the late stages of deformation, quartz porphyries are emplaced and fragments of the upper mantle are tectonically emplaced as serpentinites. During the latter part of this stage (not illustrated), sediments of the Shamvaian Group are deposited unconformably on the Bulawayan Group, and tonalitic plutons are intruded into the succession.

Tarney et al. (1976) have presented a detailed model equating Archean greenstone belts to back-arc basins, and in particular, the Rocas Verdes basinal succession of late Mesozoic age in southern Chile. The following features are shared in common between this back-arc basin succession and many Archean greenstone successions:

(1) At the margins of the basin, oceanic crust overlies older continental crust.

(2) The overall size, shape, and lithologic association of the Rocas Verdes succession is similar to those characteristic of greenstone belts.

(3) The general style of deformation is similar to that observed in some greenstone belts (i.e., southwest Rhodesia).

(4) Graywacke-argillite is the dominant sedimentary rock type.

(5) The dominant volcanic rocks are tholeiites.

(6) Tonalitic batholiths intrude the Rocas Verdes complex.

Two main lines of evidence suggest that greenstone successions are more analogous to back-arc basin sequences than to open-ocean or arc environments (Windley, 1977). Oceanic crust and lithosphere "self-destruct" by subduction and hence, are likely not to be preserved and volcanic arcs tend to be uplifted and unroofed by erosion such that only their plutonic roots are preserved. The absence of true ophiolite successions in the Archean may not be an obstacle to back-arc basin models in that a higher thermal regime existed in the Archean (Tarney et al., 1976). Because of the additional heat, back-arc spreading may produce more thinning of the crust and more rapid extrusion such that pillowed flows greatly exceed intrusive components such as gabbros and sheeted dikes which characterize Phanerozoic ophiolites. Rutland (1973) has proposed a variant of the back-arc basin model in which a multiple sag or multiple rift system develops over a descending slab and greenstone belts form in the multiple depressions.

Recently, Windley and Smith (1976), Windley (1976), and Tarney and Windley (1977) have suggested that Archean high-grade regions may represent the uplifted and eroded root zones of arcs adjacent to back-arc basins in which they propose greenstone belts form. In particular, the Mesozoic batholithic complexes in the Circum-Pacific area share in common many features with Archean high-grade terranes. Some of the more important are as follows.

(1) The most abundant rock types are tonalite-trondhjemite and granodiorite with high-K granites being uncommon and older plutons in both associations are deformed and foliated.

(2) Some young plutonic complexes (viz., the Southern California batholith) contain inclusions of cratonic sediments perhaps similar to those found in Archean high-grade terranes. However, such sediments are not characteristic of arc systems, in general, and graywackes and related sediments generally dominate in the latter.

(3) Layered mafic igneous complexes occur both as primary intrusives and as remnants in gneissic complexes. These complexes are characterized by an abundance of anorthosite with calcic plagioclase (An_{80}—An_{100}) and cumulus hornblende and appear to have crystallized from wet mafic magmas at high water pressures (Windley and Smith, 1974, 1976).

An idealized sequence of events leading to the development of a granite-

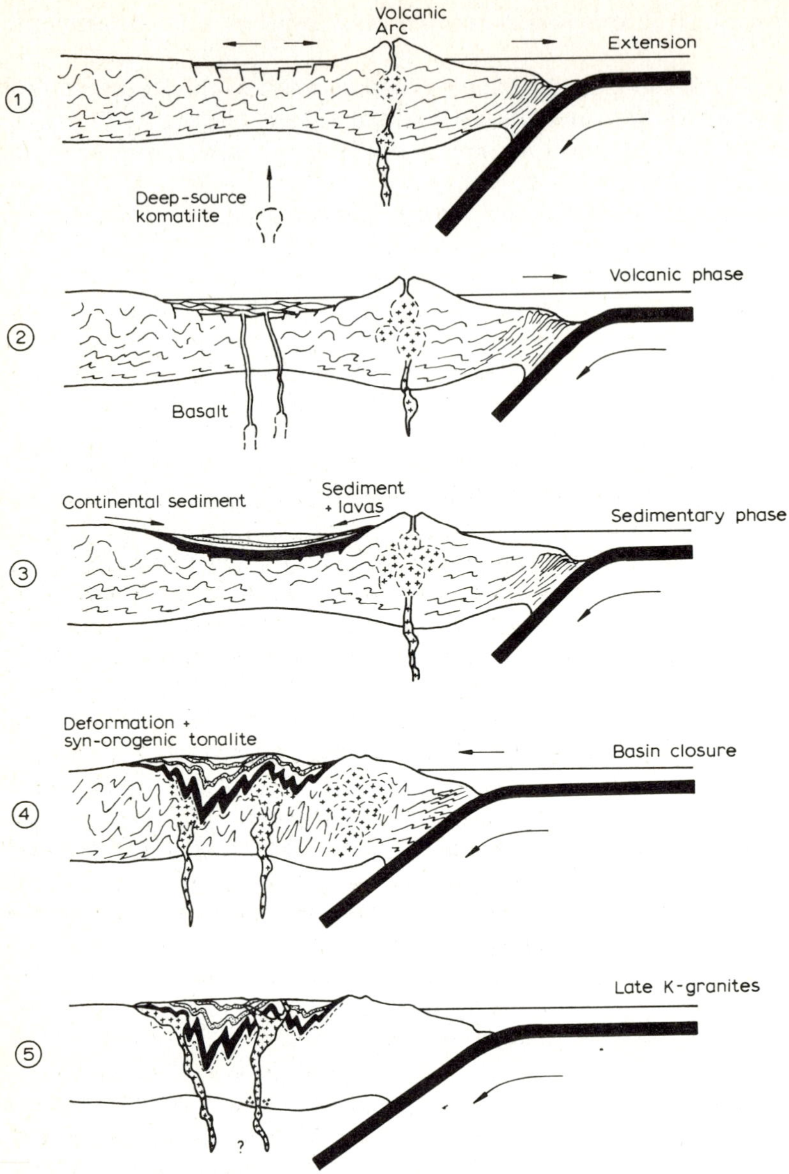

Fig. 10-21. Suggested evolution of an Archean greenstone belt in a back-arc basin tectonic setting (from Tarney et al., 1976).

greenstone terrane (in a back-arc basin) and adjacent high-grade terrane (in an arc) above a descending slab are summarized in Fig. 10-21. The model is based chiefly on the sequence of events in the Rocas Verdes basin (Tarney et al., 1976). The onset of subduction along a continental border produces an

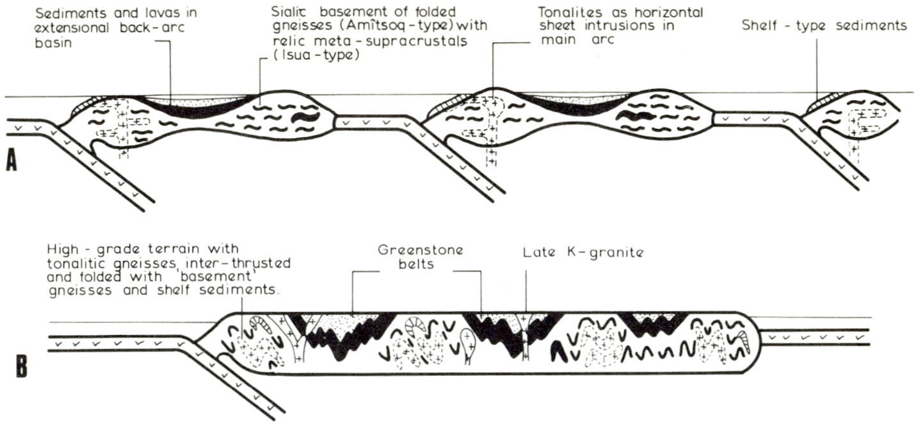

Fig. 10-22. Generalized plate tectonic model for the development of the Archean crust (from Windley, 1977).

arc and adjoining back-arc basin (stage 1). Mafic to felsic magmas are erupted and intruded both into the arc and into the back-arc basin. Ultramafic komatiites may be derived from plumes from greater mantle depths. The arc grows laterally during stage 2 as the subduction zone migrates seaward. Sediments are deposited in the basin from uplift and erosion of the arc and adjoining continent (stage 3). The back-arc basin succession is deformed and intruded with syn-tectonic tonalite-trondhjemite diapirs during an activation-type orogeny (stage 4). Late-stage, post-tectonic granites are derived from melting in the descending slab and/or in the lower crust (stage 5). Preferential uplift of the arc is then necessary to expose the high-grade metamorphic and plutonic rocks, whereas only a small amount of unroofing of the granite-greenstone terrane is allowed for its preservation.

A generalized version of this model and its application to the development of Archean cratons has been presented by Windley (1977) and is shown diagrammatically in Fig. 10-22. In this model, the arcs form before the back-arc basins and both develop on older small tonalitic-trondhjemitic sialic plates. Tonalite-trondhjemite together with mafic magmas are produced from partial melting of descending slabs. Some mafic magmas fractionate under hydrous conditions to form calcic anorthosites and related rocks. This results in thickening arcs that undergo deformation and high-grade metamorphism at depth. Extension and incipient development of back-arc basins occurs during the late stages of arc development. Ultramafic-mafic volcanics may be extruded onto thinned sialic crust in this environment or they may be erupted into a rift-opened basin. Later volcanic stages are characterized by calc-alkaline volcanism as the arc emerges above sea level. Still later sediments are derived from erosion of the arc and uplifted gneissic basement rocks. Closure of the basin produces the synformal structure of the greenstone belts. Growth of the continents is envisioned to occur by accretion at the leading edges of plates.

Arc and back-arc basin models are appealing in terms of the geological similarities of Archean granite-greenstone and high-grade terranes to modern convergent plate boundary rock assemblages developed along continental margins. Unlike the continental rift models, the formulation of the arc models has not relied strongly on geochemical and petrologic data. However, many of the geochemical features of the rift models may be equally well satisfied by the arc models. For instance, a falling geothermal gradient with time could result from decreasing rates of subduction and subduction itself provides a means of replenishing mafic mantle source rocks. There are some signficant problems with the model, however (Groves et al., 1978). First and foremost is the problem presented by the distribution and age relations of greenstone belts within and between Archean provinces. Most greenstone belts and associated granitic rocks in the Superior, Rhodesian, and Yilgarn Provinces are of the same age (2.6—2.7 b.y.) and yet cover a wide area. Does each greenstone belt represent an arc—back-arc basin couple over a descending slab or does an entire province represent multiple back-arc basins related to the same large descending slab? Neither possibility is geologically appealing. It has been suggested that the superbelts in the Superior Province grew by successive arc-arc collisions. However, the similar ages and distinctive differences in lithologic assemblages and metamorphic grade between superbelts are not explained by such a model. Paleomagnetic results seem to worsen the problem by suggesting that adjacent Archean cratons and perhaps large segments of the Archean shields were part of at most a few continents. This would appear to require vast, continuous subduction systems on a continental scale which is highly improbable in terms of heat considerations that suggest that Archean plates were small and rapidly moving.

Another problem with the model is that of obtaining komatiitic magmas which seem to require ultramafic sources at depths of $\gtrsim$ 200 km. Although one may call upon mantle plumes as in the continental rift models to produce such magmas, there is no obvious reason for such plumes to rise in front of descending slabs. Although the model provides sites for both low- and high-grade Archean terranes to form, it is not clear why the arcs should be uplifted and eroded to greater depths than the back-arc basins since both are underplated with significant amounts of sialic crust (Fig. 10-21). The presence of inclusions of cratonic sediments in the root zones of the arcs is also not consistent with modern arcs where graywackes and other immature sediments dominate in and around arcs.

The impact model

Two lines of evidence have led to the possibility that Archean greenstone belts are the result of impact melting. First, the experimental melting studies of Green et al. (1975) suggest high degrees of mantle melting (60—80%) are necessary to produce ultramafic magmas which require depths of melting > 200 km. For magmas at these depths to rise to the earth's surface without fractionating would appear to require a catastrophic event causing partial melting and rapid diapiric ascent of partially melted peridotite (D. H. Green, 1972). Also, the fact that the moon underwent severe bombardment early in its history is consistent with the earth also being bombarded. According to the impact model proposed by D. H. Green (1972), greenstone belts are interpreted as large impact scars analogous to lunar maria which were initially filled with mafic-ultramafic lavas and thereafter evolved into downfolded greenstone belts by further magmatism, sedimentation and deformation.

The proposed sequence of events in this model are summarized in Fig. 10-23. In stage 1, a large impact structure (30—50 km deep) is formed in primitive sialic crust and the floor is covered with a fall-back ejecta layer. Instantaneous unloading beneath the crater produces melting in the upper mantle; the dashed curves indicate the degree of melting. During stage 2, partially melted peridotite rises into the impact ejecta layer with an increase in the degree of melting. Both ultramafic and mafic lavas are extruded and intruded at this time, some marginal slumping and sedimentation is initiated, and the central part of the crater is uplifted. This is followed by collapse of the impact structure (stage 3) perhaps influenced by regional forces. Marginal sialic crust is remobilized and partially melted giving rise to granitic magmas which intrude the margins of the volcanic succession. During stage 4, infolding continues accentuated by diapiric intrusion of remobilized tonalitic basement. Partial remelting of the downfolded root of the greenstone belt gives rise to second-generation magmas in which calc-alkaline types may dominate.

The impact model for greenstone belt formation is beset with difficulties which render it an unlikely mechanism. The most fatal blow for the model relates to timing. The impact structures on the moon formed at ≳ 3.8 b.y., whereas greenstone belts on the earth represent younger ages (chiefly < 3.0 b.y.). It is, however, conceivable, as pointed out by Glikson (1976b) that impact structures filled with mafic-ultramafic lavas did form on the earth prior to 3.8 b.y., but that they were not preserved, or preserved only as minor inclusions in some of the old gneissic complexes. Also not consistent with an impact origin for greenstone belts is the absence of impact textures, structures, and minerals (viz., shatter cones, unique cleavages, high-pressure polymorphs of silica) and the absence of impact breccias in greenstone successions. Considering the low metamorphic grades and excellent degree of preservation of primary volcanic and sedimentary textures in greenstone belts, one would expect impact features also to be preserved.

Although it is unlikely that Archean granite-greenstone terranes were produced by impact phenomena, it is difficult to see how the earth could avoid an impact history prior to 3.8 b.y. where the evidence from the moon and other terrestrial planets (except perhaps for Venus) preserve such a clear

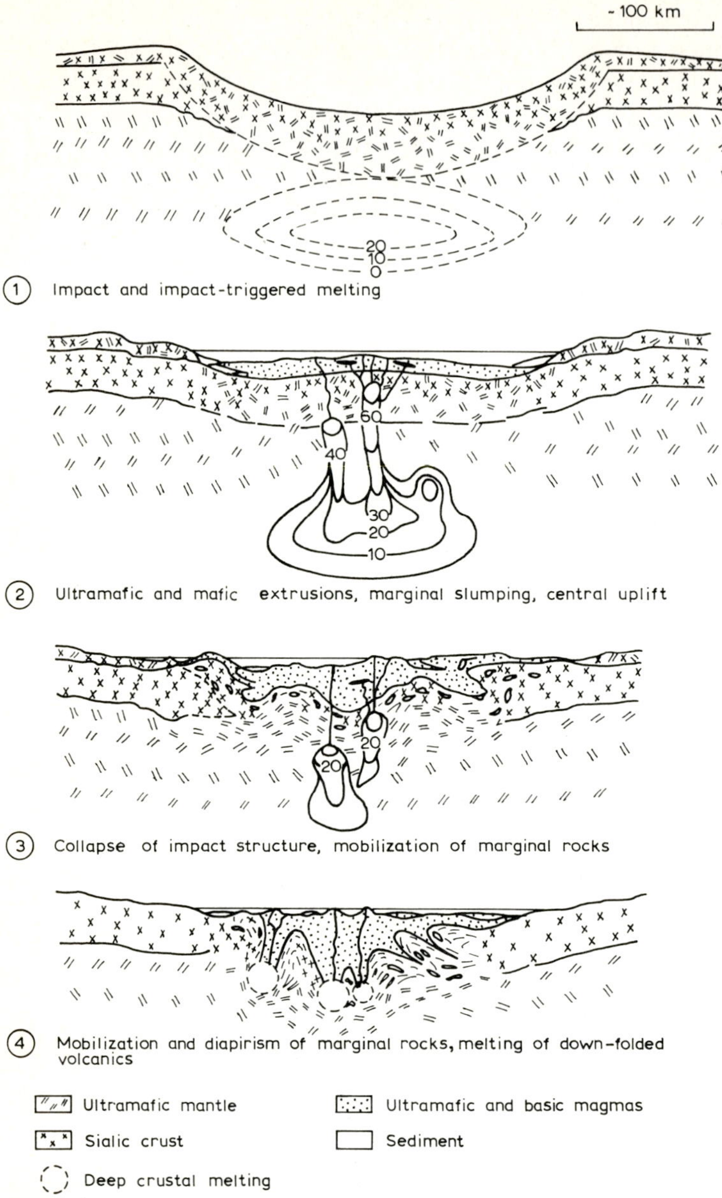

Fig. 10-23. Impact model for the origin of Archean greenstone belts (from D.H. Green, 1972).

record of early impact. The preservation of craters on these bodies may be related either to the absence of a plate tectonic stage during their early history or the cessation of this stage by 4.0 b.y.

TOWARDS AN INTEGRATED MODEL

Constraints

Most models for the origin and development of the Archean crust are constructed from only a few constraints which accounts, in part, for the diversity of published models. In the past decade, a large amount of data relevant to this subject, particularly from geochemical and isotopic studies, has become available as summarized in earlier chapters. Condie (1980a) has recently presented a model for the early history of the earth's crust based on eight likely assumptions. Employing these and other constraints as discussed in this and previous chapters, an integrated model for the origin and history of the Archean crust is now proposed. This model is based on three types of constraints: (1) factual observations from Archean granite-greenstone and high-grade terranes; (2) geochemical and experimental petrologic constraints on magma production and source; and (3) probable assumptions which are deduced from available data. The constraints on magma production and source are given in Chapter 9 and below are lists of the other two groups of constraints as extracted from discussions in previous chapters.

Factual observations

(1) Some greenstone belts were erupted, at least in part, on older sialic crust (Chapter 1).

(2) The oldest sialic rocks in granite-greenstone terranes contain supracrustal inclusions which may represent, in part, fragments of still older greenstone belts (Chapters 1 and 5).

(3) The overall synclinal structure of most greenstone belts (reflecting shortening up to 50%) appears to have developed chiefly in response to the rise of granitic diapirs (Chapter 6).

(4) Some greenstone belts show evidence of sub-horizontal compressive forces during their early stages of development (Chapter 6).

(5) Considered as a whole, granite-greenstone terranes are bimodal in character in that rocks of andesitic composition are uncommon compared to mafic and felsic end members (Chapters 1 and 5).

(6) The Superior Province is characterized by alternating superbelts of sedimentary-plutonic and volcanic-plutonic associations (Chapter 1).

(7) A minimum of two or three periods of greenstone-granite formation are recorded in many Archean provinces, where an individual period (involving magmatism, deformation, metamorphism, uplift and erosion) lasts 50—100 m.y. (Chapters 1 and 2).

(8) Two types of volcanic associations are recognised in greenstone belts: bimodal and calc-alkaline (Chapters 2 and 3). Lower greenstone belts may be either type, but upper are typically calc-alkaline.

(9) Variations in the abundances of bimodal and calc-alkaline types of greenstone belts occur between Archean provinces (Chapter 2).

(10) Ultramafic and mafic volcanic rocks dominate in greenstone successions although becoming less frequent with stratigraphic height as they are displaced with increasing proportions of calc-alkaline or felsic volcanics (Chapters 2 and 3).

(11) Although composed dominantly of subaqueous eruptive units, subaerial volcanics (chiefly pyroclastics) become increasingly important at higher stratigraphic levels in greenstone successions (Chapters 2 and 3).

(12) Members of the komatiite, tholeiite, and calc-alkaline series may be present in the same greenstone succession with their relative importances increasing with stratigraphic height in the order listed (Chapter 3).

(13) Immature sediments (dominantly graywacke-argillite) dominate in the upper parts of many greenstone successions (Chapter 4).

(14) Graywacke-argillite is deposited in a tectonically active basin by slumping and turbidity currents (Chapter 4). Provenance is dominantly volcanic although local plutonic sources may have been important.

(15) Cyclicity occurs in both sediments and volcanics in greenstone successions (Chapters 2, 3, and 4). Individual volcanic cycles show increasing proportions of calc-alkaline or felsic rocks with increasing stratigraphic height.

(16) Gneissic complexes in granite-greenstone terranes are chiefly tonalite-trondhjemite in composition and appear to have been emplaced as plutons (Chapter 5).

(17) High-K granites in granite-greenstone provinces are minor and appear to represent late, post-tectonic intrusions (Chapter 5).

(18) Metamorphic grade in greenstone belts is typically greenschist facies although often increasing to amphibolite facies near contacts with intrusive plutons (Chapter 6).

(19) Increases in metamorphic grade occur in going from the center to the margins of some granite-greenstone provinces (Chapter 6).

(20) In greenstone belts, nickel sulfide deposits are associated with the komatiite series and copper sulfide deposits with the calc-alkaline series (Chapter 7).

(21) The earliest evidences of living organisms occur in cherts and iron formation in Archean greenstone successions (Chapter 8).

(22) Many transition metals are enriched in Archean greenstone volcanics compared to Phanerozoic volcanics of similar bulk composition (Chapter 9).

(23) Most of the sedimentary supracrustal rocks in Archean high-grade terranes appear to represent cratonic sediments (Chapters 1 and 9).

(24) Layered igneous complexes in high-grade terranes are distinct from those in granite-greenstone terranes in that they contain calcic plagioclase and cumulus amphibole (Chapter 1). The high-grade complexes appear to have crystallized with higher water contents than the low-grade complexes.

(25) Deformational style in high-grade terranes reflects principally subhorizontal compressive forces and complex polyphase deformation (Chapters 1 and 6).

(26) Unlike granite-greenstone terranes, some high-grade terranes contain a complete spectrum of rocks of calc-alkaline affinities (Chapter 9).

(27) Metamorphic mineral assemblages in high-grade terranes indicate burial depths up to 30—40 km (Chapter 6).

(28) Many high-grade terranes exhibit initial $^{87}Sr/^{86}Sr$ ratios higher than granite-greenstone terranes (Chapter 9).

(29) Archean magmatism and orogeny was episodic with the major period occurring at 2.6—2.7 b.y. (Chapter 1).

Probable assumptions

(1) The early geothermal gradient in the earth was adiabatic (Chapter 10).

(2) Heat production in the earth has decreased with time such that during the Archean two to three times the present heat was being produced (Chapter 10).

(3) As a logical consequence of increased heat during the early stages of the earth's evolution, it is difficult to avoid convection (Chapter 10); convection cells occur on two scales.

(4) Archean plate tectonics, driven by viscous drag, is characterized by thin, rapidly moving plates (Chapter 10).

(5) The first stable crust was basaltic in composition (Condie, 1980a). Evidence for this comes from experimental studies that indicate mafic magmas segregate from their residual crystals between 20 and 50% melting and geochemical model studies that indicate a mafic source must have been available before sialic crust could form.

(6) The terrestrial planets, including the earth, were subjected to intense surface cratering prior to about 3.8 b.y. (Condie 1980a). Concurrent and subsequent plate tectonic processes have destroyed evidence for such cratering on the earth.

(7) Continental crust, once formed, is not recycled through the mantle to an appreciable extent (Chapter 10).

(8) The early continents grew at convergent plate boundaries by magmatic processes and arc-microcontinent collisions (Condie, 1980a).

(9) Large amounts of mafic crust must have been recycled through the upper mantle to account for the large volumes of tonalite-trondhjemite formed during the Archean (Chapter 9).

(10) At least 50% of the present continental crust had formed by 2.7 b.y. (Chapters 1 and 10).

(11) Archean continents were approximately the same thickness as present-day continents (Chapter 10).

(12) By 2.7 b.y., one supercontinent or at most a few major continents existed on the earth's surface (Chapter 10).

(13) Average continental geotherms beneath Archean continental crust were steeper than most present-day continental geotherms (Chapter 6). Geotherms (and heat flow) beneath Archean oceanic areas were also steeper than those beneath modern oceans.

(14) Archean high-grade and granite-greenstone terranes are not depth equivalents of each other, but must reflect different tectonic settings (Chapters 6 and 10).

(15) Adjacent high-grade and granite-greenstone terranes appear to have evolved simultaneously although in response to differing mantle heat sources (Chapters 6 and 10).

(16) Heat flow model studies indicate that the crust in granite-greenstone terranes is vertically zoned with respect to composition and especially with respect to LIL elements (Chapters 1, 6, and 10). Combined heat flow and heat production relationships suggest the original presence of a thin, upper layer (2—4 km) rich in radiogenic heat-producing nuclides which has been variably removed by later erosion (Jessop and Lewis, 1978).

(17) Archean provinces appear to have been uplifted to their approximate present-day erosion levels within 300—400 m.y. of their stabilization (Watson, 1976b).

(18) Three source rocks are necessary for the production of igneous rocks represented in granite-greenstone terranes: ultramafic, mafic, and intermediate (Chapter 9). Stratigraphic changes in greenstone successions and age relationships of granitic rocks in granite-greenstone terranes indicate that these three source rocks increased in importance with time in a given granite-greenstone episode in the order listed.

(19) Ultramafic komatiite liquids must be extracted before 50% melting is reached (Chapter 9). These and related mafic liquids are probably produced by successive partial melting and extraction of melts in ascending mantle plumes.

(20) Compositional changes with time in a given granite-greenstone episode imply a falling geothermal gradient (Chapter 9).

(21) Volcanic cyclicity necessitates replenishment of fertile magma source rocks during greenstone belt evolution (Chapter 9).

(22) The Archean mantle was heterogeneous despite rapid convection (Chapter 10). It was also less depleted in LIL elements than the modern mantle that serves as a source for MORB (Chapters 9 and 10).

(23) High initial $^{87}Sr/^{86}Sr$ ratios in many Archean high-grade terranes implies either a longer crustal residence time or a less-depleted mantle source than is reflected by most granite-greenstone terranes.

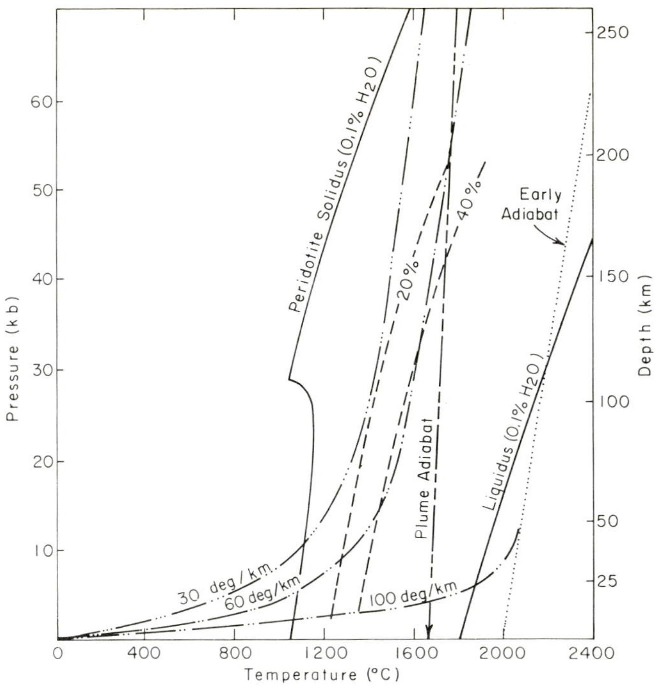

Fig. 10-24. Melting relations of hydrous peridotite (after D.H. Green, 1975). Early adiabatic gradient from Ringwood (1975). Dashed curves are contours of percent melting of peridotite.

A proposed model

The model herein described is modified after the model proposed by Condie (1980a). The earth heats rapidly during the late stages of its formation establishing an adiabatic gradient (Fig. 10-24). Interpretations of Pb and Sr isotopic data suggest that core formation was complete in less than a few hundred million years (Oversby and Ringwood, 1971; Vollmer, 1977; Vidal and Dosso, 1978). Melting of the outer part of the earth may extend to the surface as suggested by Ringwood (1975). Loss of heat by radiation and volatile escape cools the surface region rapidly and a very thin (few kilometers) crust composed chiefly of ultramafic rocks is formed (Fig. 10-25). This ultramafic layer is unstable because its density is greater than that of the underlying melted mantle and it is disrupted by rapid convection in the upper mantle. As cooling continues to gradients of about 60°/km (Fig. 10-24), voluminous tholeiitic magmas segregate and rise to the surface along rifts in the ultramafic crust (Fig. 10-25). The ultramafic crust is broken up and sinks and divergent plate boundaries are established where tholeiitic crust forms (Fig. 10-26). Volatiles escaping during volcanism begin to accumulate into an

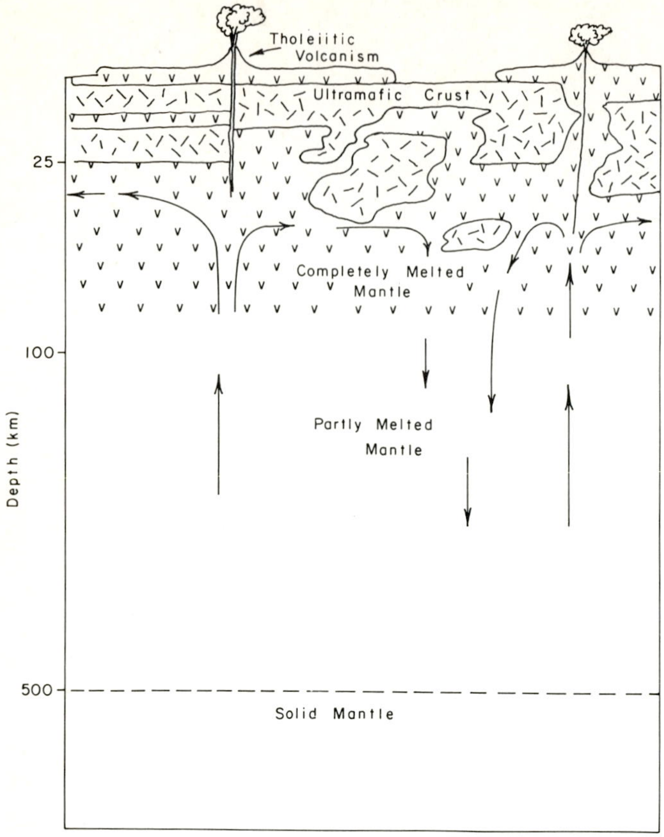

Fig. 10-25. Illustration of an early ultramafic crust being disrupted by basaltic magmatism and mantle convection. Zones of melting correspond to the 100°C/km geotherm in Fig. 10-24.

atmosphere and oceans. Some fractional crystallization may accompany production of the mafic crust producing, by analogy with the moon, gabbroic anorthosites and related rocks.

Melting relations are summarized in Fig. 10-24 for hydrous ultramafic parent rocks. As the adiabatic temperature drops, the mantle crystallizes from base upwards as the geotherm successively intersects shallower levels of the peridotite solidus. Basaltic crust thickens as cooling continues at the surface. Tholeiitic magmas, which require 20—30% melting, separate from ultramafic residue at depths of 30—50 km. Partial melting at depths between 200 and 400 km produces plumes which rise adiabatically to the base of the crust losing one or more batches of basaltic magma on route as proposed in the plume model of Naldrett and Turner (1977) (see previous discussion). Komatiitic and ultramafic magmas are residual liquid-crystal mixtures in

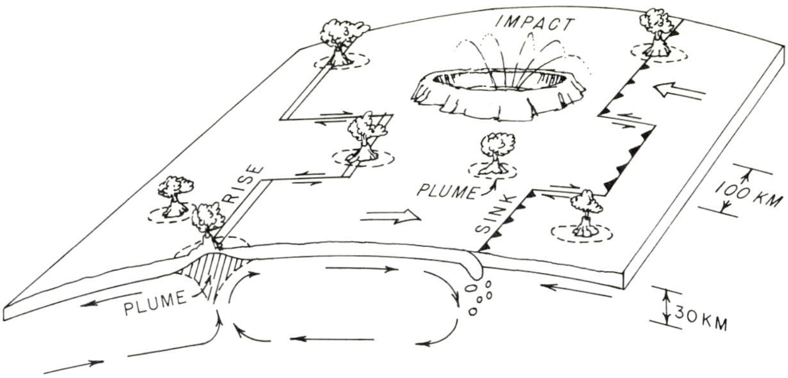

Fig. 10-26. Illustration of the development of an early Archean tholeiitic crust. Tholeiitic magma is generated at rises and in the early stages of plume ascent. Crust is completely melted and recycled in sinks.

plumes after removal of the basaltic magma and form a minor but widespread component in the early mafic crust.

The early mafic crust is produced in a complex network of rift systems, consumed at sinks, and recycled through the mantle. Duffield (1972) has proposed that plate interactions observed in Hawaiian lava lakes are naturally occurring models of global plate tectonics. In these lava lakes, a basaltic crust (< 2 cm thick) is created at ridges and consumed at sinks. The crust presumably overlies magma of the same composition but lower density. This situation may be even more analogous to the tholeiitic crust formed during the early stages of the earth's history because of the higher temperatures and greater spreading rates in the early stages of earth history. Using this model and assuming plates of a few hundred kilometers on a side and velocities of meters per year (McKenzie and Weiss, 1975), the early basaltic crust could be recycled on a time scale of a few hundred thousand years. Temperatures are high enough that such crust is completely remelted in subduction zones.

By analogy with the moon, the earth underwent a period of intense bombardment ending between 3.8 and 3.9 b.y. (Wasserburg et al., 1977). Since the bodies that produced the lunar maria during this time interval did not directly produce melting (Taylor, 1975), it is unlikely that such impacts on the earth resulted in significant melting. In this sense, the early impacts on the earth were probably passive except in promoting crustal disruption and perhaps faster recycling of basaltic crust. The production and growth of the terrestrial crust must have been controlled chiefly by internal processes. This early stage in the earth's history, as diagrammatically shown in Fig. 10-26, is considered to have lasted from the late stages of accretion (~ 4.6 b.y.) to about 4.0 b.y.

At about 4.0 b.y., the geotherms beneath at least some subduction zones decrease sufficiently to allow only partial rather than complete melting of

Fig. 10-27. Diagrammatic illustration of the formation of early Archean andesitic arc systems at convergent plate boundaries.

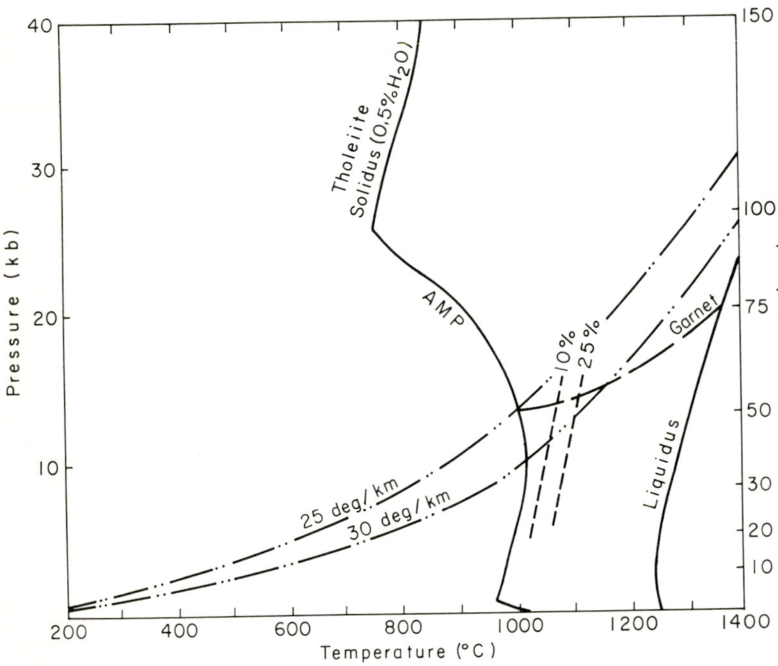

Fig. 10-28. Melting relations of hydrous tholeiite (after Wyllie, 1971) showing estimated contours of percent melting. Garnet stability after Stern, et al. (1975). *AMP* = upper stability limit of amphibole. Dashed curves are contours of percent melting.

descending mafic slabs before magma segregation (Fig. 10-27). This event occurs at different locations of the earth's surface at different times and initially only a few isolated arcs are formed, as represented, for instance, by the Isua greenstone belt in Greenland. Andesitic magmas require about 25% melting of a mafic parent rock under hydrous conditions and tonalitic magmas about 10% (Fig. 10-28). As discussed in Chapter 9, many Archean

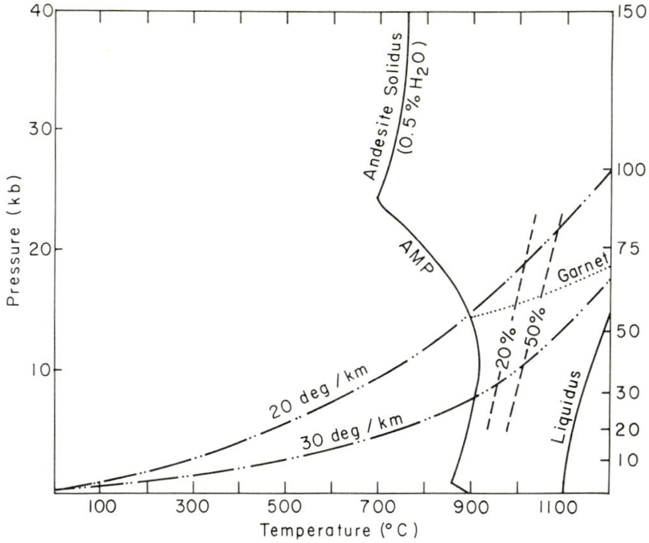

Fig. 10-29. Melting relations of hydrous andesite (after T.H. Green, 1972; Stern et al., 1975). Dashed curves are contours of percent melting. Garnet stability limit is also shown, *AMP* as in Fig. 10-28.

tonalite-trondhjemites and andesites exhibit heavy-REE-depleted patterns indicating the presence of residual garnet and/or amphibole. In systems with only small amounts of water present (such as shown in Fig. 10-28), amphibole is unlikely to form a residual phase because its upper stability limit is close to the solidus. In more water-rich systems, however, it may form a residual phase. It is clear from the figure, that for garnet to be a residual phase, the thermal gradient must be $\leqslant 30°C/km$. If the earliest andesitic-tonalitic crust was also heavy-REE depleted, a similar constraint can be placed on the geothermal gradients beneath this crust. Such gradients are consistent with those deduced from metamorphic mineral assemblages in Archean terranes as discussed in Chapter 6. As cooling continues, the rate of production of new lithosphere at ridges, plume activity, and the average degree of melting of descending mafic slabs all decrease. Thus, tonalite-trondhjemite magmas begin to dominate in the arc system provided water contents of the mafic source rocks are high. Sialic arcs locally thicken to 30—40 km as in Southwest Greenland (Wells, 1976). Andesite and some tonalite-trondhjemite are depressed into the lower crust (together with mafic remnants and some tonalites) and undergo progressive metamorphism and partial melting which leads to a compositionally zoned crust. Melting relations for hydrous andesite are shown in Fig. 10-29. Degrees of melting ranging from 20 to 50% will produce granitic magmas ranging from granodiorite to granite in composition. Because Archean granitic rocks of this composition are generally not observed to have heavy-REE depletion (Chapter 5), geothermal gradients

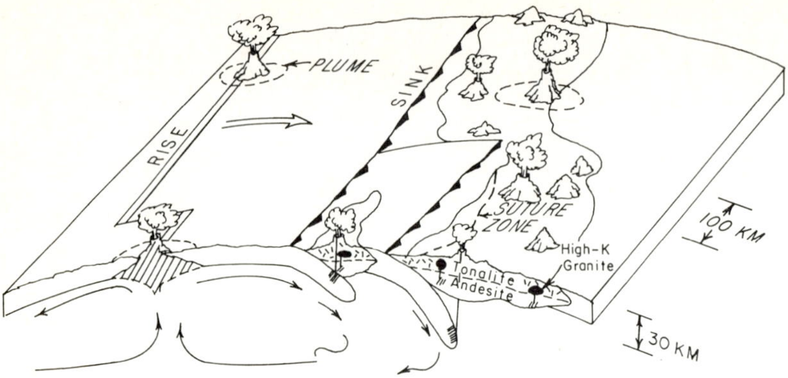

Fig. 10-30. Illustration of the growth of sialic crust by tonalite-trondhjemite plutonism and arc-arc collisions. High-K granitic plutons are produced by partial melting of andesitic rocks in the lower crust.

must not fall below about 25°C/km in order to avoid entering the garnet stability field and/or garnet must be a minor constituent in the andesitic source such that it completely melts. Geochemical model studies discussed in Chapter 9 indicate that high-K granitic magmas cannot readily be produced by partial melting of ultramafic or mafic parents. Granulite in the lower crust with andesitic bulk chemistry, however, can provide an adequate source for these magmas and is considered as the major source in the proposed model. Granitic melts, once produced, rise diapirically into the tonalitic crust (Fig. 10-30). Sialic arcs grow also by arc-arc collisions. Between 3.4 and 3.7 b.y., several continental nuclei form as illustrated on a supercontinent reconstruction in Fig. 10-31.

Any model for the origin and development of the early crust (> 3.8 b.y.) must be lead naturally into processes that give rise to the preserved Archean crust (2.5—3.8 b.y.). A diagrammatic cross-section showing various tectonic settings that may have existed on the earth between 3.8 and 2.5 b.y. is given in Fig. 10-32. In this model, greenstone belts are produced in both back-arc basin and continental rift environments in response to convergent plate boundaries and mantle plumes, respectively. High-grade Archean terranes form in sialic crust above major convective upwellings which have linear patterns in plan view. One feature of the model is that the physical characteristics of the sialic crust differ between high-grade and granite-greenstone terranes. Such crust behaves in a less brittle manner in the former and a more brittle manner in the latter. Thus rifts can form in granite-greenstone crust, but not in high-grade crust where the convective upcurrent heats the crust and causes it to deform plasticly. Cratonic sediments are deposited in basins above high-grade crust which is thinned by necking. Kaapvaal-type cratonic basins, however, are analogous to Phanerozoic trailing continental edge (miogeoclinal) assemblages. Partial melting of the sialic crust in high-grade

Fig. 10-31. Distribution of Archean crust shown on a paleomagnetic reconstruction of a Precambrian supercontinent (after Piper, 1976b). Minimum extent of Archean crust outlined by the heavy dashed line is estimated from the distribution of relict Archean ages in Proterozoic mobile belts.

areas traps basaltic magma beneath the crust by encapsulation (Fyfe, 1974). This cools and crystallizes to amphibolite or eclogite. Uplifted segments of the Archean sialic crust may preserve infolded remnants of older greenstone belts. Continental nuclei continue to grow by sialic collisions in which arcs are added to continental margins and by ~ 2.5 b.y. a supercontinent (Fig. 10-31) or several major continents exist. The alternating superbelts in the Superior Province are interpreted in terms of a somewhat modified version of the model of Goodwin (1977b) previously discussed. Volcanic-plutonic superbelts form over a linear array of plumes which give rise to a multiple-rift system. As the rift systems fill with volcanics (and some sediments) and

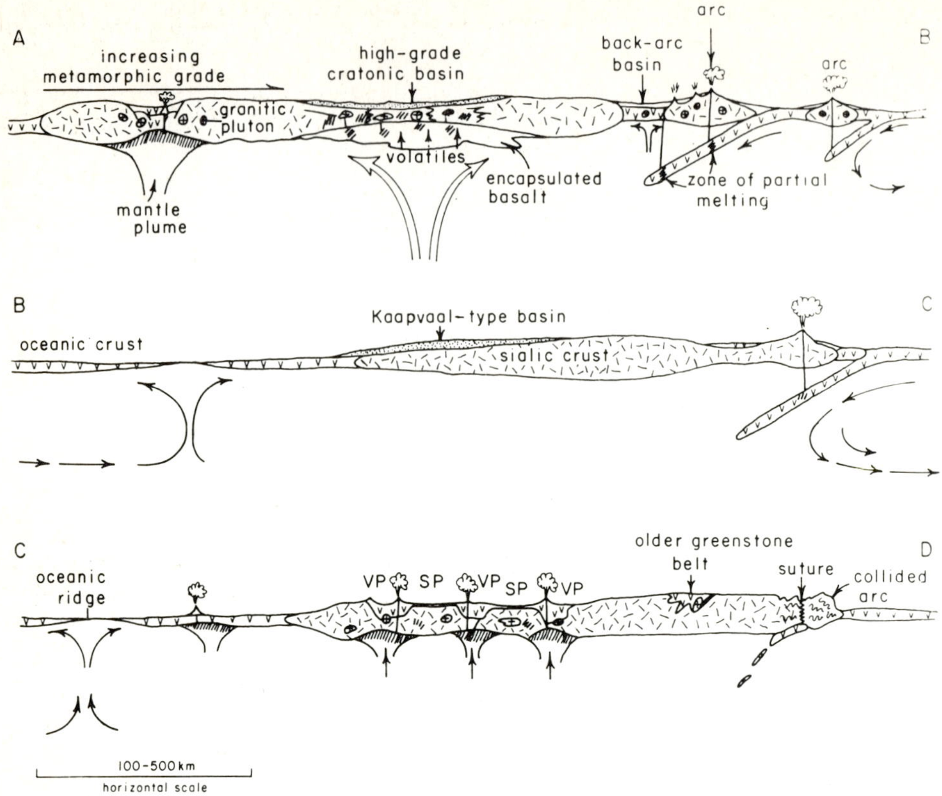

Fig. 10-32. Diagrammatic cross-sections of proposed Archean tectonic settings between 3.8 and 2.5 b.y.

are underplated with tonalite-trondhjemite, they rise isostatically and are unroofed by erosion which carries sediments into intervening basins. Linear arrays of these basins, which subside, undergo partial melting, and are intruded by granites, become the sedimentary-plutonic superbelts.

Model evaluation in terms of constraints

The proposed model accommodates most of the constraints listed above and in Chapter 9. It is worthwhile to evaluate the model in terms of specific constraints. Beginning with the factual observations, the model allows greenstone belts to be deposited, in part (around their margins) or entirely, on sialic crust in both the rift and back-arc tectonic settings (Fig. 10-32). It allows for earlier greenstone belts developed by the same mechanism, inclusions of which are preserved in deformed segments of sialic crust. The synclinal structure of most greenstone belts results from a combination of granitic diapirism and squeezing by the sialic blocks along the margins of rifts or back-arc basins. Those greenstone belts exhibiting nappes and major thrusts may result from convergence of a rift or back-arc basin that had earlier opened in response to a spreading plume or small-scale convective upcurrent in the mantle. The overall bimodal composition of granite-greenstone terranes as a whole as opposed to the continuous calc-alkaline composition of some high-grade terranes may be related to a more hydrous upper mantle beneath the former or to metamorphic processes in high-grade areas that have "homogenized" an originally bimodal crust. The latter alternative is favored for reasons discussed later. The existence of bimodal and calc-alkaline greenstone belts may be due to differing H_2O contents in the mantle on scales ranging from a few hundred (intra-province differences) to tens of thousands of kilometers (inter-province differences). Each cycle of renewed greenstone-granite production would appear to reflect either renewed plume activity or renewed subduction in a crustal segment that had been quiescent for a period of time. During this time, uplift and erosion may have occurred such that later greenstone belts may be deposited unconformably on earlier ones.

The changing proportions of volcanic rocks in greenstone successions with stratigraphic height and the inferred drop in geothermal gradient are thought to reflect subsiding plume or subduction activity which leads to increasingly smaller amounts of melting of mafic and intermediate source rocks with time. It is appropriate at this point to discuss the time dependence of source rocks (constraint 18). The dominance of ultramafic-komatiitic lavas in the lower part of many greenstone successions reflects tapping of liquids from a mantle plume that has undergone earlier stages of partial melting and magma extraction during its ascent. It is suggested, therefore, that greenstone belts with abundant ultramafic-komatiitic components develop in a plume-generated continental rift environment while those containing little if any ultramafic component develop in rift or back-arc basin environments. The increasing importance of mafic source rocks with time in the rift greenstone successions reflects plume subsidence and settling of overlying mafic rocks which invert to amphibolite (or eclogite) and undergo partial melting producing andesite and tonalite-trondhjemite magmas.

As described in the Barberton rift model (Condie and Hunter, 1976), continued plume subsidence leads to downwarping and partial melting of andesitic and tonalitic granulites in the crust to produce high-K granites which are emplaced near the end of granite-greenstone episodes. Similar thickening and sinking of lower crustal rocks in back-arc basin areas could also result in production of these types of magmas. The change from subaqueous to at least partly subaerial eruptions during the latest stages of greenstone belt volcanism would appear to reflect thickening of the crust and some isostatic uplift such that volcanic complexes are elevated, in part, above sea level. Erosion of these volcanic complexes and locally of uplifted granitic plutons gives rise to graywacke-argillite successions which are deposited by slumping and turbidity currents in tectonically active basins. These basins may be in continental rifts or in back-arc areas. Volcanic cyclicity necessitates replenishment of fertile magma source rocks (constraint 21). Such periodic replenishment is accomplished in the rift model by renewed plume activity and in the back-arc model by renewed subduction. In any given greenstone-granite episode (50—100 m.y.), however, the overall plume activity or subduction decreases with time.

The tonalite-trondhjemite plutons that dominate in granite-greenstone terranes are formed by hydrous partial melting of amphibolite (or related rocks) produced by continuing collapse of rift-generated greenstone belts or continuing subduction of oceanic crust (Fig. 10-32). Large volumes of mafic crust are recycled into the mantle by these mechanisms and provide an adequate source for the large amount of tonalite-trondhjemite observed. The large amounts of tonalite-trondhjemite in high-grade areas are produced by partial melting of the mafic parent rocks that are encapsulated as basaltic magma beneath the sialic crust. The sparsity of high-K granites in both types of Archean terranes and their general post-tectonic occurrence is related to prolonged cooling of the crust during a cycle of magmatism and deformation such that only small amounts of melt are produced (by partial melting of intermediate granulite in the lower crust) during the late stages. These granites are generally emplaced after major deformation and thus are typically post-tectonic.

The typical greenschist-facies metamorphic grade of greenstone belts indicates they were not buried very deeply and that heat from underlying plumes did not penetrate to upper crustal levels. The increase in metamorphic grade towards the margins of granite-greenstone provinces (such as the Rhodesian Province) reflects the more intense heat sources beneath adjacent high-grade areas (Fig. 10-32). The association of Ni sulfides with komatiites may be related to the melting relations of sulfides to silicates in the mantle; however, the association of Cu-Zn sulfides with calc-alkaline or bimodal volcanics is not readily explained by this mechanism. The relative enrichment of many transition trace metals in Archean volcanics may also be tied to the sulfide melting relationship since many of these elements follow sulfur (see Chapter 9).

One of the peculiar features of the model is that it provides different tectonic settings for granite-greenstone and high-grade terranes which appear not to represent depth equivalents of the same crust. The more ductile behavior of the sialic crust in high-grade areas provides a relatively stable tectonic setting at shallow depths for cratonic sedimentation while at greater depths polyphase plastic deformation and anatexis are to be expected. Compressive forces, which appear to dominate in high-grade areas, result from either the return flow of convective upwelling or movement of the crust over convective upcurrents in a manner analogous to millipede tectonics of Wynne-Edwards (1976). Adjacent high-grade and granite-greenstone terranes evolve together yet their development is controlled by different mantle heat sources (Fig. 10-32). Layered igneous complexes, requiring large water contents during their crystallization, are derived from intrusive basaltic magmas that undergo fractional crystallization. Clearly, these complexes and mafic dikes must be intruded into high-grade crust during somewhat cooler periods (caused perhaps by depressed mantle upwelling) when the sialic crust behaves more as a brittle substance and is capable of sustaining fractures. The abundant water accompanying the crystallization of high-grade layered complexes is obtained from extensive devolatilization of the mantle over the convective upcurrent. A smaller amount of water liberated into the crust from plumes and descending slabs may account for layered complexes in granite-greenstone terranes fractionating under rather dry conditions following Skaergaard-type trends.

One of the difficult features to explain in the model is the great burial depths (30—40 km) of some high-grade terranes as implied by metamorphic mineral assemblages. Of the possibilities mentioned in an earlier section of this chapter, an original crust 60—80 km thick seems unrealistic in terms of the ductile nature of crust over a mantle upcurrent. Crustal underplating accompanying rapid uplift (300—400 m.y.), however, may account for high-pressure mineral assemblages now at the surface. The underplating with tonalite-trondhjemite magmas could, in fact, be responsible for the rapid uplift in high-grade areas. The tonalite melts would be derived from partial melting of continuously supplied mafic magmas which are encapsulated and crystallize beneath the crust and later are partially melted. Alternatively, the high-grade terranes may have been emplaced at shallow levels by low-angle thrusting during the compressive phases of deformation. It is interesting, also that the convective upcurrent beneath high-grade terranes may introduce relatively undepleted mantle and derivative basaltic melts to the base of the crust. Partial melting of this basaltic material to produce the tonalite-trondhjemite components in high-grade areas may account for the relatively high initial $^{87}Sr/^{86}Sr$ ratios in many high-grade terranes.

Now let us examine some of the geochemical constraints given in Chapter 9. The mantle plumes from which one or more mafic liquids have been extracted are composed largely of refractory minerals depleted in LIL elements and especially light REE. Thus they provide an adequate source for PK (peridotitic komatiite) and BK (basaltic komatiite) groups 2 and 3. One of the problems with the Naldrett and Turner (1977) plume model is that their early mafic melt extractions do not appear to be represented in the lower part of greenstone successions; they should, in fact, precede stratigraphically PK and BK. BK1 and TH1, the latter of which comprises most of the volcanic rocks in greenstone belts, may represent some of these earlier liquid extractions; if so, however, it is difficult to explain why TH1 occurs dominantly at higher stratigraphic levels than PK and BK2 and BK3. Alternatively, BK1 and TH1 may represent partial melts of undepleted lherzolite at shallow mantle depths. In the plume model, this would occur after plume collapse, whereas in the back-arc basin model, such magmas could be produced by partial melting of lherzolite in the mantle wedge overlying the descending slab. TH2, Andesite Types I and II, high-Al_2O_3 tonalite-trondhjemite, and FI felsic volcanics can be related by varying degrees of melting of mafic rocks subsiding into the mantle as plume activity or subduction subside. New, unmelted mafic rocks are supplied by continued settling over plumes or by subduction. As mentioned above, the overall trend from mafic to felsic compositions with time reflects a falling geotherm in response either to a collapsing plume or slowing subduction rate. Granodiorite, high-K granites, and felsic volcanics FII are produced by partial melting of andesite and/or tonalite-trondhjemite granulites in the lower crust as the crust settles over collapsing plumes or slowing descending slabs. Initial strontium isotope ratios in high-K granites indicate the andesitic granulite source in the lower crust can range in age from just earlier than partial melting to several hundred million years older (Chapter 9). Late syenites and related rocks are produced from undepleted lherzolite in the mantle at depths greater than the depths that ascending plumes stop.

Turning now to the probable assumptions not already discussed, we are faced first with the existence of two scales of convection in the earth. The small scale of convection is correlated with oceanic ridges and subduction zones in Figs. 10-30 and 10-32. Although it is tempting to correlate the large scale of convection with upcurrents beneath high-grade mobile belts (Fig. 10-32) as originally suggested by Williams (1977), the sinks for the convective upcurrents do not have an obvious tectonic counterpart in the model. Perhaps some of the subduction zones served as return currents for both large- and small-scale convection, although the steep angle between the two convection systems would seem to preclude this possibility. The episodicity of magmatism in the Archean may be related to changing convective systems and, in particular, the 2.6 to 2.7-b.y. event may reflect the onset of deep-mantle convection. The model, of course, is built around plate tectonics with the plates being driven by viscous drag rather than positive buoyancy of descending slabs. The first stable crust in the model is basaltic in composition and is rapidly recycled through the mantle (Fig. 10-26). Continental crust is formed first in arc systems over descending slabs and it grows by magmatic processes and arc-arc collisions. By 2.5 b.y., a supercontinent or a few major continents are present on the earth's surface; after this time, the supercontinent or continents drift as part of the same plate or plates until major breakup begins at about 1 b.y. Compositionally zoned crust in granite-greenstone terranes, as required by heat flow models, develops in response to progressive increases in metamorphic grade with depth. Rapid uplift (300—400 m.y.) of Archean terranes results from isostatic uplift in response to the crustal thickening by tonalite-trondhjemite intrusion and as mentioned above, by rapid tonalite-trondhjemite underplating of high-grade areas.

Heterogeneity in the composition of the Archean mantle is maintained by incomplete mantle mixing, recycling of mafic rocks into parts of the upper mantle, and introduction of both more-depleted (plumes) and less-depleted (mantle upwellings beneath high-grade areas) mantle material at shallow mantle levels.

REFERENCES

Abelson, P.H., 1966. Chemical events on the primitive earth. *Proc. Natl. Acad. Sci. U.S.A.*, 55: 1365—1372.

Alderman, A.R., 1936. Eclogites from the vicinity of Glenelg, Inverness-shire. *Q. J. Geol. Soc. London*, 92: 488—533.

Allaart, J.H., 1976. The pre-3760 m.y. old supracrustal rocks of the Isua area, central West Greenland and the associated occurrence of quartz-banded ironstone. In: B.F. Windley (Editor), *The Early History of the Earth*. J. Wiley & Sons, New York, N.Y. pp. 177—189.

Allègre, C.J. and Minster, J.F., 1978. Quantitative models of trace element behavior in magmatic processes. *Earth Planet. Sci. Lett.*, 38: 1—25.

Allis, R.G., 1979. A heat production model for stable continental crust. *Tectonophysics*, 57: 151—165.

Allis, R.G. and Garland, G.D., 1977. Heat flow measurements in the Geotraverse and their constraints on crustal models. *Proc. 1977 Geotraverse Conf., Univ. Toronto*, pp. 132—133.

Anders, E., DuFresne, E.R., Hayatsu, R., Cavaille, A., DuFresne, A. and Fitch, F.W., 1964. Contaminated meteorite. *Science*, 146: 1157—1161.

Anders, E., Hayatsu, R. and Studier, M.H., 1973. Organic compounds in meteorites. *Science*, 182: 781—790.

Anderson, C.A. and Silver, L.T., 1979. Yavapai Series — a greenstone belt. *Ariz. Geol. Soc. Digest*, 10: 13—26.

Anhaeusser, C.R., 1971a. The Barberton Mountain Land, South Africa — a guide to the understanding of the Archean geology of Western Australia. *Geol. Soc. Aust., Spec. Publ.*, 3: 103—119.

Anhaeusser, C.R., 1971b. Cyclic volcanicity and sedimentation in the evolutionary development of Archean greenstone belts of shield areas. *Geol. Soc. Aust., Spec. Publ.*, 3: 57—70.

Anhaeusser, C.R., 1973a. The evolution of the early Precambrian crust of southern Africa. *Philos. Trans. R. Soc. London, Ser. A.*, 273: 359—388.

Anhaeusser, C.R., 1973b. The geology and geochemistry of the Archean granites and gneisses of the Johannesburg-Pretoria dome. *Geol. Soc. S. Afr., Spec. Publ.*, 3: 361—386.

Anhaeusser, C.R., 1974. The geology of the Sheba Hills area of the Barberton Mountain Land, South Africa. *Univ. Witwatersrand, Econ. Geol. Res. Unit, Inf. Circ.*, 94.

Anhaeusser, C.R., 1975. Precambrian tectonic environments. *Annu. Rev. Earth Planet. Sci.*, 3: 31—53.

Anhaeusser, C.R., 1976a. Archean metallogeny in southern Africa. *Econ. Geol.*, 71: 16—43.

Anhaeusser, C.R., 1976b. Geological and geochemical investigations of the Roodekrans ultramafic complex and surrounding Archean volcanic rocks, Krugersdorp district. *Univ. Witwatersrand, Econ. Geol. Res. Unit, Inf. Circ.*, 103: 16 pp.

Anhaeusser, C.R., 1976c. The nature and distribution of Archean gold mineralization in southern Africa. *Miner. Sci. Eng.*, 8: 46—84.

Anhaeusser, C.R., 1976d. The nature of crysotile asbestos occurrences in southern

Africa — a review. *Econ. Geol.*, 71: 96—116.

Anhaeusser, C.R., Roering, C., Viljoen, M.J. and Viljoen, R.P., 1968. The Barberton Mountain Land: a model of the elements and evolution of an Archean fold belt. *Geol. Soc. S. Afr., Annex. Vol.*, 71: 225—254.

Anhaeusser, C.R., Mason, R., Viljoen, M.J. and Viljoen, R.P., 1969. A reappraisal of some aspects of Precambrian Shield geology. *Geol. Soc. Am. Bull.*, 80: 2175—2200.

Archibald, N.J., Bettenay, R.A., Binns, R.A., Groves, D.I. and Gunthorpe, R.J., 1978. The evolution of Archean greenstone terrains, eastern Goldfields Province, western Australia. *Precambrian Res.*, 6: 103—131.

Armbrustmacher, T.J., 1977. Geochemistry of Precambrian mafic dikes, central Bighorn Mountains, Wyoming. *Precambrian Res.*, 4: 13—38.

Armstrong, H.S., 1960. Marbles in the Archean of the southern Canadian Shield. *Proc. 21st. Int. Geol. Congr.*, part 9: 7—20.

Armstrong, R.L., 1968. A model for Sr and Pb isotope evolution in a dynamic earth. *Rev. Geophys.*, 6: 175—199.

Armstrong, R.L. and Hein, S.M., 1973. Computer simulation of Pb and Sr isotope evollution of the earth's crust and upper mantle. *Geochim. Cosmochim. Acta*, 37: 1—18.

Arndt, N.T., 1976. Melting relations of ultramafic lavas (komatiites) at one atmosphere and high pressure. *Carnegie Inst. Washington Yearb.*, 75: 555—561.

Arndt, N.T., 1977a. Thick, layered peridotite-gabbro lava flows in Munro Township. *Can J. Earth Sci.*, 14: 2620—2637.

Arndt, N.T., 1977b. Ultrabasic magmas and high-degree melting of the mantle. *Contrib. Mineral. Petrol.*, 64: 205—221.

Arndt, N.T., Naldrett, A.J. and Pyke, D.R., 1977. Komatiitic and iron-rich tholeiitic lavas of Munro Township, northeast Ontario. *J. Petrol.*, 18: 319—369.

Arriens, P.A., 1971. The Archean geochronology of Australia. *Geol. Soc. Aust., Spec. Publ.*, 3: 11—24.

Arth, J.G., 1979. Some trace elements in trondhjemites — their implications to magma genesis and paleotectonic setting. In: F. Barker (Editor), *Trondhjemites and Related Rocks.* Elsevier, Amsterdam, pp. 123—132.

Arth, J.G. and Barker, F., 1976. Rare-earth partitioning between hornblende and dacitic liquid and implications for the genesis of trondhjemitic-tonalitic liquids. *Geology*, 4: 534—536.

Arth, J.G. and Hanson, G.N., 1972. Quartz diorites derived by partial melting of eclogite or amphibolite at mantle depths. *Contrib. Mineral. Petrol.*, 37: 161—174.

Arth, J.G. and Hanson, G.N., 1975. Geochemistry and origin of the early Precambrian crust of northeastern Minnesota. *Geochim. Cosmochim. Acta*, 39: 325—362.

Arth, J.G., Arndt, N.T. and Naldrett, A.J., 1977. Genesis of Archean komatiites from Munro Township, Ontario: trace-element evidence. *Geology*, 5: 590—594.

Arth, J.G., Barker, F., Peterman, Z.E. and Friedman, I., 1978. Geochemistry of the gabbro-diorite-tonalite-trondhjemite suite of southwest Finland and its implications for the origin of tonalitic and trondhjemitic magmas. *J. Petrol.*, 9: 289—316.

Ayres, L.D., 1969. *Early Precambrian Stratigraphy of Part of Lake Superior Provincial Park, Ontario.* Ph.D. Dissertation. Princeton University, Princeton, N.J. 399 pp. (unpublished).

Ayres, L.D., 1974. Geology of the Trout Lakes area. *Ont. Div. Mines, Geol. Rep.*, 113.

Ayres, L.D., 1977. Importance of stratigraphy in early Precambrian volcanic terranes: cyclic volcanism at Setting Net Lake, northwestern Ontario. *Geol. Assoc. Can., Spec. Paper*, 16: 243—264.

Ayres, L.D., 1978. Metamorphism in the Superior Province of northwestern Ontario and its relationship to crustal development. *Geol. Surv. Can., Paper*, 78-10: 25—36.

Baer, A.J., 1977. Speculations on the evolution of the lithosphere. *Precambrian Res.*, 5: 249—260.

Baker, D.R. and Claypool, G.E., 1970. Effects of incipient metamorphism on organic matter in mudrock. *Am. Assoc. Petrol. Geol. Bull.*, 54: 456—468.

Baragar, W.R.A., 1966. Geochemistry of the Yellowknife volcanic rocks. *Can. J. Earth Sci.*, 3: 9—30.

Baragar, W.R.A., 1968. Major-element geochemistry of the Noranda volcanic belt, Quebec-Ontario. *Can. J. Earth Sci.*, 5: 773—790.

Baragar, W.R.A., 1972. Some physical and chemical aspects of Precambrian volcanic belts of the Canadian shield. *Can. Dep. Energy, Mines, Resour., Publ. Earth Phys. Branch*, 42(3): 129—140.

Baragar, W.R.A. and Goodwin, A.M., 1969. Andesites and Archean volcanism of the Canadian Shield. *Oreg. Dep. Geol. Miner. Ind., Bull.*, 65: 121—142.

Baragar, W.R.A. and McGlynn, J.C., 1976. Early Archean basement in the Canadian Shield: a review of the evidence. *Geol. Surv. Can., Paper*, 76—14: 21 pp.

Barghoorn, E.S. and Schopf, J.W., 1966. Microorganisms three billion years old from the Precambrian of South Africa. *Science*, 152: 758—763.

Barker, F., 1979. Trondhjemite: definition, environment and hypotheses of origin. In: F. Barker (Editor), *Trondhjemites, Dacites, and Related Rocks*. Elsevier, Amsterdam, pp. 1—12.

Barker, F. and Arth, J.G., 1976. Generation of trondhjemitic-tonalitic liquids and Archean bimodal trondhjemite-basalt suites. *Geology*, 4: 596—600.

Barker, F. and Friedman, I., 1969. Carbon isotopes in pelites of the Precambrian Uncompahgre Formation, Needle Mountains, Colorado. *Geol. Soc. Am. Bull.*, 80: 1403—1408.

Barker, F. and Peterman, Z.E., 1974. Bimodal tholeiitic-dacitic magmatism and the early Precambrian crust. *Precambrian Res.*, 1: 1—12.

Barker, F., Arth, J.G., Peterman, Z.E. and Friedman, I., 1976. The 1.7- to 1.8-b.y. old trondhjemites of southwestern Colorado and northern New Mexico: geochemistry and depths of genesis. *Geol. Soc. Am. Bull.*, 87: 189—198.

Barker, F., Millard, Jr., H.T., and Lipman. P.W., 1979. Four low-K siliceous rocks of the western U.S.A. In: F. Barker (Editor), *Trondhjemites, Dacites, and Related Rocks*. Elsevier, Amsterdam, pp. 415—433.

Barley, M.E., Dunlop, J.S.R., Glover, J.E. and Groves, D.I., 1979. Sedimentary evidence for an Archean shallow-water volcanic-sedimentary facies, eastern Pilbara Block, Western Australia. *Earth Planet. Sci. Lett.*, 43: 74—84.

Barr, K.G., 1967. Upper mantle structure in Canada from seismic observations using chemical explosions. *Can. J. Earth Sci.*, 4: 961—975.

Barton, Jr., J.M., Fripp, R.E.P. and Ryan, B., 1977. Rb-Sr ages and geological setting of ancient dikes in the Sand River area, Limpopo mobile belt, South Africa, *Nature*, 267: 487—489.

Bartram, G.D. and McCall, G.J.H., 1971. Wall-rock alteration associated with auriferous lodes in the Golden Mile, Kalgoorlie. *Geol. Soc. Aust., Spec. Publ.*, 3: 191—200.

Bayley, R.W., Proctor, P.D. and Condie, K.C., 1973. Geology of the South Pass area, Fremont County, Wyoming. *U.S. Geol. Surv. Prof. Paper*, 793: 39 pp.

Beach, A. and Tarney, J., 1978. Major and trace element patterns established during retrogressive metamorphism of granulite-facies gneisses, N.W. Scotland. *Precambrian Res.*, 7: 325—348.

Beckinsale, R.D., Drury, S.A. and Holt, R.W., 1980. 3, 360-Myr-old genisses from the South Indian craton. *Nature*, 283: 469—470.

Bell, C.K., 1971. Boundary geology, upper Nelson River area, Manitoba and northwestern Ontario. *Geol. Assoc. Can., Spec. Paper*, 9: 11—39.

Bell, K., Blenkinsop, J. and Moore, J.M., 1975. Evidence for a Proterozoic greenstone belt from Snow Lake, Manitoba. *Nature*, 258: 689—701.

Berkner, L.V. and Marshall, L.C., 1965. On the origin and rise of oxygen concentration in the earth's atmosphere. *J. Atmos. Sci.*, 22: 225—261.

Berry, M.J., 1973. Structure of the crust and upper mantle in Canada. *Tectonophysics*, 20: 183—201.

Beswick, A.E. and Soucie, G., 1978. A correction procedure for metasomatism in an Archean greenstone belt. *Precambrian Res.*, 6: 235—248.

Beukes, N.J., 1973. Precambrian iron formations of southern Africa. *Econ. Geol.*, 68: 960—1004.

Bhattacharya, B.K. and Morley, L.W., 1965. The delineation of deep crustal magnetic bodies from total field aeromagnetic anomalies. *J. Geomagn. Geoelectr.*, 17: 237—252.

Bickle, M.J. and Nisbet, E.G., 1977. The chicken or the egg. *Geology*, 5: 68.

Bickle, M.J., Martin, A. and Nisbet, E.G., 1975. Basaltic and peridotitic komatiites and stromatolites above a basal unconformity in the Belingwe greenstone belt, Rhodesia. *Earth Planet. Sci. Lett.*, 27: 155—162.

Bickle, M.J., Hawkesworth, C.J., Martin, A., Nisbet, E.G. and O'Nions, R.K., 1976. Mantle compositions derived from the chemistry of ultramafic lavas. *Nature*, 263: 577—580.

Bickle, M.J., Ford, C.E. and Nisbet, E.G., 1977. The petrogenesis of peridotitic komatiites: evidence from high-pressure melting experiments. *Earth Planet. Sci. Lett.*, 37: 97—106.

Binns, R.A., Gunthorpe, R.J. and Groves, D.I., 1976. Metamorphic patterns and development of greenstone belts in the eastern Yilgarn Block, Western Australia. In: B.F. Windley (Editor), *The Early History of the Earth.* J. Wiley & Sons, New York, N.Y. pp. 303—316.

Birch, F., 1965. Energetics of core formation. *J. Geophys. Res.*, 70: 6217—6221.

Blais, S., Auvray, B., Capdevila, R., Jahn, B.M., Bertrand, J.M. and Hameurt, J., 1978. The Archean greenstone belts of Karelia (eastern Finland) and their komatiitic and tholeiitic series. In: B.F. Windley and S.M. Naqvi (Editors), *Archean Geochemistry.* Elsevier, Amsterdam, pp. 87—108.

Bliss, N.W., 1970. The geology of the country around Gatooma. *Rhod. Geol. Surv. Bull.*, 64.

Bliss, N.W. and Stidolph, P.A., 1969. A review of the Rhodesian basement complex. *Geol. Soc. S. Afr., Spec. Publ.*, 2: 305—333.

Blockley, J.G., 1975. Pilbara Block. *Geol. Surv. W. Aust., Mem.*, 2: 81—92.

Boldy, J., 1968. Geological observations on the Delbridge massive sulfide deposit. *Can. Inst. Min. Metall., Trans.*, 71: 247—256.

Bouma, A.H., 1961. *Sedimentology of Some Flysch Deposits.* Elsevier, Amsterdam, 168 pp.

Boutcher, S.M.A., Edhorn, A.S. and Moorhouse, W.W., 1966. Archean conglomerates and lithic sandstones of Lake Timiskaming, Ontario. *Proc. Geol. Assoc. Can.*, 17: 21—42.

Bowes, D.R., 1976. Archean crustal history in the Baltic Shield. In: B.F. Windley (Editor), *The Early History of the Earth.* J. Wiley & Sons, New York, N.Y., pp. 481—488.

Boyle, R.W., 1959. The geochemistry, origin, and role of CO_2, H_2O, sulfur, and boron in the Yellowknife gold deposits, Northwest Territories, Canada. *Econ. Geol.*, 54: 1506—1524.

Boyle, R.W., 1961. The geology, geochemistry, and origin of the gold deposits of the Yellowknife district. *Geol. Surv. Can., Mem.*, 310: 193 pp.

Boyle. R.W., 1976. Mineralization processes in Archean greenstone and sedimentary belts. *Geol. Surv. Can., Paper*, 75-15.

Brandt, R.T., Gross, G.A., Gruss, H., Semenenko, N.P. and Dorr, J.U.N., 1972. Problems of nomenclature for banded ferruginous-cherty sedimentary rocks and their metamorphic equivalents. *Econ. Geol.*, 67: 682—684.

Breaks, F.W., Bond, W.D. and Stone, D., 1978. Preliminary geological synthesis of the

English River Subprovince. *Ont. Geol. Surv., Misc. Paper,* 72: 55 pp.

Bridgwater, D. and Collerson, K.D., 1976. The major petrological and geochemical characters of the 3600 m.y. Uivak gneisses from Labrador. *Contrib. Mineral. Petrol.,* 54: 43—59.

Bridgwater, D. and Collerson, K.D., 1977. On the origin of early Archean gneisses: a reply. *Contrib. Mineral. Petrol.,* 62: 179—191.

Bridgwater, D., McGregor, V.R. and Myers, J.S., 1974. A horizontal tectonic regime in the Archean of Greenland and its implications for early crustal thickening. *Precambrian Res.,* 1: 179—197.

Bridgwater, D., Keto, L., McGregor, V.R. and Myers, J.S., 1976. Archean gneiss complex of Greenland. In: *Geology of Greenland.* Geological Survey of Greenland, Copenhagen, pp. 19—75.

Bridgwater, D., Collerson, K.D. and Myers, J.S., 1978. The development of the Archean gneiss complex of the North Atlantic region. In: D.H. Tarling (Editor), *Evolution of the Earth's Crust.* Academic Press, London, pp. 19—70.

Brisbin, W.C., 1971. Gravity studies in the Rice Lake—Beresford Lake region, Manitoba. *Manit. Dep. Mines, Nat. Resour., Mines Branch Publ.,* 71-1: 387—396.

Brooks, C. and Hart, S.R., 1974. On the significance of komatiite. *Geology,* 2: 107—110.

Brooks, C., Hart, S.R., Krogh, T.E. and Davis, G.L., 1969. Carbonate contents and $^{87}Sr/^{86}Sr$ ratios of calcites from Archean metavolcanics. *Earth Planet. Sci. Lett.,* 6: 35—38.

Brooks, J. and Muir, M.D., 1971. Morphology and chemistry of the organic insoluble matter from the Onverwacht Series Precambrian chert and the Orgueil and Murray carbonaceous chondrites. *Grana,* 11: 9—14.

Brooks, J. and Shaw, G., 1971. Evidence for life in the oldest known rocks — the Onverwacht cherts, Swaziland System South Africa. *Grana,* 11: 1—8.

Brooks, J., Muir, M.D. and Shaw, G., 1973. Chemistry and morphology of Precambrian microorganisms. *Nature,* 244: 215—217.

Brownell, G.M., 1941. Geology of the Falcon Lake stock, southeastern Manitoba. *Can. Inst. Min. Metall., Trans.,* 44: 230—250.

Buck, P.S., 1975. An early Precambrian caldera in the Favourable Lake metavolcanic-metasedimentary belt, N.W. Ontario. *Univ. Manit., Center Precambrian Stud., Annu. Rep.,* 1975: 108—115.

Buddington, A.F., 1959. Granite emplacement with special reference to North America. *Geol. Soc. Am. Bull.,* 70: 671—747.

Burke, K. and Dewey, J.F., 1972. Orogeny in Africa. In: T.F.J. Dessauvagie and A.J. Whiteman (Editors), *African Geology.* Ibadan University Press, Ibadan, pp. 583—608.

Burke, K. and Dewey, J.F., 1973. An outline of Precambrian plate development. In: D.H. Tarling and S.K. Runcorn (Editors), *Implications of Continental Drift to the Earth Sciences, Vol. 2.* Academic Press, London, pp. 1035—1045.

Burke, K. and Kidd, W.S.F., 1978. Were Archean continental geothermal gradients much steeper than those of today? *Nature,* 272: 240—241.

Burke, K., Dewey, J.F. and Kidd, W.S.F., 1976a. Precambrian paleomagnetic results compatible with contemporary operation of the Wilson cycle. *Tectonophysics,* 33: 287—299.

Burke, K., Dewey, J.F. and Kidd, W.S.F., 1976b. Dominance of horizontal movements, arc and microcontinental collisions during the later per mobile regime. In: B.F. Windley (Editor), *The Early History of the Earth.* J. Wiley & Sons, New York, N.Y., pp. 113—130.

Burke, K., Dewey, J.F. and Kidd, W.S.F., 1977. World distribution of sutures — the sites of former oceans. *Tectonophysics,* 40: 69—99.

Burley, A.J., Evans, R.B., Gillingham, J.M. and Masson Smith, D., 1970. Gravity anomalies in Swaziland. *Geol. Surv. Swaziland, Bull.,* 7: 4—16.

Cahen, L. and Snelling, N.J., 1966. *The Geochronology of Equitorial Africa.* North-Holland, Amsterdam, 195 pp.

Cahen, L., Delhal, J. and Lavreau, J., 1976. The Archean of Equitorial Africa: a review. In: B.F. Windley (Editor), *The Early History of the Earth.* J. Wiley & Sons, New York, N.Y., pp. 489—498.

Calvin, M., 1965. Chemical evolution. *Proc. R. Soc. London, Ser. A,* 288: 441—466.

Cameron, E.M. and Baumann, A., 1972. Carbonate sedimentation during the Archean. *Chem. Geol.,* 10: 17—30.

Campbell, F.H.A., 1971. Stratigraphy and sedimentation of part of the Rice Lake Group, Manitoba. *Manit. Dep. Mines, Nat. Resour., Mines Branch Publ.,* 71-1: 135—188.

Carey, S.W., 1958. A tectonic approach to continental drift. *Symp. Continental Drift, Hobart, Tasmania,* pp. 177—355.

Carey, S.W., 1976. *The Expanding Earth.* Elsevier, Amsterdam, 488 pp.

Carstens, H., 1963. On variolitic structure. *Nor. Geol. Unders., Arb.,* 223: 26—42.

Casella, C.J., 1969. A review of the Precambrian geology of the eastern Beartooth Mountains, Montana and Wyoming. *Geol. Soc. Am., Mem.,* 115: 53—71.

Catanzaro, E.J. and Kulp, J.L., 1964. Discordant zircons from the Little Belt, Beartooth, and Santa Catalina Mountains. *Geochim. Cosmochim. Acta,* 28: 87—124.

Catherall, D., 1973. The Shabani younger granite, Shabani, Rhodesia. *Geol. Soc. S. Afr., Spec. Publ.,* 3: 197—200.

Cavanaugh, M.D. and Seyfert, C.K., 1977. Apparent polar wander paths and the joining of the Superior and Slave Provinces during early Proterozoic time. *Geology,* 5: 207—211.

Cawthorn, R.G., 1975. Degrees of melting in mantle diapirs and the origin of ultrabasic liquids. *Earth Planet. Sci. Lett.,* 23: 113—120.

Cawthorn, R.G. and Strong, D.F., 1975. The petrogenesis of komatiites and related rocks as evidence for a layered upper mantle. *Earth Planet. Sci. Lett.,* 27: 369—375.

Cermak, V. and Jessop, A.M., 1971. Heat flow, heat generation and crustal temperature in the Kapuskasing area of the Canadian Shield. *Tectonophysics,* 11: 287—303.

Cerny, P., 1976. Pegmatite studies. *Univ. Manit., Center Precambrian Stud., Annu. Rep.* 1976: 22—44.

Chappell, B.W. and White, A.J.R., 1970. Further data on an eclogite from the Sittampundi complex, India. *Mineral. Mag.,* 37: 555—560.

Chappell, B.W. and White, A.J.R., 1974. Two contrasting granite types. *Pacific Geol.,* 8: 173—174.

Chinner, G.A. and Sweatman, T.R., 1968. A former association of enstatite and kyanite. *Mineral. Mag.,* 36: 1052—1060.

Chou, C.L., Hamud, A.J., Harris, N.B.W. and Goodwin, A.M., 1977. Geochemistry and origin of the eastern Lake Seul area, English River gneiss belt. *Proc. 1977 Geotraverse Conf., Univ. Toronto,* pp. 80—92.

Christensen, N.I., Frey, F., MacDougall, D., Melson, W.G., Peterson, M.N.A., Thompson, G. and Watkins, N., 1973. Deep-Sea Drilling Project: properties of igneous and metamorphic rocks of the oceanic crust. *Trans. Am. Geophys. Union,* 54: 972—1035.

Church, B.N., 1975. Quantitative classification and chemical comparison of common volcanic rocks. *Geol. Soc. Am. Bull.,* 86: 257—263.

Church, N. and Wilson, H.D.B., 1971. Volcanology of the Wanipigow Lake—Beresford Lake area. *Manit. Dep. Mines, Nat. Resour., Mines Branch Publ.,* 71-1: 127—134.

Church, S.E., 1973. Limit of sediment involvement in the genesis of orogenic volcanic rocks. *Contrib. Mineral. Petrol.,* 39: 17—32.

Clark, S.P., Jr., Turekian, K.K. and Grossman, L., 1972. Model for the early history of the earth. In: E.C. Robertson (Editor), *The Nature of the Solid Earth.* McGraw-Hill, New York, N.Y., pp. 3—18.

Claus, G. and Nagy, B., 1961. A microbiological examination of some carbonaceous chondrites. *Nature*, 192: 594—596.

Clifford, P.M., 1972. Behavior of an Archean batholith. *Can. J. Earth Sci.*, 9: 71—77.

Clifford, T.N., 1974. Review of African granulites and related rocks. *Geol. Soc. Am., Spec. Paper,* 156: 49 pp.

Cloud, P.E., 1968. Pre-metazoan evolution and the origins of the metazoa. In: E.T. Drake (Editor), *Evolution and Environment.* Yale University Press, New Haven, Conn., pp. 1—72.

Cloud, P., 1973. Paleoecological significance of the banded iron-formation. *Econ. Geol.* 68: 1135—1143.

Cloud, P., 1974. Evolution of ecosystems. *Am. Sci.*, 62: 54—66.

Cloud, P. and Morrison, K., 1979. On microbial contaminants, micropseudo-fossils, and the oldest records of life. *Precambrian Res.*, 9: 81—91.

Coats, C.J.A., Clark, L.A., Buchan, R. and Brummer, J.J., 1970. Geology of the Cu-Zn deposits of Stall Lake Mines Ltd., Snow Lake area, Manitoba. *Econ. Geol.*, 65: 970—984.

Coleman, R.G., 1977. *Ophiolites.* Springer-Verlag, Berlin, 229 pp.

Collender, F.D., 1964. The geology of the Cam and Motor mine, southern Rhodesia. In: S.H. Haughton (Editor), *The Geology of Some Ore Deposits in Southern Africa, Vol. 2.* Geological Society of South Africa, pp. 15—27.

Collerson, K.D. and Fryer, B.J., 1978. The role of fluids in formation and subsequent development of early continental crust. *Contrib. Mineral. Petrol.*, 67: 151—167.

Compston, W. and Arriens, P.A., 1968. The Precambrian geochemistry of Australia. *Can. J. Earth Sci.*, 5: 561—583.

Condie, K.C., 1967a. Geochemistry of early Precambrian graywackes from Wyoming. *Geochim. Cosmochim. Acta*, 31: 2135—2149.

Condie, K.C., 1967b. Composition of the ancient North American crust. *Science,* 155: 1013—1015.

Condie, K.C., 1969a. Geologic evolution of the Precambrian rocks in northern Utah and adjacent areas. In: *Guidebook, Northern Utah. Utah Geol. Mineral. Surv. Bull.*, 82: 71—95.

Condie, K.C., 1969b. Petrology and geochemistry of the Laramie batholith and related metamorphic rocks of Precambrian age, eastern Wyoming. *Geol. Soc. Am. Bull.*, 80: 57—82.

Condie, K.C., 1972. A plate-tectonic evolutionary model of the South Pass Archean greenstone belt, southwestern Wyoming. *24th Int. Geol. Congr.*, Sect. 1: 104—112.

Condie, K.C., 1973. Archean magmatism and crustal thickening. *Geol. Soc. Am. Bull.*, 84: 2981—2992.

Condie, K.C., 1975. A mantle plume model for the origin of Archean greenstone belts based on trace element distributions. *Nature*, 258: 413—414.

Condie, K.C., 1976a. *Plate Tectonics and Crustal Evolution.* Pergamon Press, New York, N.Y., 288 pp.

Condie, K.C., 1976b. The Wyoming Archean Province in the western United States. In: B.F. Windley (Editor), *The Early History of the Earth.* J. Wiley & Sons, New York, N.Y., pp. 499—510.

Condie, K.C., 1976c. Trace-element geochemistry of Archean greenstone belts. *Earth-Sci. Rev.*, 12: 393—417.

Condie, K.C., 1978. Geochemistry of Proterozoic granitic plutons from New Mexico, U.S.A. *Chem. Geol.*, 21: 131—149.

Condie, K.C., 1980a. Origin and early development of the earth's crust. *Precambrian Res.*, 11: 183—197.

Condie, K.C., 1980b. Archean andesites. In: R.S. Thorpe (Editor), *Orogenic Andesites and Related Rocks.* J. Wiley & Sons, New York, N.Y. (in press).

Condie, K.C. and Lo, H.H., 1971. Trace element geochemistry of the Louis Lake batholith of early Precambrian age, Wyoming. *Geochim. Cosmochim. Acta,* 35: 1099–1119.

Condie, K.C. and Baragar, W.R.A., 1974. Rare-earth element distributions in volcanic rocks from Archean greenstone belts. *Contrib. Mineral. Petrol.*, 45: 237–246.

Condie, K.C. and Harrison, N.M., 1976. Geochemistry of the Archean Bulawayan Group, Midlands greenstone belt, Rhodesia. *Precambrian Res.*, 3: 253–271.

Condie, K.C. and Hunter, D.R., 1976. Trace element geochemistry of Archean granitic rocks from the Barberton region, South Africa. *Earth Planet. Sci. Lett.*, 29: 389–400.

Condie, K.C. and Moore, J.M., 1977. Geochemistry of Proterozoic volcanic rocks from the Grenville Province, eastern Ontario. *Geol. Assoc. Can., Spec. Paper,* 16: 149–168.

Condie, K.C., Barsky, C.K. and Mueller, P.A., 1969. Geochemistry of Precambrian diabase dikes from Wyoming. *Geochim. Cosmochim. Acta,* 33: 1371–1388.

Condie, K.C., Macke, J.E. and Reimer, T.O., 1970. Petrology and geochemistry of early Precambrian graywackes from the Fig Tree Group, South Africa. *Geol. Soc. Am. Bull.*, 81:2759–2776.

Condie, K.C., Viljoen, M.J. and Kable, E.J.D., 1977. Effects of alteration on element distributions in Archean tholeiites from the Barberton greenstone belt, South Africa. *Contrib. Mineral. Petrol.*, 64: 75–89.

Cooke, D.L. and Moorehouse, W.W., 1968. Timiskaming volcanism in the Kirkland Lake area, Ontario, Canada. *Can. J. Earth Sci.*, 6: 117–132.

Cooper, D.G., 1964. The geology of the Bikita pegmatite. In: S.H. Haughton (Editor), *The Geology of Some Ore Deposits in Southern Africa, Vol. 2.* Geological Society of South Africa, pp. 441–461.

Cooper, J.A., Nesbitt, R.W., Platt, J.P. and Mortimer, G.E., 1978. Crustal development in the Agnew region, western Australia, as shown by Rb/Sr isotopic and geochemical studies. *Precambrian Res.*, 7: 31–59.

Cordani, U.G., Amaral, G. and Kawashita, K., 1973. The Precambrian evolution of South America. *Geol. Rundsch.*, 62: 309–317.

Cotterill, P., 1969. The chromite deposits of Selukwe, Rhodesia. In: H.D.B. Wilson (Editor), *Magmatic Ore Deposits. Econ. Geol. Monogr.*, 4: 154–186.

Coward, M.P., 1976. Archean deformation patterns in southern Africa. *Philos. Trans. R. Soc. London, Ser. A,* 283: 313–331.

Coward, M.P. and James, P.R., 1974. The deformation patterns of two Archean greenstone belts in Rhodesia and Botswana. *Precambrian Res.*, 1: 235–258.

Coward, M.P., Lintern, B.C. and Wright, L.I., 1976a. The pre-cleavage deformation of the sediments and gneisses of the northern part of the Limpopo belt. In: B.F. Windley (Editor), *The Early History of the Earth.* J. Wiley & Sons, New York, N.Y., pp. 323–330.

Coward, M.P., James, P.R. and Wright, L., 1976b. Northern margin of the Limpopo mobile belt, southern Africa. *Geol. Soc. Am. Bull.*, 87: 601–611.

Cox, K.G., 1978. Komatiites and other high-magnesia lavas: some problems. *Philos. Trans. R. Soc. London, Ser. A,* 288: 599–609.

Crawford, A.R., 1969. Reconnaissance Rb-Sr dating of Precambrian rocks in southern Peninsular India. *J. Geol. Soc. India,* 10: 117–166.

Crouse, R.A. and Cerny, P., 1972. The Tanco pegmatite at Bernic Lake, Manitoba, 1. Geology and paragenesis. *Can. Mineral.*, 11: 591–608.

Cummins, W.A., 1962. The graywacke problem. *Geol. J.*, 3: 51–72.

Davidson, A., 1972. Granite studies in the Slave Province. *Geol. Surv. Can., Paper,* 72-1A: 109–115.

Davies, D.N., 1964. The tin deposits of Swaziland. In: S.H. Haughton (Editor), *The Geology of Some Ore Deposits in Southern Africa, Vol. 2.* Geological Society of South Africa, pp. 535–541.

Davies, G.F., 1979. Thickness and thermal history of continental crust and root zones. *Earth Planet. Sci. Lett.*, 44: 231—238.

Davies, J.F. and Luhta, L.E., 1978. An Archean "porphyry-type" disseminated copper deposit, Timmons, Ontario. *Econ. Geol.*, 73: 383—396.

Davies, R.D. and Allsopp, H. L., 1976. Strontium isotopic evidence relating to the evolution of the lower Precambrian granitic crust in Swaziland. *Geology*, 4: 553—556.

Davies, R.D., Allsopp, H.L., Erlank, A.J. and Manton, W.I., 1970. Sr-isotopic studies on various layered mafic intrusions in southern Africa. *Geol. Soc. S. Afr., Spec. Publ.*, 1: 576—593.

Davis, Jr., P.A. and Condie, K.C., 1977. Trace element model studies of Nyanzian greenstone belts, western Kenya. *Geochim. Cosmochim. Acta*, 40: 271—277.

Dawson, K.R., 1966. A comprehensive study of the Preissac-Lacorne batholith, Abitibi County, Quebec. *Geol. Surv. Can., Bull.*, 142: 76 pp.

Dawson, K.R. and Whitten, E.H.T., 1962. The quantitative mineralogical composition and variation of the Lacorne, LaMotte, and Preissac granitic complex, Quebec, Canada. *J. Petrol.*, 3: 1—37.

De Almeida, F.F.M., Amaral, G., Cordani, U.G. and Kawashita, K., 1974. The Precambrian evolution of the South American cratonic margin south of the Amazon River. In: A.E. Nairn and F.G. Stehli (Editors), *The Ocean Basins and Margins, Vol. 1.* Plenum Press, New York, N.Y., pp. 411—446.

Dearnley, R., 1966. Orogenic fold belts and a hypothesis of earth evolution. In: L. Ahrens, F. Press, S.K. Runcorn and H.C. Urey (Editors), *Physics and Chemistry of the Earth, Vol. 7.* Pergamon Press, London. pp. 1—114.

De Laeter, J.R. and Blockley, J.G., 1972. Granite ages within the Pilbara block, Western Australia. *J. Geol. Soc. Aust.*, 19: 363—370.

De Laeter, J.R., Lewis, J.D. and Blockley, J.G., 1975. Granite ages within the Shaw batholith of the Pilbara block. *Geol. Surv. W. Aust., Annu. Rep.*, 1974: 73—79.

De la Hunty, L.E., 1975. Murchison Province. *Geol. Surv. W. Aust., Mem.*, 2: 54—64.

DePaolo, D.J. and Wasserburg, G.J., 1976. Nd isotopic variations and petrogenetic models. *Geophys. Res. Lett.*, 3: 249—252.

Dewey, J.F. and Bird, J.M., 1970. Mountain belts and the new global tectonics. *J. Geophys. Res.*, 75: 2625—2647.

Dickinson, B.B. and Watson, J., 1976. Variations in crustal level and geothermal gradient during the evolution of the Lewisian complex of northwest Scotland. *Precambrian Res.*, 3: 363—374.

Dickinson, W.R. and Luth, W.C., 1971. A model for plate tectonic evolution of mantle layers. *Science*, 174: 400—404.

Dimroth, E., 1975. Paleo-environment of iron-rich sedimentary rocks. *Geol. Rundsch.*, 64: 751—767.

Dimroth, E., 1976. Physical volcanology and sedimentology of the Abitibi greenstone belt, Quebec. *Geol. Surv. Can., Paper*, 76-1B: 107—111.

Dimroth, E., 1977. Models of physical sedimentation of iron formations. *Geosci. Can.*, 4: 23—30.

Dimroth, E. and Kimberley, M.M., 1976. Precambrian atmospheric oxygen: evidence in the sedimentary distributions of carbon, sulfur, uranium, and iron. *Can. J. Earth Sci.*, 13: 1161—1185.

Dimroth, E. and Demarcke, J., 1978. Petrography and mechanism of eruption of the Archean Dalembert tuff, Rouyn-Noranda, Quebec, Canada. *Can. J. Earth Sci.*, 15: 1712—1723.

Dimroth, E. and Lichtblau, A.P., 1978. Oxygen in the Archean ocean: comparison of ferric oxide crusts on Archean and Cenozoic pillow basalts. *Neues Jahrb. Mineral., Abh.*, 133: 1—22.

Dimroth, E., Boivin, P., Goulet, N. and Larouche, M., 1973. Tectonic and volcanological studies in the Rouyn-Noranda area. *Que. Dep. Nat. Resour., Open-file Ms.*, pp. 1—60.

Dimroth, E., Cousineau, P., Leduc, M. and Sanschagrin, Y., 1978. Structure and organization of Archean subaqueous basalt flows, Rouyn-Noranda area, Quebec, Canada. *Can. J. Earth Sci.*, 15: 902—918.

Dimroth, E., Cousineau, P., Leduc, M., Sanschagrin, Y. and Provost, G., 1979. Flow mechanisms of Archean subaqueous basalt and rhyolite flows. *Geol. Surv. Can., Paper*, 79-1A: 207—211.

Divis, A.F., 1977. Isotopic studies on a Precambrian geochronologic boundary, Sierra Madre Mountains, Wyoming. *Geol. Soc. Am. Bull.*, 88: 96—100.

Dodson, M.H., Gledhill, A.R., Shackleton, R.M. and Bell, K., 1975. Age difference between Archean cratons of eastern and southern Africa. *Nature*, 254: 315—318.

Donaldson, C.H., 1974. Olivine crystal types in harrisitic rocks of the Rhum pluton and in Archean spinifex rocks. *Geol. Soc. Am. Bull.*, 85: 1721—1726.

Donaldson, J.A. and Jackson, G.D., 1965. Archean sedimentary rocks of North Spirit Lake area, northwestern Ontario. *Can. J. Earth Sci.*, 2: 622—647.

Donaldson, J.A. and Ojakangas, R.W., 1977. Orthoquartzite pebbles in Archean conglomerate, north Spirit Lake, northwestern Ontario, *Can. J. Earth Sci.*, 14: 1980—1990.

Donn, W.L., Donn, B.D. and Valentine, W.G., 1965. On the early history of the earth. *Geol. Soc. Am. Bull.*, 76: 287—306.

Donnelly, T.H., Lambert, I.B., Oehler, D.Z., Hallberg, J.A., Hudson, D.R., Smith, J.W., Bavinton, D.A. and Golding, L., 1978. A reconnaissance study of stable isotope ratios in Archean rocks from the Yilgarn Block, Western Australia. *J. Geol. Soc. Aust.*, 24: 409—420.

Drury, S.A., 1977. Structures induced by granite diapirs in the Archean greenstone belt at Yellowknife, Canada: implications for Archean geotectonics. *J. Geol.*, 85: 345—358.

Drury, S.A., 1979. Rare-earth and other trace element data bearing on the origin of Archean granitic rocks from Yellowknife, Northwest Territories. *Can. J. Earth Sci.*, 16: 809—815.

Drury, S.A., Naqvi, S.M. and Hussain, S.M., 1978. REE distribution in basaltic anorthosites from the Holenarasipur greenstone belt, Karnataka, south India. In: B.F. Windley and S.M. Naqvi (Editors), *Archean Geochemistry*. Elsevier, Amsterdam, pp. 363—374.

Duffield, W.A., 1972. A naturally occurring model of global plate tectonics. *J. Geophys. Res.*, 77: 2543—2555.

Duke, J.M. and Naldrett, A.J., 1978. A numerical model of the fractionation of olivine and molten sulfide from komatiitic magma. *Earth Planet. Sci. Lett.*, 39: 255—266.

Dunbar, G.J. and McCall, G.J.H., 1971. Archean turbidites and banded ironstones of the Mt. Belches area (western Australia). *Sediment. Geol.*, 5: 93—133.

Dunlop, J.S.R., Muir, M.D., Milne, V.A. and Groves, D.I., 1978. A new microfossil assemblage from the Archean of Western Australia. *Nature*, 274: 676—678.

Dunnet, D., 1969. A technique of finite strain analysis using elliptical particles. *Tectonophysics*, 7: 117—136.

Durney, D.W., 1972. A major unconformity in the Archean, Jones Creek, western Australia. *J. Geol. Soc. Aust.*, 19: 251—259.

Dusanowskyj, T.H., 1977. A gravity study of the Sturgeon Lake metavolcanic-metasedimentary belt. *Proc. 1977 Geotraverse Conf., Univ. Toronto*, p. 2.

Eade, R.E. and Fahrig, W.F., 1971. Geochemical evolutionary trends of continental plates — a preliminary study of the Canadian Shield. *Geol. Surv. Can., Bull.*, 179: 51 pp.

Eade, R.E. and Fahrig, W.F., 1971. Geochemical evolutionary trends of continental abundances of some trace elements in the Canadian Shield. *Geol. Surv. Can., Paper*, 72-46.

Eckelmann, F.D. and Poldervaart, A., 1957. Geologic evolution of the Beartooth Mountains, Montana and Wyoming, 1. Archean history of the Quad Creek area. *Geol. Soc. Am. Bull.*, 68: 1225—1262.

Eglinton, G. and Calvin, M., 1967. Chemical fossils. *Sci. Am.*, 216: 32—41.

Egyed, L., 1956. The change in the earth's dimensions determined from paleogeographic data. *Geofis. Pur. Appl.*, 33: 42—48.

Elbers, F.J., 1973. Greenstones project. In: *Summary of Field Work, 1973. Manit. Mines Branch, Geol. Rep.*, 2/73: 12.

Elder, J.W., 1972. Dynamic control of magmatic processes. *J. Earth Sci.*, Leeds Univ., 8: 287—303.

Elliot, R.B., 1973. Chemistry of gabbro-amphibolite transitions in South Norway, I. *Contrib. Mineral. Petrol.*, 38: 71—79.

Elsasser, W.M., 1963. Early history of the earth. In: J. Geiss and E.D. Goldberg (Editors), *Earth Science and Meteoritics.* North-Holland, Amsterdam. pp. 1—30.

Engel, A.E.J., 1968. The Barberton Mountain Land: clues to the differentiation of the earth. *Geol. Soc. S. Afr., Annex. Vol.*, 71: 255—270.

Engel, A.E.J. and Engel, C.G., 1958. Progressive metamorphism and granitization of the major paragneiss, northwest Adirondak Mountains, New York, 1. Total rock. *Geol. Soc. Am. Bull.*, 69: 1369—1414.

Engel, A.E.J. and Engel, C.G., 1962. Progressive metamorphism of amphibolites, northwest Adirondak Mountains, New York. *Geol. Soc. Am., Buddington Vol.*, pp. 37—82.

Engel, A.E.J. and Kelm, D.L., 1972. Pre-Permian global tectonics: a tectonic test. *Geol. Soc. Am. Bull.*, 83: 2325—2340.

Engel, A.E.J., Nagy, B., Nagy, L.A., Engel, C.G., Kremp, G.O. and Drew, C.M., 1968. Alga-like forms in Onverwacht Series, South Africa: oldest recognized lifelike forms on earth. *Science*, 161: 1005—1008.

Eriksson, K.A., 1977. Tidal deposits from Archean Moodies Group, Barberton Mountain Land, South Africa. *Sediment. Geol.*, 18: 257—281.

Eriksson, K.A., 1979. Marginal marine depositional processes from the Archean Moodies Group, Barberton Mountain Land, South Africa: evidence and significance. *Precambrian Res.*, 8: 153—182.

Ermanovics, I.F., 1971. Granites, granite gneiss, and tectonic variation of the Superior Province in southeastern Manitoba. *Geol. Assoc. Can., Spec. Paper*, 9: 77—81.

Ermanovics, I.F. and Davison, W.L., 1976. The Pikwitonei granulites in relation to the northwestern Superior Province of the Canadian Shield. In: B.F. Windley (Editor), *The early History of the Earth.* J. Wiley & Sons, New York, N.Y., pp. 331—350.

Eskola, P.E., 1948. On the problem of mantled gneiss domes. *Q. J. Geol. Soc. London,* 104: 461—476.

Fahrig, W.F. and Wanless, R.K., 1963. Age and significance of diabase dike swarms of the Canadian Shield. *Nature*, 200: 934—937.

Fairbridge, R.W., 1972. *The Encyclopedia of Geochemistry and Environmental Science, Part IV.* Van Nostrand-Rheinhold, New York, N.Y.

Ferguson, J. and Currie, K.L., 1972. Silicate immiscibility in the ancient basalts of the Barberton Mountain Land, Transvaal. *Nature Phys. Sci.*, 235: 86—89.

Fermor, L.L., 1936. An attempt at the correlation of the ancient schistose formations of Peninsular India. *Geol. Surv. India, Mem.*, 70: 324 pp.

Field, D. and Elliott, R.B., 1974. The chemistry of gabbro-amphibolite transitions in South Norway, II. Trace elements. *Contrib. Mineral. Petrol.*, 47: 63—76.

Findlay, D.J., 1975. Lang Lake, an Archean porphyry copper-molybdenum deposit. *Univ. Manit. Center Precambrian Stud., Annu. Rep.*, 1975: 103—107.

Fox, S.W., 1965. A theory of macromolecular and cellular origins. *Nature.* 205: 328—340.

Franklin, J.M., 1978. Petrochemistry of the Sturgeon Lake volcanic belt. In: I.E.M. Smith and J.G. Williams (Editors), *Proc. 1978 Archean Geochemistry Conference.* University of Toronto Press, Toronto, Ont., pp. 161—180.

Franklin, J.M., Kasarda. J. and Poulsen, K.H., 1975. Petrology and chemistry of the alteration zone of the Mattabi massive sulfide deposit. *Econ. Geol.*, 70: 63—79.

Frarey, M.J. and Roscoe, S.M., 1970. The Huronian Supergroup north of Lake Huron. *Geol. Surv. Can., Paper,* 70-40: 143—157.

Fraser, J.A. and Heywood, W.W. (Editors), 1978. Metamorphism in the Canadian Shield. *Geol. Surv. Can., Paper,* 78-10: 367 pp.

Frey, F.A., Haskin, L.A. and Haskin, M.A., 1971. Rare-earth abundances in some ultramafic rocks. *J. Geophys. Res.*, 76: 2057—2070.

Frey, F.A., Chappell, B.W. and Roy, S.D., 1978. Fractionation of rare-earth elements in the Tuolumne intrusive series, Sierra Nevada batholith, California. *Geology,* 6: 239—242.

Frey, H., 1977. Origin of the earth's ocean basins. *Icarus,* 32: 235—250.

Fripp, R.E.P., 1976a. Gold metallogeny in the Archean of Rhodesia. In: B.F. Windley (Editor), *The Early History of the Earth.* J. Wiley & Sons, New York, N.Y., pp. 455—468.

Fripp, R.E.P., 1976b. Stratabound gold deposits in Archean banded iron formation, Rhodesia. *Econ. Geol.*, 71: 58—75.

Frith, R., Frith, R.A. and Doig, R., 1977. The geochronology of the granitic rocks along the Bear-Slave structural province boundary, northwest Canadian Shield. *Can. J. Earth Sci.*, 14: 1356—1373.

Frith, R.A., Frith, R., Helmsteadt, H., Hill, J. and Leatherbarrow, R., 1974. Geology of the Indin Lake area, District of Mackenzie. *Geol. Surv. Can., Paper,* 74-1A: 165—171.

Fryer, B.J. and Jenner, G.A., 1978. Geochemistry and origin of the Archean Prince Albert Group volcanics, Western Melville Peninsula, Northwest Territories, Canada. *Geochim. Cosmochim. Acta,* 42: 1645—1654.

Fryer, B.J., Kerrich, R., Hutchinson, R.W., Pierce, M.G. and Rogers, D.S., 1979. Archean precious-metal hydrothermal systems, Dome mine, Abitibi greenstone belt, I. *Can. J. Earth Sci.*, 16: 421—439.

Fyfe, W.S., 1973a. The granulite facies, partial melting and the Archean crust. *Philos. Trans. R. Soc. London, Ser. A,* 273: 457—461.

Fyfe, W.S., 1973b. The generation of batholiths. *Tectonophysics,* 17: 273—283.

Fyfe, W.S., 1974. Archean tectonics. *Nature,* 249: 338.

Fyfe, W.S., 1978. The evolution of the earth's crust: modern plate tectonics to ancient hot spot tectonics. *Chem. Geol.*, 23: 89—114.

Fyfe, W.S. and Henley, R.W., 1973. Some thoughts on chemical transport processes with particular reference to gold. *Mineral. Sci. Eng.*, 5: 295—303.

Fyson, W.K., 1975. Fabrics and deformation of Archean metasedimentary rocks — Ross Lake—Gordon Lake area, Slave Province, N.W.T. *Can. J. Earth Sci.*, 12: 765—776.

Fyson, W.K., 1978. Structures induced by granite diapirs in the Archean greenstone belt at Yellowknife, Canada: implications for Archean geotectonics: a discussion. *J. Geol.*, 86: 767—769.

Fyson. W.K., Herd, R.K. and Ermanovics, I.F., 1978. Diapiric structures and regional compression in an Archean greenstone belt, Island Lake, Manitoba. *Can. J. Earth Sci.*, 15: 1817—1825.

Gaál, G., Mikkola, A. and Soderholm, B., 1978. Evolution of the Archean crust in Finland. *Precambrian Res.*, 6: 199—215.

Garrels, R.M., 1960. *Mineral Equilibria.* Harper and Brothers, New York, N.Y. 254 pp.

Garrels, R.M. and MacKenzie, F.T., 1971. *Evolution of Sedimentary Rocks.* Norton, N.Y., 397 pp.

Garrels, R.M. and MacKenzie, F.T., 1974. Chemical history of the oceans deduced from post-depositional changes in sedimentary rocks. In: W.W. Hay (Editor), *Studies in Paleo-Oceanography. Soc. Econ. Paleontol. Mineral., Spec. Publ.*, 20: 193—204.

Garrett, P., 1970. Phanerozoic stromatolites: noncompetitive ecologic restriction by grazing and burrowing animals. *Science*, 169: 171—173.

Gast, P.W., 1972. The chemical composition and structure of the moon. *Moon*, 5: 121—148.

Gastil, G., 1960. The distribution of mineral dates in time and space. *Am. J. Sci.*, 258: 1—35.

Gay, N.C., 1969. The analysis of strain in the Barberton Mountain Land, eastern Transvaal, using deformed pebbles. *J. Geol.*, 77: 377—396.

Gee, R.D., 1979. Structure and tectonic style of the western Australian Shield. *Tectonophysics*, 58: 327—369.

Gelinas, L. and Brooks, C., 1974. Archean quench-texture tholeiites. *Can. J. Earth Sci.*, 11: 324—340.

Gelinas, L., Brooks, C., and Trzcienski, Jr., W.E., 1976. Archean variolites — quenched immmiscible liquids. *Can. J. Earth Sci.*, 13: 210—230.

Gelinas, L., Trzcienski, Jr., W.E. and Brooks, C., 1977a. Archean variolites and the hypothesis of quench immiscible liquids reexamined: a reply to criticisms. *Can. J. Earth Sci.*, 14: 2945—2958.

Gelinas, L., Brooks, C., Perrault, G., Carignan, J., Trudel, P. and Grasso, F., 1977b. Chemostratigraphic divisions within the Abitibi belt, Rouyn-Noranda District, Quebec. *Geol. Assoc. Can., Spec. Paper*, 16: 265—296.

Gerling, E., Kratz, K. and Lobach-Zhuckenko, S., 1968. Precambrian geochronology of the Baltic Shield. *Proc. 23rd Int. Geol. Congr.*, 4: 265—273.

Gewald, H. and Pirajno, F., 1973. The Madziwa batholiths, northeastern Rhodesia. *Geol. Soc. S. Afr., Spec. Publ.*, 3: 111—120.

Gibb, R.A. and Thomas, M.D., 1977. The Thelon Front: a cryptic suture in the Canadian Shield? *Tectonophysics*, 38: 211—222.

Gill, J.E., 1961. The origin of continents. *Proc. R. Soc. Can.*, 55: 103—113.

Gill, R.C.O., 1979. Comparative petrogenesis of Archean and modern low-K tholeiites: a critical review of some geochemical aspects. *Phys. Chem. Earth*, 11: 431—447.

Gilmour, P., 1965. The origin of the massive sulfide mineralization in the Noranda district, Quebec. *Proc. Geol. Assoc. Can.*, 16: 63—81.

Glikson, A.Y., 1970. Geosynclinal evolution and geochemical affinities of early Precambrian systems. *Tectonophysics*, 9: 397—433.

Glikson, A.Y., 1971a. Archean geosynclinal sedimentation near Kalgoorlie, Western Australia. *Geol. Soc. Aust., Spec. Publ.*, 3: 443—460.

Glikson, A.Y., 1971b. Primitive Archean element distribution patterns: chemical evidence and geotectonic significance. *Earth Planet. Sci. Lett.*, 12: 309—320.

Glikson, A.Y., 1972a. Petrology and geochemistry of metamorphosed Archean ophiolites, Kalgoorlie-Coolgardie, Western Australia. *Aust. Bur. Miner. Res. Bull.*, 125: 121—189.

Glikson, A.Y., 1972b. Early Precambrian evidence of a primitive ocean crust and island nuclei of sodic granite. *Geol. Soc. Am. Bull.*, 83: 3323—3344.

Glikson, A.Y., 1976a. Stratigraphy and evolution of primary and secondary greenstones: significance of data from shields of the southern hemisphere. In: B.F. Windley (Editor), *The Early History of the Earth*. J. Wiley & Sons, New York, N.Y., pp. 257—278.

Glikson, A.Y., 1976b. Earliest Precambrian ultramafic-mafic volcanic rocks: Ancient oceanic crust or relic terrestrial maria? *Geology*, 4: 201—206.

Glikson, A.Y., 1976c. Trace element geochemistry and origin of early Precambrian acid igneous series, Barberton Mountain Land, Transvaal. *Geochim. Cosmochim. Acta*, 40: 1261—1280.

Glikson, A.Y., 1977a. Vestiges of a beginning. *Geology*, 5: 68—71.

Glikson, A.Y., 1977b. On the origin of early Archean gneisses. *Contrib. Mineral. Petrol.*, 62: 171—178.

Glikson, A.Y., 1977c. Evidence on the radius of the Precambrian earth. *J. Aust. Geol. Geophys.*, 2: 229—232.

Glikson, A.Y., 1978. On the basement of Canadian greenstone belts. *Geosci. Can.*, 5: 3—12.

Glikson, A.Y., 1979a. Early Precambrian tonalite-trondhjemite sialic nuclei. *Earth-Sci. Rev.*, 15: 1—73.

Glikson, A.Y., 1980. Precambrian sial-sima relations: evidence for earth expansion. *Tectonophysics*, 63: 193—234.

Glikson, A.T., 1979b. The missing Precambrian crust. *Geology*, 7: 449—454.

Glikson, A.Y. and Lambert, I.B., 1973. Relations in space and time between major Precambrian shield units: an interpretation of Western Australia data. *Earth Planet. Sci. Lett.*, 20: 395—403.

Glikson, A.Y. and Lambert, I.B., 1976. Vertical zonation and petrogenesis of the early Precambrian crust in Western Australia. *Tectonophysics*, 30: 55—89.

Glikson, A.Y. and Sheraton, J.W., 1972. Early Precambrian trondhjemitic suites in western Australia and northwestern Scotland and the geochemical evolution of shields. *Earth Planet. Sci. Lett.*, 17: 227—242.

Goldich, S.S., 1972. Geochronology in Minnesota. In: P.K. Sims and G.B. Morey (Editors), *Geology of Minnesota:A Centennial Volume*. Minnesota Geological Survey, St. Paul, Minn., pp. 27—40.

Goldich, S.S. and Hedge, C.E., 1974. 3,800-myr granitic gneiss in southwestern Minnesota. *Nature*, 252: 467—468.

Goldich, S.S., Hedge, C.E. and Stern, T.W., 1970. Age of the Morton and Montevideo gneisses and related rocks, southwestern Minnesota. *Geol. Soc. Am. Bull.*, 81: 3671—3696.

Goldich, S.S., Hanson, G.N., Hallford, C.R. and Mudrey, Jr., M.G., 1972. Early Precambrian rocks in the Saganaga Lake—Northern Light Lake area, Minnesota-Ontario. *Geol. Soc. Am., Mem.*, 135: 150—177.

Goodwin, A.M., 1962. Structure, stratigraphy, and origin of iron formations, Michipicoten area, Ontario, Canada. *Geol. Soc. Am. Bull.*, 73: 561—586.

Goodwin, A.M., 1965. Mineralized volcanic complexes in the Porcupine-Kirkland Lake-Noranda region, Canada. *Econ. Geol.*, 60: 955—971.

Goodwin, A.M., 1966. Archean protocontinental growth and mineralization. *Can. Min. J.*, 87: 57—60.

Goodwin, A.M., 1968a. Evolution of the Canadian Shield. *Proc. Geol. Assoc. Can.*, 19: 1—14.

Goodwin, A.M., 1968b. Archean protocontinental growth and early crustal history of the Canadian Shield. *Proc. 23rd Int. Geol. Congr.*, 1: 69—89.

Goodwin, A.M., 1971. Metallogenic patterns and evolution of the Canadian Shield. *Geol. Soc. Aust., Spec. Publ.*, 3: 157—174.

Goodwin, A.M., 1973. Archean iron-formations and tectonic basins of the Canadian Shield. *Econ. Geol.*, 68: 915—933.

Goodwin, A.M., 1974. Precambrian belts, plumes, and shield development. *Am. J. Sci.*, 274: 987—1028.

Goodwin, A.M., 1976. Giant impacting and the development of the continental crust. In: B.F. Windley (Editor), *The Early History of the Earth*. J. Wiley & Sons, New York, N.Y., pp. 77—95.

Goodwin, A.M., 1977a. Archean volcanism in Superior Province, Canadian Shield. *Geol. Assoc. Can., Spec. Paper*, 16: 205—241.

Goodwin, A.M., 1977b. Archean basin-craton complexes and the growth of Precambrian shields. *Can. J. Earth Sci.*, 14: 2737—2759.

Goodwin, A.M., 1978. Archean crust in the Superior Geotraverse area: Geologic overview. In: I.E.M. Smith and J.G. Williams (Editors), *Proc. 1978 Archean Geochemistry Conference*. University of Toronto Press, Toronto, Ont., pp. 73—106.

Goodwin, A.M., 1979. Archean volcanic studies in the Timmons-Kirkland Lake-Noranda region of Ontario and Quebec. *Geol. Surv. Can., Bull.*, 278.

Goodwin, A.M. and Ridler, R.H., 1970. The Abitibi orogenic belt. *Geol. Surv. Can., Paper*, 70-40: 1—30.

Goodwin, A.M. and Shklanka, R., 1967. Archean volcano-tectonic basins. *Can. J. Earth Sci.*, 4: 777—795.

Goodwin, A.M. et al., 1972. The Superior Province. *Geol. Assoc. Can., Spec. Paper*, 11: 527—624.

Goodwin, A.M., Monster, J. and Thode, H.G., 1976. Carbon and sulfur isotope abundances in Archean iron formations and early Precambrian life. *Econ. Geol.*, 71: 870—891.

Gordanier, W.D., 1976. The depositional environment of Formation K, Favourable Lake volcanic complex, northwestern Ontario. *Univ. Manit., Center Precambrian Stud., Annu. Rep.*, 1976: 73—84.

Gorman, B.E., Pearce, T.H. and Birkett, T.C., 1978. On the structure of Archean greenstone belts. *Precambrian Res.*, 6: 23—41.

Gower, C.F., Westerman, C.J. and Clifford, P.M., 1978. The tectonic evolution of the western English River subprovince, northeast Ontario. In: I.E.M. Smith and J.G. Williams (Editors), *Proc. 1978 Archean Geochemistry Conference*. University of Toronto Press, Toronto, Ont., pp. 344—348.

Grant, J.A., 1972. Minnesota River Valley, southwestern Minnesota. In: P.K. Sims and G.B. Morey (Editors), *Geology of Minnesota: A Centennial Volume*. Minnesota Geological Survey, Minn., pp. 177—198.

Grant, N.K., 1973. Orogeny and reactivation to the west and southeast of the West African craton. In: A.E.M. Nairn and F.G. Stehli (Editors), *The Ocean Basins and Margins*. Plenum Press, New York, N.Y., pp. 447—492.

Green, D.C. and Baadsgaard, H., 1971. Temporal evolution and petrogenesis of an Archean crustal segment at Yellowknife, N.W.T., Canada. *J. Petrol.*, 12: 177—217.

Green, D.H., 1972. Archean greenstone belts may include terrestrial equivalents of lunar maria? *Earth Planet. Sci. Lett.*, 15: 263—270.

Green, D.H., 1975. Genesis of Archean peridotitic magmas and constraints on Archean geothermal gradients and tectonics. *Geology*, 3: 15—18.

Green, D.H., Nicholls, I.A., Viljoen, M.J. and Viljoen, R.P., 1975. Experimental demonstration of the existence of peridotitic liquids in earliest Archean magmatism. *Geology*, 3: 11—15.

Green, J.C., 1972. Ultramafic rocks in Vermilion district. In: P.K. Sims and G.B. Morey (Editors), *Geology of Minnesota:A Centennial Volume*. Minnesota Geological Survey, St. Paul, Minn., pp. 76—78.

Green, N.L., 1975. Archean glomeroporphyritic basalts. *Can. J. Earth Sci.*, 12: 1770—1784.

Green, T.H., 1972. Crystallization of calc-alkaline andesite under controlled high-pressure hydrous conditions. *Contrib. Mineral. Petrol.*, 34: 150—166.

Green, T.H. and Ringwood, A.E., 1968. Genesis of the calc-alkaline rock suite. *Contrib. Mineral. Petrol.*, 18: 105—162.

Green, T.H., Brunfelt, A.O. and Heier, K.S., 1969. Rare earth element distribution and associated high grade metamorphic rocks, Lofoten-Vestereaalen, Norway. *Earth Planet. Sci. Lett.*, 7: 93—98.

Gross, C.A., 1965. The geology of iron deposits in Canada: 1. General geology and evaluation of iron deposits. *Geol. Surv. Can., Econ. Geol. Rep.*, 22: 181 pp.

Groves, D.I., Archibald, N.J., Bettenay, L.F. and Binns, R.A., 1978. Greenstone belts as ancient marginal basins or ensialic rift zones. *Nature*, 273: 460—461.

Grubb, P.L.C., 1973. Paragenesis of spodumene and other lithium minerals in some Rhodesian pegmatites. In: L.A. Lister (Editor), *Symposium on Granite, Gneisses and Related Rocks. Geol. Soc. S. Afr., Spec. Publ.*, 3: 201—216.

Gurbuz, B., 1970. A study of the earth's crust and upper mantle using travel times and spectrum characteristics of body waves. *Bull. Seismol. Assoc. Am.*, 60: 1921—1935.

Hajnal, Z. and McClure, J.E., 1977. Seismic investigation over a segment of the Nelson River gravity trend in southeastern Saskatchewan. *J. Geophys. Res.*, 82: 4879—4892.

Hall, D.H., 1968. Regional magnetic anomalies, magnetic units, and crustal structure in the Kenora district of Ontario. *Can. J. Earth Sci.*, 5: 1277—1294.

Hall, D.H. and Hajnal, Z., 1973. Deep seismic crustal studies in Manitoba. *Bull. Seismol. Soc. Am.*, 63: 885—910.

Hall, H.T. and Murthy, V.R., 1971. The early chemical history of the earth: some critical elemental fractionations. *Earth Planet. Sci. Lett.*, 11: 239—244.

Hallberg, J.A., 1972. Geochemistry of Archean volcanic belts in the Eastern Goldfields region of Western Australia. *J. Petrol.*, 13: 45—56.

Hallberg, J.A., Johnston, C. and Bye, S.M., 1976. The Archean Marda igneous complex, Western Australia. *Precambrian Res.*, 3: 111—136.

Halls, H.C., 1978. The structural relationship between Archean granite-greenstone terranes and late Archean mafic dikes. *Can. J. Earth Sci.*, 15: 1665—1668.

Hamilton, P.J., 1977. Sr isotope and trace element studies of the Great Dyke and Bushveld mafic phase and their relation to early Proterozoic magma genesis in southern Africa. *J. Petrol.*, 18: 24—52.

Hamilton, P.J., Evensen, N.M., O'Nions, R.K. and Tarney, J., 1979. Sm-Nd systematics of Lewisian gneisses: implications for the origin of granulites. *Nature*, 277: 25—28.

Hamilton, P.J., O'Nions, R.K., Evensen, N.M., Bridgwater, D. and Allaart, J.H., 1978. Sm-Nd isotopic investigations of Isua supracrustals and implications for mantle evolution. *Nature*, 272: 41—43.

Han, J. and Calvin, M., 1969. Occurrence of fatty acids and aliphatic hydrocarbons in a 3.4-b.y.-old sediment. *Nature*, 224: 576—577.

Hancock, W., Ramsden, A.R., Taylor, G.F. and Wilmshurst, J.R., 1971. Some ultramafic rocks of the Spargoville area, western Australia. *Geol. Soc. Aust., Spec. Publ.*, 3: 269—280.

Hand, B.M. and Emery, K.O., 1964. Turbidites and topography of the north end of the San Diego trough, California. *J. Geol.*, 72: 526—542.

Hanson, G.N., 1972. Saganaga batholith. In: P.K. Sims and G.B. Morey (Editors), *Geology of Minnesota: A Centennial Volume.* Minnesota Geological Survey, St. Paul, Minn., pp. 102—107.

Hanson, G.N., 1978. The application of trace elements to the petrogenesis of igneous rocks of granitic composition. *Earth Planet. Sci. Lett.*, 38: 26—43.

Hanson, G.N. and Goldich, S.S., 1972. Early Precambrian rocks in the Saganaga Lake—Northern Light Lake area, Minnesota-Ontario, II. Petrogenesis. *Geol. Soc. Am., Mem.*, 135: 179—192.

Hanson, G.N. and Himmelberg, G.R., 1967. Ages of mafic dikes near Granite Falls, Minnesota. *Geol. Soc. Am. Bull.*, 78: 1429—1432.

Hargraves, R.B., 1976. Precambrian geologic history. *Science*, 193: 363—371.

Hargraves, R.B., 1978. Punctuated evolution of tectonic style. *Nature*, 276: 459—461.

Hargreaves, R., 1975. Basalt flow morphology at Utik Lake, Manitoba. *Univ. Manit. Center Precambrian Stud., Annu. Rep. 1975*, pp. 125—132.

Harris, N.B.W., 1976. Evolution of the eastern Lac Seul region of the English River Gneiss belt. *Geotraverse Workshop, Univ. Toronto*, pp. 1—13.

Harris, N.B.W. and Goodwin, A.M., 1976. Archean rocks from the eastern Lac Seul region of the English River Gneiss belt, northwestern Ontario, 1. Petrology, chemistry and metamorphism. *Can. J. Earth Sci.*, 13: 1202—1211.

Harrison, N.M., 1969. The geology of the country around Fort Rixon and Shangani. *Rhod. Geol. Surv., Bull.*, 61: 152 pp.

Harrison, N.M., 1970. The geology of the country around QueQue. *Rhod. Geol. Surv., Bull.*, 67.

Hart. S.R., 1969. Isotope geochemistry of crust-mantle processes. *Am. Geophys. Union Geophys. Monogr.*, 13: 58—62.

Hart, S.R. and Brooks, C., 1977. The geochemistry and evolution of early Precambrian mantle. *Contrib. Mineral. Petrol.*, 61: 109—128.

Hart, S.R., Davis, G.L., Brooks, C. and Krogh, T.E., 1970a. The K, Rb, Cs, and Sr geochemistry of Archean metavolcanics. *Carnegie Inst. Washington Yearb.*, 68: 429—434.

Hart, S.R., Brooks, C., Krogh, T.E., Davis, G.L. and Nava, D., 1970b. Ancient and modern volcanic rocks: a trace element model. *Earth Planet. Sci. Lett.*, 10: 17—28.

Hart, S.R., Erlank, A.J. and Kable, E.J.D., 1974. Sea floor basalt alteration: some chemical and Sr isotopic efforts. *Contrib. Mineral. Petrol.*, 44: 219—230.

Haughton, S.H., 1963. *The Stratigraphic History of Africa South of the Sahara.* Oliver and Boyd, Edinburgh, 365 pp.

Hawkesworth, C.J. and O'Nions, R.K., 1977. The petrogenesis of some Archean volcanic rocks from southern Africa. *J. Petrol.*, 18: 487—520.

Hawkesworth, C.J., Moorbath, S., O'Nions, R.K. and Wilson, J.F., 1975. Age relationships between greenstone belts and granites in the Rhodesian Archean craton. *Earth Planet. Sci. Lett.*, 25: 251—262.

Hedge, C.E., Marvin, R.F. and Naesser, C.W., 1975. Age provinces in the basement rocks of Liberia. *J. Res. U.S. Geol. Surv.*, 3: 425—429.

Heimlich, R.A., 1965. Petrology of the Flora Lake stock, Lake of the Woods region, Ontario, Canada. *Geol. Soc. Am. Bull.*, 76:1—26.

Heimlich, R.A., 1966. The Hope Lake stock, Lake of the Woods region, Ontario. *Can. Mineral.*, 8: 620—637.

Heimlich, R.A., 1969. Reconnaissance petrology of Precambrian rocks in the Bighorn Mountains, Wyoming. *Contrib. Geol.*, 8: 47—61.

Heimlich, R.A., 1971. Chemical data for major Precambrian rock types, Bighorn Mountains, Wyoming. *Contrib. Geol.*, 10: 131—140.

Heimlich, R.A., Gallagher, G.L. and Shatweel, L.B., 1974. Quantitative petrography of mafic dikes from the central Bighorn Mountains, Wyoming. *Geol. Mag.*, 111: 97—106.

Heinrichs, T.K. and Reimer, T.O., 1977. A sedimentary barite deposit from the Archean Fig Tree Group of the Barberton Mountain Land (S. Africa). *Econ. Geol.*, 72: 1426—1441.

Henderson, J.B., 1970. Stratigraphy of the Archean Yellowknife Supergroup, Yellowknife Bay—Prosperous Lake area, District of Mackenzie. *Geol. Surv. Can., Paper*, 70-26.

Henderson, J.B., 1972. Sedimentology of Archean turbidites at Yellowknife, Northwest Territories. *Can. J. Earth Sci.*, 9: 882—902.

Henderson, J.B., 1975a. Sedimentology of the Archean Yellowknife Supergroup at Yellowknife, District of MacKenzie. *Geol. Surv. Can., Bull.*, 246: 60 pp.

Henderson, J.B., 1975b. Archean stromatolites in the northern Slave Province. Northwest Territories, Canada. *Can. J. Earth Sci.*, 12: 1619—1630.

Henderson, J.F., 1943. Structure and metamorphism of early Precambrian rocks between Gordon and Great Slave Lakes, Northwest Territories. *Am. J. Sci.*, 241: 430—446.

Henderson, J.F. and Brown, I.C., 1966. Geology and structure of the Yellowknife greenstone belt, District of MacKenzie. *Geol. Surv. Can., Bull.*, 141: 87 pp.

Hendry, N.W., 1951. Chrysotile asbestos in Munro and Beatty townships, Ontario. *Can.*

Inst. Min. Metall., Trans., 54: 28—35.

Hensen, B.J. and Green, D.H., 1973. Experimental study of the stability of cordierite and garnet in pelitic compositions at high pressures and temperatures. *Contrib. Mineral. Petrol.*, 38: 151—166.

Hepworth, J.V., 1964. Explanation of the geology of sheets 19, 20, 28, and 29 (Southern West Nile). *Geol. Surv. Uganda, Rep.*, 10: 128 pp.

Hepworth, J.V., 1967. The photogeological recognition of ancient orogenic belts in Africa. *Q. J. Geol. Soc. London*, 123: 253—292.

Hepworth, J.V., 1972. The Mozambique orogenic belt and its foreland in northeast Tanzania: a photogeologically-based study. *J. Geol. Soc. London*, 128: 461—500.

Hepworth, J.V., 1973. A photogeological investigation of the Ancient Gneiss complex and the Mozaan of Swaziland, *U.K. Inst. Geol. Sci., Overseas Div., Rep.*, 20: 23 pp.

Hepworth, J.V. and Macdonald, R., 1966. Orogenic belts of the northern Uganda basement. *Nature*, 210: 726—727.

Herd, R.K., 1978. Notes on metamorphism in New Quebec. *Geol. Surv. Can., Paper*, 78-10: 79—83.

Hermann, A.G., Blanchard, D.P., Haskin, L.A., Jacobs, J.W., Knake, D., Koroter, R.L. and Brannon, J.C., 1976. Major, minor, and trace element compositions of peridotitic and basaltic komatiites from the Precambrian crust of southern Africa. *Contrib. Mineral. Petrol.*, 59: 1—12.

Herz, N., 1970. Gneissic and igneous rocks of the Qadrilatero Ferrifero. *U.S. Geol. Surv., Prof. Paper*, 641-B.

Hess, H.H., 1950. Vertical mineral zonation in the Great Dyke of southern Rhodesia. *Trans. Geol. Soc. S. Afr.*, 53: 159—166.

Hess, H.H., 1960. Stillwater igneous complex, Montana: a quantitative mineralogical study. *Geol. Soc. Am., Mem.*, 80: 230 pp.

Heywood, W.W. and Davidson, A., 1969. Geology of Benjamin Lake map area, District of MacKenzie. *Geol. Surv. Can., Mem.*, 361.

Hickman, A.H., 1975. Precambrian structural geology of part of the Pilbara region. *Geol. Surv. W. Aust., Annu. Rep.*, 1974: 68—73.

Hickman, A.H. and Lipple, S.L., 1975. Explanatory notes on the Marble Bar geological sheet, Western Australia. *Geol. Surv. W. Aust., Rec.*, 74/20.

Hickman, M.H., 1974. A 3500-myr-old granite in southern Africa. *Nature*, 251: 295—296.

Hickman, M.H., 1978. Isotopic evidence for crustal reworking in the Rhodesian Archean craton, southern Africa. *Geology*, 6: 214—216.

Hietanen, A., 1967. On the facies series in various types of metamorphism. *J. Geol.*, 10: 187—214.

Hills, F.A., Gast, P.W., Houston, R.S. and Swainband, I.G., 1968. Precambrian geochronology of the Medicine Bow Mountains, southeastern Wyoming. *Geol. Soc. Am. Bull.*, 79: 1757—1784.

Himmelberg, G.R. and Phinney, W.C., 1967. Granulite-facies metamorphism, Granite Falls-Montevideo area, Minnesota. *J. Petrol.*, 8: 325—348.

Hoering, T.C., 1962. The stable isotopes of carbon in the carbonate and reduced carbon of Precambrian sediments. *Carnegie Inst. Washington Yearb.*, 61: 190—191.

Hoffman, A.W. and Hart, S.R., 1975. An assessment of local and regional isotopic equilibrium in a partially molten mantle. *Carnegie Inst. Washington Yearb.*, 74: 195—210.

Holland, H.D., 1962. Model for the evolution of the earth's atmosphere. In: A.E.J. Engel, H.L. James and B.F. Leonard (Editors), *Petrologic Studies.* Geological Society of America, Boulder, Colo., pp. 447—477.

Holland, H.D., 1965. The history of ocean water and its effect on the chemistry of the atmosphere. *Proc. Natl. Acad. Sci. U.S.A.*, 53: 1173—1183.

Holland, H.D., 1972. The geologic history of seawater — an attempt to solve the problem.

Geochim. Cosmochim. Acta, 36: 637—651.
Horwitz, R.C. and Smith, R.E., 1978. Bridging the Yilgarn and Pilbara blocks, Western Australia. *Precambrian Res.,* 6: 298—322.
Hossack, J.R., 1968. Pebble deformation and thrusting in the Bygdin area, southern Norway. *Tectonophysics,* 5: 315—339.
Hubregtse, J.J.M.W., 1976. Volcanism in the western Superior Province in Manitoba. In: B.F. Windley (Editor), *The Early History of the Earth.* J. Wiley & Sons, New York, N.Y., pp. 279—288.
Huddleson, A., 1951. Geology of the Kisii district. *Geol. Surv. Kenya, Rep.,* 18.
Hughs, C.J., 1976. Archean variolites — quenched immiscible liquids: a discussion. *Can. J. Earth Sci.,* 14: 137—139.
Humphris, S.E. and Thompson, G., 1978. Trace element mobility during hydrothermal alteration of oceanic basalts. *Geochim. Cosmochim. Acta,* 42: 127—136.
Hunter, D.R., 1970. The Ancient Gneiss Complex in Swaziland. *Trans. Geol. Soc. S. Afr.,* 73: 107—150.
Hunter, D.R., 1973. The granitic rocks of the Precambrian in Swaziland. *Geol. Soc. S. Afr., Spec. Publ.,* 3: 131—147.
Hunter, D.R., 1974a. Crustal development in the Kaapvaal craton, I. The Archean. *Precambrian Res.,* 1: 259—294.
Hunter, D.R., 1974b. Crustal development in the Kaapvaal craton, II. The Proterozoic. *Precambrian Res.,* 1: 295—326.
Hunter, D.R., Barker, F. and Millard, Jr., H.T., 1978. The geochemical nature of the Archean Ancient Gneiss Complex and Granodiorite Suite, Swaziland: a preliminary study. *Precambrian Res.,* 7: 105—127.
Hurley, P.M. and Rand, J.R., 1969. Pre-drift continental nuclei. *Science,* 164: 1229—1242.
Hurley, P.M. and Rand, J.R., 1974. Outline of Precambrian chronology in lands bordering the South Atlantic exclusive of Brazil. In: A.E. Nairn and F.G. Stehli (Editors), *The Ocean Basins and Margins, Vol. 1.* Plenum Press, New York, N.Y., pp. 391—410.
Hurley, P.M., Hughes, H., Favre, G., Fairbairn, H.W. and Pinson, W.H., 1962. Radiogenic strontium-87 model of continent formation. *J. Geophys. Res.,* 67: 5315—5333.
Hurley, P.M., Leo, G.W., White, R.W. and Fairbairn, H.W., 1971. Liberian age province and adjacent provinces in Liberia and Sierra Leone. *Geol. Soc. Am. Bull.,* 82: 3483—3490.
Hurley, P.M., Pinson, Jr., W.H., Nagy, B. and Teska, T.M., 1972. Ancient age of the Middle Marker Horizon, Onverwacht Group, Swaziland Sequence, South Africa. *Earth Planet. Sci. Lett.,* 14: 360—366.
Hurley, P.M., Fairbairn, H.W. and Gaudette, H.E., 1976. Progress report on early Archean rocks in Liberia, Sierra Leone, and Guyana, and their general stratigraphic setting. In: B.F. Windley (Editor), *The Early History of the Earth.* J. Wiley & Sons, New York, N.Y., pp. 511—521.
Hurst, R.W., 1978a. Sr evolution in the West Greenland-Labrador craton: a model for early Rb depletion in the mantle. *Geochim. Cosmochim. Acta,* 42: 39—44.
Hurst, R.W., 1978b. Reply to critical comment by S. Moorbath. *Geochim. Cosmochim. Acta,* 42: 1585—1586.
Hutchinson, R.W., 1973. Volcanogenic sulfide deposits and their metallogenic significance. *Econ. Geol.,* 68: 1223—1246.
Hutchinson, R.W., Ridler, R.H. and Suffel, G.G., 1971. Metallogenic relationships in the Abitibi belt, Canada: a model for Archean metallogeny. *Can. Inst. Min. Metall., Trans.,* 74: 106—115.
Hyndman, R.D., Lambert, I.B., Heier, K.S., Jaeger, J.C. and Ringwood, A.E., 1968. Heat flow and surface radioactivity measurements in the Precambrian shield of Western Australia. *Phys. Earth Planet. Inter.,* 1: 129—135.

Innes, M.J.S., 1960. Gravity and isostasy in northern Ontario and Manitoba. *Dom. Observ. Publ.*, 21: 261—338.

Innes, M.J.S., Goodacre, A.K., Weber, J.R. and McConnell, R.K., 1967. Structural implications of the gravity field in Hudson Bay and vicinity. *Can. J. Earth Sci.*, 4: 977—993.

Irvine, T.N. and Baragar, W.R.A., 1971. A guide to the chemical classification of the common volcanic rocks. *Can. J. Earth Sci.*, 8: 523—548.

Irvine, T.N. and Smith, C.H., 1967. The ultramafic rocks of the Muskox instrusion. Northwest Territories, In: P.J. Wyllie (Editor), *Ultramafic and Related Rocks.* J. Wiley & Sons, New York, N.Y., pp. 38—49.

Irving, E. and McGlynn, J.C., 1976. Proterozoic magnetostratigraphy and tectonic evolution of Laurentia. *Philos. Trans. R. Soc. London, Ser. A*, 280: 433—468.

Irving, E. and Naldrett, A.J., 1977. Paleomagnetism in Abitibi greenstone belt and Abitibi and Matachewan diabase dikes: evidence of the Archean geomagnetic field. *J. Geol.*, 85: 157—176.

Jackson, E.D., 1961. Primary textures and mineral associations in the ultramafic zone of the Stillwater Complex, Montana. *U.S. Geol. Surv., Prof. Paper*, 358: 106 pp.

Jahn, B.M. and Condie, K.C., 1976. On the age of Rhodesian greenstone belts. *Contrib. Mineral. Petrol.*, 57: 317—330.

Jahn, B.M. and Murthy, V.R., 1975. Rb-Sr ages of the Archean rocks from the Vermilion district northeastern Minnesota. *Geochim. Cosmochim. Acta*, 39: 1679—1689.

Jahn, B.M. and Nyquist, L.E., 1976. Crustal evolution in the earth-moon system: Constraints from Rb-Sr studies. In: B.F. Windley (Editor), *The Early History of the Earth.* J. Wiley & Sons, New York, N.Y., pp. 55—76.

Jahn, B.M. and Shih, C., 1974. On the age of the Onverwacht Group, Swaziland Sequence, South Africa. *Geochim. Cosmochim. Acta*, 38: 873—885.

Jahn, B.M., Shih, C. and Murthy, V.R., 1974. Trace element geochemistry of Archean volcanic rocks. *Geochim. Cosmochim. Acta*, 38: 611—627.

Jahn, B.M., Auvray, B., Blais, S., Capdevila, R., Cornichet, J., Vidal, P. and Hameurt, J., 1979. Trace element geochemistry and petrogenesis of Finnish greenstone belts. *J. Petrol.* (in press).

Jakeš, P., 1973. Geochemistry of continental growth. In: D.H. Tarling and S.K. Runcorn (Editors), *Implications of Continental Drift to the Earth Sciences.* Academic Press, London, pp. 999—1010.

Jakeš, P. and Taylor, S.R., 1974. Excess europium content in Precambrian sedimentary rocks and continental evolution. *Geochim. Cosmochim. Acta*, 38: 739—745.

Jakeš, P. and White, A.J.R., 1972. Major and trace element abundances in volcanic rocks of orogenic areas. *Geol. Soc. Am. Bull.*, 83: 29—40.

James, H.L., 1954. Sedimentary facies of iron formation. *Econ. Geol.*, 49: 235—293.

James, H.L. and Sims, P.K., 1973. Precambrian iron formations of the world. *Econ. Geol.*, 68: 913—914.

James, R.S. and Hamilton, D.L., 1969. Phase relations in the system Ab-Or-An-Q at one kilobar water vapor pressure. *Contrib. Mineral. Petrol.*, 21: 111—141.

Jessop, A.M. and Lewis, T., 1978. Heat flow and heat generation in the Superior Province of the Canadian Shield. *Tectonophysics*, 50: 55—77.

Jolliffe, A.W., 1938. Yellowknife Bay-Prosperous Lake area N.W.T. *Geol. Surv. Can., Paper*, 38-21.

Jolliffe, A.W., 1955. Geology and iron ores of Steep Rock Lake. *Econ. Geol.*, 50: 373—398.

Jolliffe, A.W., 1966. Stratigraphy of the Steep Rock Lake Group; Steep Rock Lake, Ontario. *Geol. Assoc. Can., Spec. Paper*, 3: 75—98.

Jolly, W.T., 1974. Regional metamorphic zonation as an aid in study of Archean terrains:

Abitibi region, Ontario. *Can. Mineral.*, 12: 499—508.

Jolly, W.T., 1975. Subdivision of the Archean lavas of the Abitibi area, Canada, from Fe-Mg-Ni-Cr relations. *Earth Planet. Sci. Lett.*, 27: 200—210.

Jolly, W.T., 1977. Relations between Archean lavas and intrusive bodies of the Abitibi greenstone belt, Ontario-Quebec. *Geol. Assoc. Can., Spec. Paper*, 16: 311—340.

Jolly, W.T., 1978. Metamorphic history of the Archean Abitibi belt. *Geol. Surv. Can., Paper*, 78-10: 63—78.

Jones, L.M., Walker, R.L. and Allard, G.O., 1974. The Rb-Sr whole-rock age of major units of the Chibougamau greenstone belt, Quebec. *Can. J. Earth Sci.*, 11: 1550—1561.

Jordan, P., 1966. *The Expanding Earth*. Pergamon Press, New York, N.Y., 202 pp.

Kamineni, D.C., 1973. *Petrology and Geochemistry of Some Archean Metamorphic Rocks near Yellowknife, District of MacKenzie*. Ph.D. Dissertation, University of Ottawa, Ottawa, Ont., 228 pp. (unpublished).

Kay, R.W. and Senechal, R.G., 1976. The rare earth geochemistry of the Troodos ophiolite complex. *J. Geophys. Res.*, 81: 964—970.

Kehlenbeck, M.M., 1976. Nature of the Quetico-Wabigoon boundary in the de Courcey-Smiley Lakes area, northwestern Ontario. *Can. J. Earth Sci.*, 13: 737—748.

Kerrich, R. and Fryer, B.J., 1979. Archean precious-metal hydrothermal systems, Dome mine, Abitibi greenstone belt, II. *Can. J. Earth Sci.*, 16: 440—458.

Key, R.M., Litherland, M. and Hepworth, J.V., 1976. The evolution of the Archean crust of northeast Botswana. *Precambrian Res.*, 3: 375—413.

Kimberley, M.M., 1978. Paleoenvironmental classification of iron formations. *Econ. Geol.*, 73: 215—229.

Knauth, L.P. and Lowe, D.R., 1978. Oxygen isotope geochemistry of cherts from the Onverwacht Group, Transvaal, South Africa, with implications for secular variations in the isotopic composition of cherts. *Earth Planet. Sci. Lett.*, 41: 209—222.

Knoll, A.H. and Barghoorn, E.S., 1977. Archean microfossils showing cell division from the Swaziland system of South Africa. *Science*, 198: 396—398.

Koziar, A. and Strangway, D.W., 1978. Shallow crustal sounding in the Superior Province by audio frequency magnetotellurics. *Can. J. Earth Sci.*, 15: 1701—1711.

Kratts, K.O., Krylov, I.N. and Lobach-Zhuckenko, S.B., 1974. Geochronoligiceskie rubezi i geologiceskaja zvoljucija Baltijskogo scita. *Nov. Dannyja Absol. Geochronol. (27 Sess.), Moskva*, pp. 125—130.

Kratz, K.O. and Glebovitsky, V.A., 1972. Metamorphic belts of the U.S.S.R. *Proc. 24th Int. Geol. Congr.*, 1: 348—353.

Krogh, T.E., Harris, N.B.W. and Davis, G.L., 1976. Archean rocks from the eastern Lac Seul region of the English River Gneiss belt, northwestern Ontario, Part 2. Geochronology. *Can. J. Earth Sci.*, 13: 1212—1215.

Kröner, A., 1977a. The Precambrian geotectonic evolution of Africa: plate accretion versus plate destruction. *Precambrian Res.*, 4: 163—213.

Kröner, A., 1977b. Precambrian mobile belts of southern and eastern Africa — ancient sutures or sites of ensialic mobility? A case for crustal evolution towards plate tectonics. *Tectonophysics*, 40: 101—135.

Kuenen, P.H. and Migliorini, C.I., 1950. Turbidity currents as the cause of graded bedding. *J. Geol.*, 58: 91—127.

Kvenvolden, K.A., 1972. Organic geochemistry of early Precambrian sediments. *Proc. 24th Int. Geol. Congr.*, 1: 31—41.

Kvenvolden, K.A., 1974. Natural evidence for chemical and early biologic evolution. *Origins Life*, 5: 71—86.

Lajoie, J. and Gelinas, L., 1978. Emplacement of Archean peridotitic komatiites in LaMotte Township, Quebec. *Can. J. Earth Sci.*, 15: 672—677.

Lambert, I.B., Donnelly, T.H., Dunlop, J.S.R. and Groves, D.I., 1978. Stable isotopic compositions of early Archean sulfate deposits of probable evaporitic and volcanogenic origins. *Nature*, 276: 808—811.

Lambert, M.B., 1978. The Back River volcanic complex — a cauldron subsidence structure of Archean age. *Geol, Surv. Can., Paper*, 78-1A: 153—157.

Lambert, R. St.J., 1976. Archean thermal regimes, crustal and upper mantle temperatures, and a progressive evolutionary model for the earth. In: B.F. Windley (Editor), *The Early History of the Earth.* J. Wiley & Sons, New York, N.Y., pp. 363—376.

Lambert, R.St.J., 1978. The early thermal history of the earth. In: I.E.M. Smith and J.G. Williams (Editors), *Proc. 1978 Archean Geochemistry Conference.* University of Toronto Press, Toronto, Ont., pp. 353—354.

Large, R.R., 1977. Chemical evolution and zonation of massive sulfide deposits in volcanic terrains. *Econ. Geol.*, 72: 549—572.

Larson, E.E., Reynolds, R. and Hoblitt, R., 1973. New virtual and paleomagnetic pole positions from isotopically dated rocks from Wyoming, Montana, and Arizona. *Geol. Soc. Am. Bull.*, 84: 3231—3248.

Laubscher, D.H., 1968. The origin and occurrence of chrysotile asbestos in the Shabani and Mashaba areas, Rhodesia. *Geol. Soc. S. Afr., Annex. Vol.*, 71: 195—204.

Lavreau, J. and Ledent, D., 1975. Etablissement du cadre géochronologique du Kibalien (Zaire). *Ann. Soc. Géol. Belg.*, 98: 197—212.

Leeman, W.P., 1975. Radiogenic tracers applied to basalt genesis in the Snake River Plain—Yellowstone National Park region — evidence for a 2.7 b.y. old upper mantle keel. *Geol. Soc. Am., Abstr. Progr.*, 7: 1165.

Leggo, P.J., 1974. A geochronologic study of the basement complex of Uganda. *J. Geol. Soc. London*, 130: 263—277.

Le Roex, H.D., 1964. Nickel deposit on the Trojan claims, Bindura district, southern Rhodesia. In: S.H. Haughton (Editor), *The Geology of Some Ore Deposits in Southern Africa, Vol. 2*, Geological Society of South Africa, pp. 509—521.

Lindemann, B., 1927. *Kettengebirge, Kontinentale Zerspaltung und Erdexpansion.* Gustav Fischer, Jena.

Lo, H.H., 1970. *Geochemistry of the Louis Lake Pluton. Southwestern Wyoming.* Ph.D. Dissertation, Washington University, St. Louis, Mo. (unpublished).

Lobach-Zhuckenko, S.B., Krylov, I.N., Baykova, V.S., Chekulaev, V.P., Arestova, N.A. and Shuleshko, I.K., 1976. The geological evolution of the Karelian greenstone terrane of the Baltic Shield. *Proc. 25th Int. Geol. Congr.*, 1: 16—17.

Longstaffe, F.J., 1979. The oxygen-isotope geochemistry of Archean granitoids. In: F. Barker (Editor), *Trondhjemites and Related Rocks.* Elsevier, Amsterdam, pp. 363—400.

Longstaffe, F.J. and Schwarcz, H.P., 1977. $^{18}O/^{16}O$ of Archean clastic metasedimentary rocks: a petrogenetic indicator for Archean gneisses? *Geochim. Cosmochim. Acta*, 41: 1303—1312.

Longstaffe, F.J., Schwarcz, H.P. and McNutt, R.H., 1978. Oxygen isotope evidence for the genesis or Archean granitoids, western Superior Province. *Geol. Soc. Am., Abstr. Progr.*, 10: 447.

Lowman, Jr., P.D., 1976. Crustal evolution in silicate planets: implications for the origin of continents. *J. Geol.*, 84: 1—26.

Lubimova, E.A., 1958. Thermal history of the earth with consideration of the variable thermal conductivity of the mantle. *Geophysics*, 1: 115—134.

Ludden, J.N. and Thompson, G., 1978. Behavior of rare earth elements during submarine weathering of tholeiitic basalt. *Nature*, 274: 147—149.

Luth, W.C., Jahns, R.H. and Tuttle, O.F., 1964. The granitic system at pressures of 4—10 kb. *J. Geophys. Res.*, 69: 759—773.

MacDonald, G.J.F., 1959. Calculations on the thermal history of the earth. *J. Geophys. Res.*, 64: 1967—2000.

MacGeehan, P.J., 1978. The geochemistry of altered volcanic rocks at Matagami, Quebec: a geothermal model for massive sulfide genesis. *Can. J. Earth Sci.*, 15: 551—570.

Macgregor, A.M., 1932. Batholiths of southern Rhodesia. *Geol. Mag.*, 69: 18—29.

Macgregor, A.M., 1940. A Precambrian algal limestone in southern Rhodesia. *Trans. Geol. Soc. S. Afr.*, 43: 9—16.

Macgregor, A.M., 1951. Some milestones in the Precambrian of southern Rhodesia. *Trans. Geol. Soc. S. Afr.*, 54: 27—71.

MacLaren, A.S. and Charbonneau, B.W., 1968. Characteristics of magnetic data over major subdivisions of the Canadian Shield. *Proc. Geol. Assoc. Can.*, 19: 57—65.

MacRae, N.D., 1969. Ultramafic intrusions of the Abitibi area, Ontario. *Can. J. Earth Sci.*, 6: 281—304.

Malcuit, R.J. and Heimlich, R.A., 1972. Zircons from Precambrian gneiss, southern Bighorn Mountains, Wyoming. *Am. Mineral.*, 57: 1190—1209.

Manzer, Jr., G.K. and Heimlich, R.A., 1974. Petrology and geochemistry of mafic and ultramafic rocks from the northern Bighorn Mountains, Wyoming. *Geol. Soc. Am. Bull.*, 85: 703—708.

Manzer, Jr., G.K., Heimlich, R.A. and Ross, M.E., 1971. Mineralogic variations across Precambrian mafic dikes in the southern Bighorn Mountains, Wyoming. *The Compass*, 48: 75—83.

Marston, R.J., 1978. The geochemistry of Archean clastic metasediments in relation to crustal evolution, northeastern Yilgarn block, Western Australia. *Precambrian Res.*, 6: 157—175.

Marston, R.J. and Travis, G.A., 1976. Stratigraphic implications of heterogeneous deformation in the Jones Creek Conglomerate. Kathleen Valley, Western Australia. *J. Geol. Soc. Aust.*, 23: 141—156.

Martin, H.J., 1964. The Bikita Tinfield. *Rhod. Geol. Surv. Bull.*, 58: 114—132.

Martin, P.L., 1966. Structural analysis of the Chisel Lake orebody. *Can. Inst. Min. Metall., Trans.*, 69: 208—214.

Mascarenhas, J. de F., 1973. A geologia do centro-leste do Estado da Bahia. *S. B. G., An. 27th Congr. Braz. Geol.*, 2: 35—36.

Mason, R., 1970. *The Geology of the Country between Francistown and Madinare, North-eastern Botswana.* Ph.D. Thesis, University of Witwatersrand, Johannesburg, 186 pp. (unpublished).

Mason, R., 1973; The Limpopo mobile belt — southern Africa. *Philos. Trans. R. Soc. London, Ser. A*, 273: 463—485.

Mason, T.R. and Von Brunn, V., 1977. 3-Gry-old stromatolites from South Africa. *Nature*, 266: 47—49.

Mathur, S.P., 1974. Crustal structure in southwestern Australia from seismic and gravity data. *Tectonophysics*, 24: 151—182.

Matthews, C.N. and Moser, R.E., 1966. Prebiological protein synthesis. *Proc. Natl. Acad. Sci. U.S.A.*, 56: 1087—1094.

McCall, G.J.H., 1958. Geology of the Gwasi area. *Geol. Surv. Kenya, Rep.*, 45: 88 pp.

McCall, G.J.H., 1969. The Archean succession to the west of Lake Lefroy. *Proc. R. Soc. W. Aust.*, 52: 119—128.

McCall, G.J.H., 1971. Some ultrabasic and basic igneous rock occurrences in the Archean of western Australia. *Geol. Soc. Aust., Spec. Publ.*, 3: 429—442.

McCall, G.J.H., 1973. Geochemical characteristics of some Archean greenstone suites of the Yilgarn structural province, Australia. *Chem. Geol.*, 11: 243—269.

McCall, G.J.H. and Doepel, J.J.G., 1969. Stratiform basic bodies within the Yilgarn ophiolites, Western Australia. *Proc. Australas. Inst. Min. Metall.*, 231: 67—74.

McCall, G.J.H., Braybrooke, J.C. and Middleton, D.D., 1970. The Merougil Creek subarea. *J. R. Soc. W. Aust.*, 53: 9—32.

McCulloch, M.T. and Wasserburg, G.J., 1978. Sm-Nd and Rb-Sr chronology of continental crust formation. *Science*, 200: 1003—1011.

McElhinny, M.W. and Embleton, B.J.J., 1976. Precambrian and early Paleozoic paleomagnetism in Australia. *Philos. Trans. R. Soc. London, Ser. A*, 280: 417—431.

McElhinny, M.W. and McWilliams, M.O., 1977. Precambrian geodynamics — a paleomagnetic view. *Tectonophysics*, 40: 137—159.

McElhinny, M.W., Briden, J.C., Jones, D.L. and Brock, A., 1968. Geological and geophysical implications of paleomagnetic results from Africa. *Rev. Geophys.*, 6: 201—238.

McGlynn, J.C. and Henderson, J.B., 1970. Archean volcanism and sedimentation in the Slave structural province. *Geol. Surv. Can., Paper*, 70-40: 31—44.

McGlynn, J.C. and Henderson, J.B., 1972. The Slave Province. *Geol. Assoc. Can., Spec. Paper*, 11: 505—526.

McGrath, P.H. and Hall, D.H., 1969. Crustal structure in northwestern Ontario: regional magnetic anomalies. *Can. J. Earth Sci.*, 6: 191—207.

McGregor, V.R., 1973. The early Precambrian geology of the Godthaab district, West Greenland. *Philos. Trans. R. Soc. London, Ser. A*, 273: 343—358.

McGregor, V.R., 1979. Archean gray gneisses and the origin of the continental crust: evidence from the Godthaab region, West Greenland. In: F. Barker (Editor), *Trondhjemites and Related Rocks.* Elsevier, Amsterdam, pp. 169—204.

McIver, J.R., 1975. Aspects of some high magnesia eruptives in Southern Africa. *Contrib. Mineral. Petrol.*, 51: 99—118.

McIver, J.R. and Lenthall, D.H., 1974. Mafic and ultramafic extrusives of the Onverwacht Group in terms of the system CMAS. *Precambrian Res.*, 1: 327—343.

McKenzie, D.P., 1969. Speculations on the consequences and causes of plate motions. *Geophysics*, 18: 1—32.

McKenzie, D.P. and Weiss, N., 1975. Speculations on the thermal and tectonic history of the earth. *Geophys. J. R. Astron. Soc.*, 42: 131—174.

McKirdy, D.M., 1974. Organic geochemistry in Precambrian research. *Precambrian Res.*, 1: 75—137.

McKirdy, D.M. and Powell, T.G., 1974. Metamorphic alteration of carbon isotopic composition in ancient sedimentary organic matter. *Geology*, 2: 591—595.

McLimans, R.K., 1972. Granite-bearing conglomerates in the Knife Lake Group, Vermilion district. In: P.K. Sims and G.B. Morey (Editors), *Geology of Minnesota: A Centennial Volume.* Minnesota Geological Survey, St. Paul, Minn., pp. 91—101.

McRitchie, W.D. and Weber, W., 1971. Metamorphism and deformation of the Manigotagan gneissic belt, southeastern Manitoba. *Manit. Dep. Mines, Nat. Resour., Mines Branch Publ.*, 71-1: 235—284.

Mehnert, K.R., 1971. *Migmatites and the Origin of Granitic Rocks.* Elsevier, Amsterdam, 405 pp.

Mereu, R.F. and Hunter, J.A., 1969. Crustal and upper mantle stucture under the Canadian Shield from Project Early Rise Data. *Bull. Seismol. Soc. Am.*, 59: 147—165.

Mereu, R.F. and Jobidon, G., 1971. A seismic interpretation of the crust and Moho on a line perpendicular to the Grenville Front. *Can. J. Earth Sci.*, 8: 1553—1583.

Miller, S.L., 1953. A production of amino acids under possible primitive earth conditions. *Science*, 117: 528—529.

Minnitt, R.C.A., 1975. *The Geology of the Eastern Portion of the Murchison Range between Quagga Camp Area and the Kruger National Park.* Master's Thesis, University of Witwatersrand, Johannesburg, 191 pp. (unpublished).

Miyashiro, A., 1973. *Metamorphism and Metamorphic Belts.* J. Wiley & Sons, New York, N.Y., 492 pp.

Molnar, P. and Gray, D., 1979. Subduction of continental lithosphere: some constraints and uncertainties. *Geology*, 7: 58—62.

Monster, J., Appel, P.W.U., Thode, H.G., Schidlowski, M., Carmichael, C.M. and Bridgwater, D., 1979. Sulfur isotope studies in early Archean sediments from Isua, West Greenland: implications for the antiquity of bacterial sulfate reduction. *Geochim. Cosmochim. Acta*, 43: 405—414.

Moorbath, S., 1976. Age and isotope constraints for the evolution of Archean crust. In: B.F. Windley (Editor), *The Early History of the Earth.* J. Wiley & Sons, New York, N.Y., pp. 351—360.

Moorbath, S., 1977. Ages, isotopes and evolution of Precambrian continental crust. *Chem. Geol.*, 20: 151—187.

Moorbath, S., 1978. Critical comment on Sr evolution in the West Greenland-Labrador craton: a model for early Rb depletion in the mantle. *Geochim. Cosmochim. Acta*, 42: 1583—1584.

Moorbath, S., O'Nions, R.K. and Pankhurst, R.J., 1973. Early Archean age for the Isua iron formation, West Greenland. *Nature*, 245: 138—139.

Moorbath, S., Wilson, J.F. and Cotterill, P., 1976. Early Archean age for the Sebakwian Group at Selukwe, Rhodesia. *Nature*, 264: 536—538.

Moorbath, S., Wilson, J.F., Goodwin, R. and Humm, M., 1977a. Further Rb-Sr age and isotope data on early and late Archean rocks from the Rhodesian craton. *Precambrian Res.* 5: 229—239.

Moorbath, S., Allaart, J.H., Bridgwater, D. and McGregor, V.R., 1977b. Rb-Sr ages of early Archean supracrustal rocks and Amitsoq gneisses at Isua. *Nature*, 270: 43—45.

Moore, Jr., J.M., 1977. Orogenic volcanism in the Proterozoic of Canada. *Geol. Assoc. Can., Spec. Paper*, 16: 127—148.

Moorehouse, W.W., 1970. A comparative atlas of textures of Archean and younger volcanic rocks. *Geol. Assoc. Can., Spec. Paper*, 8.

Morey, G.B., Green, J.C., Ojakangas, R.W. and Sims, P.K., 1970. Stratigraphy of the lower Precambrian rocks in the Vermilion District, northeastern Minnesota. *Minn. Geol. Surv., Rep. Inv.*, 14: 33 pp.

Morley, L.W., MacLaren, A.S. and Charbonneau, B.W., 1967. Magnetic anomaly map of Canada. *Geol. Surv. Can., Map*,1255A.

Morrison, E.R., 1972. Corundum in Rhodesia. *Rhod. Geol. Surv., Miner. Resour. Ser.*, 16: 24 pp.

Muehlberger, W.R., Denison, R.E. and Lidiak, E.G., 1967. Basement rocks in continental interior of the United States. *Am. Assoc. Pet. Geol. Bull.*, 51: 2351—2380.

Mueller, P.A. and Rogers, J.J.W., 1973. Secular chemical variation in a series of Precambrian mafic rocks, Beartooth Mountains, Montana and Wyoming. *Geol. Soc. Am. Bull.*, 84: 3645—3652.

Muir, M.D. and Grant, P.R., 1976. Micropaleontological evidence from the Onverwacht Group, South Africa. In: B.F. Windley (Editor), *The Early History of the Earth.* J. Wiley & Sons, New York, N.Y., pp. 595—604.

Mulligan, R., 1975. Geology of Canadian tin occurrences. *Geol. Surv. Can., Econ. Geol. Rep.*, 28: 155 pp.

Murthy, V.R., 1976. Composition of the core and the early chemical history of the earth. In: B.F. Windley (Editor), *The Early History of the Earth.* J. Wiley & Sons, New York, N.Y., pp. 21—31.

Myers, J.S., 1976. The early Precambrian gneiss complex of Greenland. In: B.F. Windley (Editor), *The Early History of the Earth.* J. Wiley & Sons, New York, N.Y., pp. 165—176.

Mysen, B.O. and Kushiro, I., 1977. Compositional variations of coexisting phases with degree of melting of peridotite in the upper mantle. *Am. Mineral.*, 62: 843—865.

Naganna, C., 1971. Origin of the manganese ore deposits of Mysore State, India. *Soc. Min. Geol. Jpn., Spec. Issue*, 3: 454—459.

Naganna, C., 1976. Manganese ore deposits of Karnataka. *Proc. Symp. Geology of Ferrous and Ferro-alloy Minerals, Bangalore, 1976.* pp. 110—113.

Nagy, B. and Nagy, L.A., 1969. Early Precambrian Onverwacht microstructures: possibly the oldest fossils on earth. *Nature*, 223: 1226—1229.

Nagy, B., Nagy, L.A., Zumberge, J.E., Sklarew, D.S. and Anderson, P., 1977. Indications of a biological and biochemical evolutionary trend during the Archean and early Proterozoic. *Precambrian Res.*, 5: 109—120.

Nairn, A.E.M. and Ressetar, R., 1978. Paleomagnetism of the Peri-Atlantic Precambrian. *Annu. Rev. Earth Planet. Sci.*, 6: 75—91.

Naldrett, A.J., 1970. Ultramafic and related mafic rocks of the Abitibi orogen. *Geol. Surv. Can., Paper*, 70-40: 24—30.

Naldrett, A.J., 1973. Nickel sulfide deposits — their classification and genesis, with special emphasis on deposits of volcanic association. *Can. Min. Metall. Bull.*, 66: 45—63.

Naldrett, A.J., 1976. Ultramafic and related mafic rocks of the Abitibi orogen. *Geol. Surv. Can., Paper*, 70-40: 24—29

Naldrett, A.J. and Cabri, L.J., 1976. Ultramafic and related mafic rocks: their classification and genesis with special reference to the concentration of nickel sulfides and platinum-group elements. *Econ. Geol.*, 71: 1131—1158.

Naldrett, A.J. and Gasparrini, E.L., 1971. Archean nickel sulfide deposits in Canada: their classification, geological setting and genesis with some suggestions as to exploration. *Geol. Soc. Aust., Spec. Publ.*, 3: 201—226.

Naldrett, A.J. and Goodwin, A.M., 1977. Volcanic rocks of the Blake River Group, Abitibi greenstone belt, Ontario, and their sulfur content. *Can. J. Earth Sci.*, 14: 539—550.

Naldrett, A.J. and Mason, G.D., 1968. Contrasting Archean ultramafic igneous bodies in Dundonald and Clergue townships, Ontario. *Can. J. Earth Sci.*, 5: 111—143.

Naldrett, A.J. and Turner, A.R., 1977. The geology and petrogenesis of a greenstone belt and related nickel sulfide mineralization at Yakabindie, Western Australia. *Precambrian Res.*, 5: 43—103.

Naldrett, A.J., Goodwin, A.M. and Fisher, T.L., 1978. The sulfur content of Archean volcanic rocks and a comparison with ocean floor basalts. *Can. J. Earth Sci.*, 15: 715—728.

Nance, W.B. and Taylor, S.R., 1976. Rare earth element patterns and crustal evolution, I. Australian post-Archean sedimentary rocks. *Geochim. Cosmochim. Acta*, 40: 1539—1551.

Nance, W.B. and Taylor, S.R., 1977. Rare earth element patterns and crustal evolution, II. Archean sedimentary rocks from Kalgoorlie, Australia. *Geochim. Cosmochim. Acta*, 41: 225—231.

Naqvi, S.M., 1967. The banded pyrififerous chert from Ingaldhal and adjoining areas of Chitaldrug schist belt, Mysore. *Bull. Natl. Geophys. Res. Inst.*, 5: 173—181.

Naqvi, S.M. and Hussain, S.M., 1972. Petrochemistry of early Precambrian metasediments from the central part of the Chitaldrug schist belt, Mysore, India. *Chem. Geol.*, 10: 109—135.

Naqvi, S.M. and Hussain, S.M., 1973a. Geochemistry of Dharwar metavolcanics and composition of the primeval crust of the peninsular India. *Geochim. Cosmochim. Acta*, 37: 159—164.

Naqvi, S.M. and Hussain, S.M., 1973b. Relation between trace and major element composition of the Chitaldrug metabasalts, Mysore, India, and the Archean mantle. *Chem. Geol.*, 11: 17—30.

Naqvi, S.M. and Hussain, S.M., 1979. Geochemistry of metaanorthosites from a greenstone belt in Karnataka, India. *Can. J. Earth Sci.*, 16: 1254—1264.

Naqvi, S.M., Divakara, V. and Narain, H., 1974. The protocontinental growth of the Indian Shield and the antiquity of its rift valleys. *Precambrian Res.*, 1: 345—398.

Naqvi, S.M., Divakaro, V. and Narain, H., 1978a. The primitive crust: evidence from the Indian Shield. *Precambrian Res.*, 6: 323—345.

Naqvi, S.M., Rao, D., Hussain, S.M., Narayana, B.L., Rogers, J.J.W. and Satyanarayana, K., 1978b. The petrochemistry and geologic implications of conglomerates from Archean geosynclinal piles of southern India. *Can. J. Earth Sci.*, 15: 1085—1100.

Nath, J.S., Ramakrishnan, M. and Viswanatha, M.N., 1976. Dharwar stratigraphic model and Karnataka craton evolution. *Rec. Geol. Soc. India*, 107: 149—175.

Nautiyal, S.P., 1965. Precambrian of Mysore Plateau. *53rd Indian Sci. Congr., Calcutta, Sect. Geol. Geogr.*, pp. 1—14.

Nesbitt, R.W., 1971. Skeletal crystal forms in the ultramafic rocks of the Yilgarn block, western Australia. *Geol. Soc. Aust., Spec. Publ.*, 3: 331—350.

Nesbitt, R.W. and Sun, S.S., 1976. Geochemistry of Archean spinifex-textured peridotites and magnesian and low-magnesian tholeiites. *Earth Planet. Sci. Lett.*, 31: 433—453.

Neumann, H., Mead, J. and Vitaliano, C.J., 1954. Trace element variation during fractional crystallization as calculated from the distribution law. *Geochim. Cosmochim. Acta*, 6: 90—99.

Newton, R.C., 1978. Experimental and thermodynamic evidence for the operation of high pressures in Archean metamorphism. In: B.F. Windley and S.M. Naqvi (Editors), *Archean Geochemistry*. Elsevier, Amsterdam, pp. 221—240.

Nisbet, E.G., Bickle, M.J. and Martin, A., 1977. The mafic and ultramafic lavas of the Belingwe greenstone belt, Rhodesia. *J. Petrol.*, 18: 521—566.

Nockolds, S.R., 1954. Average chemical composition of some igneous rocks. *Geol. Soc. Am. Bull.*, 65: 1007—1032.

O'Beirne, W.R., 1968. *The Acid Porphyries and Porphyroid Rocks of the Kalgoorlie Area.* Ph.D. Dissertation, University of Western Australia, Perth. W.A.

Oberlies, F. and Prashnowsky, 1968. Biogeochemische und Elektronen-mikroskopische Untersuchung Präkambrischer Gesteine. *Naturwissenschaften*, 55: 25—28.

Oehler, D.Z., Schopf, J.W. and Kvenvolden, K.A., 1972. Carbon isotopic studies of organic matter in Precambrian rocks. *Science*, 175: 1246—1247.

Oehler, J.H., 1976. Experimental studies in Precambrian paleontology: Structural and chemical changes in blue-green algae during simulated fossilization in synthetic chert. *Geol. Soc. Am. Bull.*, 87: 117—129.

Oftedahl, C., 1953. Petrologic reconnaissance in the Precambrian of the western part of the Wind River Mountains, Wyoming. *Nor. Geol. Tidsskr.*, 32: 1—17.

O'Hara, M.J., 1968. The bearing of phase equilibria studies in synthetic and natural systems on the origin and evolution of basic and ultrabasic rocks. *Earth-Sci. Rev.*, 4: 69—133.

O'Hara, M.J., 1977. Thermal history of excavation of Archean gneisses from the base of the continental crust. *Q. J. Geol. Soc. London*, 134: 185—200.

Ojakangas, R.W., 1972. Archean volcanogenic graywackes of the Vermilion district, northeastern Minnesota. *Geol. Soc. Am. Bull.*, 83: 429—442.

Old, R.A. and Rex, D.C., 1971. Rb and Sr age determination of some Precambrian granitic rocks, S.E. Uganda. *Geol. Mag.*, 108: 353—360.

Oliver, R.L. and Ward, M., 1971. A petrological study of serpentinous rocks associated with nickel sulfide mineralization at Pioneer, Western Australia. *Geol. Soc. Aust., Spec. Publ.*, 3: 311—320.

O'Nions, R.K. and Pankhurst, R.J., 1978. Early Archean rocks and geochemical evolution of the earth's crust. *Earth Planet. Sci. Lett.*, 38: 211—236.

Oro, J., Nooner, D.W. and Zlatkis, A., 1965. Hydrocarbons of biological origin in sediments about 2 billion years old. *Science*, 148: 77—79.

Oversby, V.M., 1975. Lead isotopic systematics and ages of Archean acid intrusives in the Kalgoorlie-Norseman area, Western Australia. *Geochim. Cosmochim. Acta,* 39: 1107—1125.

Oversby, V.M., 1976. Isotopic ages and geochemistry of Archean acid igneous rocks from the Pilbara, Western Australia. *Geochim. Cosmochim. Acta,* 40: 817—829.

Oversby, V.M., 1978. Lead isotopes in Archean plutonic rocks. *Earth Planet. Sci. Lett.*, 38: 237—248.

Oversby, V.M. and Ringwood, A.E., 1971. Time of formation of the earth's core. *Nature,* 234: 463—465.

Page, R.O. and Clifford, P.M., 1977. Physical volcanology of an Archean vent complex, Minnitaki Lake area, northwestern Ontario. *Geol. Surv. Can., Paper,* 77-1A: 441—443.

Pakiser, L.C. and Zietz, I., 1965. Transcontinental crustal and upper mantle structure. *Rev. Geophys.*, 3: 505—520.

Pallister, J.W., 1971. The tectonics of East Africa. In: *Tectonique de l'Afrique. Earth Sciences, 6.* UNESCO, Paris, pp. 511—542.

Park, R.G. and Ermanovics, I.F., 1978. Tectonic evolution of two greenstone belts from the Superior Province in Manitoba. *Can. J. Earth Sci.*, 15: 1808—1816.

Parker, R.B., 1962. Precambrian agmatites of the Wind River Range, Wyoming. *Contrib. Geol.*, 1: 13—19.

Pasco, E.H., 1950. *A Manual of the Geology of India and Burma, 1.* Government of India, 3rd ed., pp. 1—483.

Paulus, G.E. and Turnock, A.C., 1971. Petrology of the Ross River Pluton. *Manit. Dep. Mines, Nat. Resour., Mines Branch Publ.*, 71-1: 215—226.

Pearce, J.A., 1976. Statistical analysis of major element patterns in basalts. *J. Petrol.*, 17: 15—43.

Pearce, J.A. and Cann, J.R., 1973. Tectonic setting of basic volcanic rocks determined using trace element analysis. *Earth Planet. Sci. Lett.*, 19: 290—300.

Pearce, T.H. and Birkett, T.C., 1974. Archean metavolcanic rocks from Thackeray Township, Ontario. *Can. Mineral.*, 12: 509—519.

Pearce, T.H., Gorman, B.E. and Birkett, T.C., 1977. The relationship between major element chemistry and tectonic environment of basic and intermediate volcanic rocks. *Earth Planet. Sci. Lett.*, 36: 121—132.

Pecerillo, A. and Taylor, S.R., 1976. Geochemistry of Eocene calc-alkaline volcanic rocks from the Kastamonu area, norther Turkey. *Contrib. Mineral. Petrol.*, 58: 63—81.

Percival, J.A., 1979. Kyanite-bearing rocks from the Hackett River area, N.W.T.: implications for Archean geothermal gradients. *Contrib. Mineral. Petrol.*, 69: 177—184.

Perry, Jr., E.C., 1967. The oxygen isotope chemistry of ancient cherts. *Earth Planet. Sci. Lett.*, 3: 62—66.

Perry, Jr., E.C., Monster, J. and Reimer, T.O., 1971. Sulfur isotopes in Swaziland system barites and the evolution of the earth's atmosphere. *Science*, 171: 1015—1016.

Perry, Jr., E.C., Ahmad, S.N. and Swulius, T.M., 1978. The oxygen isotope composition of 3800-m.y. old metamorphosed chert and iron formation from Isukasia, West Greenland. *J. Geol.*, 86: 223—239.

Peterman, Z.E., Goldich, S.S., Hedge, C.E. and Yardley, D.H., 1972. Geochronology of the Rainy Lake region, Minnesota-Ontario. *Geol. Soc. Am., Mem.*, 135: 193—215.

Peterman, Z.E. and Hildreth, R.A., 1978. Reconnaissance geology and geochronology of the Precambrian of the Granite Mountains, Wyoming. *U.S. Geol. Surv., Prof. Paper,* 1055: 22 pp.

Pettijohn, F.J., 1943. Archean sedimentation. *Geol. Soc. Am. Bull.*, 54: 925—972.

Pettijohn, F.J., 1957. *Sedimentary Rocks.* Harper & Brothers, New York, N.Y., 718 pp.

Pettijohn, F.J., 1972. The Archean of the Canadian Shield: a resume. *Geol. Soc. Am., Mem.*, 135: 131—149.

Pflug, H.D., 1966. Structured organic remains from the Fig Tree Series of the Barberton Mountain land. *Univ. Witwatersrand, Econ. Geol. Res. Unit, Inf. Circ.*, 28: 14 pp.

Pflug, H.D., 1967. Structured organic remains from the Fig Tree Series of the Barberton Mountain land. *Rev. Paleobot. Palynol.*, 5: 9—29.

Pflug, H.D. and Jaeschke-Boyer, H., 1979. Combined structural and chemical analysis of 3800-m.y.-old microfossils. *Nature*, 280: 483—486.

Phaup, A.E., 1973. The granitic rocks of the Rhodesian craton. *Geol. Soc. S. Afr., Spec. Publ.*, 3: 59—68.

Pichamuthu, C.S., 1962. Some observations on the structure, metamorphism, and geological evolution of Peninsular India. *J. Geol. Soc. India*, 3: 106—118.

Pichamuthu, C.S., 1967. The Precambrian of India. In: K. Rankama (Editor), *The Precambrian, Vol. 3.* Interscience, New York, N.Y., pp. 1—96.

Pichamuthu, C.S., 1974. The Dharwar craton. *J. Geol. Soc. India*, 15: 339—346.

Pichamuthu, C.S., 1975. Some aspects of the Precambrian geology of India. *Q. J. Geol., Min. Metall. Soc. India*, 47: 25—43.

Pichamuthu, C.S., 1976. Some problems pertaining to the Peninsular Gneissic Complex. *J. Geol. Soc. India*, 17: 1—16.

Pidgeon, R.T., 1978. 3450-m.y.-old volcanics in the Archean layered greenstone succession of the Pilbara block, Western Australia. *Earth Planet. Sci. Lett.*, 37: 421—428.

Piper, J.D.A., 1976a. Paleomagnetic evidence for a Proterozoic supercontinent. *Philos. Trans. R. Soc. London, Ser. A*, 280: 469—490.

Piper, J.D.A., 1976b. Definition of pre-2000 m.y. apparent polar movements. *Earth Planet. Sci. Lett.*, 28: 470—478.

Pirie, J. and Mackasey, W.O., 1978. Preliminary examination of regional metamorphism in parts of Quetico metasedimentary belt, Superior Province, Ontario. *Geol. Surv. Can., Paper*, 78-10: 37—48.

Pirie, N.W., 1959. Chemical diversity and the origins of life. In: A.I. Oparin (Editor), *Origin of Life on Earth.* Pergamon, London, pp. 76—83.

Platt, J.P., Allchurch, P.D. and Rutland, R.W.R., 1978. Archean tectonics in the Agnew supracrustal belt, Western Australia. *Precambrian Res.*, 7: 3—30.

Poldervaart, A., 1955a. Zircons in rocks, 1. Sedimentary rocks. *Am. J. Sci.*, 253: 433—461.

Poldervaart, A., 1955b. Chemistry of the earth's crust. *Geol. Soc. Am., Spec. Paper*, 62: 119—144.

Poldervaart, A., 1956. Zircon in rocks, 2. Igneous rocks. *Am. J. Sci.*, 254: 521—554.

Ponnamperuma, C., 1965. The chemical origin of life. *Sci. J.*, May, pp. 39—45.

Pretorius, D.A., 1974. The structural boundary between the Kaapvaal and Sonama crustal provinces. *Univ. Witwatersrand, Econ. Geol. Res. Unit, Inf. Circ.*, 88: 27 pp.

Prinz, M., 1964. Geologic evolution of the Beartooth Mountains, Montana and Wyoming, 5. Mafic dike swarms of the southern Beartooth Mountains. *Geol. Soc. Am. Bull.*, 75: 1217—1248.

Proctor, P.D. and El-Etr, H.A., 1968. Layered pegmatites, southern Wind River Mountains, Fremont County, Wyoming. *Econ. Geol.*, 63: 595—611.

Purdie, J.J., 1967. Lake Dufault Mines Ltd. *Can Inst. Min. Metall., Centennial Field Excursion Guideb.*, 1967: 52—57.

Pyke, D.R., 1975. On the relationship of gold mineralization and ultramafic rocks in the Timmins area. *Ont. Div. Mines, Misc. Paper*, 62: 23 pp.

Pyke, D.R., Naldrett, A.J. and Eckstand, O.R., 1973. Archean ultramafic flows in Munro Township, Ontario. *Geol. Soc. Am. Bull.*, 84: 955—978.

Pytkowicz, R.M., 1967. Carbonate cycle and the buffer mechanism of recent oceans. *Geochim. Cosmochim. Acta*, 31: 63—73.

Radhakrishna, B.P., 1976. Mineralization episodes in the Dharwar craton of Peninsular India. *J. Geol. Soc. India,* 17: 79—88.

Radhakrishna, B.P. and Vasudev, V.N., 1977. The early Precambrian of the southern Indian Shield. *J. Geol. Soc. India,* 18: 525—541.

Ramakrishnan, M., Viswanatha, M.M. and Nath, J.S., 1976. Basement-cover relationships of Peninsular Gneiss with high grade schists and greenstone belts of southern Karnataka. *J. Geol. Soc. India,* 17: 97—111.

Ramberg. H., 1964. A model for the evolution of continents, oceans, and orogens. *Tectonophysics,* 2: 159—174.

Ramberg, H., 1971. Model studies in relation to intrusion of plutonic bodies. In: G. Newell and N. Rast (Editors), *Mechanism of Igneous Intrusion. Geol. J., Spec. Issue,* 2: 261—286.

Ramberg, H., 1973. Model studies of gravity-controlled tectonics by the centrifuge technique. In: K.A. de Jong and R. Scholten (Editors), *Gravity and Tectonics.* J. Wiley & Sons, New York, N.Y., pp. 49—66.

Ramiengar, A.S., Ramakrishnan, M. and Viswanatha, M.N., 1978. Charnockite-gneiss complex relationship in southern India. *J. Geol. Soc. India,* 19: 411—419.

Ramsay, J.G., 1963. Structural investigations in the Barberton Mountain Land, eastern Transvaal. *Trans. Geol. Soc. S. Afr.,* 66: 353—398.

Ramsay, J.G., 1967. *Folding and Fracturing of Rocks.* McGraw-Hill, New York, N.Y. 568 pp.

Rankama, K., 1955. Geologic evidence of chemical composition of the Precambrian atmosphere. *Geol. Soc. Am., Spec. Paper,* 62: 651—664.

Rao, R.U.M. and Jessop, A.M., 1975. A comparison of the thermal characters of shields. *Can. J. Earth Sci.,* 12: 347—360.

Rao, V.D., Qureshy, M.N. and Aswathanarayana, U., 1969. Major element geochemistry of parts of the Closepet granite pluton, Mysore State, India. *Bull. Natl. Geophys. Res. Inst.,* 7: 145—157.

Rao, V.D., Aswathanarayana, U. and Qureshy, M.N., 1972. Petrological studies in parts of the Closepet granite pluton, Mysore State. *J. Geol. Soc. India,* 13: 1—12.

Rao, V.D., Naqvi, S.M., Satyanarayana, K. and Hussain, S.M., 1974. Geochemistry and origin of the Peninsular Gneisses of Karnataka, India. *J. Geol. Soc. India,* 15: 270—277.

Raudsepp, M., 1975. Differentiation in the gabbro sill complex, Favourable Lake, Ontario. *Univ. Manit., Center Precambrian Stud., Annu. Rep.,* 1975: 116—124.

Reed, Jr., J.C., 1963. Structure of Precambrian crystalline rocks in the northern part of Grand Teton National Park, Wyoming. *U.S. Geol. Surv., Prof. Paper,* 475-C: 1—6.

Reed, Jr., J.C. and Zartman, R.E., 1973. Geochronology of Precambrian rocks of the Teton Range, Wyoming. *Geol. Soc. Am. Bull.,* 84: 561—582.

Reid, R.R., McMannis, W.J. and Palmquist, J.C., 1975. Precambrian geology of North Snowy Block, Beartooth Mountains, Montana. *Geol. Soc. Am., Spec. Paper,* 157: 135 pp.

Reimer, T., 1971. Strontium depletion in early Precambrian sediments. *Neues Jahrb. Mineral. Monatsh.,* H12: 527—541.

Reimer, T., 1975a. Untersuchungen über Abtragung, Sedimentation und Diagenese im frühen Präkambrium und Beispiel der Sheba-Formation Südafrica. *Geol. Jahrb.,* 17: 3—108.

Reimer, T.O. 1975b. Paleogeographic significance of the oldest known oolite pebbles in the Archean Swaziland Supergroup (South Africa). *Sediment. Geol.,* 14: 123—133.

Richardson, S.W., 1970. The relation between a petrogenetic grid facies series and the geothermal gradient in metamorphism. *Fortschr. Mineral.,* 47: 65—76.

Ridler, R.H., 1970. Relationship of mineralization to volcanic stratigraphy in the Kirkland Lake-Larder Lakes area, Ontario. *Proc. Geol. Assoc. Can.,* 21: 33—42.

Rimsaite, J., 1974. Mineral assemblages and low-grade metamorphic-metasomatic alterations in an Archean greenstone belt, Malartic, Quebec. *Can. Mineral.*, 12: 520—526.

Ringwood, A.E., 1975. *Composition and Petrology of the Earth's Mantle.* McGraw-Hill, New York, N.Y., 617 pp.

Ringwood, A.E., 1977. The petrological evolution of island arc systems. *Q. J. Geol. Soc. London*, 130: 183—204.

Rittman, A., 1952. Nomenclature of volcanic rocks. *Bull. Volcanol.*, 12: 75—102.

Roberts, R.G., 1966. *The Geology of the Mattagami Lake Mine, Galinee Township, Quebec.* Ph.D. Thesis, McGill University, Montreal, Que. (unpublished).

Robertson, D.K., 1973. A model discussing the early history of the earth based on a study of lead isotope ratios from veins in some Archean cratons in Africa. *Geochim. Cosmochim. Acta*, 37: 2099—2124.

Robertson, I.D.M., 1968. Granulite metamorphism of the basement complex in the Limpopo metamorphic zone. *Geol. Soc. S. Afr., Annex. Vol.*, 71: 125—134.

Roddick, J.C., Compston, W. and Durney, D.W., 1976. The radiometric age of the Mount Keith granodiorite, a maximum age estimate for an Archean greenstone sequence in the Yilgarn Block, Western Australia. *Precambrian Res.*, 3: 55—78.

Rollinson, H.R., 1978. Zonation of supracrustal relics in the Archean of Sierra Leone, Liberia, Guinea, and Ivory Coast. *Nature*, 272: 440—442.

Rona, P.A., 1973. Relation between rates of sediment accumulation on continental shelves, sea-floor spreading, and eustacy inferred from the central North Atlantic. *Geol. Soc. Am. Bull.*, 84: 2851—2872.

Ronov, A.B., 1964. Common tendencies in the chemical evolution of the earth's crust, ocean, and atmosphere. *Geochem. Int.*, 4: 713—737.

Ross, J.V., 1962. Depositition and current direction within the Yellowknife Group at Mesa Lake, N.W.T., Canada. *Geol. Soc. Am. Bull.*, 73: 1159—1162.

Ross, M.E. and Heimlich, R.A., 1972. Petrology of Precambrian mafic dikes from the Bald Mountain area, Bighorn Mountains, Wyoming. *Geol. Soc. Am. Bull.*, 83: 1117—1124.

Rousell, D.H., 1965. Geology of the Cross Lake area. *Manit. Dep. Mines, Nat. Resour., Mines Branch Publ.*, 62-4.

Roy, J.L., Morris, W.A., Lapointe, P.L., Irving, E., Park, J.K. and Schmidt, P.W., 1978. Apparent polar wander paths and the joining of the Superior and Slave Provinces during early Proterozoic time: comment. *Geology*, 6: 132—133.

Roy, R.F., Blackwell, D.D. and Birch, F., 1968. Heat generation of plutonic rocks and continental heat flow provinces. *Earth Planet. Sci. Lett.*, 5: 1—12.

Roy, S., 1966. *Syngenetic Manganese Formations of India.* Jadavpur University, Calcutta, 219 pp.

Rubey, W.W., 1955. Development of the hydrosphere and atomosphere with special reference to probable composition of the early atmosphere. *Geol. Soc. Am., Spec. Paper*, 62: 631—650.

Runcorn, S.K., 1962. Convection currents in the earth's mantle. *Nature*, 195: 1248—1249.

Runcorn, S.K., 1965. Changes in the convection pattern of the earth's mantle and continental drift; evidence for a cold origin of the earth. *Philos. Trans. R. Soc. London, Ser. A*, 258: 228—251.

Runcorn, S.K., Libby, L.M. and Libby, W.F., 1977. Primeval melting of the moon. *Nature*, 270: 676—681.

Rutland, R.W.R., 1973. Tectonic evolution of the continental crust of Australia. In: D.H. Tarling and S.K. Runcorn (Editors), *Implications of Continental Drift to the Earth Sciences.* Academic Press, London, pp. 1011—1034.

Rutland, R.W.R., 1976. Orogenic evolution of Australia. *Earth-Sci. Rev.*, 12: 161—196.

Rutten, M.G., 1971. *The Origin of Life by Natural Causes.* Elsevier, Amsterdam, 420 pp.

Ryan, G.R., 1965. The geology of the Pilbara Block, Western Australia. *Proc. Australas. Inst. Min. Metall,.* 214: 61—94.

Ryan, G.R. and Kriewaldt, M., 1964. Facies changes in the Archean of the West Pilbara Goldfield. *W. Aust. Geol. Surv., Annu. Rep.,* 1963: 28—30.

Sabarg, S. and Blackburn, C.E., 1977. Granitic rocks at Meggisi Lake, northwest Ontario. *Proc. 1977 Geotraverse Conf., Univ. Toronto,* pp. 134—138.

Saggerson, E.P., 1952. The geology of the Kisuma district. *Geol. Surv. Kenya, Rep.,* 21.

Saggerson, E.P., 1973. Metamorphic facies series in Africa: a contrast. *Geol. Soc. S. Afr., Spec. Publ.,* 3: 227—234.

Saggerson, E.P. and Owen, L.M., 1969. Metamorphism as a guide to the depth of the top of the mantle in Southern Africa. In: *Upper Mantle Project. Geol. Soc. S. Afr., Spec. Publ.,* 2: 335—349.

Saggerson, E.P. and Turner, L.M., 1976. A review of distribution of metamorphism in the ancient Rhodesian craton. *Precambrian Res.,* 3: 1—53.

Sahli, E.W., 1961. Antimony in the Murchison Range of the northeastern Transvaal. *Trans. 7th Commonw. Min. Metall. Congr.,* 1: 181—199.

Salisbury, J.W. and Ronca, L.B., 1966. The origin of continents. *Nature,* 210: 669—690.

Salop, L.I., 1968. Precambrian of the U.S.S.R. *Proc. 23rd Int. Geol. Congr.,* 4: 61—73.

Salop, L.I. and Scheinmann, Y.M., 1969. Tectonic history and structure of platforms and shields. *Tectonophysics,* 7: 565—597.

Salop, L.I. and Travin, L.V., 1972. Archean stratigraphy in central part of the Aldan Shield. *Int. Geol. Rev.,* 14: 29—40.

Sanders, L.D., 1965. Geology of the contact between the Nyanza Shield and the Mozambique belt in western Kenya. *Geol. Surv. Kenya, Bull.,* 7: 45 pp.

Sangster, D.F., 1972. Precambrian volcanogenic massive sulfide deposits in Canada: a review. *Geol. Surv. Can., Paper,* 72-22: 42 pp.

Sangster, D.F. and Scott, S.D., 1976. Precambrian stratabound massive Cu-Zn-Pb sulfide ores of North America. In: K.H. Wolf (Editor), *Handbook of Stratabound and Stratiform Ore Deposits, Vol. II.* Elsevier, Amsterdam, pp. 129—222.

Satterly, J., 1952. Geology of Munro Township. *60th Annu. Rep. Ont. Dep. Mines,* 60, Rt. 8: 60 pp.

Satyanarayana, K., Divakara Rao, V., Naqvi, S.M. and Hussain, S.M., 1973. Geochemistry and petrogenesis of dikes from Nuggihalli schist belt, Mysore. *Geophys. Res. Bull.,* 11: 273—279.

Satyanarayana, K., Naqvi, S.M., Divakara Rao, V. and Hussain, S.M., 1974. Geochemistry of Archean amphibolites from Karnataka State, Peninsular India. *Chem. Geol.,* 14: 305—315.

Schau, M., 1977. Komatiites and quartzites in the Archean Prince Albert Group. *Geol. Assoc. Can., Spec. Paper,* 16: 341—354.

Schidlowski, M., 1970. Untersuchungen zur Metallogenese im südwestlichen Witwatersrand-Becken (Südafrica). *Beih. Geol. Jahrb.,* 85: 46 pp.

Schidlowski, M., Eichmann, R. and Junge, C.E., 1975. Precambrian sedimentary carbonates: carbon and oxygen isotope geochemistry and implications for the terrestrial oxygen budget. *Precambrian Res.,* 2: 1—69.

Schildowski, M., Appel, P.W.U., Eichmann, R. and Junge, C.E., 1979. Carbon isotope geochemistry of the 3.7×10^9-y.-old Isua sediments, West Greenland: implications for the Archean carbon and oxygen cycles. *Geochim. Cosmochim. Acta,* 43: 189—199.

Schopf, J.W., 1970. Precambrian microorganisms and evolutionary events prior to the origin of vascular plants. *Biol. Rev., Cambridge Philos. Soc.,* 45: 317—352.

Schopf, J.W., 1975. Precambrian paleobiology: Problems and perspectives. *Annu. Rev. Earth Planet. Sci.,* 3: 213—249.

Schopf, J.W., 1976. Evidence of Archean life: a brief appraisal. In: B.F. Windley (Editor),

The Early History of the Earth. J. Wiley & Sons, New York, N.Y., pp. 589—593.

Schopf, J.W. and Barghoorn, E.S., 1967. Alga-like fossils from the early Precambrian of South Africa. *Science*, 156: 508—512.

Schopf, J.W., Kvenvolden, K.A. and Barghoorn, E.S., 1968. Amino acids in Precambrian sediments: an assay. *Proc. Natl. Acad. Sci. U.S.A.*, 59: 639—646.

Schopf, J.W., Oehler, D.Z., Horodyski, R.J. and Kvenvolden, K.A., 1971. Biogenicity and significance of the oldest known stromatolites. *J. Paleontol.*, 45: 477—485.

Schwab, F.L., 1971. Geosynclinal compositions and the new global tectonics. *J. Sediment. Petrol.*, 41: 928—938.

Schwarcz, H.T., 1966. Chemical and mineralogic variations in an arkosic quartzite during progressive regional metamorphism. *Geol. Soc. Am. Bull.*, 77: 509—532.

Schwerdtner, W.M., 1976. Crescent-shaped granitic bodies in the Scotch Lakes area, Wabigoon subprovince. *Proc. 1976 Geotraverse Conf., Univ. Toronto*, pp. 34—41.

Schwerdtner, W.M., 1978. Archean tectonic record in Superior Province of northwest Ontario. *Geol. Soc. Am., Abstr. Progr.*, 10: 488.

Schwerdtner, W.M. and Sutcliffe, R.H., 1978. Patterns of total strain in the crestal region of immature diapirs. *Can. J. Earth Sci.*, 15: 1437—1447.

Schwerdtner, W.M., Morgan, J., Osadetz, K., Stone, D., Stott, G. and Sutcliffe, R.H., 1978. Structure of Archean rocks in western Ontario. In: I.E.M. Smith and J.G. Williams (Editors), *Proc. 1978 Archean Geochemistry Conference.* University of Toronto Press, Toronto, Ont., pp. 107—126.

Scoates, R.F.J., 1971. Ultramafic rocks of the Rice Lake greenstone belt. *Manit. Dep. Mines, Nat. Resour., Mines Branch Publ.*, 71-1: 7: 189—201.

Scott, R.B. and Hajash, A., 1976. Initial submarine alteration of basaltic pillow lavas: a microprobe study. *Am. J. Sci.*, 276: 480—501.

Semenenko, N.P., Scherbak, A.P., Vinogradov, A.P., Tougarinov, A.I., Eliseeva, G.D., Cotlovskay, F.I. and Demidenko, S.G., 1968. Geochronology of the Ukrainian Precambrian. *Can. J. Earth Sci.*, 5: 661—671.

Serebryakov, S.N. and Semikhatov, M.A., 1974. Riphean and recent stromatolites: a comparison. *Am. J. Sci.*, 274: 556—574.

Sethuraman, K. and Moore, J.M., 1973. Petrology of metavolcanic rocks in the Bishop Corners—Donaldson area, Grenville Province, Ontario. *Can. J. Earth Sci.*, 10: 589—614.

Shackleton, R.M., 1946. Geology of the Migori gold belt. *Geol. Surv. Kenya, Rep.*, 10.

Shackleton, R.M., 1973a. Problems of the evolution of the continental crust. *Philos. Trans. R. Soc. London, Ser. A*, 273: 317—320.

Shackleton, R.M., 1973b. Correlation of structures across Precambrian orogenic belts in Africa. In: D.H. Tarling and S.K. Runcorn (Editors), *Implications of Continental Drift to the Earth Sciences, Vol. 2.* Academic Press, London, pp. 1091—1095.

Shackleton, R.M., 1976. Shallow and deep-level exposures of the Archean crust in India and Africa. In: B.F. Windley (Editor), *The Early History of the Earth.* J. Wiley & Sons, New York, N.Y., pp. 317—322.

Sharpe, J.W.N., 1964. The Empress Ni-Cu deposit, Rhodesia. In: S.H. Haughton (Editor), *The Geology of Some Ore Deposits in Southern Africa, Vol. 2.* Geological Society of South Africa, pp. 497—508.

Shaw, D.M., 1970. Trace element fractionation during anatexis. *Geochim. Cosmochim. Acta*, 34: 237—253.

Shaw, D.M., 1972. Development of the early continental crust, 1. *Can. J. Earth Sci.*, 9: 1577—1595.

Shaw, D.M., 1976. Development of the early continental crust, 2. In: B.F. Windley (Editor), *The Early History of the Earth.* J. Wiley & Sons, New York, N.Y., pp. 33—54.

Sheraton, J.W., Skinner, A.C. and Tarney, J., 1973. The geochemistry of the Scourian gneisses of the Assynt district. In: R.G. Park and J. Tarney (Editors), *The Precambrian of Scotland and Related Rocks of Greenland.* University of Keele, pp. 13—30.

Simonen, A., 1971. Das finnische Grundgebirge. *Geol. Rundsch.*, 60: 1406—1421.

Sims, P.K., 1972a. Vermilion District and adjacent areas. In: P.K. Sims and G.B. Morey (Editors), *Geology of Minnesota: A Centennial Volume.* Minnesota Geological Survey, St. Paul, Minn., pp. 49—62.

Sims, P.K., 1972b. Metavolcanic and associated synvolcanic rocks in Vermilion district. In: P.K. Sims and G.B. Morey (Editors), *Geology of Minnesota: A Centennial Volume.* Minnesota Geological Survey, St. Paul, Minn., pp. 63—75.

Sims. P.K., 1976. Early Precambrian tectonic-igneous evolution in the Vermilion district, northeastern Minnesota, *Geol. Soc. Am. Bull.*, 87: 379—389.

Sims, P.K. and Mudrey, Jr., M.G., 1972. Syenitic plutons and associated lamprophyres. In: P.K. Sims and G.B. Morey (Editors), *Geology of Minnesota: A Centennial Volume.* Minnesota Geological Survey, St. Paul, Minn., pp. 140—152.

Sims, P.K. and Viswanathan, S., 1972. Giants Range batholith. In: P.K. Sims and G.B. Morey (Editors), *Geology of Minnesota: A Centennial Volume.* Minnesota Geological Survey, St Paul, Minn., pp. 120—139.

Sims, P.K., Sinclair, D. and Mundrey, Jr., M.G., 1972. Linden pluton. In: P.K. Sims and G.B. Morey (Editors), *Geology of Minnesota: A Centennial Volume.* Minnesota Geological Survey, St Paul, Minn., pp. 160—162.

Slack, P.B., 1974. Variance of terrestrial heat flow between the North American craton and the Canadian Shield. *Geol. Soc. Am. Bull.*, 85: 519—522.

Smith, C.C. and Fripp, R.E.P., 1973. The northern contact facies of the Matopos granite, Rhodesia. *Geol. Soc. S. Afr., Spec. Publ.*, 3: 447—454.

Smith, I.E.M., 1977. REE patterns in the Abitibi greenstone belt. *Proc. 1977 Geotraverse Conf., Univ. Toronto*, pp. 56—64.

Southwick, D.L., 1972. Vermilion granite-migmatite massif. In: P.K. Sims and G.B. Morey (Editors), *Geology of Minnesota: A Centennial Volume.* Minnesota Geological Survey, St Paul, Minn., pp. 108—119.

Southwick, D.L., 1978. The Vermilion granitic complex, northern Minnesota. In: I.E.M. Smith and J.G. Williams (Editors), *Proc. 1978 Archean Geochemistry Conference*, University of Toronto Press, Toronto, Ont., pp. 265—276.

Spence, C.D., 1967. The Noranda area. *Can. Inst. Min. Metall., Field Excursion Guideb.*, 1967: 36--39.

Stanton, R.L., 1960. General features of the conformable pyritic orebodies, I. — Field association; II. — Mineralogy. *Can. Inst. Min. Metall. Bull.*, 53: 24—29, 66—74.

Steiger, R.H. and Jäger, E., 1977. Subcommision of Geochronology: convention on the use of decay constants in geo- and cosmochemistry. *Earth Planet. Sci. Lett.*, 36: 359—362.

Stephenson, J.F., 1971. Gold deposits of the Rice Lake-Beresford Lake greenstone belt, southeastern Manitoba. *Manit. Dep. Mines, Nat. Resour., Mines Branch Publ.*, 71-1: 337—374.

Stern, C.R., Huang, W. and Wyllie, P.J., 1975. Basalt-andesite-rhyolite-H_2O: crystallization intervals with excess H_2O and H_2O-undersaturated liquidus surfaces to 35 kbar, with implications to magma genesis. *Earth Planet. Sci. Lett.*, 28: 189—196.

Stidolph, P.D., 1973. The granitic terrane north and east of the Shamva greenstone belt. *Geol. Soc. S. Afr., Spec. Publ.*, 3: 121—126.

Stockwell, C.H., 1933. Great Slave Lake—Coppermine River area, N.W.T. *Geol. Surv. Can., Annu. Rep.*, 1932, Part C: 37—63.

Stockwell, C.H., McGlynn, J.C., Einslie, R.F., Sanford, B.V., Norris, A.W., Donaldson, J.A., Fahrig, W.F. and Currie, K.L., 1970. Geology of the Canadian Shield. *Geol. Surv. Can., Econ. Geol. Rep.*, 1: 45—150.

Stone, D., 1976. The Sydney Lake cataclastic zone. *Proc. 1976 Geotraverse Conf., Univ. Toronto*, pp. 88—91.

Stowe, C.W., 1968a. Intersecting fold trends in the Rhodesian basement south and east of Selukwe. *Trans. Geol. Soc. S. Afr., Annex. Vol.*, 71: 53—78.

Stowe, C.W., 1968b. The geology of the country south and west of Selukwe. *Rhod. Geol. Surv., Bull.*, 59: 209 pp.

Stowe, C.W., 1973. The older tonalite gneiss complex in the Selukwe area, Rhodesia. *Geol. Soc. S. Afr., Spec. Publ.*, 3: 85—96.

Stowe, C.W., 1974. Alpine-type structures in the Rhodesian basement complex at Selukwe. *Q. J. Geol. Soc. London*, 130: 411—425.

Strakhov, N.M., 1964. Stages of development of the external geospheres and formation of sedimentary rocks in the history of the earth. *Int. Geol. Rev.*, 6: 1466—1482.

Streckeisen, A.L., 1976. To each plutonic rock its proper name. *Earth-Sci. Rev.*, 12: 1—33.

Stueber, A.M., Heimlich, R.A. and Ikramuddin, M., 1976. Rb-Sr ages of Precambrian mafic dikes, Bighorn Mountains, Wyoming. *Geol. Soc. Am. Bull.*, 87: 909—914.

Subramanian, A.P., 1967. Charnockites and granulites of southern India — a review. *Dansk Geol. Foren. Medd.*, 17: 473—493.

Sun, S.S. and Hanson, G.N., 1975. Evolution of the mantle: geochemical evidence from alkali basalt. *Geology*, 3: 297—302.

Sun, S.S. and Nesbitt, R.W., 1977. Chemical heterogeneity of the Archean mantle, composition of the earth and mantle evolution. *Earth Planet. Sci. Lett.*, 35: 429—448.

Sun, S.S. and Nesbitt, R.W., 1978. Petrogenesis of Archean ultrabasic and basic volcanics: evidence from rare earth elements. *Contrib. Mineral. Petrol.*, 65: 301—325.

Sutcliffe, R.H., 1977. Geology and structure of the Jackfish Lake pluton, a major intrusion in the Rainy Lake dome. *Proc. 1977 Geotraverse Conf., Univ. Toronto*, pp. 146—154.

Sutcliffe, R.H., 1978. Geology of the Rainy Lake granitoid complex, northwestern Ontario. In: I.E.M. Smith and J.G. Williams (Editors), *Proc. 1978 Archean Geochemistry Conference.* University of Toronto Press, Toronto, Ont., pp. 235—244.

Sutton, J., 1963. Long-term cycles in the evolution of the continents. *Nature*, 198: 731—735.

Sutton, J., 1967. The extension of the geological record into the Precambrian. *Proc. Geol. Assoc.*, 78: 493—534.

Sutton, J., 1971. Some developments in the crust. *Geol. Soc. Aust., Spec. Publ.*, 3: 1—10.

Sutton, J., 1973. Some changes in continental structure since early Precambrian time. In: D.H. Tarling and S.K. Runcorn (Editors), *Implications of Continental Drift to the Earth Sciences.* Academic Press, London, pp. 1071—1081.

Sutton, J., 1976. Tectonic relationships in the Archean. In: B.F. Windley (Editor), *The Early History of the Earth.* J. Wiley & Sons, New York, N.Y., pp. 99—104.

Sylvester-Bradley, P.C., 1975. The search for protolife. *Proc. R. Soc. London, Ser. B*, 189: 213—233.

Szewczyk, Z.J. and West, G.F., 1976. Gravity study of an Archean granitic area northwest of Ignace, Ontario. *Can. J. Earth Sci.*, 13: 1119—1130.

Talbot, C.J., 1968. Thermal convection in the Archean crust? *Nature*, 220: 552—556.

Talbot, C.J., 1973. A plate tectonic model for the Archean crust. *Philos. Trans. R. Soc. London, Ser. A*, 273: 413—428.

Taliferro, N.L., 1933. The relation of volcanism to diatomaceous and associated siliceous sediments. *Univ. Calif. Publ., Dep. Geol. Sci., Bull.*, 23: 1—55.

Tanner, J.G., 1969. *A Geophysical Interpretation of Structural Boundaries in the Eastern Canadian Shield.* Ph.D. Dissertation, University of Durham, Durham.

Tanner, P.W.G., 1973. Orogenic cycles in East Africa. *Geol. Soc. Am. Bull.*, 84: 2839—2850.

Tarling, D.H., 1978. Plate tectonics: present and past. In: D.H. Tarling (Editor),

Evolution of the Earth's Crust. Academic Press, London, pp. 361—408.

Tarney, J., 1976. Geochemistry of Archean high-grade gneisses, with implications as to the origin and evolution of the Precambrian crust. In: B.F. Windley (Editor), *The Early History of the Earth.* J. Wiley & Sons, New York, N.Y., pp. 405—418.

Tarney, J. and Windley, B.F., 1977. Chemistry, thermal gradients and evolution of the lower continental crust. *Q. J. Geol. Soc. London,* 134: 153—172.

Tarney, J., Dalziel, I.W.D. and De Wit, M.J., 1976. Marginal basin "Rocas Verdes" complex from S. Chile: a model for Archean greenstone belt formation. In: B.F. Windley (Editor), *The Early History of the Earth.* J. Wiley & Sons New York, N.Y., pp. 131—146.

Tasse, N., Lajoie, J. and Dimroth, E., 1978. The anatomy and interpretation of an Archean volcanoclastic sequence, Noranda region, Quebec. *Can. J. Earth Sci.,* 15: 874—888.

Tatsumi, T., Yoshihiro, S. and Keiichiro, K., 1970. Mineral deposits of volcanic affinity in Japan. In: T. Tatsumi (Editor), *Volcanism and Ore Genesis.* Tokyo University Press, Tokyo, pp. 3—50.

Taylor, S.R., 1967. The origin and growth of continents. *Tectonophysics,* 4: 17—34.

Taylor, S.R., 1975. *Lunar Science: A Post-Apollo View.* Pergamon Press, New York, N.Y., 370 pp.

Taylor, S.R. and Hallberg, J.A., 1977. Rare-earth elements in the Marda calc-alkaline suite: an Archean geochemical analogue of Andean-type volcanism. *Geochim. Cosmochim. Acta,* 41: 1125—1129.

Taylor, S.R. and White, A.J.R., 1965. Geochemistry of andesites and the growth of continents. *Nature,* 208: 271—273.

Taylor, S.R., Capp, A.C., Graham, A.L. and Blake, D.H., 1969. Trace element abundances in andesites. *Contrib. Mineral. Petrol.,* 23: 1—26.

Tera, F., Papanastassiou, D.A. and Wasserburg, G.J., 1974. Isotopic evidence for a terminal lunar cataclysm. *Earth Planet. Sci. Lett.,* 22: 1—21.

Thomas, M.D. and Tanner, J.G., 1975. A cryptic suture in the eastern Grenville. *Nature,* 256: 392—394.

Thompson, P.H., 1978. Archen regional metamorphism in the Slave structural Province — a new perspective on some old rocks. *Geol. Surv. Can., Paper,* 78-10: 85—102.

Thurston, P.C. and Breaks, F.W., 1978. Metamorphic and tectonic evolution of the Uchi-English River Subprovince. *Geol. Surv. Can., Paper,* 78-10: 49—62.

Towe, K.M., 1978. Early Precambrian oxygen: a case against photosynthesis. *Nature,* 274: 657—661.

Travis, G.A., Woodall, R. and Bartram, G.D., 1971. The geology of the Kalgoorlie goldfield. *Geol. Soc. Aust., Spec. Publ.,* 3: 175—190.

Trendall, A.F., 1975. Introduction. *Geol. Surv. W. Aust., Mem.,* 2: 25—32.

Turekian, K.K. and Clark, Jr., S.P., 1969. Inhomogeneous accumulation of the earth from the primitive solar nebula. *Earth Planet. Sci. Lett.,* 6: 346—348.

Turekian, K.K. and Wedepohl, K.H., 1961. Distribution of the elements in some major units of the earth's crust. *Geol. Soc. Am. Bull.,* 72: 175—192.

Turgarinov, A.I. and Bibikova, E.V., 1975. Geochronology of the oldest formations of Kola Peninsula. *Geokhimiya,* 9: 1275.

Tuttle, O.F. and Bowen, N.L., 1958. Origin of granite in light of experimental studies in the system Ab-Or-Q-H_2O. *Geol. Soc. Am., Mem.,* 74: 153 pp.

Tyndale-Biscoe, R., 1949. The geology of the country around Gwelo. *Rhod. Geol. Surv., Bull.,* 39: 145 pp.

Upadhyay, H.D., 1978. Phanerozoic peridotitic and pyroxenitic komatiites from Newfoundland. *Science,* 202: 1192—1195.

Vajner, V., 1976. Epeirogeny, orogeny and the tectonic development of the Kaapvaal basin, Southern Africa. *Geol. Rundsch.*, 65: 333—360.

Van Schmus, W.R., 1976. Early and middle Proterozoic history of the Great Lakes area, North America. *Philos. Trans. R. Soc. London, Ser. A*, 280: 605—628.

Van Schmus, W.R. and Anderson, J.L., 1977. Gneiss and migmatite of Archean age in the Precambrian basement of central Wisconsin. *Geology*, 5: 45—48.

Van Schmus, W.R. and Woolsey, L.L., 1975. Rb-Sr geochronology of the Republic area, Marquette County, Michigan. *Can. J. Earth Sci.*, 12: 1723—1733.

Veizer, J., 1976. $^{87}Sr/^{86}Sr$ evolution of seawater during geologic history and its significance as an index of crustal evolution. In: B.F. Windley (Editor), *The Early History of the Earth.* J. Wiley & Sons, New York, N.Y., pp. 569—578.

Venkatasubramanian, V.S., 1974. Geochronology of the Dharwar craton — a review. *J. Geol. Soc. India*, 15: 463—468.

Vidal, P. and Dosso, L., 1978. Core formation: catastrophic or continuous? Sr and Pb isotope geochemistry constraints. *Geophys. Res. Lett.*, 5: 169—172.

Viewing, K.A., 1968. Reginal geochemistry and the Rhodesian basement complex. *Geol. Soc. S. Afr., Annex. Vol.*, 71: 21—32.

Viewing, K.A. and Harrison, N.M., 1973. A geological reconnaissance of the Chinamora batholith near Salisbury, Rhodesia. *Geol. Soc. S. Afr., Spec. Publ.*, 3: 419—432.

Viljoen, M.J. and Viljoen, R.P., 1969a. An introduction to the geology of the Barberton granite-greenstone terrane. *Geol. Soc. S. Afr., Spec. Publ.*, 2: 9—28.

Viljoen, R.P. and Viljoen, M.J., 1969b. The effects of metamorphism and serpentinization on the volcanic and associated rocks of the Barberton region. *Geol. Soc. S. Afr., Spec. Publ.*, 2: 29—54.

Viljoen, M.J. and Viljoen, R.P., 1969c. The geology and geochemistry of the lower ultramafic unit of the Onverwacht Group and a proposed new class of igneous rock. *Geol. Soc. S. Afr., Spec. Publ.*, 2: 55—86.

Viljoen, M.J. and Viljoen, R.P., 1969d. Evidence for the existence of a mobile extrusive peridotitic magma from the Komati Formation of the Onverwacht Group. *Geol. Soc. S. Afr., Spec. Publ.*, 2: 87—112.

Viljoen, R.P. and Viljoen, M.J., 1969e. The geological and geochemical significance of the upper formations of the Onverwacht Group. *Geol. Soc. S. Afr., Spec. Publ.*, 2: 113—152.

Viljoen, M.J. and Viljoen, R.P., 1969f. A proposed new classification of the granitic rocks of the Barberton region. *Geol. Soc. S. Afr., Spec. Publ.*, 2: 153—188.

Viljoen, M.J. and Viljoen, R.P., 1969g. The chemical evolution of the granitic rocks of the Barberton region. *Geol. Soc. S. Afr., Spec. Publ.*, 2: 189—220.

Viljoen, R.P. and Viljoen M.J., 1970. The geology and geochemistry of the layered ultramafic bodies of the Kaapmuiden area, Barberton Mountain Land. *Geol. Soc. S. Afr., Spec. Publ.*, 1: 661—688.

Viljoen, R.P., Saager, R. and Viljoen, M.J., 1969. Metallogenesis and ore control in the Steynsdorp goldfield, Barberton Mountain Land, South Africa. *Econ. Geol.*, 64: 778—797.

Viljoen, M.J., Bernasconi, A., Van Coller, N., Kinlock, E. and Viljoen, R.P., 1976. The geology of the Shangani nickel deposit, Rhodesia. *Econ. Geol.*, 71: 76—95.

Villaume, J.F. and Rose, A.W., 1977. The geochemistry of some Archean ultramafic lavas. *Chem. Geol.*, 19: 43—60.

Visser, D.J.L., 1956. The geology of the Barberton area. *Geol. Surv. S. Afr., Spec. Publ.*, 15: 221 pp.

Viswanatha, M.N., 1968. Unpublished report of the Geological Survey of India.

Viswanatha, M.N. and Ramakrishnan, M., 1975. The pre-Dharwar supracrustal rocks of Sargur Schist Complex in southern Karnataka and their tectonometamorphic significance. *Indian Mineral.*, 16: 48—65.

Viswanathan, S., 1974. Basaltic komatiite occurrences in the Kolar goldfield of India. *Geol. Mag.*, 111: 353—354.

Vollmer, R., 1977. Terrestrial lead isotopic evolution and formation time of the earth's core. *Nature*, 270: 144—147.

Von Brunn, V. and Hobday, D.K., 1976. Early Precambrian tidal sedimentation in the Pongola of South Africa. *J. Sediment. Petrol.*, 46: 670—679.

Wager, L.R. and Brown, G.M., 1967. *Layered Igneous Rocks.* Freeman and Co., San Francisco, Calif., 584 pp.

Walker, D., Kirkpatrick, R.J., Longhi, J. and Hays, J.F., 1976. Crystallization history of lunar picrite basalt 12002: phase equilibria and cooling rate studies. *Geol. Soc. Am. Bull.*, 87: 646—656.

Walker, J.C.G., 1976. Implications for atmospheric evolution of the inhomogeneous accretion model for the origin of the earth. In: B.F. Windley (Editor), *The Early History of the Earth.* J. Wiley & Sons, New York, N.Y., pp. 535—546.

Walker, R.G., 1967. Turbidite sedimentary structures and their relationship to proximal and distal depositional environments. *J. Sediment. Petrol.*, 37: 25—43.

Walker, R.G., 1978. A critical appraisal of Archean basin-craton complexes. *Can. J. Earth Sci.*, 15: 1213—1218.

Walker, R.G. and Pettijohn, F.J., 1971. Archean sedimentation: analysis of the Minnitaki basin, northwestern Ontario, Canada. *Geol. Soc. Am. Bull.*, 82: 2099—2130.

Walter, M.R., 1976. *Stromatolites.* Elsevier, Amsterdam, 790 pp.

Walter, M.R., 1977. Interpreting stromatolites. *Am. Sci.*, 65: 563—571.

Wasserburg, G.J., MacDonald, G.J.F., Hoyle, F. and Fowler, W.A., 1964. Relative contributions of U, Th, and K to heat production in the earth. *Science*, 143: 465—467.

Wasserburg, G.J., Papanastassiou, D.A., Tera, F. and Huneke, J.C., 1977. Outline of a lunar chronology. *Philos. Trans. R. Soc. London, Ser. A*, 285: 7—22.

Watkinson, D.H. and Irvine, T.N., 1964. Peridotitic intrusions near Quetico and Shebandowan, northwestern Ontario: a contribution to the petrology and geochemistry of ultramafic rocks. *Can. J. Earth Sci.*, 1: 63—97.

Watson, J.V., 1973. Influence of crustal evolution on ore deposition. *Trans. Inst. Min. Metall., Sect. B*, 82: 107—113.

Watson, J.V., 1976a. Mineralization in Archean provinces. In: B.F. Windley, (Editor), *The Early History of the Earth.* J. Wiley & Sons, New York, N.Y., pp. 443—454.

Watson, J.V., 1976b. Vertical movements in Proterozoic structural provinces. *Philos. Trans. R. Soc. London, Ser. A*, 280: 629—640.

Watson, J.V., 1978. Precambrian thermal regimes. *Philos. Trans. R. Soc. London, Ser. A*, 288: 431—440.

Webber, G.R., 1962. Variation in the composition of the Lake Dufault granodiorite. *Can. Inst. Min. Metall., Trans.*, 65: 55—62.

Weber, W. and Scoates, R.F.J., 1978. Archean and Proterozoic metamorphism in the northwestern Superior Province and along the Churchill-Superior boundary, Manitoba. *Geol. Surv. Can., Paper*, 78-10: 5—16.

Wells, P.R.A., 1976. Late Archean metamorphism in the Buksefjorden region, southwest Greenland. *Contrib. Mineral. Petrol.*, 56: 229—242.

Wernick, E., 1979. The geotectonic environments of early Precambrian granites in Brazil. *Precambrian Res.*, 8: 1—17.

Wernick, E., 1980. Archean of Brazil. *Earth-Sci. Rev.*, (in press).

West, G.F., 1976. Geophysics in the Geotraverse. *Proc. 1976 Geotraverse Conf., Univ. Toronto*, pp. 17—21.

West, G.F., Szewezyk, Z., Dusanowskyj, T., Urquart, W.E., Wright, J. and Godlewski, M., 1977. Geophysical studies in the Geotraverse: gravity, magnetics, seismics. *Proc. 1977 Geotraverse Conf., Univ. Toronto*, pp. 1—9.

Wetherill, G.W., 1972. The beginning of continental evolution. *Tectonophysics*, 13: 31—45.

White, A.J.R., Jakeš, P. and Christie, D.M., 1971. Composition of greenstones and the hypothesis of sea-floor spreading in the Archean. *Geol. Soc. Aust., Spec. Publ.*, 3: 47—56.

Whitford, D.J. and Arndt, N.T., 1978. Rare earth element abundances in a thick-layered komatiite lava flow from Ontario, Canada. *Earth Planet. Sci. Lett.*, 41: 188—196.

Wildeman, T.R. and Condie, K.C., 1973. Rare earths in Archean graywackes from Wyoming and from the Fig Tree Group, South Africa. *Geochim. Cosmochim. Acta*, 37: 439—453.

Wildeman, T.R. and Haskin, L.A., 1973. Rare earths in Precambrian sediments. *Geochim. Cosmochim. Acta*, 37: 419—438.

Wiles, J.W., 1972. The geology of the country around Sipolilo. *Rhod. Geol. Surv., Bull.*, 63: 76 pp.

Willett, G., Eshuys, E. and Guy, B., 1978. Ultramafic rocks of the Widgiemooltha-Norseman area, Western Australia: petrological diversity, geochemistry, and mineralization. *Precambrian Res.*, 6: 133—156.

Williams, D.A.C., 1971. Determination of primary mineralogy and textures in ultramafic rocks from Mt. Monger, Western Australia, *Geol. Soc. Aust., Spec. Publ.*, 3: 259—268.

Williams, D.A.C. and Hallberg, J.A., 1973. Archean layered intrusions of the eastern Goldfields region, Western Australia. *Contrib. Mineral. Petrol.*, 38: 45—70.

Williams, G.E., 1973. Geotectonic cycles, lunar evolution, and the dynamics of the earth-moon system. *Modern Geol.*, 4: 159—183.

Williams, H.R., 1977. African Archean mobile belts and granite-greenstone terrane. *Nature*, 266: 163—164.

Williams, H.R., 1978. The Archean geology of Sierra Leone. *Precambrian Res.*, 6: 251—268.

Williams, H.R. and Williams, R.A., 1976. The Kasila Group, Sierra Leone, an interpretation of new data. *Precambrian Res.*, 3: 505—508.

Williams, I.R., 1973. Structural subdivision of the Eastern Goldfields Province, Yilgarn Block. *Geol. Surv. W. Aust., Annu. Rep.*, 1973: 53—58.

Williams, I.R., 1975a. Eastern Goldfields Province. *Geol. Surv. W. Aust., Mem.*, 2: 33—53.

Williams, I.R., 1975b. Southwestern Province. *Geol. Surv. W. Aust., Mem.*, 2: 65—70.

Wilson, A.F., 1969. Granulite terrains and their tectonic setting and relationship to associated metamorphic rocks in Australia. *Geol. Soc. Aust., Spec. Publ.*, 2: 234—258.

Wilson, H.B.D., Andrews, P., Moxham, R.L. and Ramlal, K., 1965. Archean volcanism in the Canadian Shield. *Can. J. Earth Sci.*, 2: 161—175.

Wilson, H.D.B. and Morrice, M.G., 1977. The volcanic sequence in Archean shields. *Geol. Assoc. Can., Spec. Paper*, 16: 355—374.

Wilson, J.F., 1964. The geology of the country around Fort Victoria. *Rhod. Geol. Surv., Bull.*, 58: 147 pp.

Wilson, J.F., 1968. The Mashaba igneous complex and its subsequent deformation. *Geol. Soc. S. Afr., Annex. Vol.*, 71: 175—188.

Wilson, J.F., 1973a. The Rhodesian Archean cration — an essay in cratonic evolution. *Philos. Trans. R. Soc. London, Ser. A*, 273: 389—411.

Wilson, J.F., 1973b. Granites and gneisses of the area around Mashaba, Rhodesia. *Geol. Soc. S. Afr., Spec. Publ.*, 3: 79—84.

Wilson, J.F., 1979. A preliminary reappraisal of the Rhodesian basement complex. *Geol. Soc. S. Afr., Spec. Publ.*, 5: (in press).

Wilson, J.F., Bickle, M.J., Hawkesworth, C.J., Martin, A., Nisbet, E.G. and Orpen, J.L., 1978. Granite-greenstone terrains of the Rhodesian Archean craton. *Nature*, 271: 23—27.

Wilson, M.E., 1962. Royn-Beauchastel map areas. *Geol. Surv. Can., Mem.*, 315.

Winchester, J.A. and Floyd, P.A., 1976. Geochemical magma type discrimination: application to altered and metamorphosed basic igneous rocks. *Earth Planet. Sci. Lett.*, 28: 459—469.

Winchester, J.A. and Floyd, P.A., 1977. Geochemical discrimination of different magma series and their differentiation products using immobile elements. *Chem. Geol.*, 20: 325—343.

Windley, B.F., 1970. Anorthosites in the early crust of the earth and moon. *Nature*, 226: 333—335.

Windley, B.F., 1973. Crustal development in the Precambrian. *Philos. Trans. R. Soc. London, Ser. A*, 273: 321—341.

Windley, B.F., 1976. New tectonic models for the evolution of Archean continents and oceans. In: B.F. Windley (Editor), *The Early History of the Earth.* J. Wiley & Sons, New York, N.Y., pp. 105—112.

Windley, B.F., 1977. *The Evolving Continents.* J. Wiley & Sons, New York, N.Y., 385 pp.

Windley, B.F. and Bridgwater, D., 1971. The evolution of Archean low- and high-grade terrains. *Geol. Soc. Aust., Spec. Publ.*, 3: 33—46.

Windley, B.F. and Smith, J.V., 1974. The Fiskenaesset complex, West Greenland, 2. General mineral chemistry from Qeqertarssuatsiaq. *Grønl. Geol. Unders., Bull.*, 108: 1—54.

Windley, B.F. and Smith, J.V., 1976. Archean high grade complexes and modern continental margins. *Nature*, 260: 671—675.

Windley, B.F. and Davies, F.B., 1978. Volcano spacings and lithospheric/crustal thickness in the Archean. *Earth Planet. Sci. Lett.*, 38: 291—297.

Windley, B.F., Herd, R.K. and Bowden, A.A., 1973. The Fiskenaesset complex, West Greenland. *Grønl. Geol. Unders., Bull.*, 106: 80 pp.

Wise, D.U., 1973. Freeboard of continents through time. *Geol. Soc. Am., Mem.*, 132: 87—100.

Wise, D.U., 1974. Continental margins, freeboard and volumes of continents and ocean through time. In: C.A. Burk and C.L. Drake (Editors), *The Geology of Continental Margins.* Springer-Verlag, Berlin, pp. 45—58.

Wolhuter, L.E., 1973a. The petrology of the Opemisca Lake pluton, Quebec, Canada. *Geol. Soc. S. Afr., Spec. Publ.*, 3: 345—360.

Wolhuter, L.E., 1973b. Major and trace elements in the Opemisca Lake granite pluton, Quebec, Canada. *Geol. Soc. S. Afr., Spec. Publ.*, 3: 387—410.

Wood, B.J., 1975. The influence of pressure, temperature, and bulk composition on the appearance of garnet in orthogneisses — an example from South Harris, Scotland. *Earth Planet. Sci. Lett.*, 26: 299—311.

Wooden, J.L., 1978. Rb-Sr isotopic studies of the Archean rocks of the eastern Lac Seul and Kenora areas, English River Subprovince, Ontario. In: I.E.M. Smith and J.G. Williams (Editors), *Proc. 1978 Archean Geochemistry Conference.* University of Toronto Press, Toronto, Ont., pp. 131—149.

Worst, B.G., 1956. The geology of the country between Belingwe and West Nicholson. *Rhod. Geol. Surv., Bull.*, 43: 218 pp.

Worst, B.G., 1958. The differentiation and structure of the Great Dyke of southern Rhodesia. *Trans. Geol. Soc. S. Afr.*, 61: 283—354.

Wright, J., 1977. *Seismic Crustal Studies in Northwestern Ontario.* Ph.D. Dissertation, University of Toronto, Toronto, Ont.

Wright, J. and West, G.R., 1976. Seismic studies in the southern Geotraverse. *Proc. 1976 Geotraverse Conf., Univ. Toronto*, pp. 1—8.

Wright, L.I., 1975. A deformation study of the Gwanda schist belt and its implications for the formation of the belt. *Univ. Leeds, Res. Inst. Afr. Geol., 19th Annu. Rep.*, pp. 59—65.

Wright, T.L. and Dohery, P.C., 1970. A linear programming and least squares computer method for solving petrologic mixing problems. *Geol. Soc. Am. Bull.*, 81: 1995—2008.

Wyllie, P.J., 1971. Experimental limits for conditions for melting in the earth's crust and upper mantle. *Am. Geophys. Union, Geophys. Monogr.*, 14: 279—302.

Wynne-Edwards, H.R., 1972. The Grenville Province. *Geol. Assoc. Can., Spec. Paper,* 11: 263—334.

Wynne-Edwards, H.R., 1976. Proterozoic ensialic orogenesis: the millipede model of ductile plate tectonics. *Am. J. Sci.*, 276: 927—953.

Yardley, B.W.D. and Blacic, J.D., 1976. Sappharine in the Sitampundi complex, India: a discussion. *Mineral. Mag.*, 40: 523—524.

Yellur, D.D. and Nair, R.S., 1978. Assigning a magmatically defined tectonic environment to Chitdradurga metabasalts, India, by geochemical methods. *Precambrian Res.*, 7: 259—281.

Young, R.A., 1979. *Crustal Structure for Explosion Seismology in the Superior Province.* Ph.D. Dissertation, University of Toronto, Toronto, Ont.

Young, R.A. and West, G.F., 1977. Seismic observations in the central Geotraverse. *Proc. 1977 Geotraverse Conf., Univ. of Toronto,* pp. 65—72.

SUBJECT INDEX